ANALOG AND MIXED-SIGNAL ELECTRONICS

ANALOG AND MIXED-SIGNAL ELECTRONICS

KARL D. STEPHAN
Texas State University, San Marcos

WILEY

Published by John Wiley & Sons, Inc., Hoboken, New Jersey
Published simultaneously in Canada

Library of Congress Cataloging-in-Publication Data:

Stephan, Karl David, 1953–
Analog and mixed-signal electronics / Karl D. Stephan.
pages cm
Includes bibliographical references and index.
ISBN 978-1-118-78266-8 (cloth)
1. Electronic circuits. 2. Mixed signal circuits. I. Title.
TK7867.S84 2015
621.3815–dc23

2014050119

Set in 10/12pt Times by SPi Publisher Services, Pondicherry, India

10 9 8 7 6 5 4 3 2 1

1 2015

CONTENTS

PREFACE

All but the simplest electronic devices now feature embedded processors, and software development represents the bulk of what many electrical engineers do. So, some would question the need for a new book on analog and mixed-signal electronics. Surely everything about analog electronics has been known for decades and can be found in old textbooks, so what need is there for a new one?

In teaching a course on analog and mixed-signal design for the past few years, I have found that as digital and software design has taken over a larger part of the electrical engineering curriculum, some important matters relating to analog electronics have fallen into the cracks, so to speak. Problems as simple as wiring up a dual-output power supply for an operational amplifier circuit prove daunting to some students whose main engineering tool up to that point has been a computer. While all undergraduate electrical engineering students master the basics of linear circuits and systems, these subjects are often taught in an abstract, isolated fashion that gives no clue as to how the concepts taught can be used to make something worth building and selling, which is what engineering is all about.

This book is intended to be a practical guide to analog and mixed-signal electronics, with an emphasis on design problems and applications. Many examples are included of actual circuit designs developed to meet specific requirements, and several of these have been lab-tested, with experimental results included in the text. While advances in analog electronics have not occurred as rapidly as they have in digital systems and software, analog systems have found new uses in concert with digital systems, leading to the prominence of mixed-signal systems in many technologies today. The modern electrical engineer should be able to address a given design problem with the optimum mix of digital, analog, and software approaches to get the job done efficiently, economically, and reliably. While most of a system's functionality may depend on software, none of it can get off the ground without power, and power supplies are largely still an analog domain.

Beginning with reviews of electronic components and linear systems theory, this book covers topics such as noise, op amps, analog filters, oscillators, conversion between analog and digital domains, power electronics, and high-frequency design. It closes with a chapter on a subject that is rarely addressed in the undergraduate curriculum: electromagnetic compatibility. Problems having to do with electromagnetic compatibility and electromagnetic interference happen all the time, however, and can be very difficult to diagnose and fix, which is why methods to detect and diagnose such problems are included. Although familiarity with standard electrical engineering concepts such as complex numbers and Laplace transforms is assumed in parts of the text, other parts can be used by those without a calculus or electrical engineering background: technicians, hobbyists, and others interested in analog and mixed-signal electronics, but who are not members of the electrical engineering profession. References for further study and a set of problems are provided at the end of each chapter, as well as an appendix describing test equipment useful for analog and mixed-signal work.

San Marcos, TX
July 3, 2014

KARL D. STEPHAN

ACKNOWLEDGMENTS

"No man is an island, Entire of itself," as John Donne's poem says, and this book is my work only in the sense that I am the medium through which it passes. Many educators, mentors, and friends contributed to the knowledge it represents. Among these, I should mention first the late R. David Middlebrook (1929–2010), whose electronics course I took as a Caltech undergraduate in the 1970s. Professor Middlebrook never met an analog circuit he couldn't analyze with nothing more than paper, pencil, and a slide rule, and his disciplined and insightful approach to analog circuit analysis is an ideal that I am sure I fall short of. I can only hope that some of the clarity and depth with which he taught shows through in this text. In my 16 years at the University of Massachusetts Amherst, I shared teaching responsibilities with my colleagues and friends Robert W. Jackson and K. Sigfrid Yngvesson. Bob Jackson in particular was never the one to let a mathematical or technical ambiguity slip by, and I thank him for the quality check he performed on any lecture material we presented jointly. A. David Wunsch, for many years a professor at the University of Massachusetts Lowell, reviewed a draft of Chapter 7 and made helpful suggestions for which I am grateful. The course entitled "Analog and Mixed-Signal Design" was developed at my present institution, Texas State University, to form part of a new Electrical Engineering program initiated in 2008. The founding Director of the School of Engineering, Harold Stern, was kind enough to give me a free hand in developing a lab-based course which has an unconventional structure, consisting of four or five multi-week projects interspersed with lectures. I thank him for creating a congenial teaching environment that helped me to develop the material that forms the basis of this text. I also thank historian of science Renate Tobies for providing information on Heinrich Barkhausen that is not generally available in English.

Finally, I express my appreciation and gratitude to my wife Pamela, whose artistic skills provided the templates for most of the illustrations. Together we can say, "Be thankful unto him, and bless his name. For the Lord is good; his mercy is everlasting, and his truth endureth to all generations."

ABOUT THE COMPANION WEBSITE

This book is accompanied by a companion website:

http://wiley.com/go/analogmixedsignalelectronics

The website includes:

- Solutions Manual available to Instructors.

1

INTRODUCTION TO ANALOG AND MIXED-SIGNAL ELECTRONICS

1.1 INTRODUCTION

"In the beginning, there were only analog electronics and vacuum tubes and huge, heavy, hot equipment that did hardly anything. Then came the digital—enabled by integrated circuits and the rapid progress in computers and software—and electronics became smaller, lighter, cheaper, faster, and just better all around, all because it was digital." That's the gist of a sort of urban legend that has grown up about the nature of analog electronics and **mixed-signal** electronics, which means simply electronics that has both analog and digital circuitry in it.

Like most legends, this one has some truth to it. Most electronic systems, ever since the time that there was anything around to apply the word "electronics" to, were analog in nature for most of the twentieth century. In electronics, an **analog signal** is a voltage or current whose value is proportional to (an analog of) some physical quantity such as sound pressure, light intensity, or even an abstract numerical value in an **analog computer**. **Digital signals**, by contrast, ideally take on only one of two values or ranges of values and by doing so represent the discrete binary ones and zeros that form the language of digital computers. To give you an idea of how things used to be done with purely analog systems, Figure 1.1 shows on the left a two-channel vacuum-tube audio amplifier that can produce about 70 W per channel.

The vacuum-tube amplifier measures 30 cm × 43 cm × 20 cm and weighs 17.2 kg (38 lb) and was state-of-the-art technology in about 1955. On its right is a solid-state class D amplifier designed in 2008 that can produce about the same amount of

Analog and Mixed-Signal Electronics, First Edition. Karl D. Stephan.

Companion Website: http://wiley.com/go/analogmixedsignalelectronics

FIGURE 1.1 A comparison: Vacuum-tube audio amplifier (left) using a design circa 1955 and class D amplifier (right) using a design circa 2008.

output power. It is a mixed-signal (analog and digital) design. It measures only 15 cm × 10 cm × 4 cm and weighs only 0.33 kg, not including the power supply, which is of comparable size and weight. The newer amplifier uses its power devices as switches and is much more efficient than the vacuum-tube unit, which is about 50 times its size and weight. So the claim that many analog designs have been made completely obsolete by newer digital and mixed-signal designs is true, as far as it goes.

Sometimes, you will hear defenders of analog technology argue that "the world is essentially analog, and so analog electronics will never go away completely." Again, there's some truth to that, but it depends on your point of view. The physics of quantum mechanics tells us that not only are all material objects made of discrete things called atoms but many forms of energy appear as discrete packets called quanta (photons, in the case of electromagnetic radiation). So you can make just as good an argument for the case that the whole world is essentially digital, not analog, because it can be represented as bits of quanta and atoms that are either there or not there at all.

The fact of the matter is that while the bulk of today's electronics technology is implemented by means of digital circuits and powerful software, a smaller but essential part of what goes into most electronic devices involves analog circuitry. Even if the analog part is as simple as a battery for the power supply, no one has yet developed a battery that behaves digitally: that is, one that provides an absolutely constant voltage until it depletes and drops abruptly to zero. So even designers of an otherwise totally digital system have to deal with the analog problem of power-supply characteristics.

This book is intended for anyone who has an interest in understanding or designing systems involving analog or mixed-signal electronics. That includes undergraduates with a basic sophomore-level understanding of electronics, as well as more advanced undergraduates, graduate students, and professionals in engineering, science, or other fields whose work requires them to learn about or deal with these types of electronic systems. The emphasis is practical rather than theoretical, although enough theory to enable an understanding of the essentials will be presented as needed throughout. Many textbooks present electronics concepts in isolation without any indication of how a component or circuit can be used to meet a practical need, and we will try to avoid that error in this book. Practical applications of the various circuits and systems described will appear as examples, as paper or computer-simulation design exercises, and as lab projects.

1.2 ORGANIZATION OF THE BOOK

The book is divided into three main sections: devices and linear systems (Chapters 2 and 3), linear and nonlinear analog circuits and applications (Chapters 4–7), and special topics of analog and mixed-signal design (Chapters 8–12). A chapter-by-chapter summary follows.

1.2.1 Chapter 2: Basics of Electronic Components and Devices

In this chapter, you will learn enough about the various types of two- and three-terminal electronic devices to use them in simple designs. This includes rectifier, signal, and light-emitting diodes and the various types of three-terminal devices: field-effect transistors (FETs), bipolar junction transistors (BJTs), and power devices. Despite the bewildering number of different devices available from manufacturers, there are usually only a few specifications that you need to know about each type in order to use them safely and efficiently. In this chapter, we present basic circuit models for each type of device and how to incorporate the essential specifications into the model.

1.2.2 Chapter 3: Linear System Analysis

This chapter presents the basics of linear systems: how to characterize a "black box" circuit as an element in a more complex system, how to deal with characteristics such as gain and frequency response, and how to define a system's overall specifications in terms that can be translated into circuit designs. The power of linear analysis is that it can deal with complex systems using fairly simple mathematics. You also learn about some basic principles of noise sources and their effects on electronic systems.

1.2.3 Chapter 4: Nonlinearities in Analog Electronics

While linear analysis covers a great deal of analog-circuit territory, nonlinear effects can both cause problems in designs and provide solutions to other design problems. Noise of various kinds is always present to some degree in any circuit, and in the case of high-gain and high-sensitivity systems dealing with low-level signals, noise can determine the performance limits of the entire system. You will be introduced to the basics of nonlinearities and noise in this chapter and learn ways of dealing with these issues and minimizing problems that may arise from them.

1.2.4 Chapter 5: Op Amp Circuits in Analog Electronics

The workhorse of analog electronics is the operational amplifier ("op amp" for short). Originally developed for use in World War II era analog computers, in integrated-circuit form the op amp now plays essential roles in most analog electronics systems of any complexity. This chapter describes op amps in a simplified ideal form and outlines the more complex characteristics shown by actual op amps. Basic op amp circuits and their uses make up the remainder of the chapter.

1.2.5 Chapter 6: The High-Gain Analog Filter Amplifier

High-gain amplifiers bring with them unique problems and capabilities, so we dedicated an entire chapter to a discussion of the special challenges and techniques needed to develop a good high-gain amplifier design. We also introduce the basics of analog filters in this section and apply them to the design of a practical circuit: a guitar preamp.

1.2.6 Chapter 7: Waveform Generation

While many electronic systems simply sense or detect signals from the environment, other systems produce or generate signals on their own. This chapter describes circuits that generate periodic signals, collectively termed *oscillators*, as well as other signal-generation devices. Because oscillators that produce a stable frequency output are the heart of all digital clock systems, you will also find information on the basics of stabilized oscillators and the means used to stabilize them: quartz crystals and, more recently, microelectromechanical system (MEMS) resonators.

1.2.7 Chapter 8: Analog-to-Digital and Digital-to-Analog Conversion

Most new electronic designs of any complexity include a microprocessor or equivalent that does the heavy lifting in terms of functionality. But many times, it is necessary to take analog inputs from various sensors (e.g., photodiodes, ultrasonic sensors, proximity detectors) and transform their outputs into a digital format suitable for feeding to the digital microprocessor inputs. Similarly, you may need to take a digital output from the microprocessor and use it to control an analog or high-power device such as a lamp or a motor. All these problems involve interfacing between analog and digital circuitry. While no single solution solves all such problems, this chapter describes several techniques you can use to create successful, reliable connections between analog systems and digital systems.

1.2.8 Chapter 9: Phase-Locked Loops

A **phase-locked loop** is a circuit that produces an output waveform whose phase is locked, or synchronized, to the phase of an input signal. Phase-locked loops are used in a variety of applications ranging from wireless links to biomedical equipment.

This chapter presents the control theory needed for a basic understanding of phase-locked loops and gives several design examples.

1.2.9 Chapter 10: Power Electronics

Most "garden-variety" electronic components and systems can control electrical power ranging from less than a microwatt up to a few milliwatts without any special techniques. But if you wish to power equipment or devices that need more than 1 W or so, you will have to deal with power electronics. Audio amplifiers, lighting controls, and motor controls (including those in increasingly popular electric or hybrid automobiles) all use power electronics. Special devices and circuits have been developed to deal with the problems that come when large amounts of electrical power must be produced in a controlled way. Because no system is 100% efficient, some of the primary input power must be dissipated as heat, and as the power delivered rises, so does the amount of waste heat that must be gotten rid of somehow in order to keep the power devices from overheating and failing. This chapter will introduce you to some basics of power electronics, including issues of heat dissipation, efficiency calculations, and circuit techniques suited for power-electronics applications. Besides conventional linear power-control circuits, the availability of fast switching devices such as **insulated-gate bipolar transistors (IGBTs)** and **power FETS** means that **switch-mode** power systems (those that use active devices as on–off switches rather than linear amplifiers) are an increasingly popular way to implement power-control circuits that have much higher efficiency than their linear-circuit relatives. For this reason, we include material on switch-mode **class D** amplifiers and switching power supplies in this chapter as well.

1.2.10 Chapter 11: High-Frequency (Radio-Frequency) Electronics

As long as no signal in a system has a significant frequency component above 20 kHz or so, which is the limit of human hearing, no special design techniques are needed for most analog circuits. However, depending on what you are trying to do, at frequencies in the MHz range the capacitance of devices, cables, and simply the circuit wiring itself becomes increasingly significant. Above a few MHz, the small amount of inductance that short lengths of wire or circuit-board traces show can also begin to affect the behavior of a circuit. At radio frequencies, which start at about 1 MHz and extend up to the GHz range, a wire is not simply a wire. It often must be treated as a **transmission line** having characteristic values of distributed inductance and capacitance per unit length, and sometimes, it can even act as an antenna, radiating some of the power transmitted along it into space.

The set of design approaches that deal with these types of high-frequency problems are known as high-frequency design or radio-frequency (RF) design. This chapter will introduce you to the basics of the field: transmission lines, filters, impedance-matching circuits, and RF circuit techniques such as tuned amplifiers.

1.2.11 Chapter 12: Electromagnetic Compatibility

Electromagnetic compatibility, **electromagnetic interference**, and **RF interference** are usually referred to just by their respective initials: **EMC**, **EMI**, and **RFI**. These phrases all refer to various problems that can arise when electromagnetic fields (electric, magnetic, or a combination) produced by a circuit disturb (or **couple** to) another circuit, usually with undesirable consequences. Of course, every time you use a mobile telephone, you employ RF coupling between the phone and the cell-tower base station that achieves the desirable purpose of making a phone call. But the same radio waves that are used in wireless and mobile equipment can also interfere with the proper operation of other electronic systems that are not necessarily designed to receive them. While reading this chapter will not make you an EMC/EMI expert, you will learn the basics of how these problems occur and some simple ways to alleviate or avoid them entirely.

Following the chapters above is an appendix containing useful information on measurement equipment for analog and mixed-signal design.

Each chapter is followed by a set of problems that range from simple applications of concepts developed in the chapter to open-ended design problems and lab projects. While paper designs and simulations using software such as National Instruments' Multisim™ are necessary steps in circuit design, the ultimate test of any design is building it. This is why we have included many lab-based projects and encourage both students and instructors to avail themselves of the opportunity of building circuits and trying them out. In this field as in many others, there is no substitute for hands-on experience.

BIBLIOGRAPHY

Analog Devices, www.analog.com. This website of a well-known analog IC manufacturer features an electronic periodical called "Analog Dialogue" as well as numerous datasheets and application notes about a wide variety of analog circuits, ICs, and applications.

Crecraft, D. I., and S. Gergely. *Analog Electronics: Circuits, Systems and Signal Processing*. Oxford, UK: Butterworth-Heinemann, 2002.

Hickman, Ian. *Analog Electronics*, Second Edition. Oxford, UK: Newnes, 1999.

Horowitz, P. and W. Hill. *The Art of Electronics*. Cambridge, UK: Cambridge University Press, 1989.

PROBLEMS

1.1. *Maximum output power for different waveforms.* One fundamental limitation of all linear amplifiers is the fact that every amplifier has a maximum voltage output limit it is capable of supplying. For example, an op amp with a dual power supply of ±15 V cannot produce a signal whose voltage exceeds about 25 V peak to peak (V_{pp}). Assuming that such an amplifier drives a 10-kΩ load resistor connected between the output and ground, calculate the maximum output power delivered to the load if

(a) the 25-V (peak-to-peak) waveform is a pure sine wave (no distortion) and

(b) the 25-V (peak-to-peak) waveform is an ideal square wave (50% duty cycle). (The **duty cycle** of a pulse is the percentage of the period during which the pulse is high.)

1.2. *Efficiency and heat sink limitations.* Suppose you are using a power device with a **heat sink** (a mechanical structure designed to dissipate heat into the surrounding air) and the heat sink can handle up to $P_{HEAT} = 20$ W of thermal power (in the form of heat) before the device becomes dangerously hot above 150°C. The **power efficiency** η of a device is defined as

$$\eta = \frac{P_{OUT}}{P_{IN}} \tag{1.1}$$

and the dissipated thermal power $P_{HEAT} = P_{IN} - P_{OUT}$. Calculate the maximum output power P_{OUT} that can be obtained from this device-and-heat sink combination if the power efficiency of the device is

(a) $\eta = 25\%$ (b) $\eta = 75\%$ (c) $\eta = 90\%$ (d) $\eta = 95\%$ Comment on why high efficiency is so important in high-power devices.

1.3. *Size of circuit compared to wavelength.* One reason high-frequency designs must use special techniques is that the signals involved take a finite amount of time to travel through the circuit. Suppose you are dealing with a sine-wave signal at a frequency of 900 MHz and the circuit you design is 25 cm long. Using the wavelength–frequency relationship

$$\lambda = \frac{c}{f}, \tag{1.2}$$

where λ is the wavelength in meters, c is the speed of light (3×10^8 m s^{-1}), and f is the frequency in Hz (= cycles s^{-1}, dimensions s^{-1}), express the length of the circuit in terms of wavelengths at 900 MHz. Any time a circuit occupies a substantial fraction of a wavelength in size (more than 10% or so), you should consider using high-frequency design techniques.

1.4. *Reactance of short thin wire at high frequencies.* If a thin wire 1 cm long is suspended at least a few centimeter away from any nearby conductors, it will have an equivalent inductance of about 10 nH (10^{-9} H). Using the inductive reactance formula $X_L = 2\pi fL$, calculate the reactance of this length of wire at (a) $f_1 = 1$ MHz and (b) $f_2 = 1$ GHz (GHz = **gigahertz**, pronounced "gig-a-hertz," = 10^9 Hz). This shows how parts of a circuit you would not normally consider important, such as wire leads, can begin to play a significant role in the circuit at high frequencies.

For further resources for this chapter visit the companion website at
http://wiley.com/go/analogmixedsignalelectronics

2

BASICS OF ELECTRONIC COMPONENTS AND DEVICES

2.1 INTRODUCTION

The term **electronic component** means any identifiable part that goes into an electronic system. Examples of components are resistors, capacitors, and inductors. An electronic **device** in the sense used in this text is a type of component, strictly speaking, but generally performs a more complex function than a plain component does. A component is usually specified by only one component value—for example, the resistance of a resistor. But devices, especially three-terminal devices such as transistors, often must be characterized by several parameters in order to predict their behavior adequately in an **equivalent-circuit model**.

To **model** a circuit means to create a mathematical structure that models or imitates the way the actual circuit will perform when built. Modeling a circuit at some level is almost a necessity, because building a circuit without first making a reasonably good mathematical estimate of how it will perform is a waste of time and money. Before the advent of digital computers, most circuit model calculations were performed by hand, and many simplifying assumptions were made. Currently, proprietary circuit simulation software packages such as PSpice™ or NI Multisim™ are typically used to model circuits. Whether the calculations are done by hand or with software, any circuit model is only as good as the equivalent-circuit models used for the devices and components in it. Simulation programs come with many device and component models already built in, but the software manufacturers cannot always keep up with the rapid development of new devices, so the

Analog and Mixed-Signal Electronics, First Edition. Karl D. Stephan.

Companion Website: http://wiley.com/go/analogmixedsignalelectronics

device you want to use may not have an exact equivalent model available in the analysis software you are using.

Many circuit models work reasonably well with **generic** device models, and those are the kind we will describe in this chapter: simple equivalent circuits with parameters that are adjustable to fit the device you are using. A generic model is not associated with any particular manufacturer's model number. Instead, it has adjustable parameters so that it can behave like any of the different specific devices in its class (another word for class is **genus**, hence the term "generic"). Generic models should be used with caution, especially in nonlinear or switching circuits. But if a reasonably close match to your device cannot be found in the software's library of customized device models, using a generic model is an alternative.

Devices are usually classed into two broad categories: **passive** and **active** devices. The dividing line between passive and active devices is fuzzy, but generally, passive devices are linear and either store or dissipate power. By contrast, active devices can **amplify** or **generate** signals when embedded in the appropriate circuit and provided with the appropriate AC and DC voltages or currents. The passive devices we will describe include resistors, capacitors, inductors, connectors, and antennas. Active devices include diodes and the various types of **small-signal** and **power** transistors.

2.2 PASSIVE DEVICES

2.2.1 Resistors

The need for a component to present a known resistance to current flow arose well before the field of electronics began. The designers of nineteenth-century electric generators found that in order to regulate the output voltage of their machines, they had to place a resistive element in series with the **field winding** that produced the magnetic field, which allowed the generator to operate. This resistive element was eventually named a **resistor** and sometimes took the form of a long piece of sheet iron bent into a zigzag shape as shown in Figure 2.1, which is where the modern symbol for resistors originated.

While resistors for power-control applications can dissipate up to several hundred watts or more, most resistors used in ordinary electronic circuits are capable of dissipating only 250 mW or less. Specialized devices called **power resistors** will be discussed in the chapter on power electronics.

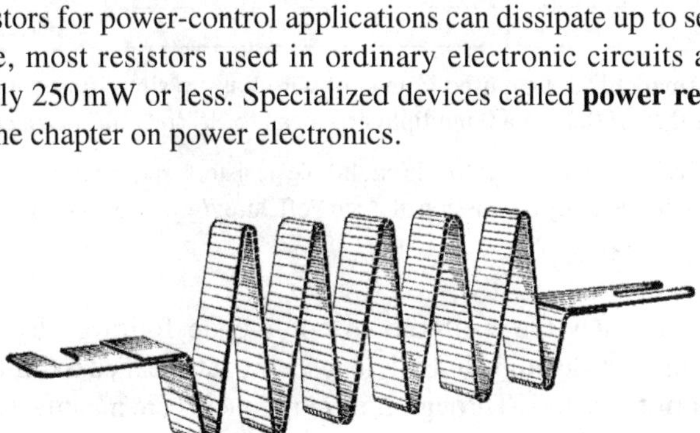

FIGURE 2.1 Early type of bent-sheet-metal current-measurement resistor (1896) from whose shape the schematic-diagram symbol for resistor was derived.

Resistor identification

The end with more bands should point left when reading colors.

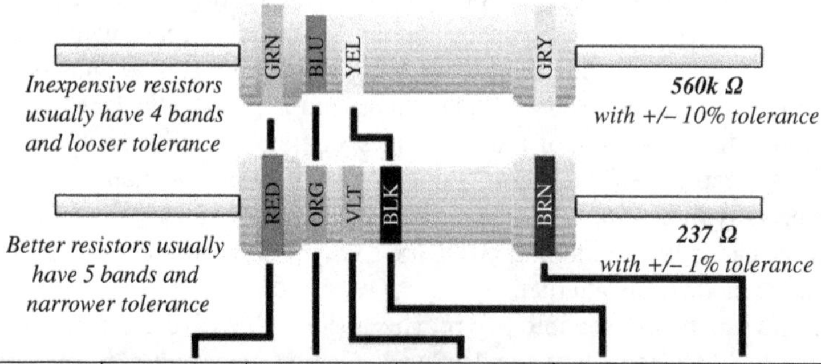

Color	1st Band	2nd Band	3rd Band	Multiplier	Tolerance
Black	0	0	0	× 1 Ω	
Brown	1	1	1	× 10 Ω	+/– 1%
Red	2	2	2	× 100 Ω	+/– 2%
Orange	3	3	3	× 1k Ω	
Yellow	4	4	4	× 10k Ω	
Green	5	5	5	× 100k Ω	+/– .5%
Blue	6	6	6	× 1M Ω	+/– .25%
Violet	7	7	7	× 10M Ω	+/– .1%
Grey	8	8	8		+/– .05%
White	9	9	9		
Gold				× .1 Ω	+/– 5%
Silver				× .01 Ω	+/– 10%

Surface-mount

Surface-mount (SMD) resistors use a similar system. Resistance is indicated by a 3-digit code like 104, sometimes followed by a letter. Rare, precision resistors have 4 digits (3+ multiplier).

1st Digit	2nd Digit	3rd Digit (rare)	Multiplier
1	0		4

(10 with 4 zeros) = **100 k Ω**

- 0 Ω resistors (marked "0") are used instead of wire links to simplify robotic assembly.
- Resistors less than 100Ω use a 0 multiplier to mean "× 1" so "100" = 10Ω, "470" = 47Ω

FIGURE 2.2 Resistor color code for through-hole resistors and marking code for surface-mount resistors. Reproduced by permission of Zach Poff, http://www.zachpoff.com/diy-resources/resistor-color-code-chart/.

Figure 2.2 illustrates the well-known color code followed by most resistor manufacturers for both **through-hole** style resistors (also called **axial-lead** resistors) and **surface-mount** resistors. Through-hole components are manufactured with wire leads that emerge from the component's body and are typically inserted into the holes in a through-hole type of **printed circuit board** (**PCB**). Through-hole PCBs were the first type of circuit board developed and are still used for some applications, though the components are typically inserted and soldered by automated machinery rather than by hand. **Surface-mount technology** (**SMT**) uses PCBs with no mounting holes for the components. Instead, the component and device terminals are formed as conductive areas on the body of the part itself, and the contact areas are soldered to corresponding areas on the SMT board surface directly, without wires. Surface-mount components can be much smaller than through-hole components because no wires are needed. Automated SMT placement machines can locate the parts with great precision, allowing them to be more closely spaced than equivalent through-hole parts. Although SMT boards can be assembled by hand for one-off prototypes, the process is very tedious and error-prone.

The physical size of a resistor determines the maximum amount of power it can dissipate, so do not expect a 1-mm^2 SMT chip resistor to be able to dissipate much more than 100 mW or so without overheating. Although values of resistors available on the market range from a few **mΩ** (**milliohm**, 10^{-3} **Ω**) (for current-sampling resistors in high-current-measurement circuits) to above 1 **GΩ** (**gigaohm**, 10^{9} **Ω**) (for converting extremely small currents into measurable voltages), most analog-circuit designs should use resistors with values between about 1 **Ω** and 100 **kΩ**. Values much smaller than 1 Ω are likely to be affected by wiring resistance, and values above 100 kΩ can show undesirable changes and instability if a circuit is used in a humid environment and a thin film of water forms in parallel with the resistor.

2.2.2 Capacitors

All capacitors are divided into two types: **nonelectrolytic** or "dry" capacitors and **electrolytic** capacitors. The two types are different enough to warrant separate discussions.

2.2.2.1 Nonelectrolytic Capacitors A capacitor consists of two conductive plates or coatings separated by a dielectric (insulator). The value of the capacitance shown by the component is directly proportional to the surface area of the dielectric and inversely proportional to the dielectric's thickness. So in order to provide the most capacitance in a given space, manufacturers try to use the thinnest dielectric possible and make its area as large as possible. Besides determining the value of capacitance, the thickness of the dielectric affects the **rated voltage** of the capacitor. Every capacitor will eventually undergo destructive **breakdown** if the total applied voltage (DC plus peak AC) exceeds its rated voltage. Many physically small nonelectrolytic capacitors have rated voltages of only 100 V or less. The ceramic or plastic dielectric is so thin that voltages in excess of 100 V will cause destructive breakdown, leading to a shorted device and possibly system failure.

Most capacitor values are marked on the body of the device either directly (e.g., "1 μF," meaning 1 **microfarad**, 10^{-6} F) or in terms of **pF** (**picofarads**, 10^{-12} F) in the same way SMT resistors are marked. For example, a capacitor marked "103" usually has a value of 10×10^{3} or 10,000 pF, which is equivalent to 10 nF (**nanofarads**, 10^{-9} F). Voltage ratings for the smaller capacitors are often not marked and must be determined from the catalog description. Nonelectrolytic capacitors are not **polarized** and can usually be reversed (terminals exchanged) without affecting the circuit performance adversely in nearly all cases. Virtually all capacitors with values less than 1 μF are nonelectrolytic capacitors.

2.2.2.2 Electrolytic Capacitors Most capacitors with values of 1 μF or larger are **electrolytic capacitors**. These capacitors are made by separating two conductive sheets of aluminum or tantalum with a thin moist electrolytic paste and then forming a dielectric layer only a few **nm** (**nanometer**, 10^{-9} m) thick on one of the plates electrolytically. Because such a thin dielectric produces a large capacitance per unit area, electrolytic capacitors can show extremely large capacitance values (from 1 μF up into the mF (**millifarad**, 10^{-3} F) range) in packages of reasonable size. However, their method of manufacture means that they can withstand their rated voltage only if the proper **polarity** is observed. This means that strictly speaking, one should not use an electrolytic capacitor in a circuit where AC only appears across it, because half the time the AC waveform will apply the wrong polarity to the capacitor and it may fail prematurely. If an electrolytic capacitor is connected backward so that a large DC voltage is applied to it with the wrong polarity, the heat generated can vaporize the water inside and cause the capacitor to explode! Besides the damage this causes, it can be really embarrassing in a lab full of other students if your circuit literally blows up.

Specially constructed capacitors called **supercapacitors** can provide even more capacitance in a given volume than conventional electrolytic capacitors, by electrochemical means that lie between the conventional dielectric-layer structure of electrolytic capacitors and the chemical processes that take place in a battery (electrochemical cell). Supercapacitors can have values in the multifarad range but tend to have low maximum rated voltages and are more costly than electrolytic units of similar ratings.

2.2.3 Inductors

Just as capacitors store energy in an electrostatic field, inductors store energy in a magnetic field. Because of the physics of magnetic fields, most inductors consist of many **turns** (loops) of insulated wire wound around a central **core**. The core may be hollow (a so-called **air-core** coil) or filled with a material such as iron, steel, or **ferrite** (a type of ceramic) that has desirable magnetic properties. The use of a magnetic core material allows greater inductance to be obtained in the same physical space.

One can always obtain more inductance in a given volume by winding more turns around the core. In order to do this in the same space, however, the wire must be made thinner, and thin wire has more resistance per unit length than thick wire does. The wire's resistance appears in the inductor's equivalent circuit along with the desired property of inductance, and too large a resistance makes the inductor less useful for most applications. Designers of inductors choose the wire size, number of turns, and core material to fit a given application, taking into consideration such factors as space available and the intended frequency range of operation.

In general, inductors are used in analog circuits only when necessary, because for values above a few **microhenries** (μH, 10^{-6} H), these components tend to be bulky, heavy, and expensive. Also, because of the many turns of fine wire in the larger values, moisture or heat can cause differential stresses in the coil that break the wire, leading to failure. Sometimes, an inductor is the best choice for a given application, however, and in some situations, there are no practical alternatives.

Transformers are simply inductors with more than one winding. A special type of transformer called an **autotransformer** uses a single winding with one or more **taps**. A tap is a point where the wire in one winding is brought out of the coil to a terminal and then returns into the winding to continue where it left off, so to speak. Transformers are frequently used in power supplies operating from the AC mains, because they allow the circuit powered to be physically isolated from the AC source for safety reasons. They also have special applications in audio-frequency equipment and radio and wireless systems.

2.2.4 Connectors

Connectors are used in nearly all electronic systems. Despite the popularity of wireless devices that need no physical connection to other devices, most electronic products use connectors internally in order to allow convenient assembly of the subsystems into a finished unit. While the consumer may never see these connectors, their design and function are essential to the operation of the system.

Simple types of connectors include **pin headers** (Fig. 2.3), which are simply conductive pins usually spaced 2.54 mm (0.1 in.) apart, that mate with sockets that are easily attached to **ribbon cables** (flat cables with equally spaced conductors across their width) by means of **insulation-displacement connectors** (**IDPs**). IDPs have sharp-edged receptacles that connect to all the conductors of the ribbon cable in one operation, avoiding the need for the assembler to solder or otherwise deal with each individual conductor. Other types of common connectors include the **D-sub** connector (Fig. 2.3), which is lockable and suitable for interfacing externally to cables outside

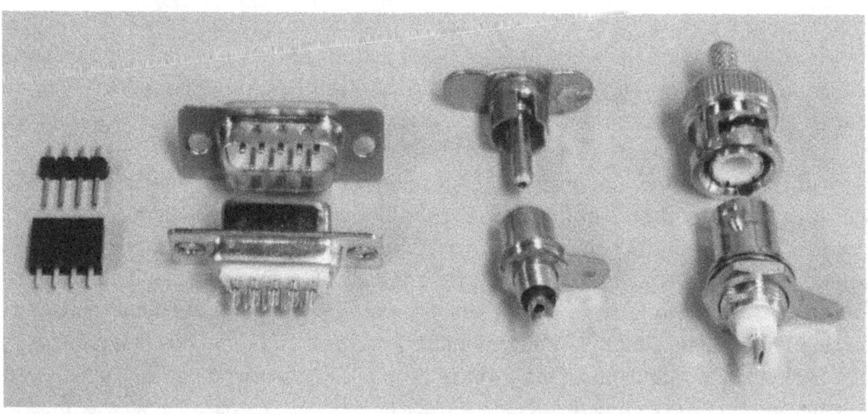

FIGURE 2.3 Common types of connectors used in electronics: pin header, 9-pin D-sub, phono, and BNC.

an equipment enclosure, and the **Universal Serial Bus** (**USB**) connector, which is used with USB-standard serial data interfaces, as well as for supplying small amounts of DC power.

When very-low-level signals or high frequencies are involved, **shielding** becomes an important feature that not all connectors provide. A shield is a conductor that encloses a housing, assembly, cable, or connector so that no electric fields can penetrate from the outside of the shield to the conductors inside. **Shielded cables** can be shielded by means of a braided metal outer conductor, which is typically grounded to the case of the system. Common types of shielded connectors include the **phono connector** (also called the **RCA connector**), the **BNC** connector (Fig. 2.3), and various types of specialized RF connectors for critical and high-power communications applications.

2.2.5 Antennas

An **antenna** provides an interface between an electrical circuit consisting of conductors that are specially designed to carry high-frequency signals (**transmission lines**) and the free space in which a radio signal is transmitted or received. Antennas typically have physical dimensions that are comparable to the wavelength of the electromagnetic wave that is transmitted. (This rule is not as important for receiving antennas, which can be much smaller than a wavelength and still operate reasonably well.) Unlike most other components, an antenna depends on its surroundings for proper operation. For example, placing an antenna inside a sealed metal box will render it useless, because the antenna's electric field cannot penetrate through the conductor. So antennas must be exposed to free space in order to work properly.

Antennas take a wide variety of forms, from the large **dish antennas** used for receiving satellite signals down to antennas small enough to be made as etched patterns on the same circuit board that holds other system components. Regardless of their form, properly designed antennas have equivalent circuits that include a resistance called the **radiation resistance**. Unlike a physical resistor that converts electrical power into heat, the radiation resistance of an antenna absorbs power that is converted into the electromagnetic radiation (radio waves) that the antenna transmits into space. This resistance is usually fairly low, on the order of $50\,\Omega$, so transmitters and receivers designed to use the antenna in question should be designed with the antenna's value of radiation resistance in mind.

With regard to how antennas are connected to a circuit, antennas come in two varieties: **balanced** and **unbalanced**. An unbalanced antenna displays a single impedance between its sole input terminal and its ground terminal, which should be connected to the ground terminal of the system. On the other hand, a balanced antenna has two terminals, neither of which should be grounded. A balanced antenna (or any type of balanced load, for that matter) presents equal impedances from each of its balanced terminals to a third ground terminal. For proper operation, a balanced antenna should be driven with opposite-polarity (180° out of phase) signals applied to its two balanced inputs. A device for converting a balanced antenna to unbalanced (or vice versa) is called a **balun**.

2.3 ACTIVE DEVICES

Active devices can be divided into two major classes: those with two terminals and those with three terminals. (Active devices having more than three terminals, such as integrated circuits, will be considered on a case-by-case basis later in the book.) Two-terminal devices include various types of **diodes** as well as specialized sensors having two terminals. The most useful three-terminal device is the **transistor**, although we will conclude this section with a brief mention of the **vacuum tube**, the transistor's main predecessor.

An intrinsic feature of most active devices is the fact that they are **nonlinear**, which means the relations between voltage and current at the various terminals cannot be expressed only with combinations of equations of the form $y=mx+b$, where m and b are constants, x is the independent variable, and y is the dependent variable. This fact must always be borne in mind when using active devices, although under certain conditions, nonlinear equivalent circuits can be replaced by linear ones.

2.3.1 Diodes

The word "diode" denotes any two-terminal device, strictly speaking, but has come to mean a device that passes current more easily in one direction than the other.

Diodes designed to process low- to medium-power signals are called **signal diodes**. Diodes designed for use in power supplies and for other medium- to high-power applications are called **rectifier diodes**. Diodes designed to emit light (either visible, infrared, or ultraviolet) are called **light-emitting diodes** or **LEDs**. (A subcategory of LEDs are **laser diodes**, which are solid-state lasers that behave electrically like diodes.) A diode designed to detect rather than emit light is called a **photodiode**. Finally, a type of diode used for voltage regulation and references is the **Zener diode**.

2.3.1.1 Signal Diodes In low-power circuits where the signals do not convey much power, signal diodes are used for such applications as **detecting** and **limiting** signals. The schematic symbol and equivalent circuits for a general-purpose diode are shown in Figure 2.4.

A semiconductor diode conducts readily when its **anode** (arrow) is made more positive than the **cathode** (bar) by a voltage that exceeds V_{FB}, the **forward-bias voltage**. (A diode is said to be **forward biased** when its anode is more positive than

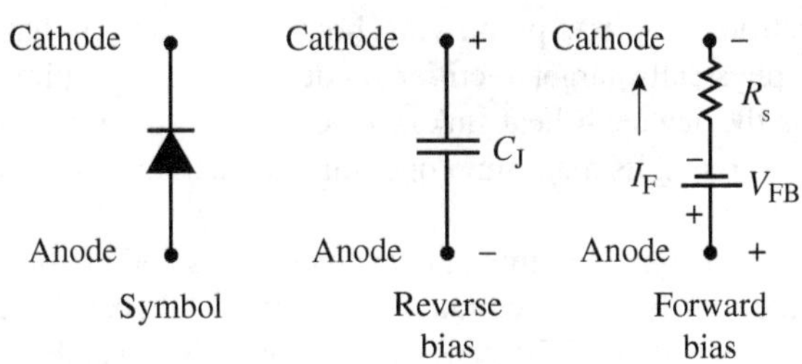

FIGURE 2.4 Diode symbol and equivalent circuits for reverse and forward bias.

its cathode and **reverse biased** when the applied voltage is of opposite polarity to forward bias). The value of the forward-bias voltage for a given current depends on the type of semiconductor that the device is made of. For **silicon** diodes (the most common semiconductor material), V_{FB} is 0.6–0.7 V, depending on the physical size of the device and other factors. For **germanium** diodes, V_{FB} can be as low as 0.3 V. Diodes made with **wide bandgap** materials such as **silicon carbide** (becoming popular in high-voltage and high-power applications) show a V_{FB} of about 0.9 V. For diodes made with **gallium arsenide** or other materials used for LEDs, V_{FB} is usually in the range of 1.6–2.0 V.

When the diode is forward biased, its simplified equivalent circuit consists of a voltage source V_{FB} in series with a resistance R_S. The forward-bias current I_F is determined by the voltage applied to the diode by the external circuit. Because this current flows *into* the positive terminal of the V_{FB} voltage source, the source *absorbs* power, as it should. For small-signal diodes, R_S typically has a value of a few ohms at most, assuming a reasonably large current of a few milliampere or more is flowing.

When the diode is reverse biased, its simplified equivalent circuit is simply a capacitor C_J, ranging in value from a few picofarads for general-purpose diodes down to 50 fF (**femtofarads**, 10^{-15} F) or less for high-frequency devices used for processing RF signals. When a forward-biased diode is suddenly placed in the reverse-bias mode, its equivalent capacitance has a charge in it that must be swept away before the reverse-bias equivalent circuit is applicable. The time it takes to sweep away this charge is expressed in terms of the **turnoff time** of the diode. A special type of diode called a **Schottky diode** is designed to have a very short turnoff time, in the **nanosecond** range (10^{-9} s).

2.3.1.2 Rectifier Diodes The main difference between small-signal diodes and rectifier diodes is that rectifier diodes are made physically larger so as to handle higher current and voltage levels. Every diode has a **reverse breakdown voltage** V_R. This is the reverse-bias voltage that may cause the device to conduct a very large current that, if not limited by the external circuit, will overheat and destroy it. Designers of power supplies and other circuits that place a large reverse-bias voltage on a diode must ensure that the diode's rated V_R is never exceeded, even for a short time.

In power supplies, rectifier diodes are often called upon to conduct large amounts of forward current I_F. The forward-current rating of a diode is not as straightforward as the reverse-voltage rating, because questions of heat dissipation and temperature are involved. We will discuss these matters in more detail in Chapter 10, which is about power electronics. At this point, you should know simply that larger forward currents require physically larger rectifier diodes that may require a **heat sink** to avoid overheating the device. A heat sink is a mechanical structure that helps remove heat from a device so that its maximum operating temperature is not exceeded.

2.3.1.3 LEDs Although experimenters as long ago as 1907 noticed that injecting a current into certain types of semiconducting crystals could cause flashes of light, it was not until the 1960s that **LEDs** became commercially available. First restricted only to the emission of red and infrared light, LEDs now can produce any color of the visible spectrum as well as white and even ultraviolet radiation. LEDs are also much more efficient than their main predecessor, the incandescent lamp. They are used solely in the forward-bias mode of Figure 2.4, and their light output is directly proportional to the applied current. This feature makes linear analog transmission of signals possible by means of LEDs and **laser diodes** operating into a fiber-optic cable link. However, most fiber-optic data is transmitted in digital form.

2.3.1.4 Photodiodes A **photodiode** is designed to detect light of a certain wavelength range. Because fiber-optic cables have their best transmission characteristics in the *near-infrared* wavelength band between 800 and 1600 nm, many photodiodes are designed to operate well in that range. Others operate with visible light as well. Most photodiodes are designed to operate with a reverse-bias voltage, and the current that results from illumination under these conditions is proportional to light intensity.

2.3.1.5 Zener Diodes The **Zener diode** is a special type of diode designed to undergo (nondestructive) breakdown at a specific voltage V_Z, which tends to stay near a constant value despite wide variations in diode current. Zener diodes are used primarily as a **voltage reference** in circuits needing a precise absolute voltage source. Because they operate in the breakdown mode, Zener diodes are used in reverse bias, with the cathode more positive than the anode. The external circuit must limit the current flow to a low enough value so that the power rating of the diode is not exceeded. The equivalent circuit of a Zener diode is the same as the forward-bias circuit of a general-purpose diode in Figure 2.4, except that the Zener voltage V_Z appears as the voltage source instead of V_{FB} and the diode is reverse biased instead of forward biased. Figure 2.5 shows the schematic symbols for the LED, the photodiode, and the Zener diode.

2.3.2 Field-Effect Transistors

The concept of the **field-effect transistor** (FET) was developed as early as the 1920s, but the technology did not then exist to construct a workable commercial device. The word **transistor** was coined by Bell Laboratories researcher John R. Pierce, who suggested the name for a solid-state amplifying device invented by his Bell Labs colleagues William Shockley, Walter Brattain, and John Bardeen in 1948. Until that

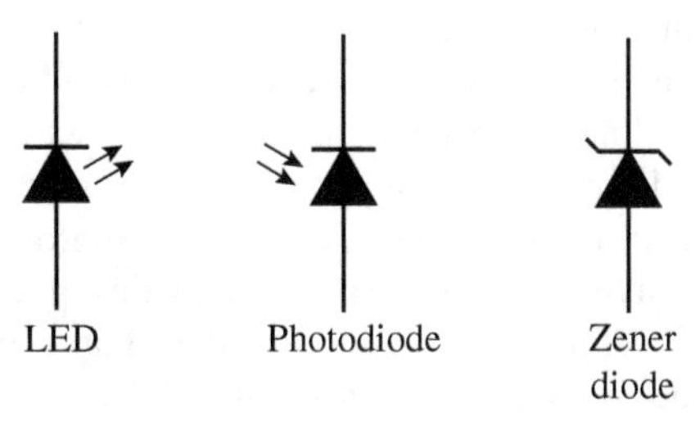

FIGURE 2.5 Schematic symbols for LED, photodiode, and Zener diode.

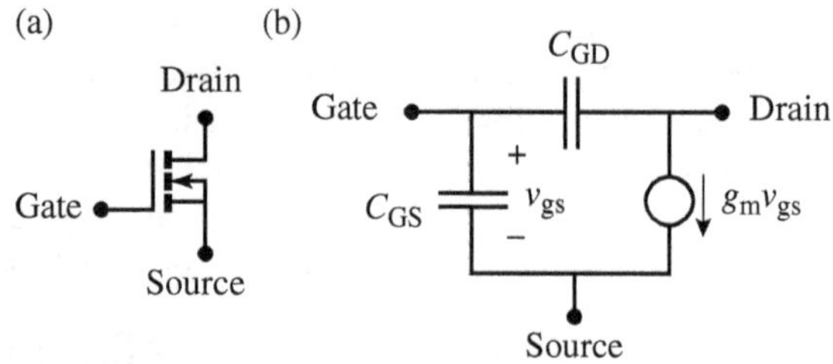

FIGURE 2.6 (a) Schematic symbol and (b) small-signal equivalent circuit for n-channel enhancement-mode MOSFET.

time, the only useful electronic amplifying device was the vacuum tube, which even in its most miniaturized form consumed several watts of power just to warm up enough to be operational and occupied several cubic centimeters of space. The earliest commercial transistors were called **point-contact** transistors, because the connections to them were made via two wire points (the **emitter** and **collector** electrodes) contacting a semiconductor crystal **base**. Even these early handmade transistors were already much more efficient and smaller than the smallest vacuum tubes available at the time. Since then, a large variety of transistor types has been developed, but we will describe only two broad categories in this chapter: **FETs** and **bipolar junction transistors** (**BJTs**).

The transistor will be the first **three-terminal device** to be described in this section. Although a two-terminal device can sometimes be used for amplification, only one voltage appears across it and only one current flows through it. The input and output signals are difficult to separate because they must appear together at both terminals. On the other hand, three-terminal devices allow the physical separation of input and output circuits, because the provision of three terminals on a device allows one terminal to be used for the input signal, one for the output signal, and one to be shared as a **common** or ground terminal. This advantage will become clearer as we examine the various equivalent circuits of the different types of transistors.

Although the BJT was the first type of transistor to be developed, the most important type currently made is the FET. The schematic symbol and one type of equivalent circuit for a common type of FET are shown in Figure 2.6.

The three terminals are called the **gate**, the **source**, and the **drain**. Typically (although not always), the gate is used as the input terminal, the drain as the output terminal, and the source as the common terminal. The complete name for this particular FET is an **n-channel enhancement-mode MOSFET**. The "MOS" in the name stands for "metal oxide semiconductor," which describes how the gate electrode was made in early versions of these devices. Although the gate is now often made of materials other than metal, the insulator between the gate and channel is a very thin layer of quartz (silicon dioxide) or other insulating oxide material. "N-channel" refers to the fact that the charge carriers that conduct current through the device's active region (the "channel") are electrons ("n" stands for "negative," which is the polarity of charge that electrons carry). An

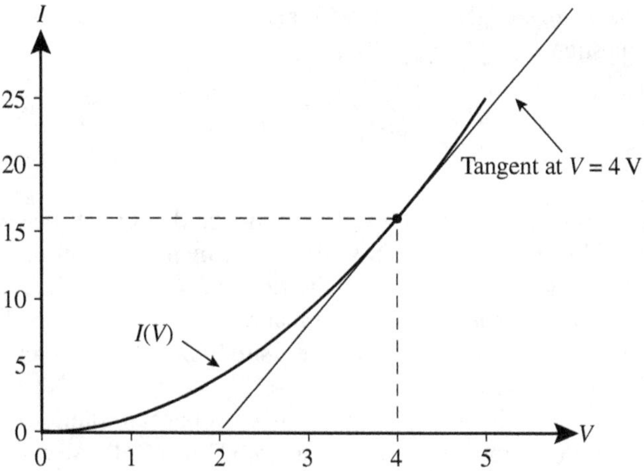

FIGURE 2.7 Parabolic I–V curve illustrating small-signal and large-signal characteristics.

enhancement-mode device does not conduct current through its channel until a suitable gate voltage is applied with respect to the source terminal. The applied voltage is denoted V_{GS} (a capital letter) for DC or total voltages or v_{gs} (a lowercase letter) if the voltage in question is a **small-signal** (low-level) AC quantity. The particular value of V_{GS} that just initiates current flow through the channel is called the **threshold voltage** V_T. In the case of an n-channel enhancement-mode FET, V_{GS} must make the gate positive with respect to the source for the device to conduct current from drain to source.

This is a good place to mention the difference between small-signal linear models and **large-signal** nonlinear models for devices. Although a full discussion of this matter will have to wait until the next chapter, it should be mentioned here. Suppose a certain imaginary nonlinear device has the current–voltage relation

$$I = kV^2. \tag{2.1}$$

where the constant $k = 1\,\mathrm{A\,V^{-2}}$. The I–V curve for this device will be a parabola, as shown by the $I(V)$ curve in Figure 2.7. If the voltage applied across the device is $V = 4\,\mathrm{V}$, the resulting current will be $(1\,\mathrm{A\,V^{-2}})(4\,\mathrm{V})^2 = 16\,\mathrm{A}$. This situation is depicted by the dot on the curve and the corresponding dashed lines indicating the voltage and current at that point.

If you want to know the total DC power dissipated by the device at this voltage, that value is $(4\,\mathrm{V})(16\,\mathrm{A}) = 64\,\mathrm{W}$. This is a large-signal nonlinear result, because it takes into account the nonlinear (quadratic) equation for the voltage–current relationship of the device. But what if you wish to know what happens to a small-signal AC voltage (e.g., 100 mV p-p) imposed on the DC voltage of 4 V? What AC current will result from this small AC voltage?

While you might be tempted to divide the total DC voltage by the total DC current at that voltage to find the device resistance, this is *not* the correct procedure to follow to find the response of the device to small voltage changes. Instead, one must calculate the *derivative* of the current with respect to voltage, evaluated at the DC voltage V in question:

$$\left.\frac{dI}{dV}\right|_V = g(V) = \frac{d}{dV}(kV^2) = 2kV \tag{2.2}$$

For $V = 4\,\text{V}$ and $k = 1\,\text{A}\,\text{V}^{-2}$, we find that the **small-signal conductance** $g(V)$ is therefore $(2)(1\,\text{A}\,\text{V}^{-2})(4\,\text{V}) = 8\,\text{A}\,\text{V}^{-1} = 8\,\text{S}$ (S stands for **siemens**, the unit of inverse resistance). This small-signal conductance is the slope of the straight line that is drawn tangent to the actual I–V curve at the point that represents the DC conditions of 4 V and 16 A, as Figure 2.7 shows. The value of the **small-signal resistance** $r(V) = 1/g(V)$ is therefore $0.125\,\Omega$ or $125\,\text{m}\Omega$.

So, for example, if we impose a 100-mV p-p AC voltage in addition to the 4-V DC voltage, the resulting current will vary from 15.603 to 16.403 A, which is a peak-to-peak variation of 0.8 A. This is almost exactly the small-signal peak-to-peak current i that will result from a linear Ohm's law calculation using the small-signal resistance r that we calculated:

$$i = \frac{v}{r} = \frac{100\,\text{mV}}{125\,\text{m}\Omega} = 800\,\text{mA}. \tag{2.3}$$

This specific example illustrates a general approach to small-signal modeling. The approach consists of first establishing what is called a **DC operating point** using the nonlinear device model and equations and then taking a derivative to find a linear equivalent value for the small-signal model. This approach does not always work as neatly as it did for the hypothetical **square-law** (quadratic) device in the example above, because usually nonlinear functions that characterize real devices have terms higher than second order in them. However, for all but the most nonlinear devices, taking derivatives in this way will produce a useful linear small-signal model that can be used for AC signals that are not too large. What "not too large" means depends on a number of factors, including the importance of distortion in the system in question.

Returning to the FET under discussion, the FET's small-signal model is shown in Figure 2.6b. This is an example of a π equivalent circuit, so called because the three elements are in the shape of the Greek letter "pi."

First, note that the equivalent circuit's gate terminal connects only to capacitances, not resistances. This means that at DC, the gate will not draw any current. This is a significant advantage in many applications, because it reduces **loading** effects on the stage preceding the device. If an AC voltage is applied between the gate and source, obviously current will flow, but because the capacitances involved tend to be small, this current can sometimes be neglected except at higher frequencies.

The capacitance that does the work of operating the device is the **gate-source capacitance** C_{GS}. An AC voltage v_{gs} imposed on this capacitance will cause an AC current $g_m v_{gs}$ to flow between the drain and the source, as the **voltage-controlled current source** between the drain and source terminals indicates. This represents the fact that the device produces a signal at its drain when a signal is imposed between the gate and source terminals. If the **gate-drain capacitance** C_{GD} were zero, one would have perfect **isolation** between the input circuit connected to the gate and the output circuit connected to the drain (assuming they share the source terminal as a common ground). Isolation is a desirable property in amplifiers, because it means that signals at the output of the device cannot travel back to the input to produce undesirable feedback effects, which can lead to loss of gain and oscillation. All real FETs have some gate-drain capacitance, however, so no FET provides perfect isolation. But if the impedance of C_{GD} is large at the frequency of operation, the degree of isolation can be very good.

The variable g_m associated with the voltage-controlled current source is called **transconductance** and has the dimensions of siemens (S) or **millisiemens** (**mS**, 10^{-3} S). The larger g_m is, the larger an AC output current will be obtained for a given AC input voltage, and this translates to larger gain in an amplifier, other things being equal. The value of small-signal transconductance in the equivalent-circuit model depends on the DC **bias** voltages and currents provided to the device by the circuit that uses it. (Bias refers to the DC conditions present in a circuit in the absence of any input signals.) In the small-signal example of Figure 2.7, the bias conditions are that the DC voltage is 4V and the DC current is 16A. Clearly, if we had chosen a different operating point (also called a **bias point** or **Q-point**), the slope of the curve would have been different, and the equivalent small-signal resistance would have changed. In general, one must first know the DC bias conditions for any nonlinear device in order to calculate or estimate its small-signal equivalent circuit.

For completeness, schematic symbols for the major important types of FETs are shown in Figure 2.8.

The n-channel enhancement-mode MOSFET we described earlier is in the center of the top row and will not conduct until the gate-source voltage is made positive. If the device is made so that the channel conducts with *zero* gate-source bias voltage, it becomes a **depletion-mode** MOSFET, and the channel is shown as a solid bar, rather than broken, to indicate the conducting channel at zero gate-source bias. Yet another type of FET is the **junction FET**, in which the gate is isolated from the channel by

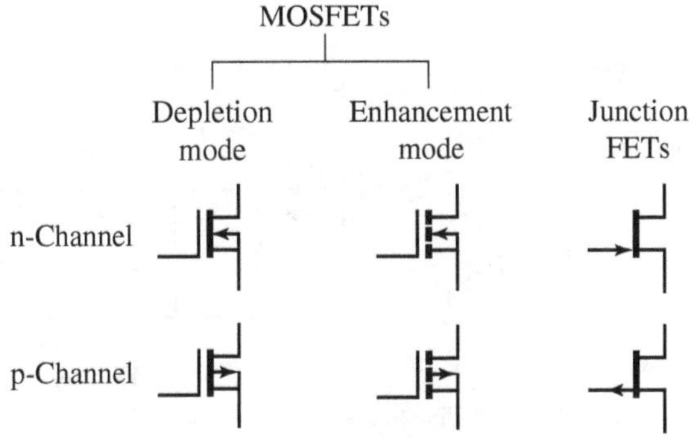

FIGURE 2.8 Schematic symbols for three types of n- and p-channel FETs: depletion-mode, enhancement-mode, and junction FETs (JFETs).

a p–n junction instead of an oxide layer. In all these devices, the n-channel type is indicated by an arrow that points toward the channel.

If p-type material is used for the channel instead of n-type, the device is called a p-channel FET. The only essential difference between p-channel and n-channel devices is that all the device polarities are reversed. So, for example, if an n-channel enhancement-mode FET normally has positive voltages on its gate and drain with respect to the source, a p-channel enhancement-mode FET will require *negative* voltages on its gate and drain with respect to its source. The availability of both p- and n-channel FETs makes for more flexible circuit designs and is the basis for the most popular type of digital logic circuitry used today, called **complementary MOS** or **CMOS**.

2.3.3 BJTs

Although BJTs were the first major type of transistor to be produced commercially, they are now less common than the FET, which excels in digital integrated-circuit applications. However, BJTs are still useful for a variety of applications in which a rugged, reliable device is needed, especially in discrete-component designs for medium and high power. While the oxide insulation used in FETs ensures that no current flows into the gate at DC, the oxide can easily be damaged by stray static discharges such as one's body can accumulate when walking across a carpet on a dry day. BJTs are much less easily damaged in this way than MOSFETs are, and the same rule applies to MOSFET ICs, which are more easily damaged than ICs that use only BJTs.

The schematic symbols and small-signal equivalent circuit for the two main types of BJTs are shown in Figure 2.9. Just as FETs come in two polarities—p-channel and n-channel—BJTs are also made as one of two types: NPN or PNP. These terms refer to the structure of semiconductor layers used to make the device. An NPN device consists of an n-type emitter, a p-type base, and an n-type collector, while with the PNP type, the layer types are reversed. Obviously, there are two p–n junctions in either device, one between the emitter and base and the other between base and collector. As with the FET symbol, the direction of the arrow (in this case on the

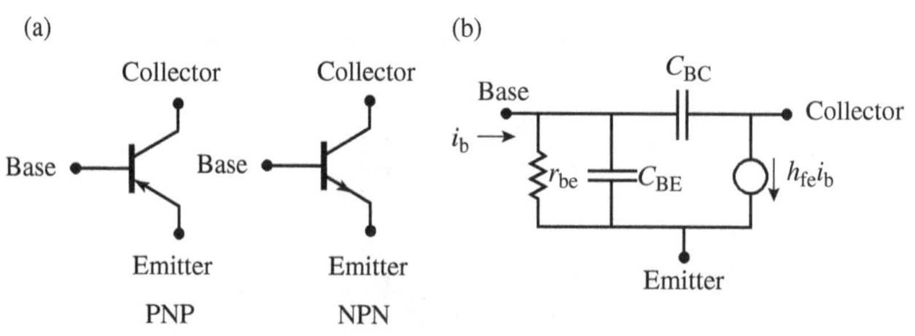

FIGURE 2.9 Bipolar junction transistors (BJTs): (a) schematic symbols for PNP and NPN types and (b) small-signal equivalent circuit.

emitter lead) indicates the type of BJT. The arrow always points in the direction of (conventional) current flow, from positive to negative. Because the emitter of an NPN BJT is always connected to the negative terminal of the power supply, the outward-pointing arrow shows it is an NPN device.

In the most common type of circuit connection using BJTs, the input signal is applied between the base and emitter terminals, and the output is taken between the collector and emitter terminals, making the emitter the common terminal. This type of connection is termed a **common-emitter** circuit. Unlike the FET, however, the BJT base must always carry some DC current I_B in order for the device to function. This current forward-biases the base–emitter p–n junction and allows current to flow from the collector to the emitter.

The small-signal equivalent circuit of the BJT is partly determined by the DC bias conditions established by the external circuit. At a given value of DC current I_B, the equivalent circuit contains a resistance r_{be} in parallel with a capacitance C_{BE}. The value of r_{be} can be estimated from the DC base current I_B as

$$r_{be} = \frac{V_T}{I_B}, \tag{2.4}$$

where V_T is the **thermal voltage**, about 25 mV at room temperature (20°C or 293°K). So, for example, if the base bias current is set at $I_B = 100\,\mu\text{A}$, the value of r_{be} is roughly $(25\,\text{mV}/100\,\mu\text{A}) = 250\,\Omega$. This equivalent resistance is in parallel with the *base capacitance* C_{BE}, whose value depends on the structure of the particular BJT used.

The BJT performs amplification by virtue of an equivalent current source whose output is proportional to the AC small-signal base current i_b. The constant of proportionality is a quantity called h_{fe}, also known as AC β. The value of h_{fe} is a differential quantity that depends on the DC bias conditions as well as the structure of the device. Generally, its value ranges from 25 to 200, meaning that a very small AC base current can cause much larger changes in the collector current. If the total (DC plus AC) collector current I_C is divided by the total base current I_B, the resulting ratio I_C/I_B is called the DC β or β_{DC}.

As with the FET, the BJT is not an ideal device in terms of input–output isolation. The presence of a base-collector capacitance C_{BC} means that some of the output voltage at the collector can be coupled to the input (base) circuit in the common-emitter connection, which shares the emitter between the input and output circuits.

For switching applications, a modified form of the small-signal model applies, although the only practical way to model a BJT's behavior accurately in large-signal and switching circuits is to use the detailed nonlinear models provided in circuit analysis software. However, it is still true even in the large-signal case that collector current is proportional to base current, as long as the device's collector-to-base reverse-bias voltage magnitude is greater than 0.5V or so. If the collector-base reverse-bias voltage falls below this value, the BJT is said to be **saturated**, and if there is no base or collector current, the device is said to be **cut off**. Otherwise, with the base–emitter junction forward biased and the collector–base junction reverse biased, a BJT is said to be in the **active region**. The active region is useful for linear amplification, while most logic circuits using BJTs (not all) switch the devices rapidly between cutoff and saturation.

Historically, the BJT was the first type of transistor used extensively in digital integrated circuits, most commonly the logic family known as **TTL** (for "transistor–transistor logic") However, when **CMOS** logic was developed using both p- and n-channel MOSFETs, the lower power consumption and higher circuit density available with CMOS ICs made most TTL ICs obsolete, although TTL circuits are still used in some applications. BJTs are useful for inexpensive medium- and high-power circuits and for high-frequency circuits operating at frequencies above 1 GHz, although various special types of FETs operate well at those frequencies too.

2.3.4 Power Devices

For handling output power levels up to 250 mW or so, the amount of heat to be dissipated is so small that it can be neglected in most designs, which can use generic or general-purpose components. However, if power in excess of 1 W must be controlled either digitally or by analog (continuous) means, special **power devices** are usually needed. This is because no circuit is 100% efficient, and as the total delivered power rises, the fraction of power dissipated as circuit losses also rises and must be dealt with by devices designed to dissipate medium to large amounts of power.

This is often done by making the devices physically larger and mounting them on thick metal substrates, which in turn must be solidly connected to heat sinks to conduct the heat away where it can be safely dissipated. For most applications, heat sinks radiate, conduct, and convect the heat directly into the surrounding air, although certain special systems use **liquid coolants** that circulate between the hot devices and the cooler **heat exchangers** that transfer the heat to the surrounding air. In the paragraphs to follow, we will describe some of the important features of the various types of power devices that are commonly used in analog and mixed-signal electronics.

2.3.4.1 Power BJTs The maximum electrical power that a BJT can safely control is determined by several factors. One of the most important is the **collector-emitter breakdown voltage** V_{CBO}. Under normal conditions, the collector–base junction is reverse biased, but like any other p–n diode, it has a maximum reverse voltage it can withstand before breakdown occurs. The circuit in which a power BJT is used must be designed so that under no circumstances is the device's breakdown voltage exceeded. If inductive loads such as motors, solenoids, or relays are used, the designer must employ protective devices such as rectifiers to make sure that short-term **voltage spikes** caused by rapid switching of currents through inductors do not reach the power BJT and break it down.

Another factor to consider when using a power BJT is the power needed to drive the base circuit. Power BJTs often show fairly low values of β_{DC} (20 or less), so if a BJT is called on to deliver 20 A at the collector, the base may require as much as 1 A to make the transistor work properly. So a power BJT needs a fairly large amount of power delivered to the base in order to work. The voltage rating (V_{CBO}) of normal BJT technology is limited to 1000 V or less. Also, the base-emitter capacitance must be

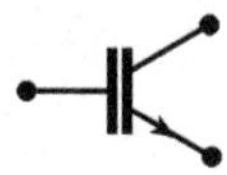

FIGURE 2.10 Schematic symbol of IGBT.

charged and discharged rapidly for fast switching, and this places additional demands on the base driver circuit. Finally, BJTs can exhibit an undesirable property called **thermal runaway**. If a BJT's temperature rises, its **collector leakage current** tends to increase, and if a large collector voltage is applied, the increase in collector current leads to greater power dissipated in the device, which leads to a further rise in temperature, which increases the current even more. A vicious circle results that can destroy the device in a very short time (a few milliseconds). For these and other reasons, power BJTs have been superseded for many applications by other power devices to be described below. However, for certain linear-amplification applications, power BJTs can deliver better performance than other devices.

2.3.4.2 IGBTs The **insulated-gate bipolar transistor** is a hybrid device that shares some characteristics of both BJTs and FETs. Like the FET, its control electrode is called a gate. The gate is insulated with an oxide layer and does not draw DC current, although it does have some capacitance with respect to the other device terminals. But like a BJT, the IGBT has p–n junctions and operates like a BJT with respect to the current that flows through it. IGBTs are normally used in switching applications because they can be made in **modules** that contain several devices connected together in one package. Some IGBT modules can withstand several kilovolts when turned off and conduct 100A or more when turned on, thus controlling a power of several kilowatts. Figure 2.10 shows the schematic symbol of an IGBT. Unlike FETs and BJTs, the IGBT comes in only one polarity and operates with the collector more positive than the emitter.

2.3.4.3 Power FETs A **power FET** has a basic internal structure that resembles a low-power FET: an insulated gate and a channel of either n- or p-type material. But a power FET is constructed so as to withstand high voltages and dissipate a fairly large amount of power while conducting large currents. Although the gate of a power FET draws no DC current, the gate capacitance can be fairly large (several hundred picofarads or more). This requires the gate driver circuit to provide a large transient current when switching the device on or off, in order to place or remove the required charge on the gate capacitance. One of the most important characteristics of a power FET is its value of $R_{DS(ON)}$, which is the equivalent resistance that appears between drain and source terminals when the FET is turned on by a suitable gate-source voltage. The value of $R_{DS(ON)}$ determines how much power is dissipated in the device when it is conducting, so lower values are more desirable. In contrast to the power BJT, power FETs do not show a tendency toward thermal runaway, because the

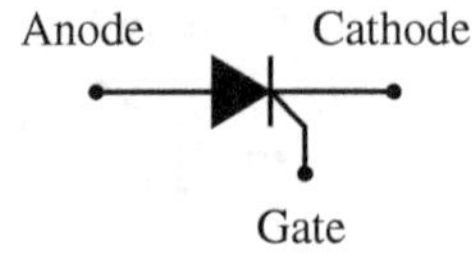

FIGURE 2.11 Schematic symbol of silicon-controlled rectifier (SCR).

channel resistance of a FET typically *increases* at higher temperatures, although other mechanisms such as voltage breakdown are present, which can cause the device to fail. The power FET is capable of faster switching than any other commonly available power device, so they are typically used in power converters with high switching frequencies above 100 kHz.

2.3.4.4 Silicon-Controlled Rectifiers The **silicon-controlled rectifier** (abbreviated **SCR**) is a switching device that is useful for controlling high-power AC circuits. The SCR is a member of a class of power devices called **thyristors**. The SCR's schematic symbol is shown in Figure 2.11. It has an anode and cathode and conducts in only one direction like an ordinary rectifier diode, except for one additional feature: a **gate** electrode, which must be made positive with respect to the cathode for the device to conduct in the forward-bias direction. Once the gate has been **triggered**, however, the gate circuit loses control, and the device continues to conduct even if the gate current falls to zero. The SCR ceases to conduct only when its anode-to-cathode voltage falls to zero. It then resets itself to a nonconducting condition and is ready for the next gate trigger signal.

SCRs find most of their uses in AC power-control circuits operating on mains-frequency (50- or 60-Hz) power. One of the first common uses for SCRs was in light dimmers for incandescent bulbs, and they are often used in industrial applications for motor control and high-power switching. A very simple circuit can provide a gate pulse to the SCR at an adjustable point along the power-line waveform, sending a variable fraction of the entire sine wave to the load and thus varying the delivered power. The zero crossing of the AC waveform automatically cuts off the SCR's current. Because of their inherent "digital" switching nature, SCRs cannot normally be used in linear analog circuits, but they are useful for power-control systems such as motor speed controls.

2.3.4.5 Vacuum Tubes Because of their size, fragility, and low efficiency compared to solid-state devices, vacuum tubes have vanished from electronic systems in all but a few specialized applications. A vacuum tube consists of an evacuated space sealed off inside an **envelope**, typically made of glass (although ceramic and metal are used in certain high-power vacuum tubes). Inside the envelope is a **cathode**, heated by a **heater**, which raises the cathode to a temperature hot enough to make it emit electrons into the surrounding vacuum. Electric fields in the tube attract electrons through a wire mesh called the **grid** to a solid conductive **plate**, where they are collected. The electrodes and heater are connected through vacuum seals to the exterior circuit. Most vacuum tubes are mounted on integral connectors that fit into

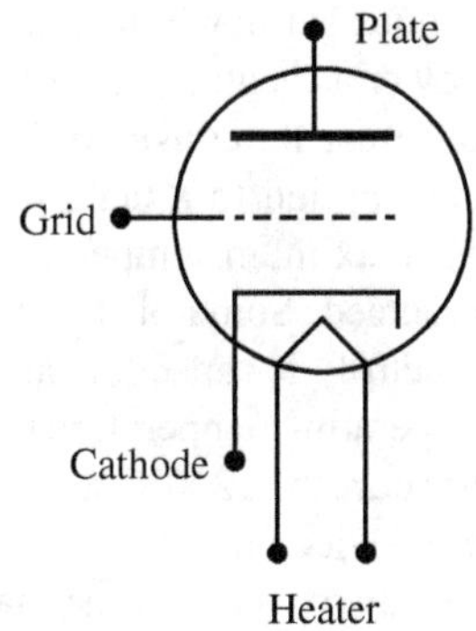

FIGURE 2.12 Schematic symbol of triode vacuum tube.

matching **tube sockets**, allowing for ease of replacement. A small voltage change on the grid varies the flow of electrons from cathode to plate, which is how the tube is used as an amplifier.

The only widespread consumer device that still uses vacuum tubes as of this writing (2013) is the **microwave oven**, which employs a diode (gridless) tube called a **magnetron** to produce 2.45-GHz energy that is absorbed by the water in foods and drinks. Other applications of vacuum tubes include very-high-power RF amplifiers (1 kW and higher), microwave amplifiers, and X-ray sources. In recent years, a small but growing number of musicians have insisted on using audio equipment such as guitar amplifiers which incorporate vacuum tubes, in the spirit of violinists who will not perform on any instrument made later than 1750.

The schematic symbol of a **triode** (three-electrode) tube is shown in Figure 2.12. Electrically, a vacuum tube behaves like a high-voltage n-channel depletion mode FET, once the heater power of a few watts is applied to make it operate.

2.3.4.6 Heat Sinks and Thermal Calculations Although **thermal design**, which means the design of a system with regard to heat flow and internal temperatures, is more mechanical than electrical engineering, electrical engineers should know enough about the subject to be aware of the basic problems that arise.

Any component that absorbs real power (as opposed to reactive power, which is stored rather than dissipated energy) typically transforms the power into heat. The exceptions are **transducers** that convert electrical energy into a form other than heat. LEDs, speakers, and motors are examples of transducers that transform at least some of electrical energy fed to them into other forms such as light waves, acoustic waves, or mechanical motion. But few transducers are 100% efficient, and the energy that is not converted into the desired form also ends up as heat.

Heat energy is measured in **joules**, which in the SI system of units are numerically the same as the **watt-seconds** familiar to electrical engineers. The *rate* of heat production is therefore measured in joules per second, or **watts**, so no conversion factors are needed in order to figure out how much heat flows from a device that is dissipating 10W of electrical power as heat: it's 10W!

The reason heat dissipation must be considered, especially with high-power equipment, is that when heat is applied to a material, its temperature rises. And electronic components all have maximum temperature ratings above which their proper performance is not guaranteed. Some of the electronic components that are most sensitive to heat include electrolytic capacitors and semiconductor devices. For silicon, the highest permissible operating temperature is about 150–200°C. Operating a silicon device at a higher temperature than 200°C will at first degrade its performance and eventually lead to premature device failure.

Fortunately, there is a simple way to calculate the maximum temperature a device will reach if you know the heat flow P_D (in W), the maximum **junction temperature** T_J that the device can tolerate (in °C), the **ambient temperature** T_A (the temperature in °C of the environment where the system will be used), and parameters relating to the thermal behavior of the device's package and the characteristics of the heat sink used. These parameters are called **thermal resistances**, because you can draw a simple circuit using thermal resistances to represent the flow of heat from the device junction to the ambient temperature. A thermal analog of Ohm's law can then be used to calculate T_J.

Figure 2.13a shows a cross section of a packaged semiconductor device (e.g., a power transistor) mounted on a heat sink. The silicon device **substrate** (the inert part of the silicon crystal that supports the device mechanically) is bonded to a **mounting tab**, typically made of plated copper, which extends beyond the plastic **encapsulation** that seals and insulates the device from the outside environment. The heat generated in the device flows through the mounting tab to the surface of the heat sink, which is usually made of a good thermal conductor such as aluminum. (Sometimes, there is an electrical connection made to the mounting tab, which means that a thin insulator must be placed between the mounting tab and the heat sink to avoid electrical shorts.) The dashed lines indicating heat flow go to the **fins** of the heat sink, which have a large surface area that allows both radiation and convection by air to carry away the heat transferred from the device. If large

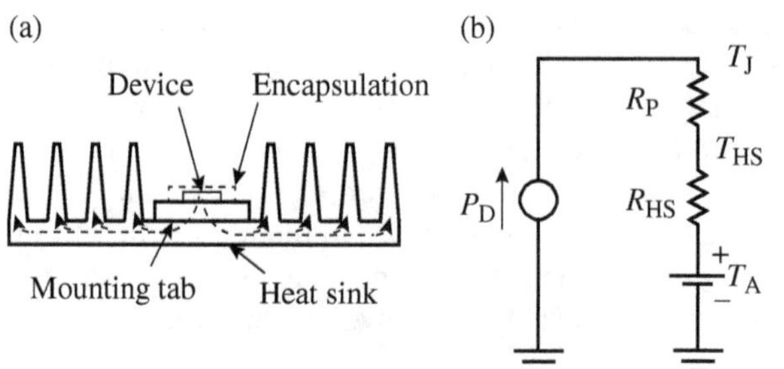

FIGURE 2.13 (a) Cross section of device mounted on heat sink, showing heat flow path (dashed lines). (b) Equivalent thermal circuit for junction temperature analysis.

amounts of power are involved, **forced-air** cooling using one or more fans can be used. If no fans are involved, the fins should be oriented vertically so that the heated air can rise past them freely as it expands for best heat transfer.

The heat "circuit" of this situation is shown in Figure 2.13b. The heat power developed by the device is represented as P_D (measured in W) but is treated as an ideal current source in the circuit. This is because the heat flow equivalent of Ohm's law is

$$T = PR \tag{2.5}$$

where T is the *temperature drop* across an element in the path of heat flow (in °C), P is the heat flow (in W), and R is the thermal resistance of the element (in $°C\,W^{-1}$). As you can see, T is like a voltage drop, P is like current flow, and R is like electrical resistance. Given the simplifying assumptions that enable this type of analysis, it can be reasonably accurate and useful for more complex thermal "circuits" than the simple one shown in Figure 2.13.

Here is an example of how to use this analysis method. Suppose a certain device has to dissipate $P_D = 60\,W$ while staying below a maximum junction temperature $T_J = 150°C$. The system that uses the device may be required to operate in an environment as hot as $T_A = 40°C$ (104F). The device's specification sheet states that the thermal resistance of the case is $R_C = 1.5°C\,W^{-1}$. What is the maximum thermal resistance R_{HS} that the heat sink can have and still not exceed the maximum junction temperature when the ambient temperature is 40°C?

From the maximum junction temperature $T_J = 150°C$, we know the "voltage" at the top of the two "resistors" in series. We also know the "voltage" at the bottom, namely, $T_A = 40°C$. The "current" flow (heat flow, really) is $P_D = 60\,W$. As we are given the value of R_C, the problem is to find the value of R_{HS}. The solution is

$$R_{HS} = \frac{T_J - T_A}{P_D} - R_C = \frac{(150-40)°C}{60\,W} - \frac{1.5°C}{W} = 0.33°C\,W^{-1} \tag{2.6}$$

The value of $0.33°C\,W^{-1}$ is a rather low number that might be difficult to attain with a heat sink operating by free-air convection alone, although a forced-air heat sink cooled by a fan would probably work well. If the maximum junction temperature were higher (200°C is sometimes used) or the maximum ambient temperature lower (30°C instead of 40°C is more applicable to devices used only in air-conditioned spaces), the allowable heat sink thermal resistance R_{HS} could be larger, meaning that a smaller heat sink would work. Manufacturers of heat sinks usually provide a thermal-resistance figure for their products or else describe how to calculate it under a given set of conditions.

BIBLIOGRAPHY

Platt, C. *Encyclopedia of Electronic Components*, 1 and 2. Sevastopol, CA: Maker Media, Inc., 2012.

PROBLEMS

2.1. *Footprint of surface-mount resistors.* Suppose the "footprint" (area covered) of a 1/8-W through-hole resistor is 5.8 mm long by 1.8 mm wide. If 30 such resistors fit into a given area of through-hole circuit board, about how many surface-mount resistors having a footprint of 1 mm × 1 mm will fit into the same area?

2.2. *Comparative volumes of conventional and electrolytic capacitors.* Here is a problem that shows why electrolytic capacitors can have much larger capacitance per unit volume than nonelectrolytics can. The basic equation for the capacitance C of a **parallel-plate capacitor** with a dielectric of area A, thickness t, and **relative dielectric constant** ε_r (a material property) is

$$C = \frac{(8.854\times10^{-12}\ \mathrm{F\ m^{-1}})\varepsilon_r A}{t} \tag{2.7}$$

(a) What is the volume V of the dielectric layer in a 1-μF capacitor using a plastic-sheet dielectric that is $t_1 = 0.15$ mm thick and has a relative dielectric constant $\varepsilon_{r1} = 2.7$? (Calculate the area A_1 required, and then multiply by t to obtain the volume.) Call this volume V_1. The thickness of the conductive sheets on either side of the dielectric layer can be neglected because they are typically evaporated films only a few nm thick.

(b) Now assume the capacitor is an electrolytic unit with a dielectric whose thickness is only $t_2 = 100$ nm and whose relative dielectric constant $\varepsilon_{r2} = 9.6$ (aluminum oxide). Assume the conductive metal foils on either side of the dielectric are the same total thickness as the plastic film used in (a) above, namely, 0.15 mm. Calculate the area A_2 required to make a 1-μF capacitor this way, and then calculate its volume by multiplying A_2 by 0.15 mm. Then find the volume V_2 of the second (electrolytic) capacitor, and finally, calculate the ratio V_2/V_1 to see how much smaller the electrolytic is than the plastic-film capacitor for the same value of capacitance.

2.3. *Parameters of inductor design.* You must evaluate two different designs for an inductor. Design *A* uses 20 m of #18 AWG wire and occupies 20 cm^3. Design *B* uses 20 m of (much thinner) #32 AWG wire and occupies only 0.8 cm^3 yet has the same total inductance. Suppose the circuit in which the inductor will be used can tolerate a maximum of only 3 Ω of equivalent series resistance in the inductor's equivalent circuit. Assuming the series resistance is solely due to the DC resistance of the wire, which, if either, design can be used? (The DC resistance of #18 wire (1.04 mm dia.) is 21 mΩ m^{-1}, and #32 wire (0.2 mm dia.) has 538 mΩ m^{-1} resistance.)

2.4. *Saturation current and AC resistance of diode.* Although we showed two different equivalent circuits in Figure 2.4—one for reverse bias and one for forward bias—there is a single (nonlinear) equation that gives a fairly good approximation of a diode's current as a function of applied voltage over a wide range of both positive and negative applied voltages. If the applied voltage is V, the diode's physical temperature is T, and its "saturation current" (a device parameter) is I_S, the diode's current is approximated by

$$I(V) = \mathrm{I_S}\left(e^{\frac{qV}{k_B T}} - 1\right) \tag{2.8}$$

where q=charge on the electron (1.6×10^{-19} C), k_B=Boltzmann's constant (1.38×10^{-23} J K^{-1}), and T is the absolute temperature in degrees Kelvin (K=C+273.15).

Suppose a certain diode conducts a current I=3.5 mA at a voltage V=0.67 V, when the diode is at room temperature (T=20°C).

(a) Neglecting the −1 term in parentheses, find the diode's saturation current I_S.

(b) What is the small-signal AC resistance r of the diode at this voltage and current? (*Hint*: Take the derivative $dI(V)/dV$ to find the small-signal conductance g, and invert to find r.)

2.5. *LED bias design.* A certain type of white LED requires 2.2 V at 10 mA to emit the amount of light required for a specific application. If the only power supply available provides 5 VDC, what is the value R_{LED} of a *current-limiting resistor* required in series between the power supply and the LED? How much power P_R is dissipated in the resistor?

2.6. *Capacitance of diode at high frequencies.* At high frequencies, the reverse-bias capacitance of a signal diode can limit its performance as a *detector* (rectifier for small signals). Suppose a signal source at a frequency f=2.4 GHz has an *output impedance* of Z_{OUT}=50 Ω. If you model a detector diode as a simple capacitor with junction capacitance C_J:

(a) Draw the equivalent circuit of the signal source as a voltage source V_S in series with a resistance Z_{OUT}, and connect it to the junction capacitance C_J that returns the signal to ground.

(b) Calculate the maximum value of C_J that will allow 70.7% of V_S to appear across the diode capacitance. This problem shows why a high-frequency signal diode must have a small junction capacitance.

2.7. *Transition frequency of FET.* Suppose a FET is connected to an ideal small-signal voltage source v_S at the gate-source terminals and a zero-resistance current ammeter that measures the small-signal output current i_D from drain to source (see Fig. 2.14).

Calculate the frequency f_T at which the magnitude of the current I_G flowing into the gate terminal equals the magnitude of the equivalent current-source current $g_m v_{gs}$. Express your answer in terms of the parameters C_{GD}, C_{GS} and g_m. This frequency is called the *transition frequency*, because it is an estimate of the frequency

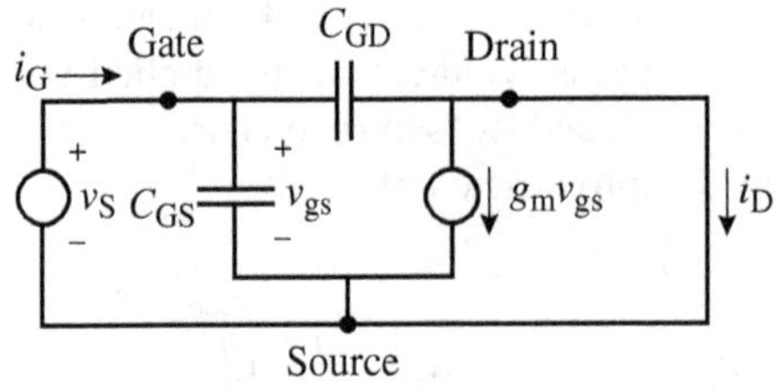

FIGURE 2.14 Circuit for measurement of FET transition frequency.

at which the device's current gain transitions to less than unity. The device usually cannot work as an amplifier for frequencies above f_T.

2.8. *Heat loss and efficiency of power FET.* Suppose you are designing a system in which a power FET is on for 50% of the time and off for the other 50% of the time, with a switching frequency between these two states of 1 kHz. When the device is off (no current flow), the applied voltage is 550 V. When the device is on, it conducts a current of 45 A. The $R_{DS(ON)}$ of the device is specified to be 50 mΩ.

(a) What is the average power $P_{D(AVG)}$ dissipated in the device?

(b) If the total thermal resistance R_{TH} from the device junction at temperature T_J to the ambient temperature T_A is 1.2°C W^{-1}, how hot will the junction become if $T_A = 40$°C?

(c) If the total applied voltage of 550 V is reduced by the voltage drop through the device while it is on, calculate the average power P_L delivered to the load. What is the system efficiency, measured as the fraction P_L/P_{IN}, where P_{IN} is the total average input power?

For further resources for this chapter visit the companion website at
http://wiley.com/go/analogmixedsignalelectronics

3

LINEAR SYSTEMS ANALYSIS

3.1 BASICS OF LINEAR SYSTEMS

Most electronic systems involving analog circuitry can be modeled mathematically to a high degree of accuracy with the aid of a discipline called **circuit theory**. In the previous chapter, we used circuit theory when we described equivalent circuits for various types of devices. In circuit theory, one assumes that a system is adequately described when it is treated as a **network** consisting of **nodes** interconnected by **branches**. Figure 3.1 shows the FET equivalent circuit of Figure 2.6 reproduced as Figure 3.1a.

The geometric essentials of this circuit are captured by its **network graph** shown in Figure 3.1b. In circuit theory, a node is any point where a unique voltage is defined (with respect to any other node), and a branch is a path between any two nodes through which a unique current is defined. If we know the relative voltages of all the nodes and the currents through all the branches, we know everything there is to know about the network, and it is said to be completely *analyzed*.

Not all circuits can be treated adequately by circuit theory. As mentioned in Chapter 1, if the frequencies involved are so high that the physical size of the circuit is no longer small compared to a wavelength, it is not necessarily true that, for example, a wire (representing a branch) carries the same current throughout its entire length, as circuit theory assumes. In such situations, the conventional node-and-branch circuit model is not applicable, and at least some parts of the circuit must be analyzed with more sophisticated techniques described in Chapter 11, which covers

Analog and Mixed-Signal Electronics, First Edition. Karl D. Stephan.

Companion Website: http://wiley.com/go/analogmixedsignalelectronics

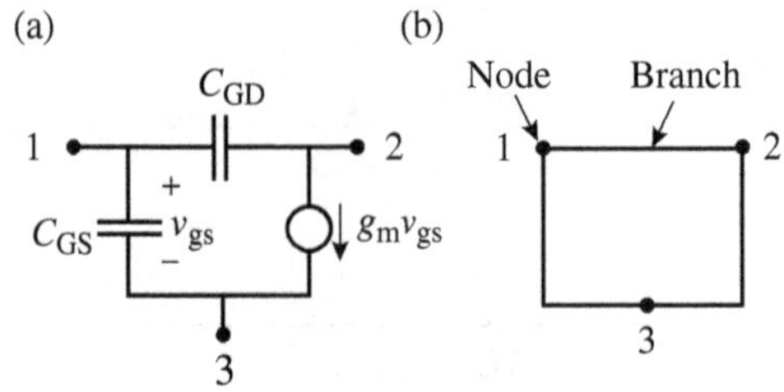

FIGURE 3.1 (a) FET equivalent circuit from Figure 2.6. (b) Network graph version showing nodes 1, 2, and 3 and branches connecting nodes.

high-frequency electronics. But with that caution in mind, we can proceed under the assumption that any circuit we deal with in this chapter is well modeled by circuit-theory analysis.

Before a complete network can be analyzed, it is necessary to know the current–voltage relationships that apply to each of its branches. In the example of Figure 3.1a, for instance, two capacitors and one voltage-dependent current source comprise the branches between the nodes, and the nature of those components establishes the relationships among the voltage differences between each node and the currents through each branch. If mathematical functions can be found to express the relations between voltage and current in each branch, one can solve for the currents and voltages everywhere in the network, at least in principle.

However, unless the mathematical functions are **linear** ones, the resulting set of equations involves nonlinear terms. Except for some simple cases, it is not possible to solve nonlinear sets of equations **analytically**—that is, in a way that allows you to write down an equation without doing numerical calculations first. Fortunately, computer software has been developed to solve systems of nonlinear equations. When you draw a diagram and indicate the components and their interconnections, the software uses circuit-theory principles to create a large matrix of nonlinear equations and then solves the matrix. While this usually works, it does not always give the user a detailed understanding of how the circuit operates.

That is why we will present mainly linear analysis in this chapter. Linear systems of equations can be solved in a straightforward way, leading to explicit equations that are easy to write down, at least for simple circuits. So while most actual circuits involve nonlinear elements, they can often be analyzed with linear approximations that give good results over a wide range of operating conditions. And the linear analysis is much easier to comprehend and to perform.

3.1.1 Two-Terminal Component Models

At low frequencies—say, below 10 kHz or so—most passive components such as resistors, capacitors, and inductors can be modeled well by a single-element equivalent circuit. Before we describe the specific equivalent circuits appropriate for passive components, we will establish some general principles about how to express the relationship between voltage and current in a two-terminal component.

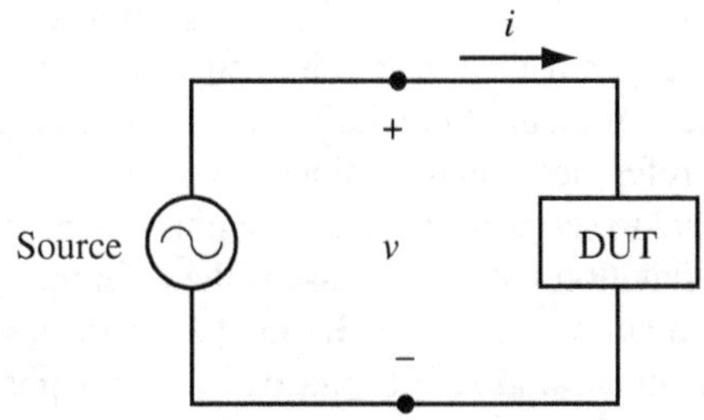

FIGURE 3.2 Conceptual measurement of two-terminal device's current–voltage characteristics. "DUT" stands for "device under test."

We can imagine characterizing such a component by imposing a voltage on it from an **ideal voltage source** (a mathematical fiction whose voltage is constant no matter what is connected across it) and measuring what the resulting current is. Because any periodic waveform can be expressed as a sum of sine waves at different frequencies that represent the **Fourier-series** terms of the sum, all we need to know about a two-terminal component in order to model it is how it responds to sine waves of various frequencies. Figure 3.2 shows one conceptual way this response can be measured for all two-terminal devices (except for ideal short circuits).

In Figure 3.2, a sinusoidal voltage $v(t)$ is applied to the **device under test (DUT)**, and a current $i(t)$ results. (We use lowercase letters for sine-wave AC quantities and capital letters for DC or total quantities, remember.) If the source is a sinusoidal function of time with peak value V_0 at a radian frequency ω (rad s^{-1}), its voltage can be written as

$$v(t) = V_0 \cos(\omega t) \tag{3.1}$$

In that case, the resulting current can be written as

$$i(t) = I_0 \cos(\omega t + \phi_0) \tag{3.2}$$

where I_0 is the peak value of the resulting current and ϕ_0 is the phase of the current waveform with respect to the voltage waveform.

A more concise way to represent these quantities is by means of **phasor notation**. **Euler's formula** $e^{jx} = \cos(x) + j\sin(x)$ (where j is the square root of -1) provides a way to describe both the phase and the amplitude of a sinusoidal wave by means of a *single* (complex) number called a **phasor**. In this book, we will distinguish phasor quantities from other types of variables, when necessary, with a mark called a **right harpoon**, which looks like a half arrow: $\vec{v}$ is the phasor form of the voltage $v(t)$, for example.

All voltages and currents that you can actually measure are real, not imaginary or complex, so how does using complex numbers help? Here is how. If we let Re(z) mean the **real part** of the complex number z, we can express the real voltage $v(t)$ in Equation 3.1 as the real part of a complex number:

$$v(t) = \mathrm{Re}(\mathrm{V}_0 e^{j\omega t}) \tag{3.3}$$

(The real voltage magnitude V_0 is multiplied by the complex exponential whose real part is $\cos(\omega t)$, and we get back Eq. 3.1 immediately). Because the phase of a sinusoidal wave is always measured in relation to a **reference phase** (just as voltage is always relative to a reference voltage point), if you do an analysis using phasors, you must choose a signal to serve as the reference phase, which is zero by definition. In our case here, the excitation voltage's phase is the reference phase.

To express the information in Equation 3.1 in phasor form, we must indicate both the **magnitude** and the **phase angle** of the phasor. Using the phasor $\vec{v}$ as an example, we can express Equation 3.1 as

$$\vec{v} = V_0 \angle 0 \tag{3.4}$$

The leading variable V_0 is the magnitude of the complex number, and the $\angle$ symbol means "having an angle of." Because $\vec{v}$ is the reference phase, its phase angle is zero. We can now express Equation 3.2 in phasor form as

$$\vec{i} = I_0 \angle \phi_0, \tag{3.5}$$

which is shorthand for

$$\vec{i} = I_0 e^{j(\omega t + \phi_0)} \tag{3.6}$$

Phasors are most useful when all the voltages and currents have the same frequency ω. In that case, the $e^{j\omega t}$ factor is shared by all the phasors and can be suppressed (omitted).

With phasor notation well in hand, we can now use it to discuss the characterization of a two-terminal network.

If a component is linear, the ratio of voltage to current will always be a constant, no matter what the actual magnitudes of voltage or currents are (assuming it is small enough to stay within the device's rated limits, of course). The only item of interest about the component from a circuit point of view is the *ratio* of voltage to current, which in the case of phasors is called the **impedance Z**.

Impedance is defined as

$$Z \equiv \frac{\vec{v}}{\vec{i}}, \tag{3.7}$$

and because phasors are complex numbers in general, impedance can also be complex. Because it is the ratio of a voltage to a current, impedance has the dimensions of ohms. Like all other complex numbers, any impedance has both a real and an imaginary part. The real part is termed the **resistive component**, and the imaginary part is termed the **reactive component** of the impedance. If we denote the resistive component by the variable R and the reactive component by the (real) variable X, impedance can be written as the sum:

$$Z = R + jX \tag{3.8}$$

Of course, either or both of the terms in the sum may be zero. However, an important point to note concerns the direction of **power flow**. The way the voltage polarity and the current direction in Figure 3.2 are shown, if R is *positive* in Equation 3.8, the component *absorbs* power. That is, power flows from the circuit into the component and doesn't come back again. Conversely, if R is *negative*, the component *delivers* power to the rest of the circuit (in this case, to the voltage source that it is connected to). Unless a component receives energy from the outside world (e.g., as a **transducer** or **generator** does) or *stores* energy somehow over the long term (e.g., as a **storage battery** does), it cannot deliver average power to the circuit, and its real part must therefore be positive or zero. Even transducers and batteries usually have a positive AC equivalent resistance for sine-wave excitation, so impedances with negative real parts are rarely encountered.

On the other hand, the X part of the impedance can be either positive or negative. A positive X is termed **inductive**, because an ideal inductor's reactance is $+jX$, where X is a positive (real) number. A negative X is termed **capacitive** for a similar reason. These are straightforward consequences of the facts that for sinusoidal excitation in an ideal inductor, the current waveform lags the voltage waveform by $\pi/2$ rad (90°), and in an ideal capacitor, the current leads the voltage by 90°.

Just to wrap up our discussion of terminology, an impedance that is purely imaginary (real part equal to zero) is called a **reactance**, while one that is purely real (no imaginary part) is called a **resistance** or "resistive impedance." Just as all bugs are insects but not all insects are bugs, all reactances are impedances, but not vice versa.

3.1.1.1 Equivalent Circuit of Capacitor At low frequencies, the two-terminal equivalent circuit of a capacitor is simply that—a capacitor whose value is the **nominal** value of the component in question. (*Nominal* means "pertaining to a name" so the nominal value is what the label says it is.) However, at higher frequencies, some capacitors begin to behave mysteriously, as though their capacitance value was increasing. At sufficiently high frequencies, every capacitor will cease to show capacitive reactance and instead will begin to show inductive reactance.

This is because it is impossible to build a capacitor without leads, and all leads have some inductance. This inductance is most conveniently shown as an inductor in series with the original nominal-value capacitor in the equivalent circuit.

Finally, although the losses in most capacitors are very low, any current flowing in a capacitor encounters some resistance in the conductors of which the device is composed. Also, it is possible that the dielectric material between the capacitor plates absorbs some of the electric-field energy, especially at high frequencies, even though the material's DC conductivity is zero. All these **losses** are usually modeled by an **equivalent series resistance** (abbreviated as **ESR**), which appears in series with the equivalent capacitance and inductance. Because the inductance is an undesired "parasite" on the desired property of capacitance, it is called **parasitic inductance**. Figure 3.3 shows the equivalent circuit of a capacitor with parasitic inductance L_{PAR} and ESR R_{PAR}.

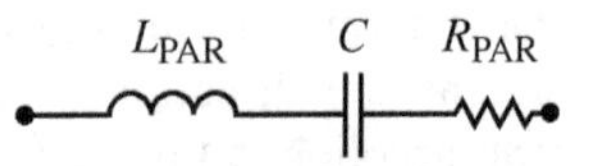

FIGURE 3.3 Equivalent circuit of capacitor showing parasitic inductance L_{PAR} and equivalent series resistance R_{PAR} in series with capacitance C.

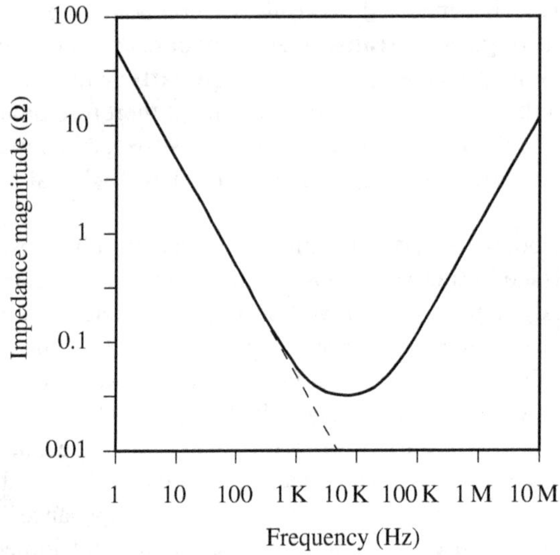

FIGURE 3.4 Impedance magnitude of capacitor equivalent circuit of Figure 3.3 with $C = 3.2\,\text{mF}$, $L_{PAR} = 187\,\text{nH}$, and $R_{PAR} = 33\,\text{m}\Omega$. Dashed line indicates ideal capacitor impedance.

The impedance Z_{CAP} of this equivalent circuit is easily found, because it is the series combination of the impedance of each element:

$$Z_{CAP}(\omega) = j\omega L_{PAR} + R_{PAR} + \frac{1}{j\omega C} \tag{3.9}$$

Clearly, the impedance of this equivalent circuit will resemble that of a capacitor only at low frequencies, when the capacitive term dominates the resistive and inductive terms. In Figure 3.4, we have plotted the theoretical impedance magnitude $|Z_{CAP}|$ for a capacitor modeled with these values: $C = 3.2\,\text{mF}$ (*millifarads*, 10^{-3} F), $L_{PAR} = 187\,\text{nH}$, and $R_{PAR} = 33\,\text{m}\Omega$.

These values were measured from an actual electrolytic capacitor whose nominal value was 3.3 mF. As you can see, at frequencies below about 100 Hz, the impedance magnitude of the model tracks that of an ideal capacitor (dashed line) almost perfectly. But above 100 Hz, the model's curve begins to deviate upward from the ideal curve and reaches a minimum at about 6.5 kHz. This frequency f_{SR} is called the **self-resonant frequency** of the capacitor and can be found from the familiar resonant-frequency formula:

$$f_{SR} = \frac{1}{2\pi\sqrt{L_{PAR}C}} \tag{3.10}$$

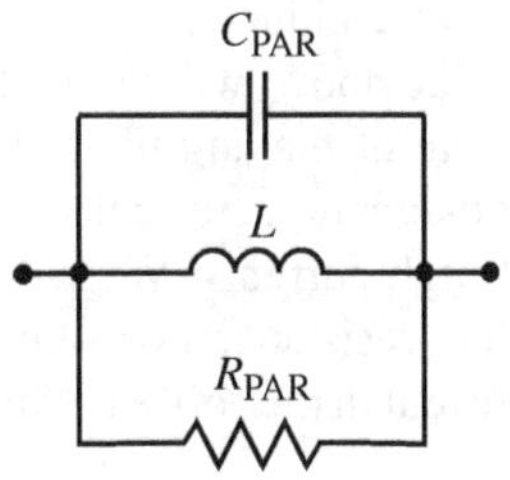

FIGURE 3.5 Equivalent-circuit model of inductor with inductance L, parasitic capacitance C_{PAR}, and parasitic resistance R_{PAR}.

This is the frequency at which the capacitive reactance of C equals the inductive reactance of L_{PAR}. The two opposite-sign reactances cancel out, leaving only the resistance R_{PAR} to be seen by the external circuit.

Above f_{SR}, the capacitor behaves like a lossy inductor, and the impedance rises with frequency. Nevertheless, the absolute magnitude of the impedance is low for a broad range of frequencies. For example, if this capacitor is used for **bypassing** a power-supply line, which requires only a low impedance to ground, it may operate as intended if its impedance magnitude is less than a certain value, say, 1 Ω. The capacitor model shown in Figure 3.4 meets that criterion for the frequency range of 50 Hz to 800 kHz. If effective bypassing is needed at higher frequencies as well, the best thing to do is to put a much smaller capacitor (say, 10 nF) in parallel with the large electrolytic one. That way, at frequencies where the larger capacitor acts like an inductor, the smaller capacitor will still behave like a capacitor, and the overall bypassing function will not be compromised.

While the values of an equivalent-circuit model for a given capacitor will vary a great deal, most capacitors are well modeled by the circuit in Figure 3.3. All capacitors have a series-resonant frequency beyond which they cease to act as capacitors, although they may be useful for some purposes above f_{SR}.

3.1.1.2 Equivalent Circuit of Inductor Although inductors are not as common as capacitors in most analog circuits, their behavior is such that an equivalent-circuit model is usually needed to take into account losses and a feature of all inductors called **parasitic capacitance**. Figure 3.5 shows an inductor equivalent circuit, which is usually adequate to model behavior over a wide range of frequencies.

Besides the inductance L, any coil's windings have the property that differences of potential between the turns give rise to electric fields between the turns. Although the effects of these electric fields can be very complex in general, their net effect for coils that are not too large can usually be adequately modeled by a single parasitic capacitance C_{PAR} in parallel with the equivalent inductance L.

All coil windings have some resistance (even **superconducting** coils that have to be cooled to liquid nitrogen temperatures or below have resistance at AC), and this resistance can be modeled in at least two ways: as a resistor in parallel with the inductance or in series with it. In Figure 3.5, we have chosen to place the resistance

R_{PAR} in parallel with L and C_{PAR}. Some inductors use a core made of magnetic material that increases the inductance value above what would be obtained with a nonmagnetic core. This has the advantage of making the inductor physically smaller for a given value of inductance. But magnetic cores can have losses of their own and can even produce nonlinearities at high currents. We will assume that whatever losses occur in either the coil's winding resistance or its core are adequately modeled by R_{PAR} in Figure 3.5 and that the current in the coil is not high enough to cause significant nonlinear effects.

The parallel equivalent circuit of Figure 3.5 is best treated as an **admittance** Y_{IND}. Admittance is the mathematical inverse of impedance: $Y=1/Z$. The advantage of using admittance to express the behavior of a parallel circuit is that admittances add in parallel. So the admittance of the circuit in Figure 3.5 is simply

$$Y_{IND}(\omega) = j\omega C_{PAR} + \frac{1}{R_{PAR}} + \frac{1}{j\omega L} \tag{3.11}$$

The reader will notice a formal similarity between the admittance Equations 3.9 and 3.11, which gives the impedance of a capacitor's equivalent circuit. Just as the capacitance term dominated Equation 3.9 at low frequencies, the inductance term is going to dominate the admittance in Equation 3.11 at low frequencies.

To show how this equivalent circuit behaves over a wide range of frequencies, we will assume some typical values for a 20-μH inductor. This particular inductor will have a self-resonant frequency f_{SR} of 17.3 MHz, which means (using the resonant-frequency formula in Eq. 3.10) that the parasitic capacitance C_{PAR} is 4.22 pF. And we will suppose that the equivalent parasitic resistance R_{PAR} is 18.8 kΩ.

We plotted the magnitude $|Z_{IND}(\omega)| = 1/|Y_{IND}(\omega)|$ that results from these values in Figure 3.6. If an inductor is used as a **choke coil** to "choke off" undesired high-frequency currents, a high impedance magnitude is desirable in the frequency ranges where the currents appear. As you can see, the impedance of the inductor does indeed rise to high values, going above 1 kΩ at about 6 MHz. The behavior of an ideal inductor (with no parasitics or losses) is shown as the straight dashed line in the figure. Around 10 MHz, the equivalent-circuit model's impedance rises *above* what the ideal inductor would show and peaks at the self-resonant frequency of 17.3 MHz. So far so good, but once we are above f_{SR}, things go to pot fairly quickly. The impedance falls below 1 kΩ at 44 MHz and keeps falling.

While this is an oversimplified version of how a real inductor behaves, it does show that assuming an inductor is an inductor regardless of frequency is not correct beyond a certain frequency range. More sophisticated inductor equivalent-circuit models use as many as three resistors, one of which varies with frequency (see the reference to www.coilcraft.com at the end of this chapter).

You will notice that the impedance-versus-frequency curve in Figure 3.4 for the capacitor's equivalent circuit has a broad minimum, while the curve for the inductor's equivalent circuit in Figure 3.6 has a sharply peaked maximum. The reason for this concerns a circuit parameter known as the **Q**, which stands for **quality factor**.

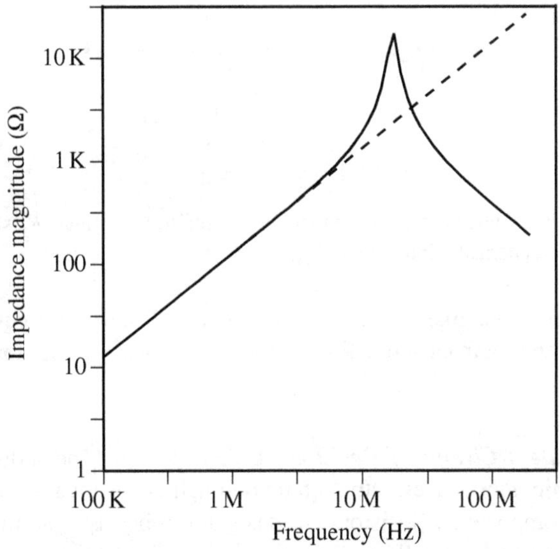

FIGURE 3.6 Impedance magnitude of inductor equivalent circuit of Figure 3.5 with $L = 20\,\mu\text{H}$, $C_{\text{PAR}} = 4.22\,\text{pF}$, and $R_{\text{PAR}} = 18.8\,\text{k}\Omega$. Dashed line indicates ideal inductor impedance.

Fundamentally, Q for a component excited by a sine-wave voltage or current is defined as 2π times the ratio of energy stored in a circuit at its resonant frequency to energy dissipated in it per cycle. In a series circuit such as the capacitor equivalent circuit of Figure 3.3, the expression for Q is

$$Q(\text{series}) = \left|\frac{X_{\text{C}}}{R_{\text{PAR}}}\right| = \frac{1}{\omega_0 C R_{\text{PAR}}}, \tag{3.12}$$

in which ω_0 is 2π times the **self-resonant frequency** f_{SR}. For the values given in the capacitor equivalent circuit of Figure 3.3 and plotted in Figure 3.4, the Q is only about 0.23. In low-Q resonant circuits, the impedance phase and magnitude change relatively slowly with frequency.

On the other hand, the inductor equivalent circuit shown in Figure 3.5 is a **parallel-resonant circuit** (as opposed to the **series-resonant circuit** in Fig. 3.3), so the equation for the inductor's Q is

$$Q(\text{parallel}) = \left|\frac{R_{\text{PAR}}}{X_{\text{L}}}\right| = \frac{R_{\text{PAR}}}{\omega_0 L} \tag{3.13}$$

For the example values chosen, the Q is 10, much higher than that of the capacitor's equivalent circuit. This accounts for the sharp, narrow peak in the impedance-magnitude plot in Figure 3.6, because as Q rises, the impedance magnitude and phase versus frequency near resonance change much more rapidly with frequency.

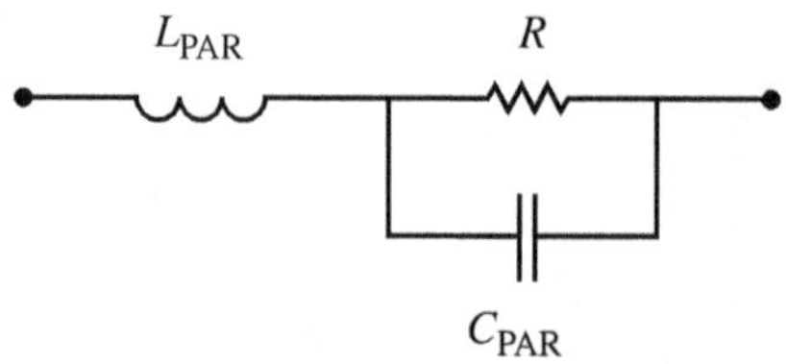

FIGURE 3.7 One equivalent-circuit model for a resistor of value R, showing parasitic capacitance C_{PAR} and parasitic inductance L_{PAR}.

Resonance phenomena like this are important in all sorts of frequency-selective analog circuits ranging from active filters to radio-frequency oscillators, transmitters, and receivers.

3.1.1.3 ***Equivalent Circuit of Resistor*** Like capacitors and inductors, resistors also have parasitic impedances, although modeling them is not always as simple as it is for reactive components. Resistors are designed to be lossy, so they do not typically show a well-defined self-resonant frequency. One resistor equivalent circuit that can be used over a limited range of frequencies is shown in Figure 3.7. The importance of the parasitic impedances depends on the value of the resistor. In low-resistance units, 1 kΩ or less, the parasitic inductance L_{PAR} begins to show up initially at high frequencies, raising the component's impedance above the value that R alone would suggest. For high-resistance units, in the range of 100 kΩ or greater, the parasitic capacitance C_{PAR} will tend to make its presence known first, causing the component's impedance to fall below its nominal value at high frequencies.

3.1.2 Two-Port Matrix Analysis

When the component in question has only two terminals, it can be treated as a single branch in a network with a uniquely defined voltage and current. From the viewpoint of linear circuit analysis, the component is completely defined by its complex impedance $Z(\omega)$ as a function of radian frequency ω. But as we have seen, many important devices in analog electronics have three terminals, not two. An extension of the impedance concept is therefore needed to deal with devices with three or more terminals. This extension is called the **impedance matrix**.

An impedance matrix expresses the relationships among the voltages and currents that exist at the two **ports** of a four-terminal or **two-port** network. (The term "port" refers to a location in a circuit where a signal can enter and leave on a given pair of wires or terminals. Four terminals means there are two ports.) A canonical or general two-port network is illustrated in Figure 3.8.

You should note that the upper terminal at each port is defined as positive, while the current associated in each port is defined as entering the device. In general, each terminal of a four-port can be unique, so it is important to be clear about which terminal is which, both in calculations and in actual laboratory work. For example, connecting the four terminals of a power transformer incorrectly can cause serious problems!

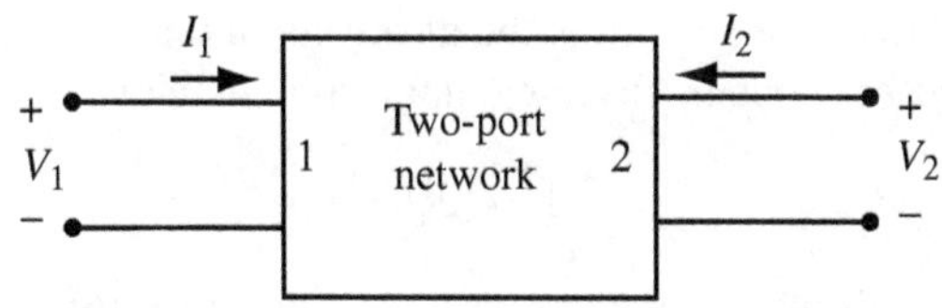

FIGURE 3.8 Two-port network showing voltage polarities and current directions at ports 1 and 2.

Given the currents and voltages defined in Figure 3.8 and assuming the circuit in question is linear, we can write two equations that completely describe its behavior:

$$V_1 = Z_{11}I_1 + Z_{12}I_2 \tag{3.14}$$

$$V_2 = Z_{21}I_1 + Z_{22}I_2 \tag{3.15}$$

Notice that instead of a single complex impedance value Z, we now need *four* complex numbers: Z_{11}, Z_{12}, Z_{21}, and Z_{22}. This is because, in general, a current I_1 at port 1 can give rise to a voltage at either port 1 or port 2 (or both) and similarly for a current I_2 at port 2. To cover all the possibilities, we need four impedance variables, not just one.

A system of simultaneous equations can be expressed compactly in terms of a single **matrix equation**. For example, if we define a **voltage vector** [*V*] as

$$[V] = \begin{bmatrix} V_1 \\ V_2 \end{bmatrix} \tag{3.16}$$

and a **current vector** as

$$[I] = \begin{bmatrix} I_1 \\ I_2 \end{bmatrix} \tag{3.17}$$

(both of these are **column vectors**), then Equations 3.14 and 3.15 can be written as one **matrix equation**:

$$[V] = [Z][I], \tag{3.18}$$

in which the **impedance matrix** is

$$[Z] = \begin{bmatrix} Z_{11} & Z_{12} \\ Z_{21} & Z_{22} \end{bmatrix} \tag{3.19}$$

The **elements** of the impedance matrix are complex numbers with the dimensions of ohms, because they are all ratios of voltages to currents. Most, but not all, two-port circuits have a finite impedance matrix, meaning that none of the matrix elements is infinite. Sometimes, however, either the impedance matrix is not defined, or it is more convenient to express the terminal currents in terms of the voltages presented to the terminals. In that case, one uses the **admittance matrix** $[Y]$, whose elements have the dimensions of siemens (inverse ohms), and the relevant matrix equation is

$$[I] = [Y][V] \tag{3.20}$$

For those familiar with matrix algebra, you will not be surprised to learn that the admittance matrix is the inverse of the impedance matrix:

$$[Y] = [Z]^{-1} \tag{3.21}$$

(Recall that a matrix multiplied by its own inverse produces the identity matrix, which leaves all vectors it multiplies unchanged.) The elements of the admittance matrix are called **admittance parameters**.

A third type of parameter (besides impedance and admittance parameters) you may encounter in transistor datasheets is the category of **hybrid parameters**. They are called "hybrid" because one of the four matrix elements has the dimensions of ohms, one has the dimension of siemens, and two are dimensionless. So they are neither impedance nor admittance parameters, but a hybrid of both. Here is the defining equation for hybrid parameters:

$$\begin{bmatrix} V_1 \\ I_2 \end{bmatrix} = \begin{bmatrix} h_{11} & h_{12} \\ h_{21} & h_{22} \end{bmatrix} \begin{bmatrix} I_1 \\ V_2 \end{bmatrix} \tag{3.22}$$

When a two-port analysis is applied to a common-emitter transistor model and the relevant quantities are small-signal voltages and currents, the dimensionless parameter h_{21} is also referred to as h_{fe}, the AC β referred to in the previous chapter. Various other types of two-port parameters are used at high frequencies, including a wave-based system called **S-parameters**.

If you know the two-port parameters for a given device, you can determine certain important facts about it. And conversely, if you wish your two-port to have certain properties, you can specify these characteristics in terms of the two-port parameters. Among these properties are **reciprocity**, **losslessness**, and **passivity**.

3.1.2.1 Reciprocity **Reciprocity** is a property that nearly all linear passive circuits show. A "seat-of-the-pants" definition of reciprocity is that a signal can travel either forward (from port 1 to port 2) or backward (from port 2 to port 1) and the network will treat it the same. Specifically, it is always true that for a reciprocal two-port network, $z_{21} = z_{12}$, and similarly for the y-matrix elements: $y_{21} = y_{12}$. Any circuit you build with linear passive components such as resistors, capacitors, and inductors will be reciprocal. This means that if you want a signal to travel in only one direction in a circuit (say, from input to output and not vice versa), you will need something more than a passive, reciprocal circuit to do the job.

Note that while $z_{12} = z_{21}$ means a two-port is reciprocal, the other two elements z_{11} and z_{22} will in general be different. If the circuit is **physically symmetrical** (i.e., if the circuit is built so that it looks the same to the outside world no matter which port is labeled 1 and which is labeled 2), then it will be true that $z_{11} = z_{22}$ as well.

Active devices such as transistors and amplifiers, when treated as two ports, are *not* reciprocal in general. This means that you cannot interchange the input and the output (e.g., port 1 and port 2) and expect the circuit to behave the same as before.

3.1.2.2 Losslessness **Losslessness** means that whatever power goes into the two ports of a device comes back out again, at least averaged over one cycle of the sine-wave excitation that we assume we are dealing with. No power is lost in the two-port itself. It turns out that all the parameters of a lossless two-port are imaginary numbers. So if you have a network composed entirely of lossless capacitors, for example, all four of the z-parameters will be imaginary.

3.1.2.3 Passivity A **passive** two-port is a net *absorber* of power. That is, if you add up the (real) power going into each port, you will get a positive number. All circuits composed of passive elements are also passive. Passivity is related to the question of **stability**, which in this context means whether a system will respond to a transient signal stimulus by eventually going back to its steady-state condition (a stable system) or whether the stimulus will excite an oscillation that in principle will grow without limit (an unstable system). The general question of stability is complicated and will be addressed later in Chapter 7. However, if one considers the case of a linear, reciprocal two-port that can be characterized by an impedance matrix (z parameters), one can show that it is passive if and only if its z-parameters meet all the following conditions, where Re(z) means "the real part of z":

1. $\mathrm{Re}(z_{11})>0$.
2. $\mathrm{Re}(Z_{22})>0$.
3. $\mathrm{Re}(Z_{11})\,\mathrm{Re}(Z_{22})>[\mathrm{Re}(Z_{21})]^2$.

Otherwise, the circuit can be a net producer of power, at least for some combinations of current phases at the two ports.

3.1.2.4 Impedance Matrices of Simple Two-Port Circuits It is fairly simple to derive the impedance or admittance matrix for two types of circuits: the t-network and the π-network. The names of these circuits are derived from their shapes, as Figures 3.9 and 3.10 show.

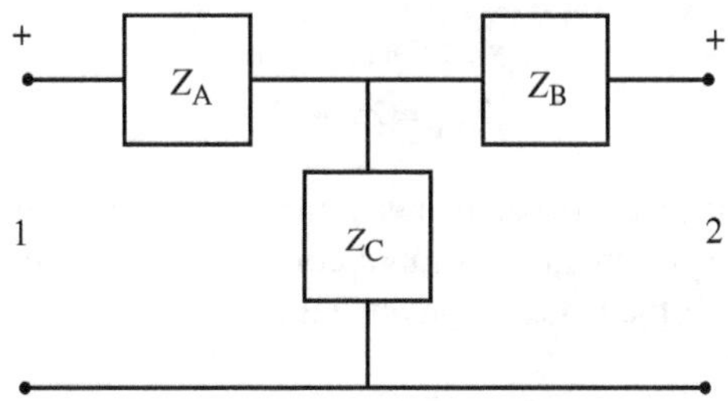

FIGURE 3.9 Diagram of t-network composed of impedance circuit elements Z_A, Z_B, and Z_C.

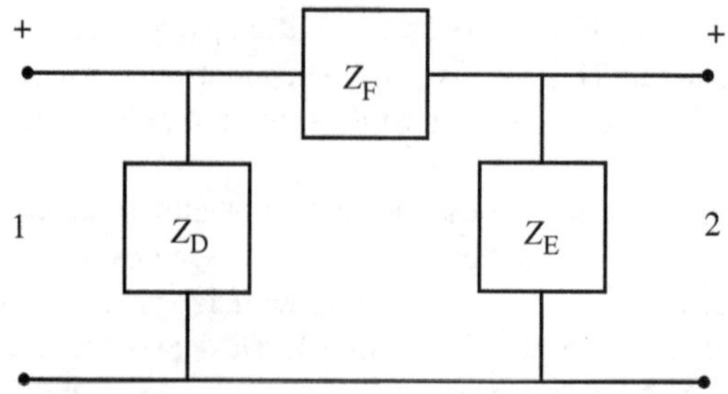

FIGURE 3.10 Diagram of π-network composed of impedance circuit elements Z_D, Z_E, and Z_F.

We can perform simple thought experiments to find the values of the four impedance-matrix elements for the t-network of Figure 3.9. By definition, each of the matrix elements can be found by injecting a known current into one of the ports and measuring a voltage at one of the ports while keeping the other port open circuited. Specifically, we find these expressions for the four matrix elements this way:

$$Z_{11} = \left.\frac{V_1}{I_1}\right|_{I_2=0} \tag{3.23}$$

$$Z_{12} = \left.\frac{V_1}{I_2}\right|_{I_1=0} \tag{3.24}$$

$$Z_{21} = \left.\frac{V_2}{I_1}\right|_{I_2=0} \tag{3.25}$$

$$Z_{22} = \left.\frac{V_2}{I_2}\right|_{I_1=0} \tag{3.26}$$

You can imagine actually performing these measurements with an ideal current source and an ideal voltmeter. No matter how they are determined, once we have these voltage and current values for a two-port, the results give expressions for the impedance-matrix elements in terms of the physical impedances of the three components in the t-network of Figure 3.9:

$$Z_{11t} = Z_A + Z_C \tag{3.27}$$

$$Z_{12t} = Z_{21t} = Z_C \tag{3.28}$$

$$Z_{22t} = Z_B + Z_C \tag{3.29}$$

In the event that you measure the elements of the impedance matrix and wish to know the t-network equivalent circuit, it is more convenient to solve for the three circuit impedances in terms of the matrix Z parameters:

$$Z_C = Z_{12t} = Z_{21t} \tag{3.30}$$

$$Z_B = Z_{22t} - Z_{21t} \tag{3.31}$$

$$Z_A = Z_{11t} - Z_{21t} \tag{3.32}$$

The π-network expressions are more complex, because there are elements in series and parallel.

The impedance-matrix elements in terms of the three π-network impedances Z_D, Z_E, and Z_F in Figure 3.10 are:

$$Z_{11\pi} = \frac{Z_D(Z_E + Z_F)}{Z_D + Z_E + Z_F} \tag{3.33}$$

$$Z_{12\pi} = Z_{21\pi} = \frac{Z_D Z_E}{Z_D + Z_E + Z_F} \tag{3.34}$$

And $Z_{22\pi}$ can be found by interchanging Z_D and Z_E in Equation 3.33. The expressions for the π-network's admittance-matrix elements are simpler. If $Y_{D,E,F} = 1/Z_{D,E,F}$, then

$$Y_{11} = \left.\frac{I_1}{V_1}\right|_{V_2=0} = Y_D + Y_F \tag{3.35}$$

$$Y_{21} = \left.\frac{I_2}{V_1}\right|_{V_2=0} = Y_{12} = \left.\frac{I_1}{V_2}\right|_{V_1=0} = -Y_F \tag{3.36}$$

$$Y_{22} = \left.\frac{I_2}{V_2}\right|_{V_1=0} = Y_E + Y_F \tag{3.37}$$

As you can see, the expressions for the admittance-matrix elements of the π-network are as simple as those for the impedance-matrix elements of the t-network. Note that the transfer admittances Y_{12} and Y_{21} are the *negative* of the component Y_F that connects the two ports. The negative sign is simply a function of the way we chose to define current directions into each port and does not mean that the individual component admittance is negative. So, other things being equal, it's better to use a t-network equivalent circuit with impedance parameters and a π-network with admittance parameters.

Because several of the device equivalent circuits that were described in Chapter 2 use a π-network, we can now show what the admittance-matrix elements are for them. For example, the FET equivalent circuit of Figure 2.6, reproduced here as Figure 3.11, has now become a two-port, with the gate-source pair of terminals serving as port 1 and the drain–source pair as port 2. This diagram therefore shows the *common-source* connection of the FET, which is only one of three ways it can be connected as a two-port.

The t- and π-network equations for impedance- and admittance-matrix values have an important limitation: they assume that each circuit element is a *passive* component whose behavior is *independent* of all other components and circuit variables, except for the voltage and current that applies to that individual component.

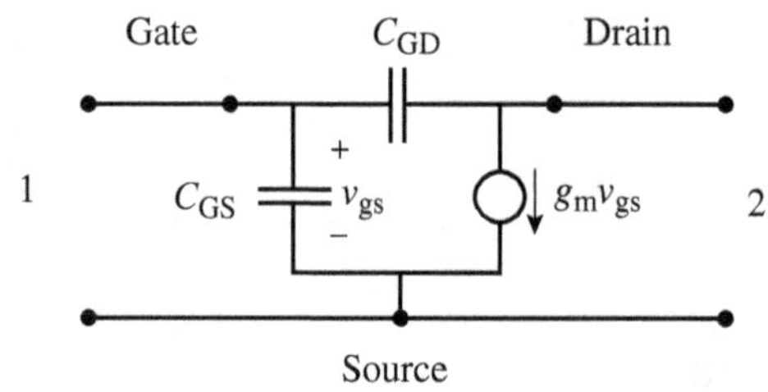

FIGURE 3.11 FET equivalent circuit for y-parameter evaluation.

If a circuit element *depends* on variables elsewhere in the circuit, as the voltage-dependent current source does in Figure 3.11, then Equations 3.35–3.37 cannot be used, and we have to look at the circuit using the basic y-parameter definitions of Equation 3.20. This gives the following equations for the FET equivalent circuit's y-parameters (they are lowercase letters because these are *differential* small-signal quantities):

$$y_{11} = j\omega(C_{\mathrm{GS}} + C_{\mathrm{GD}}) \tag{3.38}$$

$$y_{12} = j\omega C_{\mathrm{GD}} \tag{3.39}$$

$$y_{21} = g_{\mathrm{m}} - j\omega C_{\mathrm{GD}} \tag{3.40}$$

$$y_{22} = j\omega C_{\mathrm{GD}} \tag{3.41}$$

There are several important things to note about these equations. First, because $y_{12} \neq y_{21}$, the model is not **reciprocal**. So it behaves differently depending on which pair of terminals you use for the input and which for the output. In particular, the parameter y_{21}, which represents the **forward transfer admittance** of the device, has the voltage-dependent current source coefficient g_{m} in it.

Second, the gate–drain capacitance C_{GD} leads to a potentially undesirable **reverse transfer admittance** y_{12} that is not zero. If C_{GD} is small compared to the other circuit admittances, its effects can be incorporated into the overall circuit design, but one of the main goals of device designers is to make g_{m} as large as possible while making C_{GD} as small as possible. This makes the device approach the ideal of a **unilateral** device that provides perfect **isolation** between the output circuit at port 2 and the input circuit at port 1. Such isolation prevents undesirable feedback and greatly simplifies circuit designs.

3.2 NOISE AND LINEAR SYSTEMS

Most electronic analog or mixed-signal systems deal with a signal of some kind: a changing voltage or current whose changes represent information. Ideally, a linear system's output is proportional to the input and nothing else: no input would mean no output. Unfortunately, real analog systems only approach this ideal, because every circuit introduces a certain amount of **noise** along with the signal.

The term "noise" can be taken broadly to mean any undesired signal or waveform, whether produced internally by the system in question or produced by an external source of interference. For example, a high-gain public-address system can pick up the high-frequency impulses produced by a lamp dimmer circuit when it switches line-frequency power rapidly once or twice each cycle. This type of noise is periodic, somewhat predictable, and more properly referred to as **interference**, because one waveform produced in one system for a desirable purpose (dimming lights) happens to interfere with the operation of another system (the public-address system).

The type of noise we will discuss in this section is not predictable in an instantaneous sense. Rather, it is **random noise** that we are interested in. Random noise behaves in a way that is generally not possible to predict moment by moment, although its statistical behavior averaged over a long enough period of time can sometimes be predicted. As we will see, every electronic system generates a certain amount of random noise that is mixed in with the signal being processed. Depending on the application, this added noise component may be trivially unimportant, or it may establish the performance limits for the entire system. Whether or not noise is significant in your design, you should know some basic concepts relating to noise in electronic systems so you can deal with it when it causes problems.

3.2.1 Sources of Noise

3.2.1.1 Thermal Noise The most fundamental and widespread source of noise in electronic systems arises from **ohmic losses**. Any circuit whose model includes one or more resistors has ohmic losses and will contribute noise to the circuit that it is a part of. The physical reason is that a lossy element (resistor) represents a connection between the electrical power in the circuit and heat energy in the physical circuit itself. Not only can you produce physical heat in a resistor by applying a voltage to it and allowing the resulting current to flow; the heat in a resistor (as long as it is hotter than 0 K, which most resistors are!) causes random motion of the electrons in the resistor, which produces electronic noise. This is a fundamental feature of the physics of electronic devices and is required by the laws of thermodynamics. So there is no getting around the so-called thermal noise, unless you want to go to the trouble of *cooling* your noise-sensitive system a good bit below room temperature. (Thermal noise is sometimes called **Johnson noise** because an engineer named Johnson at Bell Laboratories discovered it experimentally.) For special applications such as **radioastronomy**, in which noisier receivers mean more hours spent using expensive radiotelescopes, the **low-noise** amplifiers used are often cooled to below the temperature of liquid helium (4.2°K). But this is an expensive and inconvenient thing to do, so most of us deal with whatever noise arises from having the electronic system we design at room temperature, about 300°K. (To convert temperatures in Celsius (°C) to Kelvin (°K), use the formula K = 273.15 + C.)

Like other sources of power, thermal noise sources can be modeled with equivalent circuits. Figure 3.12a shows a resistor R that produces noise because it is at a certain temperature T. The equivalent-circuit model of this resistor is shown in Figure 3.12b, in which a **noiseless** resistor of the same value R as the real (physical) resistor is

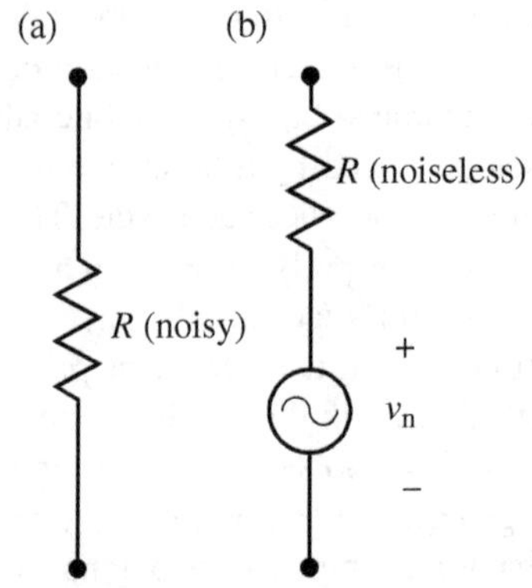

FIGURE 3.12 (a) Noisy resistor and (b) equivalent-circuit model of noisy resistor showing noiseless resistor in series with noise voltage source.

connected in series with a **noise voltage source** v_n. In this way, we can model the noise output of a physical resistor by imagining it as an ideal (noiseless) resistor in series with a noise voltage source. It turns out that this model works quite well as long as the noise voltage is chosen properly.

It would be nice if we could write an explicit function of time for the noise voltage, just as we write $v_S = V\sin(\omega t)$ for a sinusoidal signal of peak amplitude V. But if we could do that, we could predict the future! Remember, the exact value of a noise voltage at a particular instant cannot be predicted in advance. However, thermal noise *does* have a well-defined characteristic: **average power**.

You can show by thermodynamic arguments that the largest amount of power you can extract from a single two-terminal lossy component through its terminals is

$$P_n = k_B TB, \tag{3.42}$$

in which P_n is the noise power (W), k_B is **Boltzmann's constant** ($1.38\times10^{-23}\,\mathrm{J\,K^{-1}}$), T is the temperature of the component in K, and B (Hz) is the **effective noise bandwidth** (also called **equivalent noise bandwidth**) of the system used to measure the noise power. You can think of B as the upper frequency limit of a power meter, for example, although the exact definition of B for a system with an amplitude response $A(f)$ is

$$B \equiv \frac{1}{|A_{\mathrm{MAX}}|^2}\int_0^\infty |A(f)|^2\,df \tag{3.43}$$

where A_{MAX} is the maximum value of the response for any frequency. If the system's frequency response is an ideal "brick-wall" function that is constant from zero to B and then falls abruptly to zero, the 3-dB down frequency is of course equal to B. Frequency responses of practical circuits do not have this ideal characteristic, and Equation 3.43 can be used to calculate or measure B for a particular case.

In any event, unless B is very large or T is very high, the amount of power in thermal noise is tiny. For example, suppose a video-frequency amplifier has an effective noise

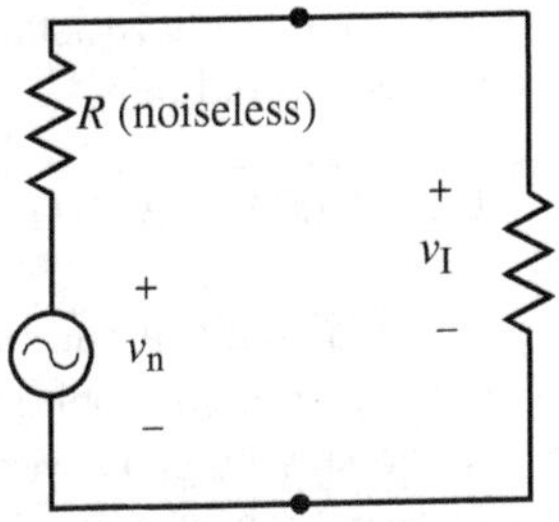

FIGURE 3.13 Resistor noise equivalent circuit of Figure 3.12b loaded by resistor R_L.

bandwidth $B=5$ MHz and a resistor heated to 1000 K is connected to it. The available noise power from such a component at that temperature is

$$P = (1.38\times10^{-23}\ \mathrm{J\,K^{-1}})(1000\ \mathrm{K})(5\times10^{6}\ \mathrm{Hz}) = 69\times10^{-15}\ \mathrm{W} \tag{3.44}$$

or 69 **femtowatts** (fW, 10^{-15} W).

What voltage does that power represent? Note that we said nothing about the resistance of the resistor to this point, because it doesn't matter when only power is involved. But once you ask about voltage or current, the resistor's value in ohms enters in. You probably know that for a **Thévenin equivalent circuit** such as the one in Figure 3.12b, the maximum available power is extracted from the circuit when a load resistance R_L is connected to the terminals and made equal to the source resistance R. This situation is illustrated in Figure 3.13, and if $R=R_L$, we are extracting power $P=k_BTB$ from the resistor noise equivalent circuit.

While we still don't know what v_n is instantaneously, we do know a couple of other things. Because we have formed a voltage divider, $v_l=v_n/2$, and since we know the power P delivered to the load R_L, we know the **root-mean-square (RMS)**[1] voltage $v_{l(RMS)}$ *at the load* (*not* the open-circuit source voltage):

$$v_{l(RMS)} = \sqrt{PR_L} = \sqrt{k_BTBR} \tag{3.45}$$

We can then work backward to find the RMS value of the *noise voltage source* in the equivalent circuit model, which is thus

$$v_{n(RMS)} = \sqrt{4k_BTBR} \tag{3.46}$$

Equation 3.46 is the universal expression for the equivalent noise voltage in a bandwidth B for a resistor R at temperature T. Returning to the example of a resistor

[1]In general, RMS voltages must be calculated numerically, because RMS literally means the square root of the mean (average) of the square of a quantity, where the mean is taken over a suitable time interval. For a purely sinusoidal AC wave (no DC), it is easy to show that the relations among the RMS voltage V_{RMS}, the peak voltage V_p, and the peak-to-peak voltage V_{PP} are $V_{RMS}(\text{sine}) = V_P/\sqrt{2} = V_{PP}/2\sqrt{2}$.

heated to 1000 K, if it was possible to maintain a resistance of 1 MΩ at such a high temperature, the equivalent noise voltage would be

$$v_{n(\mathrm{RMS})} = \sqrt{4(1.38\times10^{-23}\,\mathrm{J\,K^{-1}})(1000\,\mathrm{K})(5\,\mathrm{MHz})(10^{6}\,\Omega)} = 525\,\mu\mathrm{V} \qquad (3.47)$$

or about 0.5 mV. While such a voltage is small, it can be comparable to the signal levels of certain types of transducers such as microphones. At lower temperatures and resistance values, the equivalent noise voltage is even smaller, but sometimes, it cannot be neglected in critical low-noise, high-gain amplifier systems.

3.2.1.2 Shot Noise If current flow were truly continuous, shot noise would not exist. But as you know, current in electronic devices consists of the flow of discrete charge carriers (usually electrons, although positively charged **holes** carry the current in p-type semiconductors). When a charge carrier moves from one region to another, such as when an electron crosses the junction of a p–n junction diode, a graph of the current flow through the device shows a small impulse, whose area corresponds to the charge that the electron carried across the junction. These impulses can be compared with the sound you would hear if someone poured a cup of buckshot onto a tin roof: a rush of randomly timed impulses that is called *shot noise*. As you might expect, the noise power in shot noise is directly proportional to the current flow through the device. It is best modeled as a random noise *current source* whose RMS current is

$$i_{n(\mathrm{RMS})} = \sqrt{2qBI}, \qquad (3.48)$$

in which $q = 1.6\times10^{-19}$ C (the charge on a single electron), B is the effective noise bandwidth, and I is the current (A). Unlike thermal noise, there is no fixed amount of power associated with a given current that causes shot noise—the power delivered depends on the details of the circuit in which the noise occurs. Shot noise is small but not negligible in some circuits. For example, suppose $B = 5$ MHz and $I = 10$ mA. The shot-noise current associated with this level of DC current is

$$i_{n(\mathrm{RMS})} = \sqrt{2(1.6\times10^{-19}\,\mathrm{C})(5\,\mathrm{MHz})(10\,\mathrm{mA})} = 126.5\,\mathrm{nA} \qquad (3.49)$$

While such a current is small, it can be comparable to low-level signals in sensitive systems and sometimes must be taken into account in designs.

3.2.1.3 Flicker or 1/f Noise A type of noise that is increasingly significant at lower frequencies is **flicker** or **1/*f*** noise. This type of noise is also typically associated with semiconductors and has to do with the appearance and disappearance of charge carriers into **traps**. A trap is basically an energy level that a mobile charge carrier (electron or hole) falls into and becomes, well, trapped. The effect of traps is to cause current to vary in an unpredictable stepwise pattern. Because traps can hold charges for a long time, the energy in 1/*f* noise tends to become larger at lower frequencies, say, below about 100 Hz. There is no straightforward equation available to calculate 1/*f* noise, so typically, manufacturers of low-noise devices will characterize the $1/f$ noise of their products experimentally. The main thing you should know about $1/f$ noise is that it can interfere with attempts to amplify very slowly changing signals. A useful way around this problem is to **chop** a slowly varying signal with an automatic switch called a **chopper**, which is really a kind of modulator. Chopping transforms a low-frequency signal into modulation sidebands around the chopper's frequency, which can be much higher (e.g., in the kHz range) and avoids or minimizes problems with $1/f$ noise.

3.2.2 Noise in Designs

The design of low-noise systems is an art of its own, and we can only scratch the surface of the topic here. You should know how a noise-sensitive design is influenced both by the noise coming from the signal source and any **internal noise** generated by the system itself.

Devices such as **operational amplifiers** that we will discuss in more detail in Chapter 5 are sometimes characterized for noise in specification sheets. Typically, such a specification will list an equivalent **input noise voltage density** in units of V $Hz^{-1/2}$ (volts per square root of Hz). This seems like an odd unit until you recall that in the equation for thermal noise (Eq. 3.42), noise power P_n is directly proportional to bandwidth B. Because power goes as the square of voltage, noise voltage is therefore proportional to the square root of frequency. To see how this unit is used, it is best to consider a simplified example.

Figure 3.14 shows the equivalent circuit of a signal source whose output resistance is R. We will model the thermal noise of this resistance with a noise voltage source v_{n0}, and the signal itself is represented by a signal voltage source providing RMS voltage v_S. The signal source is connected to amplifier 1, whose equivalent input noise voltage is v_{n1}. This amplifier is connected in **cascade** (output to input) to amplifier 2, whose noise voltage is represented by v_{n2}. Finally, the second amplifier's

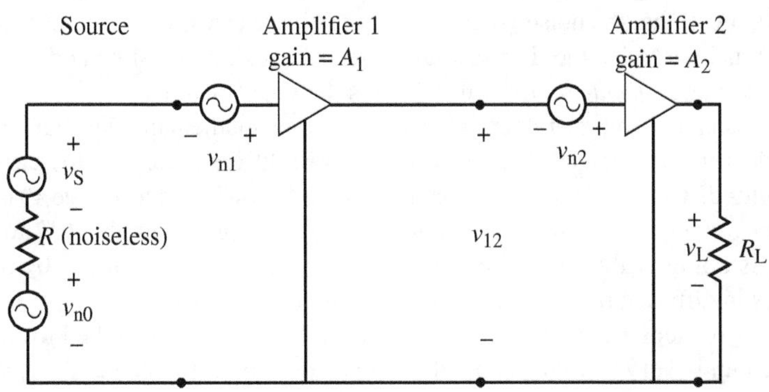

FIGURE 3.14 Two-stage amplifier design with noise sources considered. Voltage v_{n0} is the thermal noise of signal source, v_{n1} is the equivalent input noise of amplifier 1, and v_{n2} is the equivalent input noise of amplifier 2.

output is connected to a load resistance R_L, whose internal thermal noise we will neglect in this analysis.

Let us assume that the input impedance of both amplifiers is infinite, so that the amplifier inputs cause no **loading** effects. Given the voltage gains A_1 and A_2 (numeric ratios), it is easy to calculate the expression for the output voltage v_L across the load resistor R_L:

$$v_L = A_1A_2(v_S + v_{n0} + v_{n1}) + A_2v_{n2} \tag{3.50}$$

As it stands, Equation 3.50 is exact for instantaneous voltages. We will assume the amplifiers have a "brick-wall" frequency response that limits their bandwidth to B, but as long as the signal and noise voltages have no frequency components higher than B, the equation is correct.

In noise calculations, it is best to express results in terms of *power*, not voltage, for a reason that will be apparent shortly. For a given signal voltage v_S, the signal power at the load will be

$$P_{S(OUT)} = \frac{(A_1A_2v_S)^2}{R_L} \tag{3.51}$$

The signal voltage has been amplified by both amplifiers in cascade, so it is much larger than its original amplitude. But in sensitive systems, the factor that often determines the smallest signal that can be detected is not simply the overall gain of the system, but the **power signal-to-noise ratio**, sometimes abbreviated as **S/N**. This is because if a weak signal is contaminated by circuit noise at the input, no amount of amplification will remove the noise later, and simply adding more gain will not improve the situation. So let's calculate the signal-to-noise ratio S/N and calculate it as a function of the signal level v_S, the amplifier gains, and the various noise sources.

You might think this would be as easy as simply squaring the right-hand side of Equation 3.50 to obtain the squared voltage at the load, but a word of caution is in order. While the signal voltage may be in principle completely predictable (e.g., a sinusoid), recall that the noise voltages are not. All we can say for sure about a noise voltage v_n is its RMS value. For several noise voltages that must be added together, the noise voltage's *mean-square* value $\langle v_n^2\rangle$ is the quantity to use.

The reason for this is rather involved to show mathematically, but basically, independent noise voltages (i.e., noise voltages in different parts of the circuit) are **uncorrelated**. That is a fancy word that means the following: if a voltage V_A is uncorrelated to a voltage V_B, it is not possible to use any information about V_A to predict what V_B is doing at the same time. And if two voltages are produced by different structures in different parts of a system, that only makes sense.

When two uncorrelated voltages are connected in series and applied to a resistor, what adds up is not the voltages but the *powers* due to each voltage. If the voltages are **correlated** (e.g., two sine waves at the same frequency but differing by a constant phase shift), the proper thing to do is to add the phasor voltages directly. But we cannot do that with uncorrelated noise voltages, so we add the mean-square quantities and calculate the *power* in the load due to each source.

When we do that, we find that the noise power $P_{\mathrm{N(OUT)}}$ at the load in Figure 3.14 is

$$P_{\mathrm{N(OUT)}} = \frac{A_1^2 A_2^2 \left(\left\langle v_{\mathrm{n0}}^2 \right\rangle + \left\langle v_{\mathrm{n1}}^2 \right\rangle \right) + A_2^2 \left\langle v_{\mathrm{n2}}^2 \right\rangle}{R_{\mathrm{L}}} \tag{3.52}$$

Finally, we can calculate the power signal-to-noise ratio S/N at the output across the load resistor R_{L}:

$$\frac{\mathrm{S}}{\mathrm{N}} = \frac{P_{\mathrm{S(OUT)}}}{P_{\mathrm{N(OUT)}}} = \frac{(A_1 A_2 v_{\mathrm{S}})^2}{A_1^2 A_2^2 \left(\left\langle v_{\mathrm{n0}}^2 \right\rangle + \left\langle v_{\mathrm{n1}}^2 \right\rangle \right) + A_2^2 \left\langle v_{\mathrm{n2}}^2 \right\rangle} \tag{3.53}$$

Dividing numerator and denominator by the squared amplitude term $(A_1A_2)^2$ gives

$$\frac{\mathrm{S}}{\mathrm{N}} = \frac{(v_{\mathrm{S}})^2}{\left\langle v_{\mathrm{n0}}^2 \right\rangle + \left\langle v_{\mathrm{n1}}^2 \right\rangle + \left\langle v_{\mathrm{n2}}^2 \right\rangle / A_1^2} \tag{3.54}$$

The significance of the terms in Equation 3.54 is best appreciated with a calculation using typical numerical values.

Suppose we use amplifiers whose equivalent input noise voltage source has a noise voltage density of $\rho_{\mathrm{n}} = 8.7\,\mathrm{nV\,Hz^{-1/2}}$. We use this amplifier in a circuit that has an effective noise bandwidth B of 1 MHz, and suppose that the equivalent resistance of the signal source is $R = 100\,\mathrm{k\Omega}$. We also design each amplifier to have a voltage gain $A_1 = A_2$ of 10. The source itself is at room temperature ($T = 300\,\mathrm{K}$). If the RMS signal amplitude v_{S} is $100\,\mu\mathrm{V}$ (e.g., typical of signals in many areas of biomedical engineering), what is the signal-to-noise ratio at the load?

The first step is to calculate the mean-square voltages of each noise source. The equivalent-circuit noise from the source itself is found from Equation 3.46 (squared) to be

$$\left\langle v_{\mathrm{n0}}^2 \right\rangle = 4k_{\mathrm{B}}BTR = 1.66 \times 10^{-9}\,\mathrm{V}^2 \tag{3.55}$$

We do a similar calculation for the amplifier noise voltages, which are the same for the two amplifiers:

$$\left\langle v_{\mathrm{n1}}^2 \right\rangle = \left\langle v_{\mathrm{n2}}^2 \right\rangle = \rho_{\mathrm{n}}^2 B = (8.7 \times 10^{-9}\,\mathrm{V\,Hz^{-1/2}})^2 (1\,\mathrm{MHz}) = 75.7 \times 10^{-12}\,\mathrm{V}^2 \tag{3.56}$$

Now that we have these quantities, we insert them into Equation 3.54 to find S/N:

$$\frac{\mathrm{S}}{\mathrm{N}} = \frac{10 \times 10^{-9}\,\mathrm{V}^2}{(1.66 \times 10^{-9} + 75.7 \times 10^{-12} + 75.7 \times 10^{-12}/10^2)\mathrm{V}^2} = \frac{10 \times 10^{-9}\,\mathrm{V}^2}{1.7365 \times 10^{-9}\,\mathrm{V}^2} = 5.76 \tag{3.57}$$

This is an adequate signal-to-noise ratio for many purposes. But if the signal were a single-frequency sine wave displayed on an oscilloscope, the wave would have clearly visible "fuzz" on it from the noise voltage, most of which is coming from the 100-kΩ equivalent source resistance itself. If the source resistance was much lower—100 Ω, say—then the dominant noise contribution would be the equivalent noise voltage of the first amplifier stage. Notice that the equivalent noise voltage of the *second* amplifier appears in Equation 3.54 only after it is *divided* by the gain of the first stage. This means that the second-stage amplifier's noise power could be 100 times higher (in this example) before it would equal the noise contribution made by the first stage.

This is a specific example of a general rule. In a well-designed low-noise amplifier system, the first stage of amplification should have low noise and enough gain to "take over" the noise of all following stages. Low-noise amplifiers tend to be more expensive than "garden-variety" types, so it is good design practice to use them only when necessary. This example shows that in a good design, you can get by with noisier amplifiers after the first stage, but you had better be sure your first stage has low enough noise not to overwhelm the lowest expected input signal.

One way to express the noise behavior of a system is to calculate its **noise floor**, which we will take to mean the signal level that produces an output equal to the no-signal noise output. At the noise floor, the S/N ratio is 1, which means it would be hard to see the signal in the noise on an oscilloscope, for example. So calculating the noise floor is one way to state the approximate lower limit for the input signal level before it is overwhelmed by the noise in the system.

You can find the noise-floor signal level by setting S/N = 1 in Equation 3.54 and solving for $\langle vs^2 \rangle$. In general, the noise floor will depend on the source circuit as well as the system itself, so strictly speaking, you cannot discuss a noise floor for a system independently of what it is connected to. For the system shown in the previous example, the noise-floor signal voltage is $\sqrt{1.7365 \times 10^{-9}\ \text{V}^2} = 41.7\,\mu\text{V}$. Again, this is limited by the noisy 100-kΩ source resistance. The lower limit due to the amplifier noise alone is $\sqrt{76.46 \times 10^{-12}\ \text{V}^2} = 8.74\,\mu\text{V}$. This is a small voltage, but many signals of interest are much smaller than this, in the nV or pV range. So there are plenty of opportunities for low-noise amplifier designers in analog electronics.

BIBLIOGRAPHY

Coilcraft Inc., http://www.coilcraft.com/models.cfm

Guillemin, E. A. *Introductory Circuit Theory*. New York: Wiley, 1953.

Sze, S. M. and K. K. Ng. *Physics of Semiconductor Devices*, 3rd ed. Hoboken, NJ: Wiley-Interscience, 2007.

Van der Ziel, A. *Noise*. Englewood Cliffs, NJ: Prentice-Hall, 1954.

Van Valkenburg, M. E. *Network Analysis*, 3rd ed. Englewood Cliffs, NJ: Prentice-Hall, 1974.

PROBLEMS

Note: Problems of above-average difficulty are marked with an asterisk (*).

3.1. *Self-resonant frequency of capacitor.* Suppose a capacitor has the following values for the equivalent circuit shown in Figure 3.4: $C = 100\,\text{pF}$, $L_{\text{PAR}} = 25\,\text{nH}$, and $R_{\text{PAR}} = 2.3\,\Omega$. Calculate the self-resonant frequency f_{SR} and the circuit Q at that frequency.

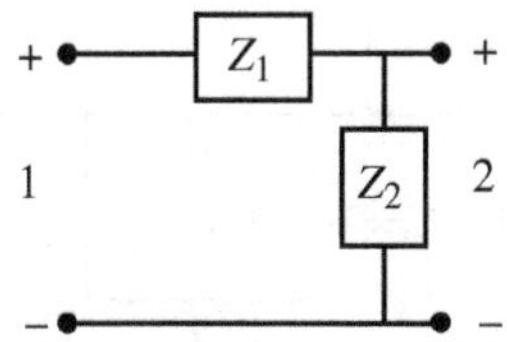

FIGURE 3.15 *L*-network circuit for Problems 3.3 and 3.4.

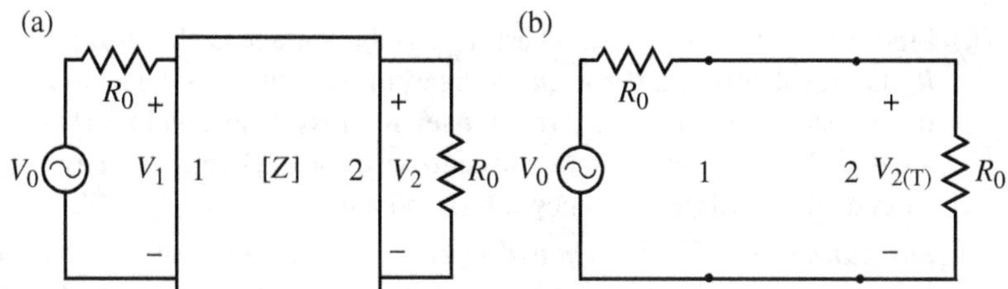

FIGURE 3.16 (a) Two-port inserted between source with resistance R_0 and load resistance R_0. (b) Through connection as reference for insertion loss test circuit (a).

3.2. *Impedance of inductor with parasitic capacitance.* In a certain application, a 25-mH choke coil is required to have an impedance magnitude of greater than 1 kΩ over a frequency range of 7–100 kHz. What should the parasitic capacitance C_{PAR} be so that the self-resonant frequency falls at the **geometric mean** of these two frequencies? (The geometric mean of A and B is $(AB)^{1/2}$.)

3.3. *Z-matrix of L network.* Figure 3.15 shows an **L network** connecting ports 1 and 2, which consists of a series element Z_1 and a shunt element Z_2 across port 2 only. Find the Z-matrix of this two-port when (a) $Z_1 = R_1$ and $Z_2 = R_2$, (b) $Z_1 = R_1$ and $Z_2 = 1/j\omega C$, (c) $Z_1 = j\omega L$ and $Z_2 = R_2$, and (d) $Z_1 = j\omega L$ and $Z_2 = 1/j\omega C$. In (d), check to make sure all elements are imaginary, as they should be.

3.4. *Y-matrix of L network.* Using the *L* network of Figure 3.15 and the element values given in 3.3 (a)–(d), find the *Y*-matrix for each case.

***3.5.** *Insertion loss of two-port network.* Figure 3.16a shows a two-port to which are connected a signal source V_0 in series with a source resistance R_0 at port 1 and an identical resistance R_0 connected to port 2. As long as $Z_{21} \neq 0$, a voltage V_2 will appear at port 2 in response to V_0.

(a) Show that the ratio V_2/V_0 can be expressed as

$$\frac{V_2}{V_0} = \frac{Z_{21}R_0}{(Z_{11} + R_0)(Z_{22} + R_0) - Z_{12}Z_{21}} \tag{3.58}$$

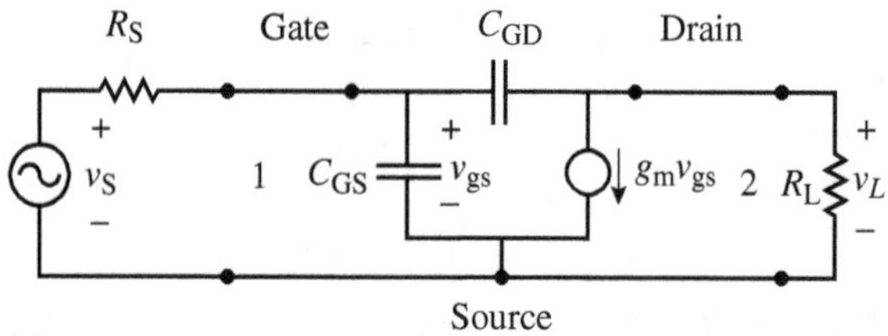

FIGURE 3.17 FET circuit model with signal source and load for Problem 3.11.

(b) Find the ratio $V_2/V_{2(T)} = \text{IL}$, where $V_{2(T)}$ is the voltage at the load resistance R_0 that results from a **through connection** straight from the source circuit to the load. The ratio IL is termed **insertion loss**. Insertion loss is an easily measured quantity that is often used to describe the behavior of everything from cables to high-frequency filters and amplifiers.

3.6. *Two-port parameters of FET equivalent-circuit model.* One of the main uses of two-port models of active devices is to derive equivalent-circuit models of amplifier circuits using the devices. Figure 3.17 shows the two-port common-source FET model of Figure 3.11 with added components. Port 1 (the gate circuit) is connected to a signal-source equivalent circuit that has a voltage generator v_S in series with resistance R_S. Port 2 (the drain circuit) is connected to a load resistor R_L.

(a) Assuming $C_{GD}=0$, find an algebraic expression for the (phasor) voltage gain $v_L/v_S = A_{V1}$ as a function of R_S, C_{GS}, g_m, R_L, and radian frequency ω.

(b) If $C_{GS}=2.0\,\text{pF}$, $R_S=1\,\text{k}\Omega$, $g_m=3\,\text{mS}$, and $R_L=5\,\text{k}\Omega$, what is $|A_V|$ (the voltage gain magnitude) at $\omega/2\pi=f=1\,\text{kHz}$, at $f=80\,\text{MHz}$, and at $f=800\,\text{MHz}$?

***(c)** Now, assume that $R_S=C_{GS}=0$, and at a frequency ω, assume the reactance $(1/\omega C_{GD}) \gg R_L$. Show that port 1 behaves like a capacitor whose value is $C_{GD}(1+g_m R_L)$. Because the low-frequency voltage gain of the FET in a common-source amplifier circuit is approximately $g_m R_L$, it is often much larger than 1. So the gate–drain capacitance of a common-source amplifier appears *larger* at the gate than it really is. This is an example of a circuit phenomenon called the *Miller effect* and is a significant limitation on the high-frequency response of amplifiers.

3.7. *Noise output of device equivalent-circuit model.* Suppose the equivalent circuit of a certain device is shown in Figure 3.18, which shows a (noiseless) resistor R, a thermal noise voltage source v_{nT} associated with R, and a shot-noise current source i_{nS}. The noise voltage source and the noise current source are completely *uncorrelated.*

(a) If the thermal noise comes from resistor $R=1\,\text{M}\Omega$ at a temperature $T=300\,\text{K}$, what is the RMS value of the thermal noise v_{nT} (only) if bandwidth $B=100\,\text{MHz}$?

(b) What DC current I through the shot-noise current source will be necessary to make the shot-noise RMS voltage produced across resistor R equal to the thermal noise voltage v_{nT} for the same bandwidth B?

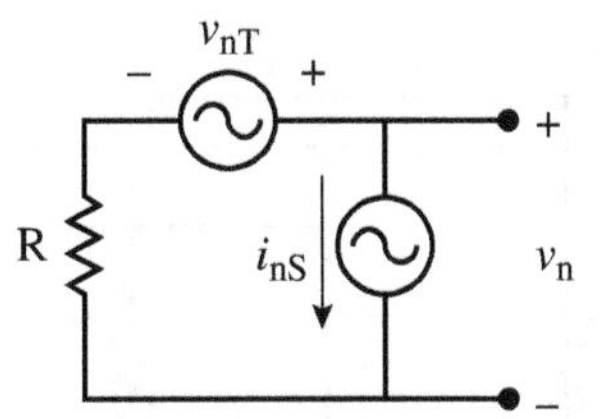

FIGURE 3.18 Noise-source equivalent circuit for Problem 3.7.

(c) If the two noise voltages (v_{nT} and the voltage produced by i_{nS} across R) have equal RMS values, what is the RMS value of the total noise voltage v_n present at the output terminals?

PROJECT PROBLEM: MEASUREMENT OF INDUCTOR CHARACTERISTICS

This project requires access to an automatic Bode plotter such as the one available in the NI-ELVIS™ modular engineering educational laboratory platform. The measurements required may be performed manually with a function generator and oscilloscope, but manual measurements are difficult and tedious to perform with good accuracy. Alternatively, the problem can be done as a simulation exercise using only a circuit simulation software package such as Multisim, although the simulation and the experiment will be the same activity.

EQUIPMENT AND SUPPLIES

1. Bode plotter capable of measuring amplitude and phase response with greater than 60 dB dynamic range from 1 Hz to 20 kHz (alternative: function generator and oscilloscope)
2. Inductor with nominal value in the range of 1–100 mH (Inductors outside this range will be difficult to measure in the audio-frequency range.)
3. 1-Ω resistor, 47-Ω resistor, and assortment of capacitors that will resonate with the inductor in the frequency range of 1–10 kHz

DESCRIPTION

An equivalent-circuit model need be only as detailed as the application it is intended for. For this exercise, it is desired to develop an equivalent-circuit model that provides a reasonably good agreement between the model and experimental data obtained from the actual component over a frequency range of 100 Hz to 10 kHz, a

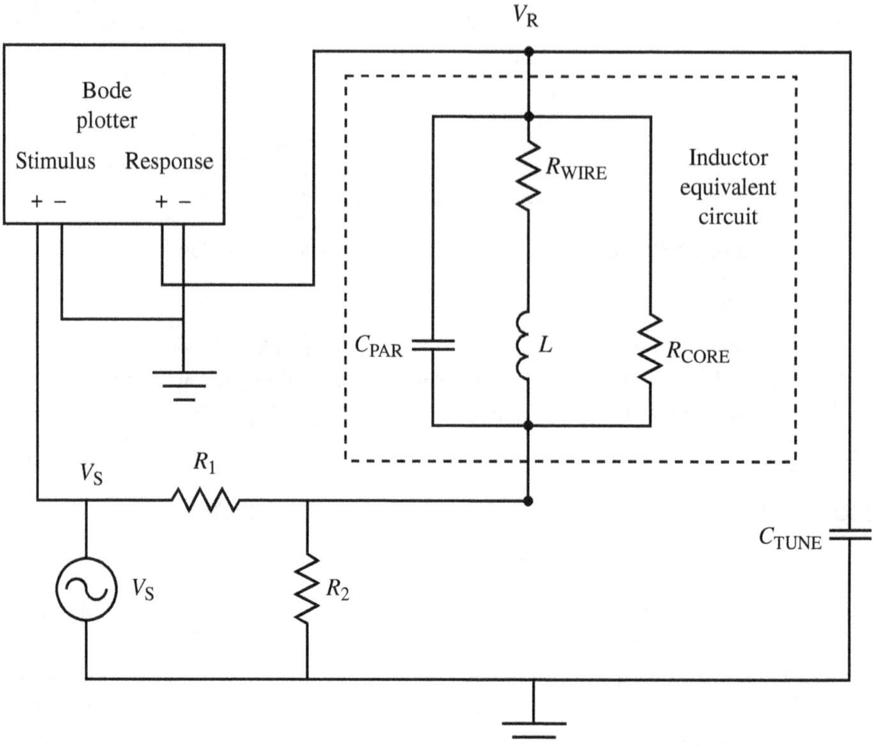

FIGURE 3.19 "Q-meter" circuit for measurement of inductor parameters.

range of two decades. An inductor model that is more than adequate for this application is shown in Figure 3.19 in the dashed-line box. The winding inductance is L. Resistive losses in the inductor's winding are represented by R_{WIRE}, while core losses (if any) are modeled by R_{CORE}. The winding's parasitic capacitance is C_{PAR}.

The circuit surrounding the inductor equivalent circuit is termed a "Q-meter" system because it resembles the circuit of a specialized instrument of that name formerly manufactured for the measurement of Q in radio-frequency inductors. In operation, voltage source V_S drives a 47-Ω resistor R_1 connected to 1-Ω resistor R_2. The Thévenin equivalent circuit of these three components is a voltage source V_T reduced from V_S by the voltage-divider ratio, in series with a resistor that is less than 1 Ω. The user chooses by trial and error a tuning capacitor C_{TUNE} that resonates with the inductance L of the device under test somewhere near the center of the range of frequency measurement (in this case, near 1–3 kHz). A Bode plotter is connected so that the source voltage V_S is the stimulus and the voltage V_R at the inductor–capacitor junction is the response. (For information on Bode plots, see Chapter 5.)

If the parasitic capacitance C_{PAR} and core losses R_{CORE} are neglected (set equal to 0), the Bode plotter response $A(\omega) = V_R/V_S$ is

$$A(\omega) = \frac{1}{j\omega C_{\mathrm{TUNE}} R_{\mathrm{S}}[1 + jQ_{\mathrm{S}}(\omega/\omega_{\mathrm{S}} - \omega_{\mathrm{S}}/\omega)]} \tag{3.59}$$

where the following parameters are defined in terms of the inductance L, Thévenin equivalent resistance of the source R_{T}, C_{TUNE}, and R_{WIRE}:

$$\omega_{\mathrm{S}}^2 = \frac{1}{LC_{\mathrm{TUNE}}} \tag{3.60}$$

$$R_{\mathrm{S}} = R_{\mathrm{T}} + R_{\mathrm{WIRE}} \tag{3.61}$$

$$Q_{\mathrm{S}} = \frac{\omega_{\mathrm{S}} L}{R_{\mathrm{S}}} \tag{3.62}$$

At resonance ($\omega = \omega_{\mathrm{S}}$), the magnitude of A attains a maximum and is equal in magnitude to Q (hence the name Q-meter). In principle, measuring the resonant frequency ω_{S} and Q should provide enough data to calculate both L and R_{WIRE}, given a knowledge of C_{TUNE}.

In your experiment, set up the system as shown in Figure 3.19 and take data with a suitable value of C_{TUNE} so that the resonant frequency is in the neighborhood of 1–3 kHz. Calculate L and R_{WIRE} from your measurement of Q and ω_{S}, and then model the test circuit. You can model it with Multisim™ software or by calculating values obtained from Equation 3.62 with Excel or MATLAB software. Begin your modeling efforts with the values for L and R_{WIRE} obtained from your initial measurements of Q and ω_{S}. If the agreement between the measurement data and model data is not very good, feel free to adjust the parameters of your inductor model so that both the phase and the amplitude curves agree near resonance. It will be more difficult to obtain good agreement over the entire frequency range, and you may find it helpful to include additional components such as R_{CORE} and possibly C_{PAR} to obtain better agreement.

In your report, describe the device under test as accurately as possible (model number, terminals used). Then describe your experimental setup and how you took the data, and explain the mathematics and analytical techniques used in enough detail so that the motivated reader could duplicate your results.

For further resources for this chapter visit the companion website at
http://wiley.com/go/analogmixedsignalelectronics

4

NONLINEARITIES IN ANALOG ELECTRONICS

4.1 WHY ALL AMPLIFIERS ARE NONLINEAR

Almost everything we have discussed up to now is based on the assumption that the system in question is **linear**. While this assumption makes the math easier and allows us to model many situations in real-life circuits, it also has its limitations.

For example, no real circuit that provides power gain can be linear under all possible input conditions, including extremely large inputs. Suppose there is an amplifier whose transfer function is almost perfectly linear with a sinusoidal input signal of $V_{IN} = 1\,V$ peak to peak. If the voltage gain is 10, for example, suppose the exact voltage $V_{OUT} = 10\,V_{IN}$ is delivered to a power-absorbing load to within the accuracy of available measuring equipment. Unless infinite power is available from the amplifier's power supply—which is impossible—as you increase the input voltage, sooner or later, the output will fail to follow the increase in input voltage as the power supply's limitations make themselves felt. When that happens, the linear expression $V_{OUT} = 10\,V_{IN}$ will no longer apply, and the amplifier is no longer described by a linear equation. This shows that for sufficiently large input signals, no real system that delivers power gain can be linear under all circumstances.

As this example implies, linearity is a condition that many systems approximately meet as long as the inputs are limited to a certain range. What that range is depends on how (nearly) linear the system is required to be. Power systems to drive motors, for example, may not be linear at all. But even a very slight nonlinearity can cause problems in a system such as a cable-TV distribution network, in which a complex

Analog and Mixed-Signal Electronics, First Edition. Karl D. Stephan.

Companion Website: http://wiley.com/go/analogmixedsignalelectronics

wideband signal is amplified repeatedly many times. We will show what some of the effects of nonlinearity are with a simple example in the next section.

4.2 EFFECTS OF SMALL NONLINEARITY

4.2.1 Second-Order Nonlinearity

Mathematically, the "gentlest" kind of nonlinearity is the presence of a **second-order** term in a system's transfer function: namely, a squared voltage or current. For simplicity, we will ignore any frequency dependence and assume the system has the following transfer function for the entire frequency range of interest:

$$v_{OUT} = a_1 v_{IN} + a_2 v_{IN}^2 \tag{4.1}$$

In Equation 4.1, v_{IN} is the input voltage, v_{OUT} is the output voltage (both are measured in terms of peak AC amplitudes), and coefficients a_1 and a_2 represent the (linear) voltage gain and the (nonlinear) **second-order transfer coefficient**, respectively. (The term a_2 is called "second-order" because it multiplies the input voltage raised to the second power.) While a_1 has no dimensions, a_2 must have the dimensions of V^{-1} for the dimensions to come out right. Right away, this ties a_2 to a particular voltage scale, which means that any calculations using Equation 4.1 must be numerically accurate and proportioned to the actual conditions in the circuit.

Before we proceed further, this is a good point at which to introduce the concept of the **decibel**, abbreviated as **dB**. The dB is a dimensionless unit used in referring to large ratios of voltage, current, or power. In the early days of telephone engineering, the dB was invented to ease calculations that involved large ratios of voltage at the input of a long telephone line to the greatly attenuated voltage at its output. The concept was so useful that it spread to many other fields in electronic engineering, which is why we are describing it now.

Suppose a (linear) amplifier has a **power gain** of 3, so that for $P_{IN} = 1$ watt applied to the input, it delivers $P_{OUT} = 3$ watts at its output. The dimensionless **numeric ratio** expressing this power gain is $A_P = 3$.

The dB version of this same amount of power gain is defined as

$$A_{P(dB)} \equiv 10\log_{10}\left(\frac{P_{OUT}}{P_{IN}}\right) = 10\log_{10}(3) = 4.77\,\text{dB} \tag{4.2}$$

The usefulness of decibel units is most apparent when a large ratio such as 10 million is involved. In dB, a power ratio of 10,000,000 is $10\log_{10}(10^7) = 70\,\text{dB}$.

Decibels are just as useful with voltage or current magnitude ratios. While the original definition of dB pertained to power ratios, the concept is easily extended to voltage or current ratios, with the implicit assumption that both quantities are measured with respect to the same resistance. Under this assumption, power varies as voltage squared, so we find that the dB gain ratio of a voltage V_{OUT} to another voltage V_{IN} is

$$A_{V(dB)} = 10\log_{10}\left(\frac{V_{OUT}^2}{V_{IN}^2}\right) = 20\log_{10}\left(\frac{V_{OUT}}{V_{IN}}\right) \tag{4.3}$$

And the same goes for a current ratio I_{OUT}/I_{IN}, so the general expression for either current or voltage ratios is

$$\text{Amplitude ratio (dB)} = 20\log_{10}\text{ (voltage or current ratio)} \tag{4.4}$$

while for power ratios,

$$\text{Power ratio (dB)} = 10\log_{10}\text{ (power ratio)} \tag{4.5}$$

Another nice thing about expressing ratios in dB shows up when you wish to calculate the gain of a **cascaded** series of two ports. If we cascade two or more two-port circuits (in an amplifier system, they are called **stages**), it means that the output of one is connected to the input of the next one in "daisy-chain" fashion, as shown in Figure 4.1, which shows four such cascaded circuits *A*, *B*, *C*, and *D*, each with a different *in-circuit* voltage gain. (The phrase "in-circuit" means the gain that it shows when it is connected to the other circuits with which it is used. This may or may not be the same gain the circuit shows when isolated on a test bench, for example.) Along the top of Figure 4.1, we show how you would calculate the overall voltage gain by multiplying together the numeric ratios of each stage's gain. The result is 1219.9, which in dB is $20\log_{10}(1219.9) = 61.72$ dB. Along the bottom, we have shown the dB equivalents of each numeric-ratio gain (rounded to the 1st decimal place). Because of the nature of logarithms, the *product* of several numeric ratios equals the *sum* of their logarithms. That means to find the gain in dB for the overall cascaded circuit, all we have to do is *add* the individual circuit gains (measured in dB), rather than multiplying them. With dB's, it is easy to see that, for example, the gain of stage *B* (+12.0 dB) is almost exactly canceled out by the *loss* (negative gain, in dB), namely, −12.8 dB, in stage *C*. It is not so easy to see that the numeric ratio of 0.23 is almost exactly the inverse of the numeric ratio of 4. The overall gain calculated with dB's—61.6 dB—is close enough to the exact answer of 61.72 dB to satisfy all but the most fastidious engineers. The difference of 0.12 dB represents an error of only about 1.4%, which is

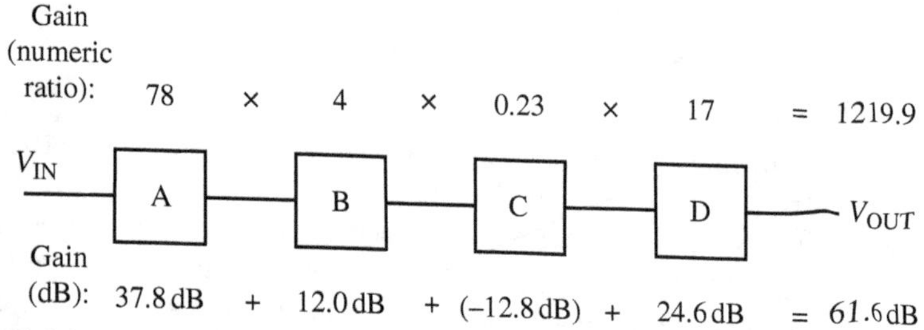

FIGURE 4.1 Cascaded series of amplifier stages with various voltage gains, shown both in numeric ratio and dB.

quite acceptable for most applications. This example shows the usefulness of dB units in calculating the overall gain or response of a long cascade of stages.

Returning to the example of a gentle nonlinearity, suppose in Equation 4.1 the constants are $a_1 = 10$ and $a_2 = 1\,\text{V}^{-1}$. What happens when we apply a sine wave of peak amplitude V_0 to this system? If $v_{\text{IN}} = V_0 \sin(\omega t)$, the expression for v_{OUT} can be expanded as follows:

$$v_{\text{OUT}} = a_1 V_0 \sin(\omega t) + a_2 [V_0 \sin(\omega t)]^2 = \frac{a_2 V_0^2}{2} + a_1 V_0 \sin(\omega t) - \frac{a_2 V_0^2}{2} \sin(2\omega t) \quad (4.6)$$

If the circuit were linear ($a_2 = 0$), the output would contain only the linear term beginning with a_1. But the second-order nonlinearity term gives rise to *two* new terms in the output: a *constant* (DC) term and a term whose frequency is *twice* the original, 2ω instead of ω.

In general, these additional terms are undesirable in a supposedly linear amplifier. The DC term can be eliminated by means of AC coupling through a capacitor, for example, but the **second harmonic** term at 2ω is not so easily dealt with. Let's see just how large the second harmonic's amplitude is with respect to the **fundamental** frequency term at ω. And it turns out using dB's will be useful in examining this question. (Harmonics are integral multiples of the fundamental frequency, numbered in engineering by the multiplier used, so that is why 2ω is called the second harmonic of the fundamental frequency ω.)

In order to plot the amplitudes of these terms using dB's, we need to decide on a **reference level** for the plot. Ordinarily, the dB concept is used to compare two different amplitudes or powers. But if you wish to express *absolute* amplitude or power using dB's, you need to choose a reference level having the same units as the quantity you are measuring. To indicate that the dB figure you come up with is not relative, but is absolute with respect to a reference level, a third letter is added after "dB" to indicate what the reference level is. There is no standard for this, but some common reference levels are 1 watt (dBW), 1 milliwatt (dBm), and 1 volt (dBV). We will use 1 volt (peak) as our reference level for an absolute voltage in dBV as we examine the second-order nonlinearity problem.

Let's call A_1 (dBV) the amplitude of the fundamental signal in Equation 4.6 and A_2 (dBV) the amplitude of the second harmonic. We have

$$A_1(\text{dBV}) = 20\log_{10}\left(\frac{a_1 V_0}{1V}\right) \quad (4.7)$$

$$A_2(\text{dBV}) = 20\log_{10}\left(\frac{a_2 V_0^2}{2(1V)}\right) \quad (4.8)$$

With specific values for a_1 and a_2, we can plot these amplitudes as a function of V_0 or, even better, as a function of A_0 (dBV), which is the dBV version of V_0. When we do this, an interesting fact emerges. Figure 4.2 shows these output amplitudes of the system represented by Equation 4.1 for the values $a_1 = 10$ and $a_2 = 1\,\text{V}^{-1}$.

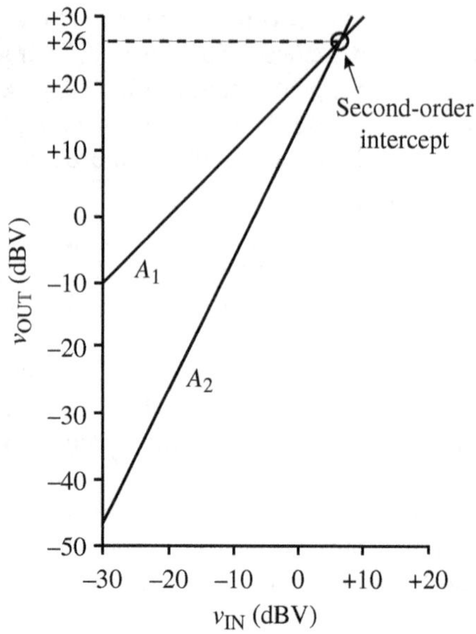

FIGURE 4.2 Fundamental (A_1) and second-harmonic (A_2) amplitudes for transfer function of Equation 4.1 for $a_1 = 10$, $a_2 = 0.1\,V^{-1}$, all in dBV.

First, note that the original amplified signal represented by A_1 has a slope of 1, meaning that a 1-dB increase in the input signal level causes a 1-dB increase in the fundamental component of the output signal. It also shows that the amplifier has a linear gain (for the fundamental component) of +20 dB, because, for example, when the input level is A_0 (dBV) = 0 dBV, the output level A_1 (dBV) = +20 dBV, and the difference 20 − 0 = 20 dB represents the linear gain of the amplifier (10 in numeric-ratio form). However, because the input amplitude V_0 is *squared* in the second-order term of Equation 4.1, the A_2 line has a slope of 2 dB/dB. Because the two lines have different slopes, they are not parallel, and the A_2 line will **intercept** the A_1 line at a specific output signal level. Because the line A_2 that intercepts the fundamental-frequency output curve A_1 arises from a second-order term, this point is called the **second-order intercept point**. With the values of a_1 and a_2 we have chosen, the second-order intercept output level is +26 dBV, which is about 20 V. We found this by equating the linear-amplitude term $a_1 V_0$ to the second-harmonic amplitude term $a_2 V_0^2/2$ in Equation 4.6, solving for V_0, and expressing it in terms of dBV. The graph tells us that if we provide an input signal with a peak level of +6 dBV, the amplitudes of the fundamental and the second-harmonic component will be equal at the output.

Clearly, if you want only the fundamental amplified and don't wish any harmonics to be generated in the system, this is a problem. And a sine wave is the cleanest sort of input one can deal with, meaning that it has the fewest harmonics to begin with. A more complex waveform will have many sine-wave components at various frequencies, and each will produce its own second harmonic. Those harmonics can interact with each other to produce yet more harmonics, and the situation gets complicated very fast. Unless one's goal is to generate harmonics (and sometimes this is desirable), having a second-order nonlinearity in a transfer function is a problem. But note that the problems occur at frequencies that are generally twice the original input frequency, which is often enough of a frequency separation to be eliminated by filtering. That is not the case for the next type of nonlinearity we will consider, however.

4.2.2 Third-Order Nonlinearity

The effects of second-order nonlinearity are not as bad as what happens with a **third-order nonlinearity**. Many linear amplifiers operate at high frequencies in the radio-frequency (RF) range, where it is fairly easy to design narrow-band frequency-selective **filters** that pass signals whose frequencies lie in a narrow band while rejecting all signals whose frequencies lie outside this band. If the output of the nonlinear system in question can be filtered this way, then harmonics at twice the fundamental frequency can usually be successfully filtered away, and no serious problem results. However, some problems caused by third-order nonlinearities cannot be solved this way, as we will now show.

Let's suppose that a particular electronic system has a linear gain term as in Equation 4.1 but instead of a second-order nonlinearity, it has only a third-order nonlinearity characterized by the coefficient a_3:

$$v_{\mathrm{OUT}(3)} = a_1 v_{\mathrm{IN}(3)} + a_3 (v_{\mathrm{IN}(3)})^3 \tag{4.9}$$

(We have designated these voltages with the (3) subscript to distinguish them from the earlier variables of the second-order Eq. 4.1.) Now instead of a single sine wave at a frequency ω, let's suppose we have *two* signals of peak amplitudes V_1 and V_2 at closely spaced frequencies ω_1 and ω_2, where $\omega_2 > \omega_1$. The output $v_{\mathrm{OUT}(3)}$ is then expressed as

$$v_{\mathrm{OUT}(3)} = a_1[V_1 \sin(\omega_1 t) + V_2 \sin(\omega_2 t)] + a_3[V_1 \sin(\omega_1 t) + V_2 \sin(\omega_2 t)]^3. \tag{4.10}$$

If we write out all the terms of this equation, it will be very messy and contains terms at the second and third harmonics of each frequency. But what is relevant to this analysis is that with a third-order nonlinearity, you will produce spurious (false) signals that are *very close in frequency* to the original two signals, so close that they cannot usually be filtered out by practical means.

Suppose we call the frequency spacing between the two original signals $\delta\omega = \omega_2 - \omega_1$. It can be shown that among the terms of the expanded version of Equation 4.10 are the following two terms, which we will call v_{LOW} and v_{HIGH}:

$$v_{\mathrm{LOW}} = -\frac{3a_3 V_1^2 V_2}{4}\cos(\omega_1 - \delta\omega)t \tag{4.11}$$

$$v_{\mathrm{HIGH}} = -\frac{3a_3 V_1 V_2^2}{4}\cos(\omega_2 + \delta\omega)t \tag{4.12}$$

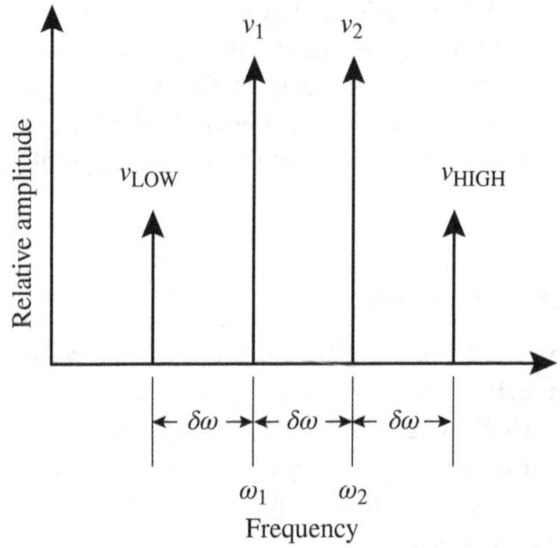

FIGURE 4.3 Plot of two original tones (frequency components) v_1 and v_2 spaced $\delta\omega$ apart in frequency, which produce two third-order intermodulation products v_{LOW} and v_{HIGH} at frequency spacings shown.

A qualitative *spectrum* of these two signals and their frequency relationship to the original two signals V_1 and V_2 is shown in Figure 4.3. (A frequency spectrum is nothing more than a graph showing the relative amplitude of various frequency components as a function of frequency.) Assuming the two original signals have the same amplitude, we see them displayed at frequencies ω_1 and ω_2. But now in addition to the two original signals, the third-order nonlinearity has produced *two more signals* that are spaced only $\delta\omega$ away from the first two, one on either side. These spurious signals are called **intermodulation products** because they result from one signal "modulating" (changing the amplitude of) the other. (Intermodulation is sometimes abbreviated as *IM*.)

In communications systems, the **channel spacing** is the frequency spacing between adjacent independent signals. In some RF systems, the channel spacing can be as close as 10 kHz or less. So if V_1 and V_2 represent two different signals in a communications system, the difference $\delta\omega$ could be only 10 kHz. And if the signals themselves have frequencies in the range of many MHz, it is very difficult to filter out signals that are spaced apart by only a very small percentage of the signal frequency. In other words, once the IM products v_{LOW} and v_{HIGH} show up, you are stuck with them. So the best thing to do is to keep them from appearing in the first place.

In practice, the way that third-order IM products are measured is straightforward. Two pure signals (sine waves) of equal amplitude (called **tones**) separated by a small difference in frequency $\delta\omega$ are sent to the input of the system to be tested. A high-quality **spectrum analyzer** or receiver that can separate out each individual signal by itself is connected to the system's output, and a plot is made of the level of both the original input tones and the level of the spurious IM tones as a function of the input level of the original tones. If this is done correctly, the levels of the 3rd-order IM tones will increase at a rate of 3 dB for every 1-dB increase in the original signal

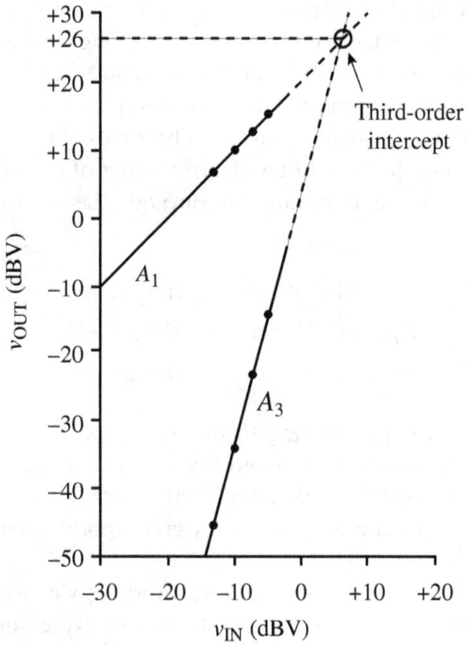

FIGURE 4.4 Hypothetical third-order intermodulation data for an amplifier showing a third-order intercept point at +26 dBV.

output level. Usually, the system will completely overload long before the third-order IM products are equal in magnitude to the original tones, so the two lines (original tones and IM product tones) are **extrapolated** to the point where they intersect, as shown in Figure 4.4. A_1 in this case is the amplitude of each of the two original tones, and A_3 is the amplitude of each third-order IM product.

The point where the two lines would cross if extrapolated is called the **third-order intercept** and is expressed in terms of an absolute output (or input) level. The higher this level is, the more input signal it will take to produce a given level of intermodulation, and the more linear the amplifier is, other things being equal. Of course, linearity is only one of many desirable qualities. Efficiency is another, and it turns out that efficiency and linearity tend to be opposing qualities, although there are some design approaches that can deliver both good linearity and good efficiency. But that is not easy to do.

4.3 LARGE-SCALE NONLINEARITY: CLIPPING

While some circuits can be designed to produce a nonlinearity that is only second order or third order, a much more common type of nonlinearity that a wide variety of circuits show is called **clipping**. Clipping (also sometimes referred to as **limiting**, especially when it occurs as a desirable feature) means that a circuit's output follows the exact shape of the input waveform up to a maximum amplitude and then stops. A clipped waveform looks as though someone came along with a pair of hedge trimmers and clipped off the top and bottom at a constant level, hence the term. When most amplifiers that use constant-voltage power supplies reach the limits of their output voltage capability, the output waveform becomes clipped.

We can see the consequences of an idealized type of clipping by starting with a sine-wave voltage V_{IN} and running it through the following mathematical operation:

$$V_{OUT} = \begin{cases} +1V & \text{if } V_{IN} > +1V \\ V_{IN} & \text{if } -1V \leq V_{IN} \leq +1V \\ -1V & \text{if } V_{IN} < -1V \end{cases} \tag{4.13}$$

The **transfer function** corresponding to this operation is shown in the graph in Figure 4.5. Again, we assume there is no frequency dependence and also that the corners on the transfer-function curve are perfectly sharp. While this is an idealization of what real electronic systems do, it is a fairly good approximation and behaves like a real circuit would in many ways.

Figure 4.6 shows the shape of the output waveform as we increase the signal level past 0 dBV, which we have chosen to use as the way of expressing the signal level in dB. Because the actual frequency of the signal is not important in this example, we have chosen to label the horizontal (time) axis in degrees rather than seconds. We have plotted curves for three input levels above 1 V: (a) +0.14 dBV (V_{IN} = 1.02 V peak), (b) +2.06 dBV (V_{IN} = 1.02 V peak), and (c) +14 dBV (V_{IN} = 5 V peak). With a clipping level of 1 V peak, signal (a)'s clipped portion is barely visible as a slight flattening at the very top and bottom of the waveform. As the input amplitude increases, the output spends a larger fraction of its time at either +1 V or −1 V, and the rest of the waveform becomes steeper. For signal (b), the sine-wave curvature is still visible but a good fraction of the waveform is now clipped. For signal (c), which

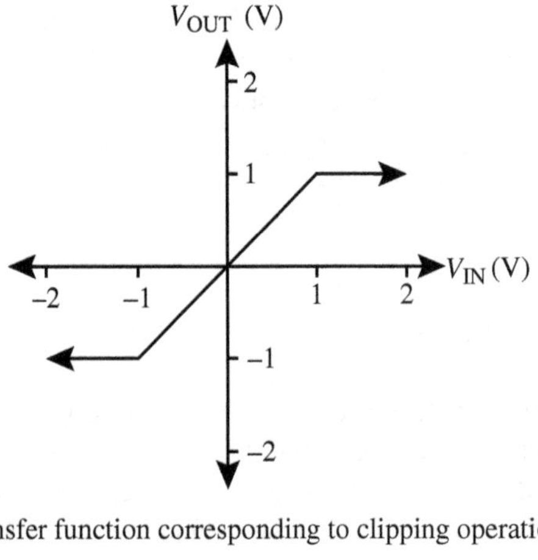

FIGURE 4.5 Transfer function corresponding to clipping operation of Equation 4.13.

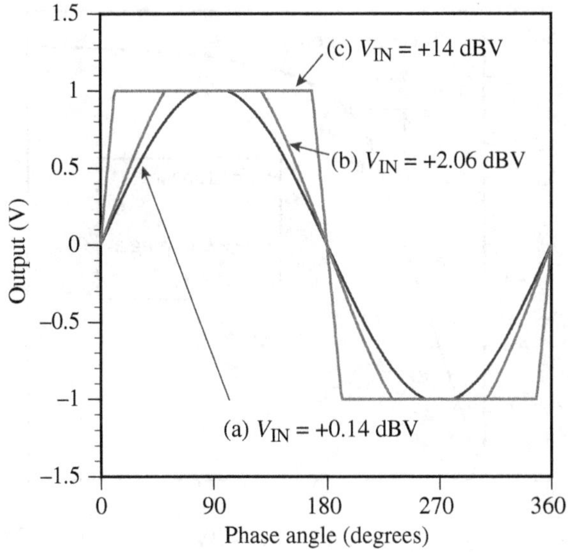

FIGURE 4.6 Waveforms of sine wave clipped at 1 V (peak) amplitude for input voltages (a) +0.14 dBV (V_{IN} = 1.02 V peak), (b) +2.06 dBV (V_{IN} = 1.268 V peak), and (c) +14 dBV (V_{IN} = 5 V peak).

corresponds to overloading the circuit by a factor of five, the waveform begins to resemble a square wave with nonzero rise and fall times. (The **rise time** of a square wave is conventionally defined as the time it takes to get from 10 to 90% of its final (peak) value from its initial value, and the **fall time** definition uses the same levels but applies to waveforms that are going down instead of up.)

A term you may encounter in dealing with analog systems is **overloading**. When a system is either driven by a much larger input than it was designed for or loaded by a much greater load than it was designed for, it is said to be *overloaded*. There is no exact definition for overloading, because the criterion depends upon the quality of performance desired from the system. Generally speaking, however, when a signal is large enough to produce more than the type of clipping associated with case (b), it is safe to say that it is overloaded. And case (c) in which the input is five times what is required to initiate clipping would be termed severe overloading for a linear system.

What does clipping do to the output power of a circuit? Recall that a linear circuit will show a 1-dB rise in output power for a 1-dB rise in input power. Because the output level is limited to the −1 V-to-+1 V range, we expect that a plot of output power versus input power will start to fall away from the ideal linear slope when clipping occurs. Figure 4.7 shows that this indeed happens. We have plotted the output power in dBV, taking into account all the power in the waveform (the fundamental frequency plus all the harmonics). Just as you would expect, at 0 dBV (1 V peak), the actual output begins to fall away from the dashed line indicating what a linear circuit would do.

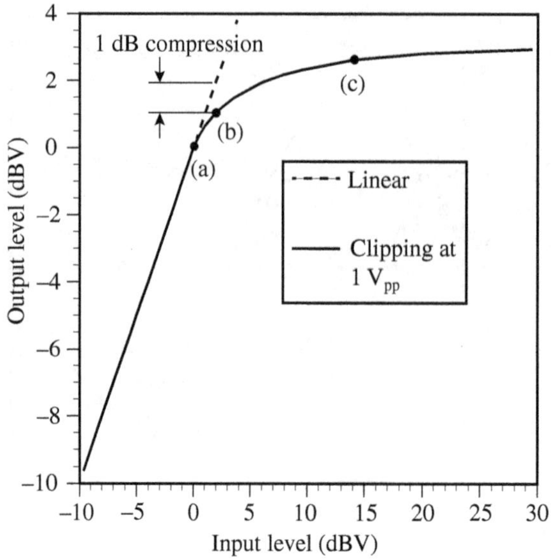

FIGURE 4.7 Output power (in dBV) versus input power for the idealized clipping operation of Figure 4.5. Input levels indicated by dots are (a) +0.14 dBV (V_{IN} = 1.02 V peak), (b) +2.06 dBV (V_{IN} = 1.268 V peak), and (c) +14 dBV (V_{IN} = 5 V peak).

A point of particular interest in Figure 4.7 is the input level at which the output falls below the ideal linear curve by 1 dB. This level is called the *1-dB compression* input level, because the output is "compressed" or clipped so that it has fallen 1 dB below what a linear system would produce. Note that for the ideal symmetrical clipping we have assumed, 1-dB compression will occur when the input sine-wave voltage is 2.06 dB (a numerical factor of 1.286) higher than it takes for clipping to begin. Generally speaking, if any pretense to linearity is desired, one should not drive an amplifier or analog system beyond the 1-dB compression point. As the input level rises farther, the actual output level flattens out and **asymptotically** (almost but never quite) approaches a constant level, which corresponds to the power contained in an ideal square wave whose amplitude is 1 V peak.

What is even more interesting than the output power curve is an analysis of the harmonics generated by clipping. Because the type of clipping we have chosen as an example is perfectly **symmetrical** (i.e., it does the same thing to the positive part of the waveform as it does to the negative part), there will be only *odd* harmonics in the output. (This is a consequence of the way Fourier-transform mathematics works.) Figure 4.8 shows the levels of these odd harmonics up to the 35th multiple of the fundamental frequency. (The reference 0-dB level has been chosen for each case to be the amplitude of the fundamental component, which changes somewhat with input level.) As you would expect, the very small amount of clipping encountered when the input level is (a), +0.14 dBV, produces low levels of harmonics—the largest one is at −50 dB or, as it is sometimes termed, "50 dB down." The harmonics for the 1-dB compression case (b), +2.06 dBV, are considerably higher: about 20 dB down for the worst case, which is the third harmonic. For the severely overloaded case (c), +14 dBV input, the third harmonic

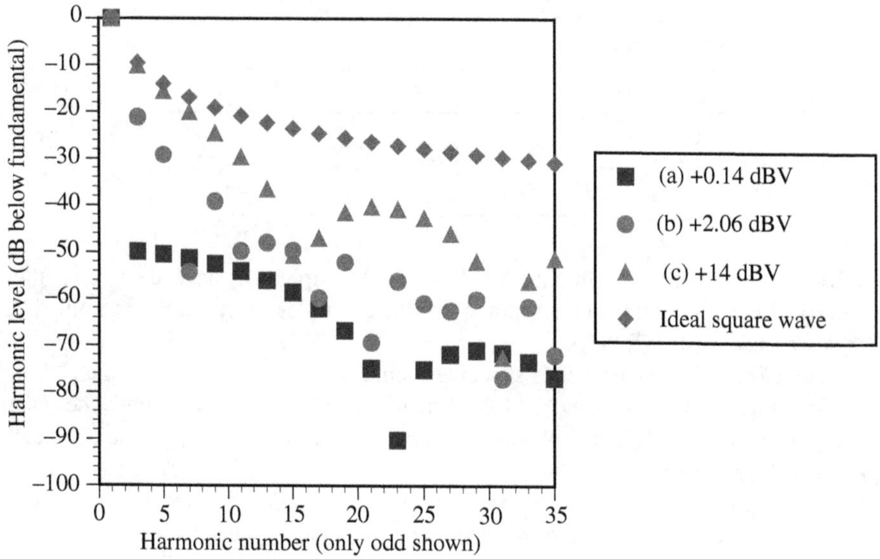

FIGURE 4.8 Harmonics generated by ±1-V peak clipping of input levels (a) +0.14dBV (V_{IN}=1.02V peak), (b) +2.06dBV (V_{IN}=1.268V peak), and (c) +14dBV (V_{IN}=5V peak).

is only about 10dB down, and higher-order harmonics ones are not much lower. As a kind of limiting case, we have also plotted the theoretical harmonic levels of an ideal square wave, which fall off rather slowly, as the graph shows.

If the system in question was being used for audio-signal amplification, most people could clearly tell when the signal becomes clipped enough to be compressed by 1dB. This is because the harmonics generated are easily heard unless the fundamental frequency is very high (above 4kHz or so). Surprisingly, analog signals representing the human voice can tolerate a rather large amount of clipping before intelligibility is impaired, but the same is not true of music.

While analyzing specific harmonic content is possible with simple input signals such as sine waves, it is sometimes more convenient to have a single number that characterizes the distortion due to all nonlinearities at a particular signal level. One such measure is called **total harmonic distortion**, abbreviated as **THD**. THD compares the power (or voltage) contained in all the harmonics to the power (or voltage) contained in the fundamental and expresses this ratio as a percentage. Possibly because the voltage ratio is the square root of the power ratio and thus smaller than the power ratio, equipment manufacturers prefer to express THD in voltage ratios, which we will denote as THD(V). The accepted definition is

$$\mathrm{THD(V)} = \frac{\sqrt{V_2^2 + V_3^2 + V_4^2 + \cdots + V_\infty^2}}{V_1} \tag{4.14}$$

where V_1 is the amplitude of the fundamental-frequency (sine-wave) component at the level specified for the test and V_2, V_3, …, V_∞ are the second, third, and so on harmonics, up to the last significant harmonic, which is designated V_∞. To show how

TABLE 4.1 Total Harmonic Distortion for Various Clipping Levels

Input level (dBV)	THD(V) (%)
+0.14	0.64
+2.06	9.52
+14	37.9

THD(V) varies with the clipping level of our example, we have calculated this quantity (again, out to the 35th harmonic) for the examples shown in Figures 4.6, 4.7, and 4.8 and show it in Table 4.1.

From a listener's perspective, most people cannot hear a THD level of less than 3% or so unless their ears are sensitive and the listening conditions are optimum. The wide variety of small portable sound-making electronic devices currently available (e.g., cell phones, portable music players, etc.) means that users have often been trained to accept a relatively high level of distortion, because of the limitations of inexpensive audio components. But distortion is also important in other contexts such as instrumentation, sensors, and digital data transmission, so it is something you will do well to understand.

4.4 THE BIG PICTURE: DYNAMIC RANGE

We have seen how noise is an unavoidable feature of analog systems and is always present to some degree. We have also seen that every supposedly linear analog system has a maximum output level beyond which nonlinearity begins to limit the system's ability to track the input level faithfully. Together, these effects determine a characteristic of an analog system called **dynamic range**. The dynamic range of a system, roughly speaking, is the ratio of the maximum input signal it can handle with acceptable linearity to the minimum signal it can accept in the presence of typical noise levels. It is the input signal range over which the system is useful, limited at the low end by noise considerations and at the high end by nonlinearity considerations. Because this ratio is often quite large, it is customary to cite it in dB.

Figure 4.9 is a plot of input signal level versus output signal level for the input stage of a hypothetical microphone preamplifier with 10 dB of gain (all levels are in dBV). The equivalent input noise level of a good low-noise preamp can be below 1 μV, so we have assumed that the internal noise of the preamp produces an output noise of −130 dBV. When we *refer* this output noise level to the input, it is reduced by 10 dB to −140 dBV, meaning that the equivalent input noise level is −140 dBV. To translate this level into an absolute number, one multiplies the reference level (1 V peak) by 10 raised to the power of the dBV divided by 20:

$$V\,(\text{absolute}) = (1\ \text{V peak}) \times 10^{\frac{(\text{dBV})}{20}}, \tag{4.15}$$

which in the case of −140 dBV gives $V_{IN}=0.1\,\mu V$. If you connect this system to a low-level signal source and reduce the signal input level to 0.1 μV, the signal level at the output

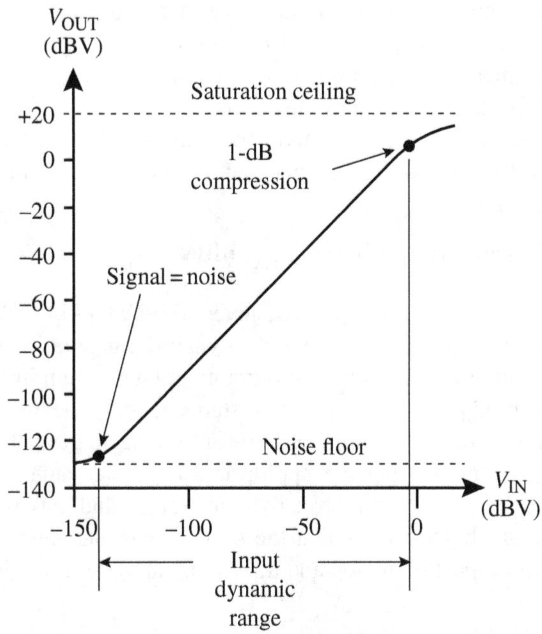

FIGURE 4.9 Input level versus output level for hypothetical microphone preamp input stage having a gain of 10 dB. Solid curve is input–output relationship; dashed lines indicate noise floor and saturation ceiling.

will be equal to the noise level at the output. While the minimum usable signal-to-noise (S/N) ratio varies from system to system, for audio-signal applications, S/N = 1 is barely usable and is often used as the lower practical limit of the input signal level. Note that the total output curve deviates from linearity and gradually flattens out to the noise-floor level at very low signal input levels. This is because the noise and the signal *powers* add, and well below a signal input of 0.1 μV, the noise will dominate the output. So for a S/N of 1, the input level is −140 dBV, which is termed the **noise floor** for the system, because an input signal level below this will not be amplified with sufficient S/N ratio to be useful.

We will choose the 1-dB compression point as the upper limit of the amplifier's dynamic range. As we have seen earlier, that point implies that THD is at least 9% or so if the reason for nonlinearity is clipping, and this amount of distortion is easily audible. In the example shown, the input level to produce 1 dB of compression at the output is about −3 dBV, or 0.707 V peak. This is an unusually high output level for a microphone unless it is reproducing a nearby thunderclap, but some types of microphones can deliver voltages in this range.

Figure 4.9 shows a **saturation ceiling**, which is the ultimate output the system can provide for indefinitely large inputs. In the clipping example shown earlier, the saturation ceiling results when the output is an ideal square wave.

"Saturation" is one of the most overused terms in engineering. In this context, a saturated amplifier is one that is producing its maximum output, and further increases in input level will no longer raise the output level at all. This is not a mode of operation to use for linear audio amplifiers, so we have established the upper dynamic range limit well before the output reaches the saturation ceiling. Other systems, such as certain high-frequency RF amplifiers, actually *should* be operated in saturation, but that topic will be addressed later in the book.

In the example shown for the lower and upper limits we have determined, the total dynamic range in dB is the difference between the input at the upper limit and the input at the lower limit:

$$\text{Dynamic range (dB)} = V_{\text{IN(MAX)}}(\text{dBV}) - V_{\text{IN(MIN)}}(\text{dBV}) \tag{4.16}$$

Doing the calculation for our example case gives $-3\,\text{dBV} - (-140\,\text{dBV}) = 137\,\text{dB}$ for the system's dynamic range. In addition to the dynamic-range ratio, it is important to know the actual absolute values for the minimum and maximum input levels so that the system may be used properly with other systems or components, which have their own dynamic-range limits. The more systems are connected together, the harder it sometimes is to determine what the appropriate dynamic range is for the entire assembly. But every analog system has a dynamic range, and making sure it is large enough to handle all the situations that the system will encounter in practice is a subtle but important aspect of analog and mixed-signal systems engineering.

BIBLIOGRAPHY

Hagen, Jon B. *Radio-Frequency Electronics: Circuits and Applications*, 2nd edition. Cambridge, UK: Cambridge University Press, 2009.

Pedro, J. C. and N. B. Carvalho. *Intermodulation Distortion in Microwave and Wireless Circuits*. Boston, MA: Artech House, 2003.

PROBLEMS

Note: Problems of above-average difficulty are marked with an asterisk (*).

4.1. *Second- and third-order intercept of nonlinear amplifier.* A certain amplifier has the following transfer function: $v_{\text{OUT}} = a_1 v_{\text{IN}} + a_2 v_{\text{IN}}^2 + a_3 v_{\text{IN}}^3$. Suppose $a_1 = 40$, $a_2 = 2\,\text{V}^{-1}$, and $a_3 = 0.1\,\text{V}^{-2}$.

- **(a)** What is the linear gain of this amplifier in dB?
- **(b)** What is the second-order intercept level at the output? (Express it in terms of dBV (peak) at the output).
- **(c)** If two single-frequency signals at f_1 = 2.09 MHz and f_2 = 2.12 MHz are applied to the input, at what frequencies f_{LOW} and f_{HIGH} will the third-order IM products appear?
- ***(d)** What is the third-order intercept level at the output, expressed in terms of dBV (peak)? Recall that this is the level of each of two tones (in dBV peak) that will (theoretically) produce a third-order IM product equal in amplitude to one of the original tones.

4.2. *Absolute dB quantities.* Suppose at the output of a particular system, you measure a sine wave whose peak voltage is 470 mV. This voltage appears across a load resistance of 600 Ω.

(a) Express this output level in dBV (with respect to 1 V peak):

(b) Express this output level in dBW (with respect to 1 watt).

(c) Express this output level in dBm (with respect to 1 milliwatt).

4.3. *Power delivered as amplifier saturates.* Suppose a system with a gain of 32 dB is linear until the output reaches a level of either +12 or −13 V, beyond which limits the output is clipped.

(a) If a symmetrical AC sine wave (no DC) is provided to the system, what is the maximum peak input voltage v_{IN}(max) the system can receive before it begins to clip on at least one half of the waveform?

(b) When the system is driven so hard that its output is a rectangular wave going from +12 to −13 V (equal times + and −), what is the average output power delivered to a 1-kΩ load, in milliwatts? Assume the output is DC-coupled to the load.

4.4. *Noise power of amplifier.* A certain amplifier is connected to a source whose internal equivalent resistance is $R_S = 100\,\text{k}\Omega$. Its output is connected to a 4-Ω load. The system's internal noise is low enough so that the source's noise dominates the internal noise and determines the noise floor of the system. The source behaves like its internal resistance is at a temperature of 30,000 K. The system has a voltage gain of 70 dB when connected to its source and load.

(a) What is the equivalent input noise voltage v_N of the source in a bandwidth $B = 20\,\text{kHz}$? Express your answer in terms of RMS voltage.

(b) What is the noise power delivered to the load? Assume that the amplifier's RMS output voltage is 70 dB higher than v_N.

(c) If the amplifier clips in an ideal mathematical fashion at an output voltage of ±5 V, what is the RMS input voltage v_{IN}(1-dB) that will cause 1-dB compression at the output?

(d) Defining the maximum input to be v_{IN}(1-dB) and the minimum input for S/N = 1 to be v_N, find the amplifier's dynamic range $DR_{(dB)}$ in dB.

For further resources for this chapter visit the companion website at **http://wiley.com/go/analogmixedsignalelectronics**

5

OP AMP CIRCUITS IN ANALOG ELECTRONICS

5.1 INTRODUCTION

An *operational amplifier* (hereafter "op amp") is a specially designed amplifier that can be used in a huge number of analog-circuit applications. It is fair to say that almost anything you can do in analog electronics without op amps is easier, smaller, and better performing if done with op amps. And some things are almost impossible to do without op amps or their functional equivalents.

Why are op amps so important to analog electronics (and mixed-signal electronics as well)? Because they make the use of negative feedback easy and straightforward and confer its blessings on any circuit where they are used. To understand the importance of negative feedback for op amps, a little history is helpful.

Although the concept of **negative feedback** was studied by nineteenth-century physicists such as James Clerk Maxwell, negative feedback as used in electronics was developed by a Bell Labs engineer named Harold S. Black, who in the 1920s was struggling with the problem of how to improve the repeater amplifiers the Bell System was trying to use to span the US continent with long-distance telephone lines. Nonlinearities were a significant problem, because in traveling through repeaters spaced every 25 miles or so along a cable, signals encountered distortion that was tolerable in one amplifier, but which accumulated to unacceptable levels as the signals passed through dozens of amplifiers in cascade. Black realized that if he built an amplifier with much more gain than necessary and then reduced the gain by "feeding back" a portion of its output signal to the input with the proper polarity, nonlinear distortion

Analog and Mixed-Signal Electronics, First Edition. Karl D. Stephan.

Companion Website: http://wiley.com/go/analogmixedsignalelectronics

would decrease roughly in proportion to the amount of sacrificed gain. Gain, while not free, was relatively easy to obtain simply by cascading stages, and it turned out to be a good engineering trade-off to sacrifice inexpensive gain for the much-sought-after reduction in distortion. It turned out that negative feedback conferred other desirable properties such as reduced noise, improved frequency response, and greater stability against changes in component values and characteristics as well.

Throughout the 1930s, the good news of negative feedback traveled through the electronics world, and when World War II began, engineers charged with the task of developing fire-control systems for ships turned to negative-feedback amplifiers to perform mathematical operations such as addition and multiplication in analog computers. John Ragazzini, a professor of electrical engineering at Columbia University, was the first person to use the phrase **operational amplifier** to refer to these specialized computing amplifiers, in a paper published in 1946. But it was a student of his, named Loebe Julie, who apparently devised the first op amp that had all the essential characteristics of today's op amps: a high gain, a dual-polarity power supply, and a pair of **differential** inputs. A voltage amplifier with differential inputs responds only to the difference in voltage between the inputs, and not just to the absolute signal value. Another colleague of Ragazzini's named George A. Philbrick founded a firm that introduced the first commercially available op amp in 1952 (see Fig. 5.1). In 1968, Fairchild Semiconductor Corporation released the μA741, the first popular integrated-circuit

FIGURE 5.1 The first commercially available operational amplifier used two vacuum tubes and cost $24 in 1952. (From www.PhilbrickArchive.org. Reproduced by permission of Joe Sousa.)

(IC) op amp. Many hundreds of op amp types are now available, but most of them share certain basic characteristics that we will now discuss.

5.2 THE MODERN OP AMP

5.2.1 Ideal Equivalent-Circuit Model

Designers of op amps use various circuit approaches to approximate an ideal op amp, which we will describe first. The ideal model is greatly simplified, but in complex systems, the ideal model is simple and convenient to use and will often give results that are a fairly good approximation to what the actual system will do. However, the designer should always bear in mind that real op amps fall short of the ideal model in various ways, as we shall see later.

Figure 5.2a shows the standard schematic-diagram symbol for an op amp. The right-pointing triangle symbolizes an amplifier with inputs on the left (wide) side and output at the tip on the right. Op amps have two inputs: a **noninverting** input whose voltage is designated V_+ and an **inverting** input designated V_-. The names are chosen the way they are because a signal fed to the inverting input is inverted (changed in sign) when it appears at the output, while a signal fed to the noninverting input does not change its sign. So if a voltage at the noninverting input rises (goes more positive), the output voltage will go more positive too. But if a voltage at the inverting input rises, this will cause the output voltage to fall. Keeping these polarities straight in your mind is vital to understanding op amp circuits.

Most (but not all) op amps have a single output lead, which carries the output voltage V_0. In addition, most op amps require **dual** DC power supplies, designated $+V_S$ and $-V_S$. Typically, these voltages are equal in magnitude and opposite in sign (e.g., ±15V). Not shown in the circuit, but very important, is the fact that each of the two power supplies has one terminal connected to **ground** and the other terminal connected to its respective op amp power input terminal. Because the load connected

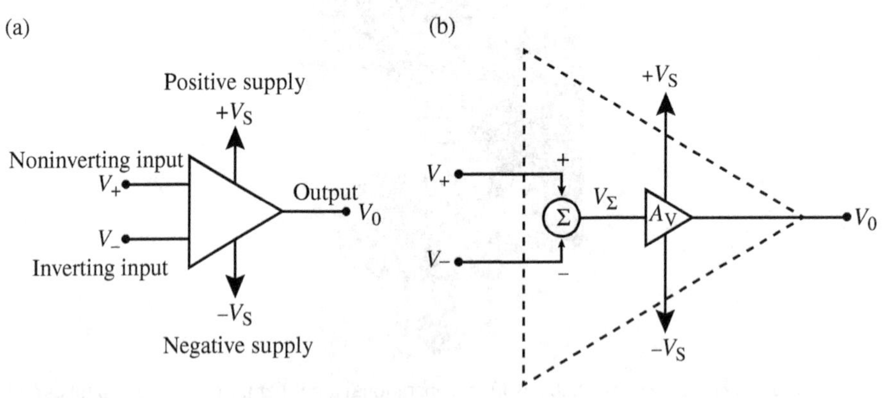

FIGURE 5.2 (a) Schematic symbol of op amp. (b) Equivalent circuit of ideal op amp.

to the op amp output is typically connected between the output and ground, the power supplies must also be grounded in order for the circuit to be completed from the supply, through the op amp, through the load, and back to ground. This is an elementary point that is, however, often overlooked by beginners.

In Figure 5.2b, we show the internal equivalent circuit of an ideal op amp. The two input voltages from the differential inputs go to a **summing junction** (indicated by the Greek letter *sigma*: Σ), which takes the mathematical difference of the two voltages. This difference, denoted V_Σ, goes to a high-gain amplifier with voltage gain A_V. Although not infinite, A_V is assumed to be high enough to satisfy certain approximations in feedback equations that we will describe later in this chapter. The output voltage V_0 of the op amp is therefore

$$V_0 = A_V(V_+ - V_-) \tag{5.1}$$

Even for an ideal op amp, Equation 5.1 is valid only if the value of V_0 lies between the negative and positive power-supply voltages: $-V_S < V_0 < +V_S$. Outside this range, the op amp **clips**, an action also known as **saturating** or **hitting the power-supply rails**, and the output fail to follow the input voltage changes beyond those limits.

The ideal op amp's input terminals draw *no current* from the source voltages connected to them. This means that the ideal op amp has **infinite input impedance**. Real op amps only approach this ideal, but some designs feature input resistances in excess of $10^{10}\,\Omega$. Also, the ideal op amp's output voltage V_0 remains the same no matter how much current is drawn from it (and also from the power supplies, of course). This means the ideal op amp has **zero output impedance**. Typical small-signal op amps can supply up to 10 mA or so before protective current-limiting circuits go into action, and specialized **power op amps** can deliver 1 A or more to the load. While the output impedance of a real op amp is low, often less than 10 Ω, it is not zero.

5.2.2 Internal Block Diagram of Typical Op Amp

Unless you are undertaking the formidable task of designing an op amp IC yourself, you need not be concerned with all the details of a typical op amp's inner workings. But in order to use these devices intelligently, the designer should know some basics of how it works.

Figure 5.3 shows a very simplified block diagram of a generic 741-type op amp.

It uses BJTs exclusively and begins with the **input stage** on the left. The two inputs, V_+ and V_-, are applied to two **matched** input transistors Q_1 and Q_2. By "matched," we mean that the devices are designed to have as nearly identical characteristics as possible. The more closely the devices are matched, the closer the input stage approaches the ideal of a truly differential amplifier. With modern IC fabrication techniques, it is fairly easy to produce two transistors on the same chip which are very closely matched.

Note that the junction of the emitters of Q_1 and Q_2 is driven by a bias-current source providing a constant bias current I_B. To the degree that this current is truly

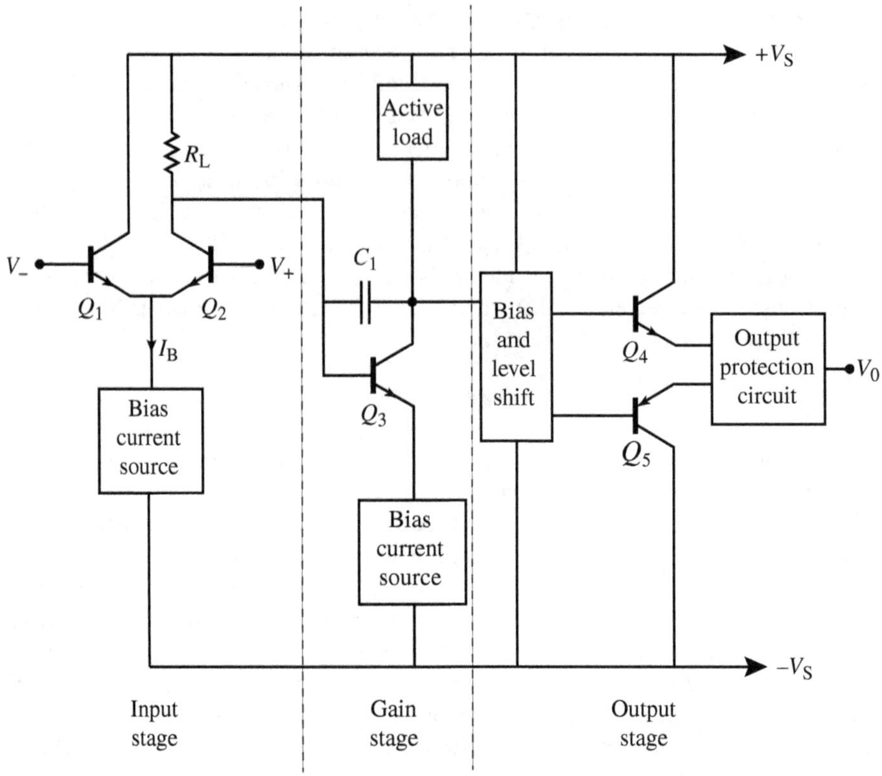

FIGURE 5.3 Simplified block diagram of typical integrated-circuit op amp of the generic "741" type.

constant (and good designs can make it nearly so), the only way that current can change in load resistor R_L at Q_2's collector is if the *division* of current between Q_1 and Q_2 changes. This division is determined exclusively by the *difference* in voltage between V_1 and V_2. If the difference is constant, the division of current will also be constant, and the current through R_L will be constant.

This will remain true even if V_+ and V_- change *together*. This type of change is termed a **common-mode voltage**. Recall the discussion of voltage vectors from Chapter 3. If we consider the pair of input voltages V_+ and V_- as a voltage vector $[V_{IN}]$:

$$[V_{IN}] = \begin{bmatrix} V_+ \\ V_- \end{bmatrix} \tag{5.2}$$

we can *transform* it by a pair of equations

$$V_\Sigma = \frac{V_+ + V_-}{2} \tag{5.3}$$

$$V_\Delta = \frac{V_+ - V_-}{2} \tag{5.4}$$

into another voltage vector $[V_{\Sigma\Delta}]$:

$$[V_{\Sigma\Delta}] = \begin{bmatrix} V_\Sigma \\ V_\Delta \end{bmatrix} \tag{5.5}$$

All of the information in the first pair of voltages is also contained in the second pair; it has simply been expressed in different mathematical form.

There are terms for the two elements of the new voltage vector $V_{\Sigma\Delta}$. The sum voltage V_Σ is called the **common-mode voltage**, while the difference V_Δ is called the **differential-mode voltage**. With these definitions in place, we can state the behavior of an ideal differential amplifier quite simply: an ideal differential amplifier responds only to V_Δ (the differential-mode voltage) and not at all to V_Σ (the common-mode voltage). If, in a pair of input signals, we hold V_Δ (the voltage difference) constant while varying the sum V_Σ of the two voltages, the division of current between Q_1 and Q_2 will not change. The voltage across load resistor R_L will also not change, and the amplifier will not respond to common-mode voltages at all. In other words, the ideal op amp will *reject* common-mode signals *perfectly*. No real op amp does this, of course, but an important op amp specification is the **common-mode rejection ratio** (sometimes abbreviated as "**CMRR**"), which is the ratio of the amplifier's response to the differential-mode signal to its response to the common-mode signal. Normally, an op amp will show very high gain for a differential-mode input and much lower gain for a common-mode input, so the CMRR is usually a large number, 70 dB or more.

The main job of the input stage is to perform common-mode rejection and to transform the differential-mode input into a so-called **single-ended** form. Differential signals such as the V_+-V_- pair are typically conveyed by pairs of wires, each of which carries a voltage with respect to ground. The voltage signal across load resistor R_L carries the differential-mode information in a single voltage value, so it is termed a "single-ended" output, again with reference to ground. Although a typical op amp does not have a terminal that must be connected directly to ground, all voltages in the circuit are indirectly referenced to ground by the fact that both constant-voltage power supplies are connected to ground.

In connection with ground returns, it is important to note that both input transistors need a certain minimum amount of DC base bias current to operate. This current is called the **input bias current**, and it is drawn automatically from the external circuit. It is typically less than 1 μA, but the designer *must* make provision in the external circuit for this DC current to flow to both inputs. A common mistake is to connect only a capacitor to an op amp input without any DC path to ground through a resistor. This AC-only connection will disable its input and cause the op amp's output to saturate at one rail or the other. If the external circuit does not naturally provide a DC path to each input, the designer must add one by means of a large-value resistor to ground.

The single-ended signal leaves the input stage without much (if any) voltage amplification. Voltage gain is the task of the gain stage that uses Q_3, shown in the middle section of Figure 5.3. This is a single transistor operating into an **active load**. If a transistor's transconductance is g_m and its load resistance is R_L, it is easy to find the voltage gain A_V (at low frequencies) for the device when connected to R_L:

$$A_V = \frac{v_o}{v_i} = \frac{-g_m R_L v_i}{v_i} - g_m R_L \tag{5.6}$$

So other things being equal, the voltage gain is proportional to the load resistance R_L. An active load uses a transistor or transistors to synthesize a load that shows a very high resistance for changing (AC) signals, while keeping the DC bias voltages within the limits of the DC power supplies. In this way, a single amplifier stage can produce a voltage gain of 10^5 (100 dB) or higher.

You will also note that a capacitor C_1 is connected between the collector and base of the gain-stage device Q_3. This capacitor produces a **dominant pole** (a lowpass filter effect) at a low frequency, typically in the range of 1 Hz, by means of the **Miller effect**. The Miller effect makes a small capacitor (C_1 is typically 30 pF or less) appear much larger when it is connected to the output of a high-gain inverting amplifier, such as C_1 is.

Why is a dominant pole at 1 Hz needed? Because the typical op amp of the 741 type is what is termed **unity-gain compensated**. That is a fancy phrase meaning that you can connect the device's output directly to its inverting input without encountering **oscillation** (the generation of a periodic output with no input) or other instability problems. It turns out that the dominant pole is a good way to achieve this desirable property of unity-gain compensation, as we will explain later.

Now that the signal has been amplified sufficiently, it needs to be sent through a final output stage, which is capable of delivering the current required by the load. The output of the gain stage is at a very high impedance, so we need a circuit with a high input impedance and an output impedance that is as low as possible. The general BJT configuration known as the **common-collector** connection (also termed an **emitter follower**) has these properties. After some additional bias circuitry and **level shifting** (changing the DC value of a signal without altering its AC content), the signal is fed to a pair of **complementary** BJTs, Q_4 and Q_5. As you will note, Q_4 is an NPN device, while Q_5 is a PNP device. "Complementary" means that, except for the polarity reversal encountered in going from an NPN to a PNP transistor, Q_4 and Q_5 have very similar characteristics. This makes it easier to design a circuit that will behave the same way for both positive and negative signals.

Because of the dual power-supply rails, the circuit can draw current from either a positive or a negative source to provide the proper polarity of output. When the circuit's output voltage V_o is greater than zero, Q_4 is active and supplies current through its collector to its emitter, which is controlled by the much smaller current into its base. During this time, Q_5 is basically inactive. But once the required output voltage goes negative, NPN device Q_4 becomes dormant and PNP device Q_5 jumps into action, providing negative current from the negative supply rail. In this way, the amplifier can either **source** or **sink** current, depending on the polarity required at the output. The output transistors are fairly large physically so as to handle the nontrivial amount of power they must dissipate in supplying up to 10 mA to the output terminal. In the 741 class of circuits, an output-protection circuit is interposed between the output devices and the output terminal. Normally, it is inactive, but when the external circuit asks for more than the 10 mA or so that the device is designed to supply, the protection circuit goes into action and **limits** the maximum current to a level that will not destroy the chip. While this is effective in many situations, it is nonetheless possible to damage op amp ICs by other means, such as connecting the power supplies incorrectly or by connecting an external voltage source to the output terminal.

5.2.3 Op Amp Characteristics

There is a bewildering variety of op amps available, and an exhaustive discussion of all the various specifications would be prohibitively long. Nevertheless, certain important characteristics make a particular op amp suitable for a particular application, and we will discuss those that the designer is most likely to be concerned with. These characteristics include the number and type of power supplies needed, input and output characteristics regarding impedances and current capability, gain, bandwidth, slew rate (SR), and others.

5.2.3.1 Power-Supply Requirements As we have seen, the useful range of output voltages an op amp can supply is limited by the number and voltage levels of its DC power supplies. No normal op amp can supply a voltage beyond the range established by the highest and lowest power-supply voltages provided to it. In fact, because of internal circuit limitations, the usable output range before clipping occurs is often a volt or two less than the actual power-supply range. For example, the generic 741 type of op amp, when connected to dual supplies at ±15 V, can typically swing only from −13 V or so to +12 V. Other op amp designs are not limited in this way and can provide outputs much closer to the **rails** (power-supply limits). It is also possible to augment the output of an op amp with external power devices and higher-voltage supplies connected to the output of the device.

Besides limiting the output voltage swing, the power-supply voltages limit the allowable input voltage ranges as well. This limitation applies to each input individually, but in practice applies mainly to the common-mode voltage, because in normal operation, the voltages applied to the two differential inputs of an op amp are very close to each other. The circuit will cease to amplify properly if the input's common-mode voltage goes outside specified limits. Some op amp designs can handle inputs whose voltage goes all the way to a positive or negative supply voltage level, however, if circumstances require it.

The absolute difference between the positive and negative supply voltages, called the total applied voltage, is also an important specification. In order for the bias circuitry to operate properly, the total applied voltage must exceed a certain minimum value. With the increasing use of low-power battery-operated equipment, both the minimum required voltage and the minimum current consumption have moved lower in new op amp designs, so that at least one type of op amp currently available can operate from a single 3-V supply. Of course, its output voltage range is similarly limited, so the choice of power-supply voltage range must be dictated by the particular application in mind.

5.2.3.2 ***Input and Output Impedance*** Some applications of op amps require that almost no current be drawn from the signal source. High-impedance sensors such as electrostatic-field sensors and condenser microphones can supply very little current to the load. In these situations, an op amp with a **FET input** is called for. FET-input op amps can have input resistance in the 10^{12}-Ω range and draw input bias currents measured in **picoamps** (pA, 10^{-12} A). Nevertheless, even such high-impedance inputs must have a DC path to ground to avoid cutting off the input bias current and disabling the op amp's proper operation.

Garden-variety BJT op amps such as the 741 series have lower input resistances (~1 MΩ) and higher bias currents (10–100 nA) than FET-input devices. These values are still high enough for many purposes, but should be taken into account when using these types of op amps with high-impedance sources. Certain types of feedback circuits can increase the effective input impedance of an amplifier circuit to levels far above the op amp's own input impedance level.

The output resistance of most op amps is typically below 100 Ω and can be made artificially lower (within limits) by the proper application of feedback as well. This low output impedance is available only if the output current is limited to less than the maximum value established by the output-protection circuit, if any.

5.2.3.3 ***Gain and Bandwidth*** We will discuss these specifications together because they are best considered as two parts of the most fundamental specification, the **gain–bandwidth product (GBP)**. As we will see when we begin discussing circuits that use op amps, there are ways of trading gain for bandwidth. But for a given op amp, the product of gain (the numeric ratio, not in dB) and the bandwidth (meaning the frequency at which the gain is 3 dB down from its DC value) is a constant, which we denote as the **GBP**.

Because most commonly used op amps are unity-gain compensated, their **open-loop** gain (the gain from one input to the output with no feedback of any kind) is accompanied by a very low open-loop bandwidth, typically 1–3 Hz. (Unity-gain compensation enables the designer to use any amount of feedback up to *unity* (1).) This frequency, which we will call f_0, is termed the **3-dB-down frequency**, which is also sometimes referred to as the **cutoff frequency** or the **corner frequency**. Because the standard definition of bandwidth is that range of frequencies over which the gain decreases from maximum by no more than 3 dB, f_0 equals the bandwidth in the case where the gain goes all the way to DC (zero frequency). Beyond that frequency, the gain falls off at a 20 dB per decade rate until it goes below unity. Above the 3-dB-down frequency, the gain is inversely proportional to frequency, so the product of frequency and gain at that frequency is a constant, termed the GBP. Values of the GBP for typical op amps range from 300 kHz to 5 MHz or more.

Just as important as the GBP is the DC or low-frequency voltage gain A_V. For the approximation of "high gain" to be true in most applications, A_V needs to be 100,000 or more. The actual gain realized in a circuit using such an op amp cannot be more than a small fraction of this if the other advantages of negative feedback are to be realized. This means that a single stage of amplification using an op amp with a gain of 100,000 should not provide a usable gain of more than about 100 or so.

5.2.3.4 ***Slew Rate (SR)*** Strictly speaking, ***SR*** is a dynamic nonlinear effect that occurs when an op amp is commanded to change its output voltage abruptly over a large range of several volts. The bandwidth-limiting capacitor in Figure 5.3 has to charge or discharge before this can happen, and the internal bias current available limits the maximum rate at which the output voltage can change, even for an instantaneous change in input voltage. SR limits are typically stated in terms of $\mathrm{V\,\mu s^{-1}}$. SR limits become important when a device is called on to amplify a large output voltage at a high frequency. Signals that may not cause a problem at lower levels can run into the device's SR limit if either the amplitude or the frequency increases. Because of SR limits, op amp circuits are not particularly suitable for the amplification of large-value digital waveforms such as square waves and pulses, which typically change very fast over a large voltage range.

5.2.3.5 ***Other Op Amp Specifications*** We have already mentioned input bias currents, which can be very small but cannot be neglected in the design of the input circuit. Manufacturers usually specify a maximum input bias current, which will cause a small constant voltage to appear at the inputs when this current flows through the internal DC resistance of the source. For this reason, when circuits designed to amplify small DC voltages are designed, it is good practice to make sure that the total DC resistance connected to each of the two inputs is approximately the same. If this is not done, equal bias currents will cause unequal bias voltages to appear across the input terminals, leading to a spurious (false) voltage that can interfere with the desired signal.

Even if input bias currents flow through balanced resistances, an op amp will not produce exactly zero volts out for zero differential volts at the inputs. Typically, an op amp with exactly zero volts on its inputs will produce a small DC output voltage $V_0(0)$. This zero input–output voltage, when divided by the amplifier's differential voltage gain A_V, yields a number called the **input offset voltage** V_{OS}:

$$V_{OS} = \frac{V_o(0)}{A_V}. \tag{5.7}$$

The input offset voltage is the differential input voltage that would be required to produce the measured nonzero output if the amplifier were ideal. You can think of it as a small voltage source inserted in series with the actual input voltages. Manufacturers specify a maximum magnitude for V_{OS}, which is typically in the millivolt to microvolt range. Unless your application requires handling DC or slowly varying input voltages in this range, a small offset voltage does not usually cause problems.

Finally, op amps all have definite frequency and phase response characteristics. Unity-gain compensation, in which the response is dominated by a single pole at a low frequency, is only one type of compensation available. Other types of frequency response may be obtained in other types of op amps by the adjustment of a capacitor external to the IC or by other means. In the following sections of this chapter, we will always assume that the op amps used are unity-gain compensated.

5.3 ANALOG CIRCUITS USING OP AMPS

The great usefulness of op amps in analog electronic circuits comes from the fact that negative feedback can be used to design circuits whose critical characteristics such as gain and frequency response are easily calculated and determined almost entirely by values of passive components. It is difficult even today to produce discrete active devices such as transistors that have precisely controlled characteristics. For example, the BJT current-gain parameter called β can vary over a 4-to-1 range and still be within typical manufacturer's specifications. Obviously, if an amplifier design's voltage gain is proportional to the β of the transistor used, the circuit's gain will vary over an equally wide range for different transistors.

As we will show, the use of large amounts of negative feedback will shift the control of circuit characteristics from the properties of the active devices to the values of passive components such as resistors and capacitors. It is very easy to make passive components with precise values—resistors with tolerances of 0.1% are available commercially—and so circuits using op amps with negative feedback can be designed to deliver precise and repeatable characteristics. We will show how this happens in the following simplified analysis of a negative-feedback circuit that will apply to most of the op amp circuits to be described in the remainder of Section 5.3.

Figure 5.4 shows the simplified block diagram of an op amp taken from Figure 5.3, together with an external feedback network characterized by the coefficient β. (For the moment, we will assume that β is a real positive constant between 0 and 1, although we will lift this restriction later.) The input signal voltage v_I goes into the positive input of the summing junction. A portion β of the output signal v_O is fed back to the negative input of the summing junction. The coefficient β is called the **feedback factor**,

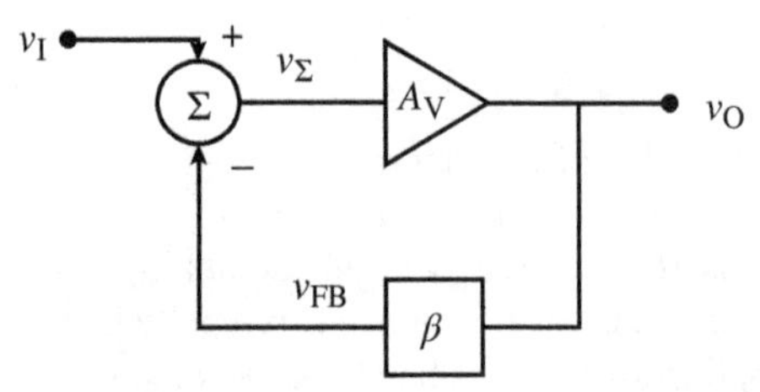

FIGURE 5.4 Generalized op amp feedback circuit with passive feedback network.

feedback coefficient, or sometimes simply the **amount of feedback**. The fed-back voltage is termed v_{FB} and is given by $v_{FB}=\beta v_O$.

The algebra to solve for the output voltage v_O in terms of v_I and the circuit constants is straightforward:

$$v_O = A_V(v_I - v_{FB}) = A_V(v_I - \beta v_O) \tag{5.8}$$

$$v_O = \frac{A_V v_I}{(A_V\beta + 1)} \tag{5.9}$$

If we denote the overall circuit's transfer function as $H=v_O/v_I$, we find that

$$H = \frac{A_V}{(A_V\beta + 1)} \tag{5.10}$$

The term $(A_V\beta+1)$ is very significant for this type of feedback system. Here is why.

Suppose, as is typical, that A_V is very large, say, 10^5. Then for any value of β greater than 10^{-3} (0.001), the product $A_V\beta$ will be more than 100 times the second term, namely, 1. So within 1% accuracy (which is sufficient for all but the most exacting instrumentation applications), we can neglect the 1 and approximate $A_V\beta+1$ as simply $A_V\beta$.

When we do that, a great simplification occurs in the math. We find that the overall transfer function becomes

$$H \approx \frac{A_V}{A_V\beta} = \frac{1}{\beta} \tag{5.11}$$

As long as A_V (the op amp's voltage gain) is large enough to cause the product $A_V\beta$ to be much greater than 1, the A_V term *cancels out* of the transfer-function equation, leaving only $1/\beta$. In other words, as long as A_V is large enough, its exact value has little influence on the transfer function. This is great news, because it is easy to devise circuits with only passive components to derive a fraction β of the output voltage to use in negative feedback. And because the circuit's transfer function now depends only on β, which can be determined by precisely valued passive components, we don't have to worry about exactly what the amplifier's gain A_V is, as long as it is sufficiently high.

Equations 5.10 and 5.11 show why high-gain op amps are useful in a wide variety of analog electronic circuits. In general, the feedback term β can be a complex function of frequency, but as long as the magnitude of the product $A_V\beta$ is much larger than 1, Equation 5.11 is still an accurate expression of the circuit's transfer function.

What if we now allow the amplifier gain A_V to be a function of frequency? Specifically, we will assume there is a dominant low-frequency pole (typically 1–3 Hz) as is the case with unity-gain-compensated op amps. (The term **pole** simply means a value for a variable that makes a function go to infinity.) We will model this pole with the following expression:

$$A_V(s) = \frac{A_{V0}}{(s/\omega_0)+1} \tag{5.12}$$

where A_{V0} is the DC gain and ω_0 is the pole frequency (in radians per second). Substituting the expression in Equation 5.12 for the constant A_V in Equation 5.10 gives the following expression for $H(s)$, the transfer function as a function of (complex) frequency $s=j\omega$:

$$H(s) = \frac{(A_{V0}/(s/\omega_0)+1)}{(\beta A_{V0}/(s/\omega_0)+1)+1} = \frac{A_{V0}}{\beta A_{V0}+(s/\omega_0)+1} \tag{5.13}$$

$$H(s) = \frac{A_{V0}/A_{V0}\beta+1}{(s/(\omega_0(A_{V0}\beta+1)))+1} \tag{5.14}$$

We recognize that Equation 5.14 has a single pole at a new, higher frequency that we will call ω_{FB}, the bandwidth with feedback:

$$\omega_{FB} = \omega_0(A_{V0}\beta+1) \tag{5.15}$$

The original bandwidth (1–3 Hz) has been multiplied by a factor similar to the one we encountered in Equation 5.10, namely, $(1+A_{V0}\beta)$. If the product of the feedback factor β and the DC gain A_{V0} is much greater than 1, the 1 can be neglected, and we find the following expression for the transfer function with feedback:

$$H(s) \approx \frac{(1/\beta)}{(s/\omega_{FB})+1}. \tag{5.16}$$

The net effect of applying an amount of feedback β is to *increase* the bandwidth from the very low figure ω_0 to the higher bandwidth $\omega_{FB}=\omega_0(1+A_{V0}\beta)$. This is easier to see in a type of frequency-response graph called a **Bode plot**, named for Hendrik W. Bode (1905–1982), a Bell Laboratories scientist who first popularized them. A Bode plot uses a horizontal axis that shows frequency plotted *logarithmically*, so that every major division along the frequency axis means that frequency is multiplied by 10. The vertical axis of a Bode plot is in dB, which is already a logarithmic function, so amplitudes expressed in dB are plotted linearly. In this way, a Bode plot of reasonable size can cover ranges of amplitude and frequency that would be difficult or impossible to distinguish with linear scales.

Figure 5.5 shows the magnitude $|H(f)|$ of the no-feedback response function (where $f=\omega/2\pi$) for an op amp having a DC gain $A_{V0}=100{,}000$ (100 dB) and a low-frequency pole at $f_0=1$ Hz. As you can see, well below the pole frequency, the magnitude approaches 100 dB **asymptotically**, meaning that it gets arbitrarily close to 100 dB the lower the frequency goes, but never quite reaches it. We have

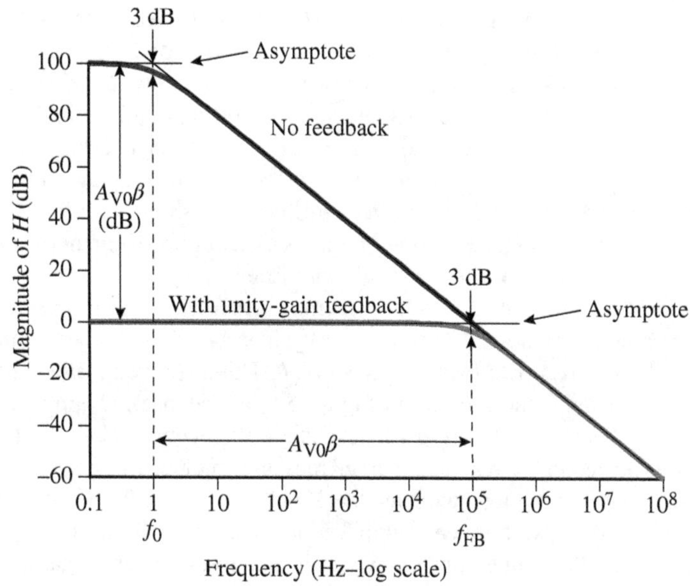

FIGURE 5.5 Bode plot of op amp response without and with unity-gain feedback.

drawn a horizontal straight-line **asymptote** that is at exactly 100 dB to show how the actual magnitude behaves in comparison to its asymptotic value.

Going up in frequency toward 1 Hz, the no-feedback curve falls to a value of 0.707 times its low-frequency asymptotic value at f_0. This is because the real and imaginary parts of the denominator term in Equation 5.16 are exactly equal when $\omega=\omega_0$ (recall that $s=j\omega$) and the magnitude of the complex number that results is $1/\sqrt{2}=0.707$ times the value at zero frequency. Expressed as a dB voltage ratio, $1/\sqrt{2}$ is −3 dB (−3.0103 dB, to be exact), and we have also indicated this fact at the 1-Hz frequency point for the no-feedback curve.

Once the frequency rises far above f_0, the real 1 in the denominator of Equation 5.16 becomes negligible compared to the imaginary part of the expression, and the magnitude becomes approximately inversely proportional to frequency. Here is where the beauty of the Bode plot becomes evident. Whereas on a linear plot, an inverse function traces out a curve called a **hyperbola**, on a Bode plot a function that varies as the inverse of frequency is a straight line with a slope of exactly −20 dB per decade (or about −6 dB per **octave**, which is a factor of two in frequency). Just as with the DC gain at low frequencies, the actual magnitude of H never quite reaches the perfect inverse line, which we have shown with another straight-line asymptote drawn with a slope of −20 dB per decade next to the curve. The important thing about these asymptote lines is that they intersect at the actual 3-dB-down frequency f_0. In fact, after some experience with Bode plots, you can simply draw the asymptotes first so that they intersect at the known frequencies of importance (called **poles** and **zeroes**), and then sketch in the actual curve so that it goes through the 3-dB-down point under the intersection of the asymptotes.

What happens when unity-gain feedback ($\beta = 1$) is applied to the amplifier? In that case, the factor $(1 + A_{V0}\beta)$ is simply A_{V0} to a very good approximation, and Equation 5.16 applies. The 3-dB-down frequency is now $\omega_{FB}/2\pi = f_{FB}$, which according to Equation 5.15 is approximately $A_{V0}f_0$. Putting numbers in, that gives us $f_{FB} = (100{,}000)(1\,\text{Hz}) = 100\,\text{kHz}$. And that is what the exact calculations show, because the response curves in Figure 5.5 are plots of the exact magnitudes with no approximations. The response plot of $|H|$ with feedback shows that the 3-dB-down frequency is now at 100 kHz, a bandwidth that is a great improvement over the 1-Hz bandwidth of the **open-loop** (no-feedback) amplifier.

Figure 5.5 also illustrates a general principle: when an amount of feedback β is applied to an amplifier whose DC (asymptotic) gain is a sufficiently large number A_{V0}, the bandwidth increases by a factor of about $A_{V0}\beta$. This is proved mathematically by Equation 5.15, but you can see that in Figure 5.5, as feedback is applied, the low-frequency gain becomes $1/\beta$, or 0 dB, and stays there until the 0-dB asymptote intersects the no-feedback response curve. It is the intersection between the low-frequency-response asymptote with feedback and the no-feedback asymptote that determines the 3-dB-down frequency with feedback, namely, 10^5 Hz or 100 kHz. If we had chosen a smaller value for β, say, 0.01, the low-frequency gain would be $20\log_{10}(1/0.01) = +40\,\text{dB}$ (a numerical ratio of 100). If you draw a straight asymptote out from the +40-dB point on the gain axis in Figure 5.5, it will intersect the no-feedback curve at a frequency of 10^3 Hz or 1 kHz. Finally, note that the product of the low-frequency (asymptotic) gain and the 3-dB-down bandwidth is the *same* for all these examples, namely, the GBP of the amplifier: $100{,}000 \times 1\,\text{Hz} = 100 \times 1\,\text{kHz} = 1 \times 100\,\text{kHz} = 100\,\text{kHz} = \text{GBP}$. This shows that the GBP is a constant that is characteristic of the amplifier and cannot be changed by feedback.

On the Bode plot, the no-feedback response curve forms a kind of boundary that the response can never cross (assuming only negative feedback is used). Every point on the downward-sloping part of the no-feedback curve has the same GBP, which cannot be exceeded for a given device. Feedback does allow you to trade lower gain for increased bandwidth, however, and that is in fact how most op amps are employed to obtain useful bandwidths.

These examples also show over what frequency range it is correct to assume that the gain with feedback is simply $1/\beta$. With unity-gain feedback ($\beta = 1$), you are safe in assuming this all the way up to 100 kHz, or nearly so. But as less feedback is applied and as the gain with feedback approaches the gain A_{V0} without feedback, the bandwidth falls, and you should be increasingly aware that it is doing so. Most of the time we will assume that we are operating at frequencies well below f_{FB}, the 3-dB-down frequency with feedback. But if you devise a new design, it will be your responsibility to check and make sure this condition is valid.

5.3.1 Linear Op Amp Circuits

With due attention to gain and bandwidth limitations, we will now describe a number of linear op amp circuits in common use and will employ the approximation of Equation 5.11 in calculating their transfer functions. But the designer should remember the conditions under which the approximation is valid and make sure these conditions are met in a given application.

5.3.1.1 Voltage Follower The simplest linear op amp circuit is a **voltage follower**, whose name derives from the fact that the output voltage follows the input voltage almost exactly. When the feedback factor β is 1, Equation 5.11 implies that the circuit's transfer function is $H=1$ also. Figure 5.6 shows the diagram of a voltage follower using an op amp and indicates that the output of the op amp is connected directly to the inverting input. We have adopted a simplified schematic symbol for the op amp in Figure 5.5, denoting the inverting and noninverting inputs by the symbols – and +, respectively, and omitting the power-supply connections. However, the reader must remember when actually building such circuits that all op amps need power supplies, whether or not the connections are explicitly shown.

The usefulness of an amplifier circuit with a gain of 1 is not perhaps immediately obvious. The voltage-follower circuit is sometimes called a **buffer**, because it isolates any circuit connected to its input from loading effects of a low-impedance load. For an example of this, we show a simple voltage-follower application in Figure 5.7: a buffer preamp for a **condenser microphone**. This is not the only way to design an op amp buffer amplifier, but it demonstrates the usefulness of the voltage-follower circuit.

A condenser microphone works by placing a high voltage V_{DC} (or its functional equivalent, a high electric field provided by a type of dielectric called an **electret** having a permanent electric field) across the plates of a capacitor C_{MIC} (**condenser** is an older term for capacitor). One plate of the capacitor is the microphone's diaphragm, which vibrates in response to sound waves impinging upon it. As the diaphragm moves, the capacitance C_{MIC} changes. If we assume a sine wave of frequency ω impinges on the microphone, we can express the capacitance as a function of time as

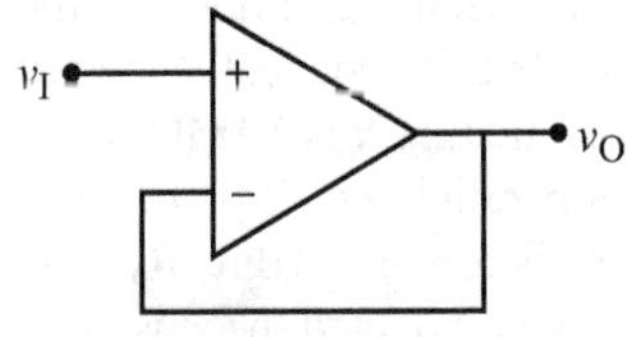

FIGURE 5.6 Voltage-follower circuit using op amp.

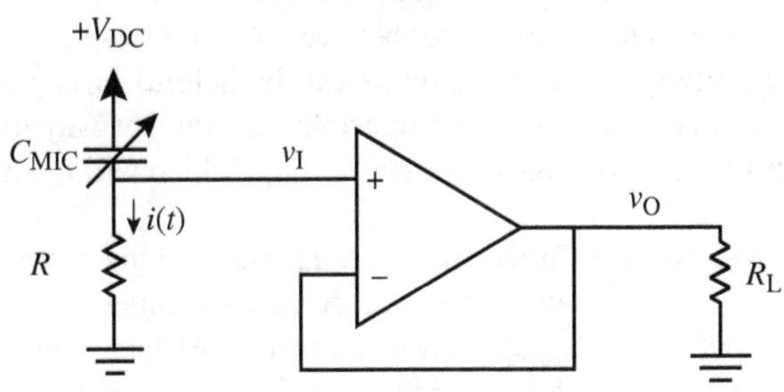

FIGURE 5.7 Voltage follower used as condenser microphone buffer preamp.

$$C_{\text{MIC}}(t) = C_0 + \Delta C \sin(\omega t) \tag{5.17}$$

where ΔC is the change in capacitance for a given amplitude of sound wave and C_0 is a constant. The charge on the capacitor $Q(t)$ can be expressed as the product of $C_{\text{MIC}}(t)$ and the voltage $V_C(t)$ across the capacitor. Taking the derivative of $Q(t)$ gives

$$\frac{dQ(t)}{dt} = C_0 \frac{dV_C(t)}{dt} + V_0 \frac{dC_{\text{MIC}}(t)}{dt}. \tag{5.18}$$

You can show that if

$$\frac{1}{2\pi RC_{\text{MIC}}} \ll f_{\text{MIN}}, \tag{5.19}$$

where f_{MIN} is the lowest sound frequency to be received (typically about 20 Hz), then the charge on the capacitor is approximately constant: $dQ/dt \sim 0$. Because the capacitance C_{MIC} is small (<100 pF), meeting this requirement needs a very large resistance value for R, in the tens to hundreds of megohms or more. Assuming that condition is met, we equate the two remaining terms in Equation 5.18 and obtain this expression for the output voltage V_I:

$$V_I(t) = \frac{V_{\text{DC}} \Delta C}{C_0} \sin(\omega t). \tag{5.20}$$

But this voltage can be obtained from the circuit only if the amplifier connected to it has an input impedance that is even higher than the value of resistor R.

If we assume that the load resistance R_L driven by the op amp output v_O is a typical value for audio circuitry, say, 10 kΩ, the usefulness of the unity-gain buffer amplifier becomes apparent. Suppose we tried to connect R_L directly to the microphone capacitor in parallel with R, and assume that R is 10 MΩ. Immediately, the effective resistance to ground is R and R_L in parallel, which is slightly less than 10 kΩ. The change in output voltage when the load is connected directly is the ratio of R_L to R, or −60 dB, and the frequency response is also affected, because the condition of Equation 5.19 is no longer met. Assuming the input impedance of the op amp is much greater than 10 MΩ (e.g., by using a FET-input op amp), we realize a voltage increase of +60 dB simply by interposing a unity-gain buffer between R and R_L, as shown in Figure 5.7. This shows how unity-gain op amp circuits can be helpful in interfacing between high-impedance sources and lower-impedance loads. Another way to do such interfaces is with a **current-to-voltage converter** circuit, which will be described next.

5.3.1.2 Current-to-Voltage Converter A better way to convert small currents into voltages under some circumstances is with a current-to-voltage converter, shown in Figure 5.8. The input current I_{IN} appears at the inverting input, which is assumed to have an input impedance much greater than feedback resistor R. Therefore, essentially all the current flows through R.

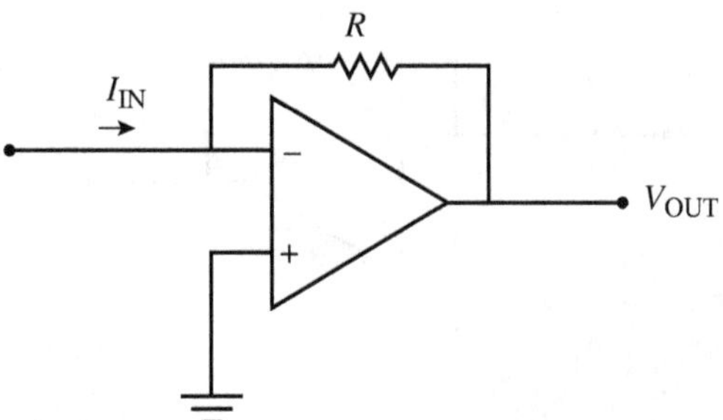

FIGURE 5.8 Current-to-voltage converter using op amp.

At this point, we will apply a simple approximate principle that is of great help in deriving the transfer function of many op amp circuits. The principle, which we will call the **negative-feedback principle** for lack of a better term, is this:

If an op amp's output is connected to its inverting input in a negative-feedback path, the output voltage v_O *will take on a value that forces the inverting and non-inverting inputs to the same voltage.*

This principle embodies the assumption that the product $A_V\beta$ is much greater than 1 and also assumes that the op amp is operating well within its bandwidth with feedback. But given these assumptions, the principle allows you to write down an op amp circuit's transfer function almost by inspection, as we will now show.

Applying this principle to the circuit shown in Figure 5.8, the noninverting input is held to 0V because it is grounded. The question then becomes, "What must V_{OUT} be in order for the voltage at the inverting input to be zero?" Given that essentially all of the input current I_{IN} flows through resistor R, if we assume the inverting input is at zero volts, the voltage drop across the resistor must be $V_{OUT}I_{IN}$. Observing the direction of current flow, we simply write down that

$$V_{OUT} = -I_{IN}R. \tag{5.21}$$

One can characterize this circuit by a **transresistance**, which is the ratio of output voltage to input current. Ignoring the minus sign, the transresistance of the current-to-voltage converter is simply R, the feedback resistor. The same voltage would be obtained simply by sending the input current through the same resistor, but you could not connect anything to that voltage without loading problems, unless the load had a much larger resistance than R. The addition of the op amp allows you to obtain the same voltage as with R alone and also to draw load current from the circuit without concern that the signal source is being affected by the load.

Unlike the buffer or voltage-follower circuit, the current-to-voltage converter circuit has an input impedance of approximately zero. It would be exactly zero only if the op amp had infinite gain, but in practice, the voltage at the inverting input is only a few millivolts at most, as long as the op amp works within its linear range.

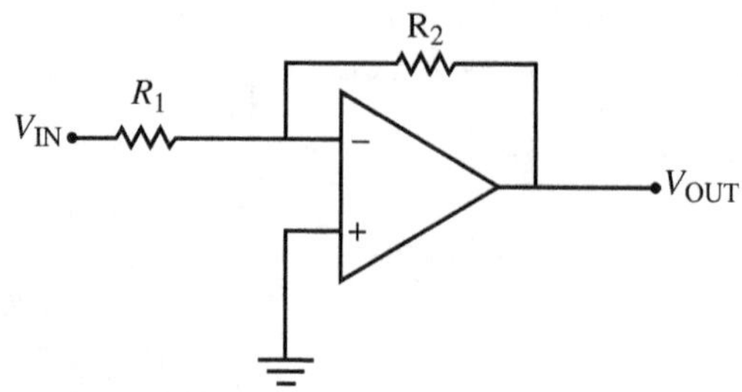

FIGURE 5.9 Inverting amplifier using op amp.

Nevertheless, the approximation is a good one and creates at the inverting input what is known as a **virtual ground**. A virtual ground is a node that is artificially maintained at zero volts, although it has no direct physical connection to ground. Virtual grounds can provide good isolation between multiple inputs of an amplifier, as we shall see when we study **summing amplifiers** in the following text.

5.3.1.3 Inverting Amplifier The basic principle behind the current-to-voltage converter is that whatever current flows into the node attached to the inverting input becomes a voltage when it flows out through the feedback resistor. If we connect a resistor between the virtual-ground inverting input and an input voltage, the input voltage is converted to current and then back to a voltage. The result is the **inverting amplifier** shown in Figure 5.9.

The input voltage V_{IN} produces a current $I = V_{IN}R_1$, which is then transformed back to a voltage. Using the current-to-voltage analysis of Equation 5.21, it is easy to see that the transfer function V_{OUT}/V_{IN} for the inverting amplifier is a constant that we will call A_V(inverting), the **voltage gain**:

$$\frac{V_{OUT}}{V_{IN}} = A_V(\text{inverting}) = -\frac{R_2}{R_1}. \tag{5.22}$$

Again, you should remember that Equation 5.22 is an approximation under the condition that the signal frequency is well below the circuit's bandwidth limit. The bandwidth limit can be found by using Equation 5.15 and setting the feedback factor equal to

$$\beta(\text{inverting}) = \frac{R_1}{R_1 + R_2}. \tag{5.23}$$

Beyond the bandwidth limit f_{FB}, the inverting amplifier's gain will fall at a rate of 20 dB per decade.

5.3.1.4 Inverting Amplifier Application: Unbalanced-to-Balanced Conversion One useful application of an inverting amplifier is in performing the conversion of an **unbalanced** signal source to a form that will drive a **balanced** load. Many audio systems

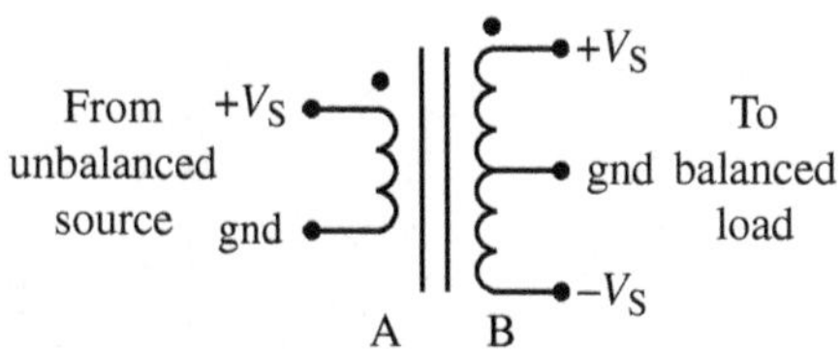

FIGURE 5.10 Transformer used to allow an unbalanced source to drive a balanced load. The dots beside the terminals indicate the primary and secondary terminals that have the same polarity.

employ balanced cables, which consist of two inner conductors surrounded by a common grounded shield. The impedance from each inner conductor to ground is the same, and the cable is designed to be driven by equal-amplitude, opposite-polarity signals. Although balanced cables are more costly, they are popular for low-level circuits such as microphones because a balanced cable is less prone than a single-conductor (unbalanced) shielded cable to pick up electromagnetic interference such as **hum** at the power-line frequency induced by nearby current-carrying extension cords, for example.

One way to convert an **unbalanced** source (single-terminal output and ground, also called **single-ended**) to drive a balanced load is with a type of transformer called a **balun**. The term "balun" is a contraction of "balanced to unbalanced" and is used mostly to refer to such devices in the radio-frequency (RF) range. Another term for a similar device used at audio frequencies is a **direct box**, which can also isolate ground shields to eliminate problems called **ground loops** (further information on ground loops is found in Chapter 12 on electromagnetic interference). A simple direct box can be made with a transformer wired as shown in Figure 5.10. The terminals labeled "gnd" go to the ground leads of the respective cables, which may or may not be connected together depending on the direct-box settings.

We have not discussed transformers up to now, but the basic operating principles are simple. A transformer consists of two or more coils of wire wound around a common magnetic core. The coils are electrically insulated from each other so that the main coupling between them takes place through their common magnetic field. The AC voltage applied to coil *A* also appears across coil *B* in proportion to the **turns ratio** of the two coils. For example, if coil *A* has 100 turns and coil *B* has 200 turns, applying 5 V_{PP} across coil *A* will cause $(200/100)\times 5=10$ V_{PP} to appear across coil *B*, assuming there is no load connected and the transformer is operating within its specified frequency range.

The transformer in Figure 5.10 works just like this. Coil *B* has twice as many turns as coil *A* and in addition is **center tapped** (has a third wire going to the center of its winding), with the center tap going to ground. So when a voltage $+V_S$ appears at coil *A* (termed the **primary** winding), the same voltage appears across the upper half of coil *B* (called the **secondary** winding). But because of the way transformers work, an equal-magnitude but opposite-polarity voltage $-V_S$ appears across the lower end of the secondary winding as well. This is exactly what is needed to drive a balanced load, and before the advent of electronic amplifiers, transformers were the only way to achieve this conversion.

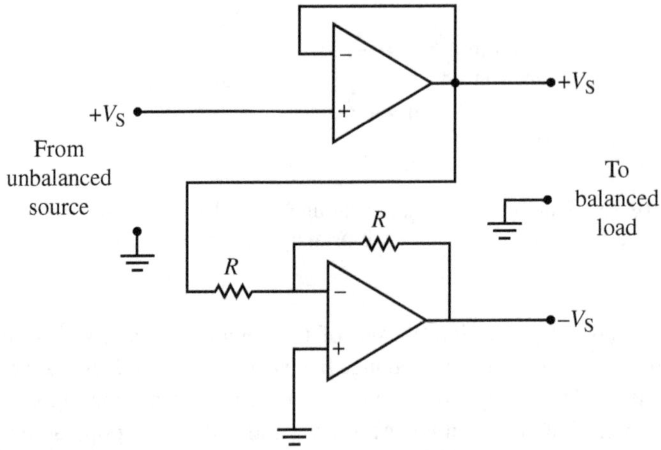

FIGURE 5.11 Voltage-follower and inverting op amp circuits to perform electronic unbalanced-to-balanced conversion.

While transformers still have their uses, they can be bulky, costly, and unreliable under some conditions, and their coils can pick up stray magnetic fields unless they are well shielded. With op amps, the same unbalanced-to-balanced function can be performed in a much more compact circuit that can also deal with a wider range of impedances. Figure 5.11 shows how one can use a voltage follower and an inverting amplifier with a gain of exactly −1 to perform the unbalanced-to-balanced conversion. DC power must be supplied to the op amp circuit, while the transformer circuit of Figure 5.10 needs no power other than that provided by the signal source. But the op amp circuit has advantages that the transformer circuit does not share. It turns out that the transformer circuit is somewhat critical as to impedances. It is most efficient when the impedance from each balanced output load to ground equals the source's impedance. If the source and load impedances do not obey this equality, using a transformer will result in some amount of **mismatch loss**, although there are special transformers available with turns ratios designed to perform impedance matching as well as balanced-to-unbalanced conversion.

The active op amp circuit of Figure 5.11 avoids these problems by presenting a very high impedance to the source and a low and matched impedance to the load. The only caution to be observed in the design of Figure 5.11 concerns the values of the resistors R. Because the gain is exactly −1 only if the resistors are perfectly matched, the degree of balance between the two balanced outputs is affected by how closely the values of the two resistors agree. Using 1% tolerance resistors would take care of this issue. And because the inverting input of the lower op amp is a virtual ground, the input impedance of the inverting amplifier is equal to the input resistor, namely, R. So R should not be so low as to cause significant loading on the voltage-follower amplifier.

5.3.1.5 Summing Amplifier By adding more than one input resistor to the inverting amplifier of Figure 5.9, we obtain a circuit called a **summing amplifier**.

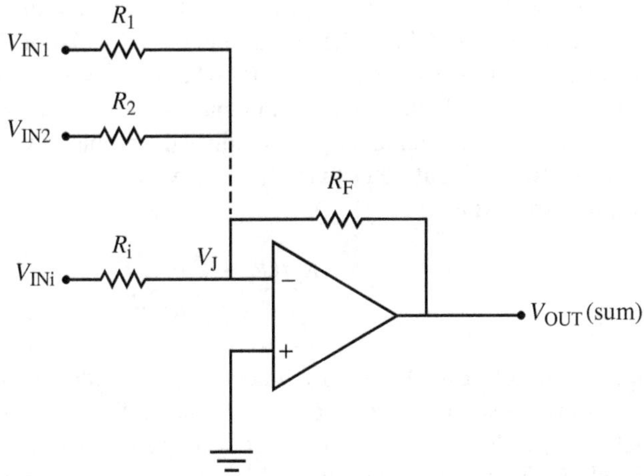

FIGURE 5.12 Summing amplifier with inputs numbered 1 to *i*.

Figure 5.12 shows the diagram of a generic summing amplifier with a total of *i* inputs, where *i* is an integer. Under the idealized assumption that the inverting input is a virtual ground and has infinite input impedance, we can write the output voltage V_{OUT}(sum) as

$$V_{OUT}(\text{sum}) = -\left(V_{IN1}\frac{R_F}{R_1} + V_{IN2}\frac{R_F}{R_2} + \cdots + V_{INi}\frac{R_F}{R_i}\right). \tag{5.24}$$

The algebraic addition of multiple input voltages is a handy operation to have for applications ranging from audio and video analog-signal **mixing** to the correction or insertion of DC offsets into AC signals and various mathematical operations that involve sums and differences. As with any inverting amplifier op amp circuit, the input resistance at each terminal is equal to the respective input resistor, for example, the input resistance at the input labeled V_{IN1} is R_1. The sum in Equation 5.24 is a **weighted sum** in which the weighting factors are the various ratios of the feedback resistor R_F to the individual input resistors R_1, R_2, ... R_i. If the weighting factors exceed a 100:1 range or more, the feedback resistor R_F must be carefully chosen so that the lower-value input resistors are not too low, nor the high-value ones too high.

5.3.1.6 Summing Amplifier Application: Audio Mixing and Isolation In many applications where several signal inputs must be linearly combined into one output, an operation termed mixing is performed. (There is also an RF operation that is also called "mixing" but that is a nonlinear operation discussed later in Chapter 11.) Audio mixing is the addition of two or more audio signals with *independent* control of each signal's contribution to the total output signal. We stress "independent" because it is important to be able to make changes and combinations among signals without allowing **crosstalk**, which is the undesired transfer of one signal into the pathway of a second independent signal.

Suppose two independent signal sources are connected to two of the inputs of the summing amplifier in Figure 5.11. What is the degree of isolation between the inputs? We can measure this by asking about the ratio between the summing-junction voltage V_J (which should ideally be zero) and an input, say, V_{IN1}, under the condition that the op amp's gain is a finite number A_V. To simplify the problem, we will assume nothing is connected to any input other than V_{IN1}, and we will call the ratio A_{ISOL}. It is straightforward to show that

$$A_{ISOL} \equiv \frac{V_J}{V_{IN1}} = \frac{(R_F/(R_1+R_F))}{1+A_V(R_1/(R_1+R_F))}. \tag{5.25}$$

The denominator is our old friend $(1+A_V\beta)$, where β is the feedback factor for this circuit if only terminal 1 is connected to a voltage source. So the result is that the degree of isolation depends on the amount of feedback used.

A typical level of isolation can be calculated for a *unity-gain* summing amplifier in which $R_1=R_F$, $A_V=100{,}000$, and $\beta=0.5$. The result is that A_{ISOL} is -100 dB, which would be less if one tried to obtain more gain from the amplifier by decreasing β. But in audio systems, isolation of -60 dB or better results in crosstalk that is inaudible except under unusual conditions.

5.3.1.7 Noninverting Amplifier The next amplifier circuit using op amps is a variation on the voltage follower. If, instead of feeding back all the output signal to the inverting input, one uses a resistive voltage divider to feed back only a fraction β of the output, one obtains the **noninverting amplifier** circuit shown in Figure 5.13.

Because the input signal V_{IN} is connected to the noninverting input, the sign (or phase for an AC wave) will be the same at the output as at the input. (In general, this is an easy way to tell whether an op amp circuit is noninverting or inverting: the name of the op amp terminal that the input signal is connected to is the same as the type of amplifier.) Because most op amps have a high input impedance, the input impedance

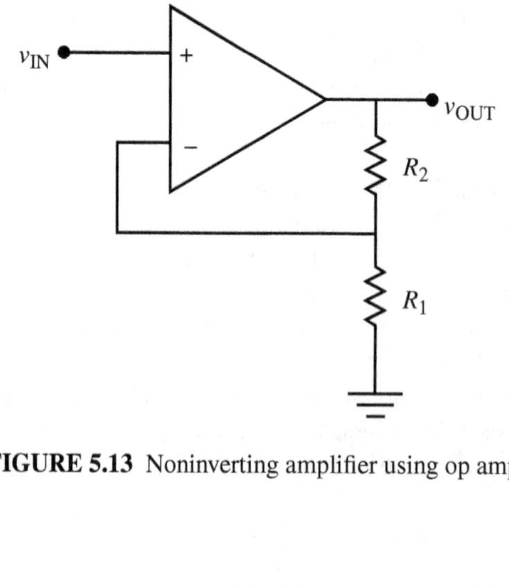

FIGURE 5.13 Noninverting amplifier using op amp.

of the noninverting amplifier circuit will also be high. This is a significant advantage claimed by the noninverting amplifier over the inverting amplifier, whose input impedance is the same as the input resistor used.

Applying the negative-feedback principle to this amplifier results in the following expression for its voltage gain A_V(noninverting):

$$A_V(\text{noninverting}) = \frac{R_2}{R_1 + R_2}. \tag{5.26}$$

For this amplifier only, the feedback factor β used in calculations of bandwidth is equal to the voltage gain: $A_V = \beta$. The noninverting amplifier is one of the most popular circuits that use op amps, because its combination of high input impedance, straightforward gain expression, and low output impedance makes it easy to use in a variety of applications.

5.3.1.8 Noninverting Amplifier Application: Instrumentation Amplifier The noninverting amplifier is a single-ended circuit; that is, its input is a single-terminal voltage referenced to ground. In many applications, a truly differential amplifier circuit is needed that will have the large input impedance of the noninverting amplifier while also showing high common-mode rejection for improved signal-to-noise ratio in systems where interfering common-mode noise is present. A circuit that will provide all of these features is the **instrumentation amplifier** circuit shown in Figure 5.14.

The requirement for high input impedance at both input terminals is met by using two op amps A_1 and A_2 in noninverting configurations. Note that the two otherwise independent amplifier circuits share a common resistor R_G. We can analyze this circuit for its behavior with an arbitrary pair of input signals by studying

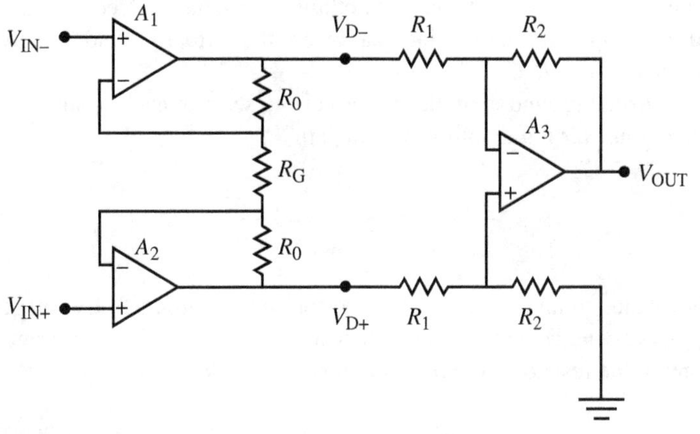

FIGURE 5.14 Instrumentation amplifier circuit with differential inputs V_{IN+} and V_{IN-}, output V_{OUT}, and intermediate voltages V_{D+} and V_{D-}.

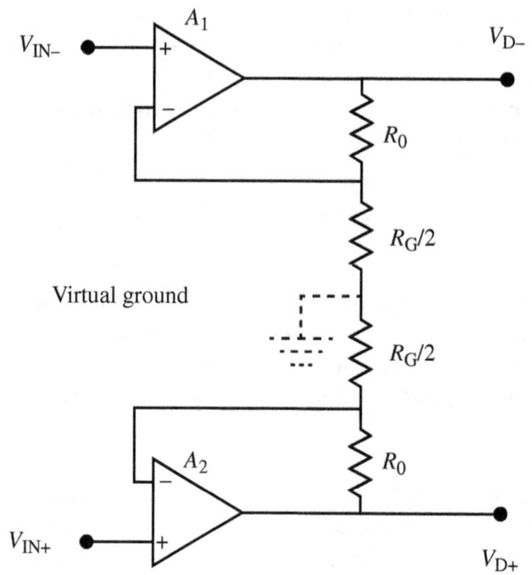

FIGURE 5.15 Equivalent circuit of input stage of instrumentation amplifier shown in Figure 5.14 for differential-mode input voltage vector where $V_{IN+} = -V_{IN-}$.

its response to a differential-mode input and then to a common-mode input and adding the two responses by virtue of the principle of **linear superposition**. The operation of this circuit for differential-mode signals can be understood by referring to Figure 5.15.

By symmetry, if a purely differential-mode input signal is present, the inputs are equal in magnitude and opposite in sign ($V_{IN+} = -V_{IN-}$), and so the center point of the common resistor R_G will be at zero volts, constituting a virtual ground. Connecting a virtual ground to a real ground changes nothing, so we have added a ground symbol in dashed lines to Figure 5.15 to indicate where the virtual ground is, between two resistors each of value $R_G/2$.

Once the virtual ground is in place, it is easy to see that each op amp at the input is a standard noninverting amplifier with a gain

$$\left.\frac{V_{D+}}{V_{IN+}}\right|_{DM} = \left.\frac{V_{D-}}{V_{IN-}}\right|_{DM} = 1 + \frac{2R_0}{R_G} \tag{5.27}$$

What about the common-mode gain of the input circuits? By inspection, if $V_{IN}+ = V_{IN}-$, every node in the circuit will be at the same voltage, no current will flow through any of the resistors, and the common-mode gain will be

$$\left.\frac{V_{D+}}{V_{IN+}}\right|_{CM} = \left.\frac{V_{D-}}{V_{IN-}}\right|_{CM} = 1 \tag{5.28}$$

So the input circuit can show a large gain for differential-mode signals, but will always show unity gain for common-mode signals. Note that the input-stage gain of the system is determined by the value of a single resistor R_G, rather than the need for providing two closely matched resistors if one wishes to change the circuit's gain, as would be the case for two completely independent noninverting amplifiers.

The differential part of the instrumentation amplifier is the circuit using amplifier A_3 in Figure 5.14. Using the negative-feedback principle and a little algebra, it is easy to show that the output voltage V_{OUT} is

$$V_{OUT} = (V_{D+} - V_{D-})\frac{R_2}{R_1} \tag{5.29}$$

Equations 5.28 and 5.29 together yield an expression for the overall instrumentation amplifier transfer function:

$$V_{OUT} = \left(1 + \frac{2R_0}{R_G}\right)\frac{R_2}{R_1}(V_{IN+} - V_{IN-}) \tag{5.30}$$

The designer can construct an instrumentation amplifier from precision-matched resistors and discrete op amps, but a better choice is to purchase an instrumentation amplifier IC. Such chips incorporate the needed op amps internally and often feature **laser-trimmed** resistors that have been matched at the factory to a higher degree of accuracy than is possible with manual matching of discrete resistors. Instrumentation amplifiers can exhibit a very high CMRR and are often used in critical systems where accurate and low-noise amplification of sensor signals is required.

5.3.1.9 Integrator Op amps can be used to perform a variety of time- and frequency-dependent functions such as filtering, differentiation, and integration. A simple integrator circuit using an op amp is shown in Figure 5.16.

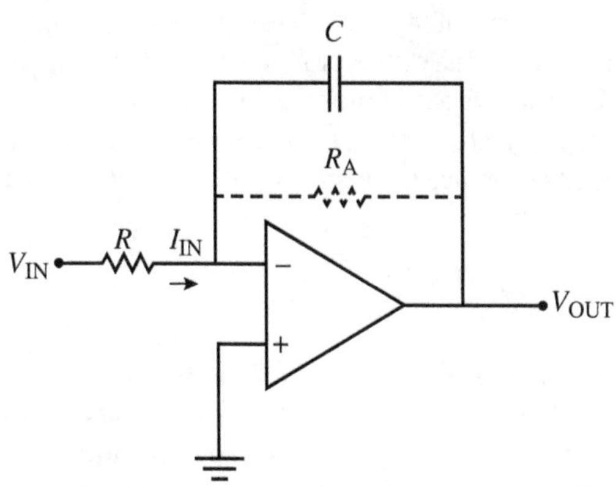

FIGURE 5.16 Integrator circuit using op amp.

First, assume that the auxiliary resistor R_A in dashed lines is not present. If that is the case, we can assume that at $t=0$, the capacitor C has no charge and the op amp is ideal, which means that all current I_{IN} entering the input terminal also enters the capacitor. We can then write the voltage $V_-(t)$ at the op amp's inverting input terminal as

$$V_-(t) = V_{OUT}(t) - \frac{1}{C}\int_{t'=0}^{t} I_{IN}(t')dt'. \tag{5.31}$$

Under the assumption that the op amp does whatever it has to in order to keep the inverting input's voltage equal to the noninverting input (namely zero), we can find $I_{IN}(t)$:

$$I_{IN}(t) = \frac{V_{IN}(t)}{R}. \tag{5.32}$$

Equations 5.31 and 5.32 together give an expression for the output voltage $V_{OUT}(t)$ as a function of the input voltage $V_{IN}(t)$ and the circuit constants:

$$V_{OUT}(t) = -\frac{1}{RC}\int_{t'=0}^{t} V_{IN}(t')dt' \tag{5.33}$$

The circuit behaves as an ideal integrator with a leading constant that is inversely proportional to the time-constant product RC. Integrator circuits like this one are very useful in timing and waveshaping applications and are essential in the design of **active filters**, as we will show in the following text.

A practical difficulty with the circuit in Figure 5.16 without auxiliary resistor R_A is that unless the circuit is part of a larger feedback loop that maintains its output voltage within the linear range of the amplifier, even a very small DC offset voltage at the V_{IN} terminal will eventually integrate to such a large number that the op amp output will saturate (go to one of the power-supply rails) and stay there, rendering the circuit nonfunctional. To prevent this, or at least to reduce the chances of this happening, you will often see an auxiliary resistor R_A connected in parallel with the capacitor C. It is easiest to see what R_A does in the frequency domain.

If we ask what the integrator's phasor voltage gain $H_I(s)$ is as a function of complex frequency $s=j\omega$, we can use the negative-feedback principle again to find that

$$H_I(s) = \frac{v_{OUT}(s)}{v_{IN}(s)} = -\frac{1}{sRC}. \tag{5.34}$$

(This can be obtained by substituting $1/sC$ for R_2 in the equation for inverting amplifier gain, Equation 5.22). Equation 5.34 neglects the presence of resistor R_A. You may know that the Laplace transform for the integration operation in the time domain is dividing by s in the frequency domain, and that is confirmed by Equation 5.34, with the RC scale factor present that makes the dimensions come out right.

With R_A added, the feedback network in Figure 5.16 becomes a parallel combination of C and R_A. The transfer function $H_{IA}(s)$ for the integrator with the auxiliary resistor R_A added is

$$H_{IA}(s) = -\frac{R_A}{R}\frac{1}{1+sCR_A} = -\frac{R_A}{R}\frac{1}{1+(s/\omega_A)}. \tag{5.35}$$

The new transfer function with the auxiliary resistor now has two terms: a constant term $-R_A/R$, and a one-pole lowpass transfer function with a 3-dB-down radian frequency ω_A, where

$$\omega_A = \frac{1}{CR_A}. \tag{5.36}$$

Suppose the lowest-frequency component ω of the input signal is well above ω_A, so that the fraction ω/ω_A is much greater than 1. The transfer function $H_{IA}(\omega)$ of Equation 5.35 then approximates the ideal integration transfer function of Equation 5.34. But for lower frequencies and DC, where $\omega/\omega_A << 1$, the magnitude of $H_{IA}(\omega)$ levels out at a constant, namely, $-R_A/R$, rather than rising to very high values. As long as any DC input is so small that its value times the gain magnitude R_A/R is still small compared to the op amp's output voltage range, adding resistor R_A of a suitable value will prevent the integrator from saturating, as it inevitably will without R_A unless a secondary feedback circuit prevents it.

5.3.2 Nonlinear Op Amp Circuits

Up to this point, all the circuits we have described that use op amps have linear transfer functions in the ideal case. As explained earlier, no amplifier circuit is perfectly linear, and even these "linear" circuits will show a certain amount of harmonic distortion before the output reaches its maximum or minimum voltage level and begins to clip. But it turns out that the application of negative feedback to an amplifier reduces the nonlinear distortion compared to the level it would have without feedback, so most linear op amp circuits that employ a significant amount of feedback show nonlinear distortion so small that it is difficult to measure.

However, there are many applications that use op amps to perform nonlinear operations on signals. Examples of these operations are **detection**, **rectification**, **limiting**, **logarithmic amplification**, and **comparison** to a reference voltage or voltages. We will discuss each of these applications now in turn, giving some examples of how they are used along the way.

5.3.2.1 Precision Rectifier **Rectification** is the conversion of an AC waveform to a DC waveform, usually with the aid of diode **rectifiers** that pass current in only one direction. While a single diode will rectify a waveform, there are several drawbacks to using a diode alone. A semiconductor diode requires a minimum forward-bias voltage to conduct appreciable current. In silicon diodes, this voltage is about 0.6V, so signals with a peak amplitude less than that will not be rectified at all, and larger signals will

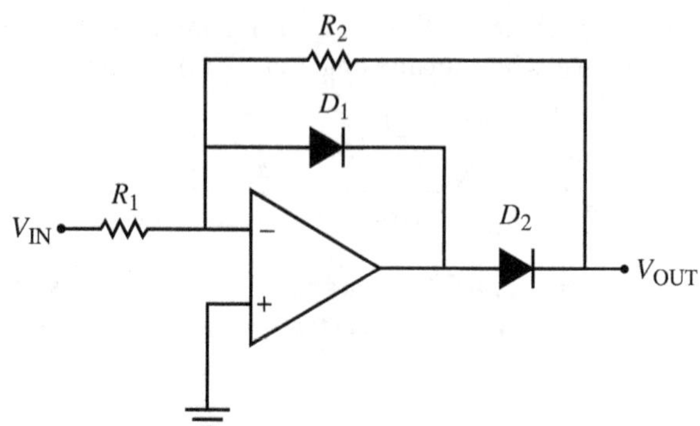

FIGURE 5.17 Precision half-wave rectifier circuit using op amp.

encounter a **dead zone** in which no rectification takes place, before the voltage is large enough so that the diode conducts. If the purpose of the rectifier circuit is to produce a DC voltage proportional to the peak amplitude of the AC input signal, the dead zone renders the circuit insensitive to low-level signals, and a plot of AC input amplitude versus DC output amplitude is very nonlinear below a peak AC input of about 1 V.

If one incorporates diodes in a feedback loop around an op amp, these difficulties can be largely overcome, and the resulting circuit is called a **precision rectifier**. A practical version of a precision rectifier is shown in Figure 5.17.

As you will see, the circuit in Figure 5.17 is a **half-wave** circuit that produces an inverted and amplified version of the *negative* parts of the input voltage wave V_{IN}. Assuming V_{IN} is negative, the voltage at V_{OUT} that will maintain the inverting input of the op amp at 0 V is

$$V_{OUT} = -\frac{R_2}{R_1} V_{IN} \tag{5.37}$$

This is the same form as the gain equation for the inverting amplifier (Eq. 5.22). And as long as the input voltage is negative, the op amp performs as a standard inverting amplifier and produces a positive output at V_{OUT}, although the voltage at the op amp output terminal must be one forward voltage diode drop higher than V_{OUT} in order to turn on diode D_2. Diode D_1 is always reverse biased under these conditions and conducts no current. So the net result so far is that for values of V_{IN} less than zero, an amplified and inverted version of V_{IN} appears at V_{OUT}.

Now, suppose that V_{IN} goes positive with respect to ground. When the op amp output voltage goes negative in response, diode D_2 becomes reverse biased and cuts off, and D_1 becomes forward biased. The op amp's output will stabilize at about one diode voltage drop negative in order to maintain the inverting input at 0 V. This voltage is also what will appear at V_{OUT}, because the only path open to the circuit's output terminal is through R_2, which goes back to the inverting input, now maintained at 0 V. Therefore, the voltage V_{OUT} will be zero for positive inputs. The resulting

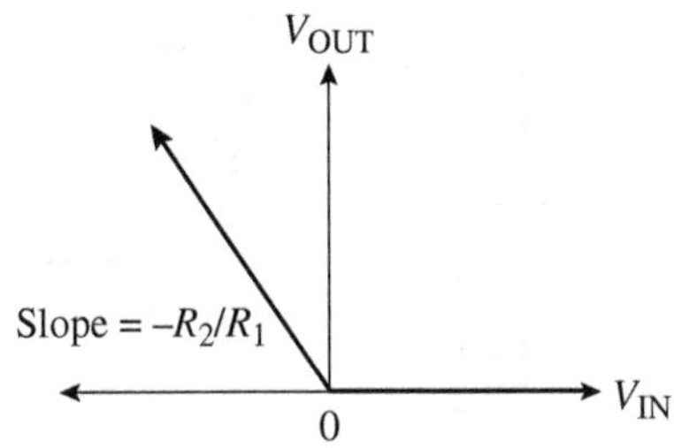

FIGURE 5.18 Transfer function of precision half-wave rectifier shown in Figure 5.17.

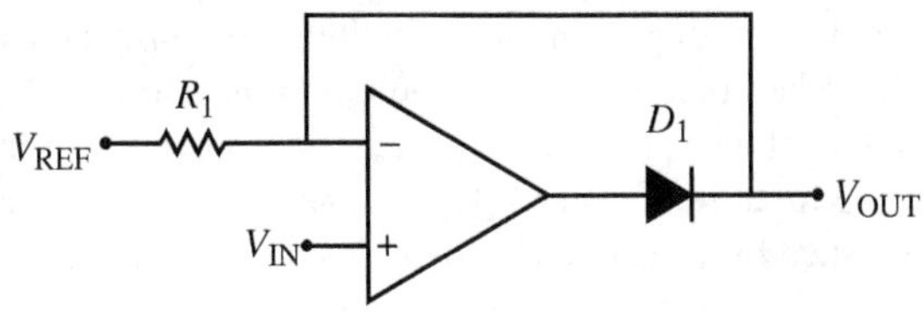

FIGURE 5.19 Precision limiter using op amp.

transfer function for the precision half-wave rectifier of Figure 5.17 is shown in Figure 5.18. The curve has a negative slope equal to the inverting amplifier gain for negative inputs and flattens sharply to zero for positive inputs. With this transfer function, the negative halves of a sine wave will appear inverted at the circuit's output, but the positive halves will not. A precision rectifier for positive inputs can be obtained from the circuit of Figure 5.17 by reversing all diode polarities.

5.3.2.2 Precision Limiter A diode can act as a **limiter**, which is a circuit that allows an input signal to pass unaltered until it exceeds (or falls below) a fixed DC **limiting voltage**. When an input signal undergoes limiting, the limited output remains at the limiting voltage until the input signal once again falls below (or exceeds) the limit set by the circuit. In this sense, an op amp behaves as a limiter when the output voltage reaches the limits dictated by the op amp's power-supply voltages. However, this "rough-and-ready" limiting action cannot be easily controlled, so special limiting circuits are usually designed to perform this function when specific limiting values are required.

While diodes can be used as limiters, the same problems that arise when they are used as rectifiers also pose issues for limiters. The turn-on voltage of the diode must be taken into account, and the turn-on is gradual rather than abrupt, leading to poorly defined limiting voltages. If abrupt limiting is desired, the **precision limiter** shown in Figure 5.19 can be used.

The limiting voltage level is determined by a fixed DC voltage source at a voltage V_{REF}. There are two types of limiting: "ceilings," which establish a maximum voltage above which the output cannot pass, and "floors," which establish a minimum voltage below which the output cannot pass. The polarity of diode D_1 in Figure 5.19 determines whether the circuit is a ceiling or floor limiter, as we will now show.

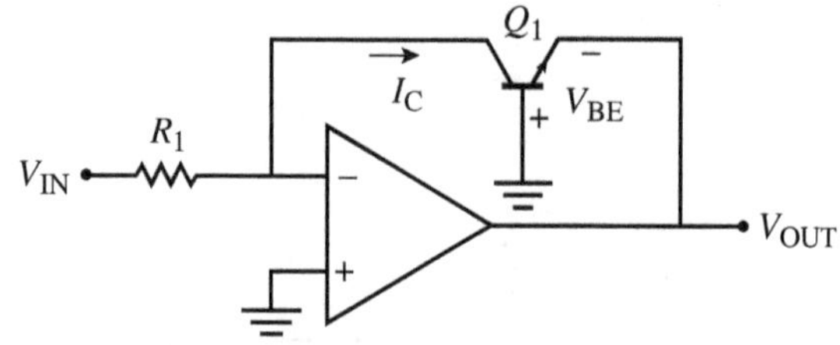

FIGURE 5.20 Log amp using op amp and transistor in feedback loop.

For all voltages $V_{IN} > V_{REF}$, the op amp's output voltage will be positive and will turn on D_1, allowing the op amp to maintain its inverting input at a voltage approximately equal to V_{IN}. When that is the case, the circuit acts as a voltage follower and $V_{OUT} = V_{IN}$. If V_{IN} falls below V_{REF}, however, the diode prevents reverse current from flowing, the inverting input remains at V_{REF}, and the output voltage stops at the floor voltage value established by V_{REF}. Because there is no ground reference involved, V_{REF} can be either positive or negative with respect to ground. To provide a ceiling (upper limiting level) instead of a floor, the diode polarity should be reversed. Two such circuits can be cascaded to provide both an upper and a lower limit.

5.3.2.3 Logarithmic Amplifier Sometimes, an analog system must process signals whose amplitude varies over a wide dynamic range while neither overloading nor allowing the signal to fall below the system's noise floor. Acoustic signals from microphones and radar signals sometimes vary over ranges of 80 dB or more. If the signal can be detected or otherwise processed so that its amplitude is always of the same sign (positive, for instance), a circuit called a **logarithmic amplifier** (or **log amp** for short) can compress a signal with an amplitude dynamic range of several thousand or more into a range of a factor of 10 or less. In a radar display, this processing allows echoes of very different amplitudes to be displayed in a way that shows their relative strengths without either concealing the weak returns altogether or causing the strong returns to have all the same brightness, which would occur without log amp processing. The circuit does this by producing an output that is the mathematical logarithm of the input voltage value.

Figure 5.20 shows one form of log amp circuit using an op amp. (There are other types that use discrete devices or special custom-designed IC configurations.) The NPN transistor in the feedback loop between the op amp output and the inverting input has the following relationship between the base–emitter voltage V_{BE} and the collector current I_C (assuming that V_{IN} is greater than about 0.2 V):

$$I_C = I_S e^{\frac{V_{BE}}{V_T}} - 1 \approx I_S e^{\frac{V_{BE}}{V_T}} \tag{5.38}$$

The quantity I_S is the emitter–base junction's **saturation current**, which is typically a very small value (10^{-9} A or less), so the exponential must be very large to sustain reasonable collector currents, justifying the approximation that the −1 term can be ignored. V_T is a constant called the **thermal voltage**, which is about 25 mV at room temperature and proportional to the absolute temperature of the device.

Because I_C is the current flowing through input resistor R_1, and $V_{BE} = -V_{OUT}$, we can write

$$\frac{V_{IN}}{R_1 I_S} = e^{\frac{-V_{OUT}}{V_T}}. \tag{5.39}$$

Taking the log of both sides leads directly to the following expression for the output voltage as a function of the input voltage:

$$V_{OUT} = -V_T \ln\left(\frac{V_{IN}}{R_1 I_S}\right). \tag{5.40}$$

To give an example of how a wide-ranging input signal is compressed to a smaller range by this circuit, suppose $R_1 = 10\,\text{k}\Omega$, $I_S = 10^{-10}\,\text{A}$, $V_T = 25\,\text{mV}$, and V_{IN} ranges from 100 μV to 20 V, a range of 106 dB. The output voltage corresponding to these input voltages ranges from −115 to −420 mV, which would probably need to be scaled up by a following DC amplifier to be usable. This shows the ability of the log amp to translate extremely wide ranges of input signal levels into a much smaller range that nevertheless preserves the relative amplitudes of each signal.

5.3.2.4 Comparators Although op amps are not designed to operate as comparators, they can be used for this important mixed-signal function if nothing better is available. Comparator circuits resemble op amps in some ways, so we will discuss the use of circuits designed to be comparators in this section as well.

A **comparator** does what its name implies: it compares two input voltages and provides an output indicating which voltage is higher. The output of a comparator is designed to produce only one of two **digital logic levels** designated HI (or 1) and LO (or 0) and typically is connected directly to digital logic circuitry. Comparators are similar to op amps in that they have differential inputs and a single output and often show very high gain. But comparators are not designed to be used with direct negative feedback. If you attempt to apply negative feedback to a comparator, it will quite possibly oscillate because no attempt has been made to tailor the device's phase response to prevent oscillation. On the other hand, a comparator can change its output state very fast, providing a useful change in output within microseconds or less of the time that a voltage change occurs at the inputs. The time it takes for the input transition to cause the output transition is called the **propagation delay**. A low value of propagation delay is especially important for high-speed comparators dealing with rapidly changing signals, because the output due to one transition must appear before the next input transition occurs. A comparator with a long propagation delay will therefore not be able to handle input transitions at a frequency higher than about twice its propagation delay, which may be a serious limitation.

The schematic symbol and ideal transfer characteristic of a comparator are shown in Figure 5.21. The symbol is the same as the one for a conventional op amp except for the miniature transfer-function symbol that looks like a squared-off *S*, which is sometimes omitted. As the transfer function in Figure 5.21 shows, the output voltage changes abruptly between V_{LO} (indicating a digital 0) and V_{HI} (indicating a digital 1) when the voltage *difference* between the V_{IN} input and the V_{REF} input crosses zero.

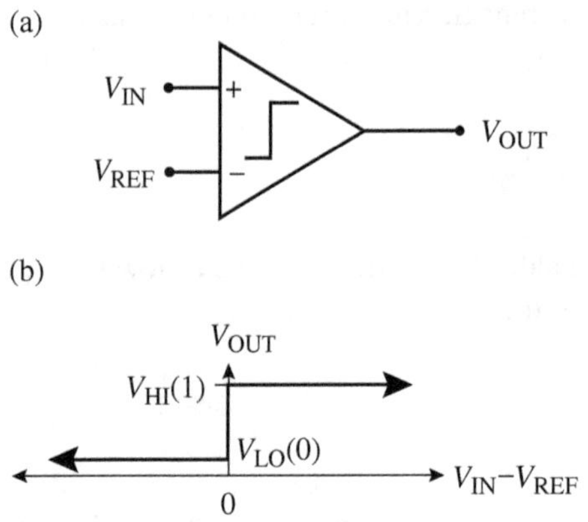

FIGURE 5.21 (a) Schematic symbol for comparator and (b) ideal transfer function.

Comparator circuits are one of the most important ways to interface between analog and digital signals. Because the inputs of a comparator form a differential analog input circuit, a comparator can deal with a wide range of analog input levels and frequencies. Because the output is designed to be fully compatible with digital circuits, it can send signals directly to digital logic circuits. While it is true that a comparator handles only one bit at a time, there are various ways of extracting a great deal of information about the analog input signal from even a single comparator circuit, as we will see in Chapter 8 on analog-to-digital conversion.

As we mentioned, in a pinch, an ordinary op amp can be used as a comparator, but it is not a good idea to do this. The op amp's output levels are near its power-supply rails and typically need further processing to be compatible with digital circuitry. Also, the unity-gain compensation included in many op amp designs means that the device switches at a relatively slow SR, which limits its ability to deal with rapidly changing signals.

Comparator design is a trade-off between high speed and power consumption. Faster comparators typically require more power, while very low-power comparators cannot respond rapidly to high-frequency inputs. One should use a comparator that is fast enough to perform adequately in a specific application, but no faster.

5.3.2.5 Schmitt Trigger Circuit Using Comparator In 1934, Otto H. Schmitt was a student studying nerve impulses in squids. While nerve impulses can be measured electrically, they tend to have random electrical noise associated with them. Schmitt came up with a circuit that processed the noisy impulses and produced a clean, well-behaved output signal that was suitable for counting or other digital operations. Once Schmitt published his circuit, it became known as a **Schmitt trigger** and has found thousands of applications in signal processing over the years since.

The basic problem addressed by the Schmitt trigger circuit is illustrated in Figures 5.22 and 5.23. In Figure 5.22, we show a hypothetical signal that rises from 0 to 5 V in about 20 μS and 30 μS later falls back to 0 V at the same rate. This signal could represent a single nerve impulse, or a single photon counted by a photon detector. As it stands, a signal having V_{HI} of 5 V and V_{LO} of 0 V is suitable to be fed

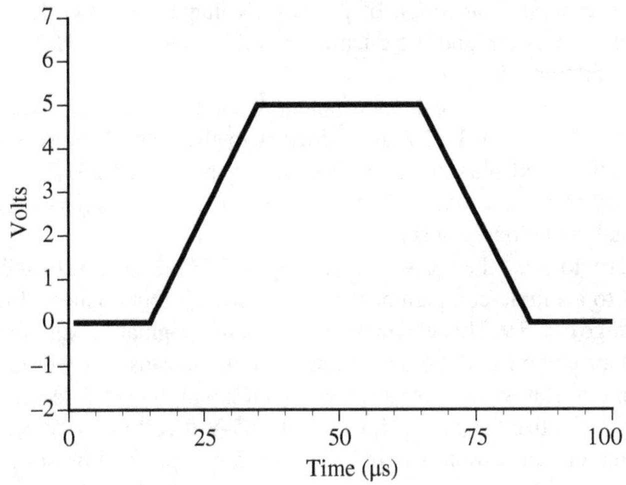

FIGURE 5.22 Hypothetical signal pulse before noise added.

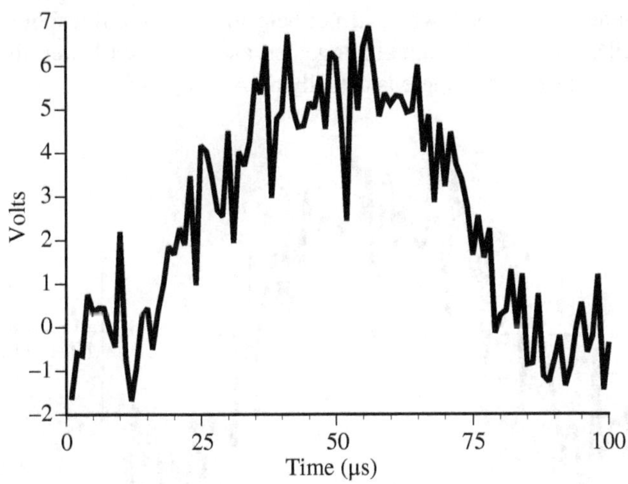

FIGURE 5.23 Signal of Figure 5.22 with $\sigma = 1.25\,\text{V}$ Gaussian noise added.

directly to a type of digital circuit that has **TTL-compatible** inputs, because these are standard voltage levels for that family of logic circuitry.

However, in many situations involving sensors or direct connections to complex systems (such as squids!), a good amount of noise is picked up along with the signal. To model this for the example, in Figure 5.23, we have added a type of noise called **Gaussian noise**. Gaussian noise has a **Gaussian probability distribution**, meaning that the likelihood of any voltage sample having an amplitude of x volts is

$$f(x) = \frac{1}{\sigma\sqrt{2\pi}} e^{-\frac{x^2}{2\sigma^2}}, \tag{5.41}$$

where σ is the *standard deviation* of the noise voltage (in volts). It turns out that thermal noise is Gaussian and the quantity σ is the same as the RMS noise voltage defined in Equation 3.45.

To simulate a noisy electrical environment, we mathematically created a random noise voltage with a σ of 1.25V and added it to the "clean" noise-free signal of Figure 5.22. The signal-plus-noise result is shown in Figure 5.23. (In order to make the graphs easy to read, voltage values are given at a sample interval of 1 μS.) The resulting signal is obviously noisy.

One could try to make the noisy signal of Figure 5.23 presentable to a digital system by sending it to a simple comparator with an inverting input connected to a constant reference voltage of 2.5V. This would work fine for the original "clean" signal, because its transitions are smooth and there is no random noise to cause early or late transitions of the comparator. But when noise is added to the signal, Figure 5.24 shows the result of attempting to digitize the pulse with a simple 2.5-V **threshold**, as it is called.

If the digital circuit responds to pulses as short as 1 μS (and most of them do), it will detect not one pulse, but *five* pulses as the noise "riding" on the slope of the original signal causes the comparator to switch HI and LO several times. This is a fundamental problem in converting noisy pulses to digital form, and moving the threshold voltage higher or lower will not help much. If the threshold is set much higher, it is likely to miss the signal altogether, and if it is set lower, the number of **false positives** (pulses when there is actually no signal present) will increase.

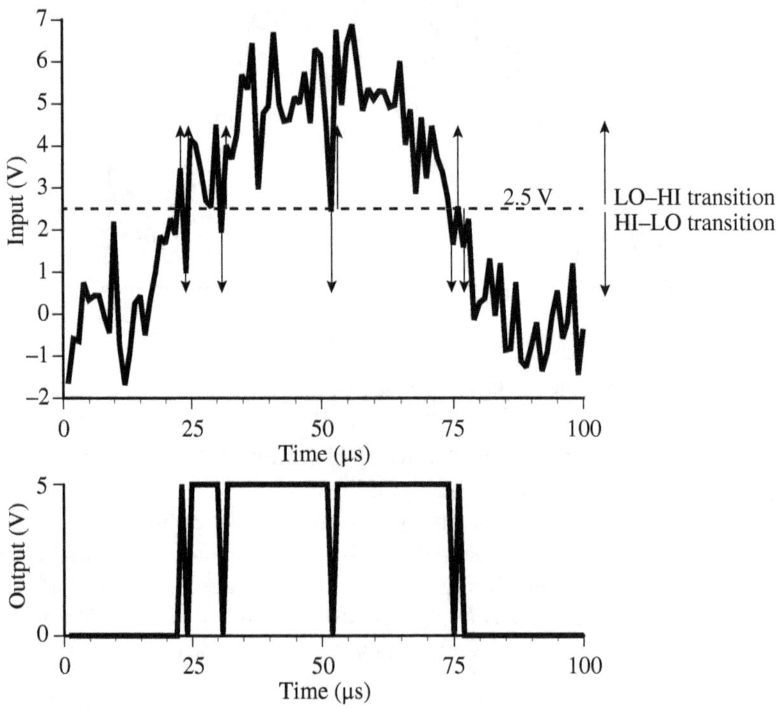

FIGURE 5.24 Result of processing signal + noise of Figure 5.23 through simple comparator with threshold of 2.5V.

At this point, the Schmitt trigger comes to the rescue. A Schmitt trigger circuit using a comparator with output voltages of $V_{HI}=5$V and $V_{LO}=0$V is shown in Figure 5.25a.

Notice that the output terminal is connected to the *noninverting* input of the comparator through resistor R_2, thus providing *positive* feedback. The positive feedback used in Schmitt trigger circuits makes the system transition faster from one state to the other and reduces the chances that noise voltages will cause multiple triggers near a transition time.

Suppose the voltage V_{IN} in Figure 5.25 is initially at 0V, and suppose V_{OUT} is at 0V as well. This is consistent because the noninverting input's voltage of 0V is less than the inverting input's (constant) voltage of 2.5V, and so the output will be LO, which is 0V. As V_{IN} goes positive from 0V, the R_1–R_2 voltage divider will present a scaled-down version of this change to the noninverting input. The scaling is arranged so that when $V_{IN}=4$V, the noninverting input will just barely rise to 2.5V.

When that happens, the output voltage V_{OUT} begins to rise, and this positive change is fed back to the noninverting input. Because of this feedback, the input signal loses control once the **positive-going threshold** of 4V is reached, and the Schmitt trigger rapidly **triggers** so its output goes HI, to 5V.

With the output HI, the situation at the noninverting input is changed so that a small excursion below 4V will do *nothing* once triggering has occurred. Once V_{OUT} rises to 5V, V_{IN} must now *fall* all the way to 1V in order to reach the **negative-going threshold** of 1V and trigger the circuit back to a LO output.

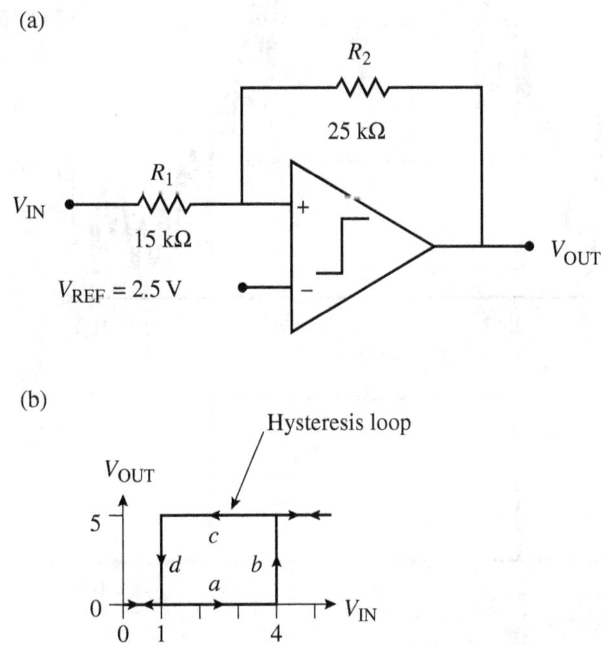

FIGURE 5.25 (a) Schmitt trigger with positive-going threshold of 4V and negative-going threshold of 1V, using comparator with $V_{HI}=5$V and $V_{LO}=0$V. (b) Transfer-function curve of Schmitt trigger showing hysteresis loop.

When the input and output voltages are plotted on a transfer-function graph such as is shown in Figure 5.25b, we find that the output voltage is no longer a **single-valued function** of the input. That is, for a particular input voltage (between 1 and 4V, anyway), we can have either of *two* output voltages: 0 or 5V. Which voltage is actually present depends on both the *present* value of the input voltage and the *history* of the circuit's input in the past. Thus, a Schmitt trigger circuit has **memory**, and you cannot tell what its output will be based simply on what its input is at a given time. You must also know what the input was for some time into the past. This type of behavior is called **hysteresis**.

If you follow the sequence of input voltages described earlier in the transfer-function graph of Figure 5.25b, you will see as the input voltage rises, the output voltage follows segment *a* and transitions at the positive-going threshold at *b* to 5V. Once the circuit has triggered HI, the input voltage must fall along segment *c* to 1V, whereupon the output falls through segment *d* back to the LO level. The open shape or "loop" in the transfer function is called a **hysteresis loop** and is always present in Schmitt trigger circuits.

The usefulness of the Schmitt trigger for processing noisy pulses is now easy to see. In Figure 5.26, we have plotted the exact same signal-plus-noise voltage versus time that was shown in Figure 5.25. Only this time, we suppose that it is sent to the Schmitt

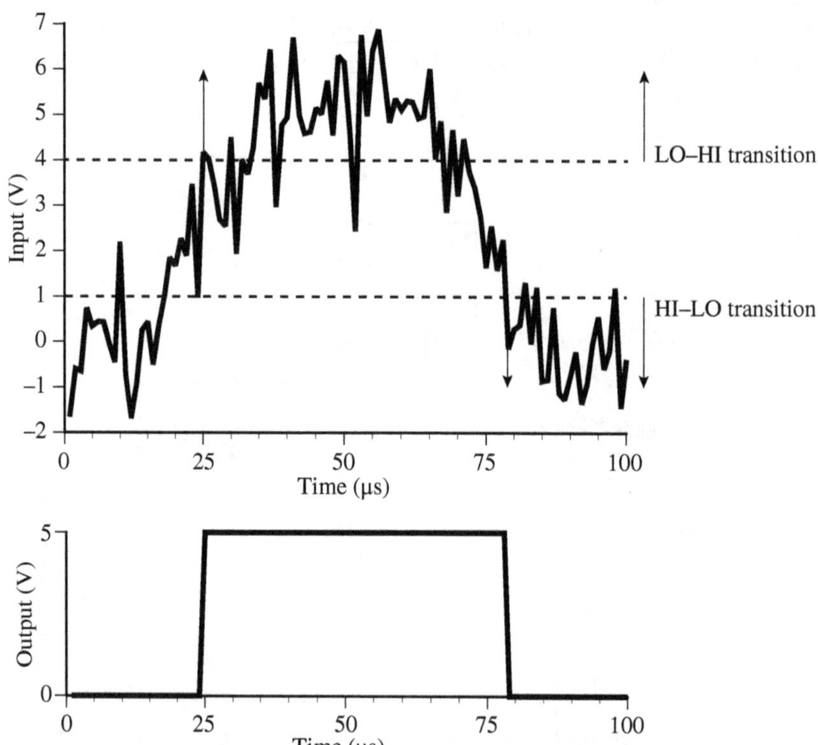

FIGURE 5.26 Noisy signal processed by Schmitt trigger with positive-going threshold of 4V and negative-going threshold of 1V.

trigger circuit with its positive-going threshold of 4V and negative-going threshold of 1V. As you can see, the output of the Schmitt trigger is a single, clean pulse whose timing is reasonably close to that of the original pulse.

If we run this experiment many times with different noise voltages, you will find that once in a while, even the Schmitt trigger circuit will produce an extra pulse. But statistically, the average number of pulses processed with the Schmitt trigger will be much closer to the true number of signal pulses than the simple comparator circuit would produce.

BIBLIOGRAPHY

Fiore, J. M. *Op Amps and Linear Integrated Circuits*. Albany, NY: Delmar, 2001.

Jung, W. *Op Amp Applications Handbook* (Newnes, 2004), also available (at this writing in 2013) for free downloading at http://www.analog.com/library/analogdialogue/archives/39-05/op_amp_applications_handbook.html

Rybin, Y. K. *Electronic Devices for Analog Signal Processing* (Springer, New York, 2012).

Sedra, A. S. and K. C. Smith, *Microelectronic Circuits*, 6th edition (Oxford University Press, New York, 2009).

PROBLEMS

Note: Problems of above-average difficulty are marked with an asterisk (*).

5.1. *Measurement of op amp CMRR*. The standard way to measure the common-mode rejection ratio (CMRR) of an op amp is to use it in a differential amplifier circuit such as the one shown in Figure 5.27. Ideally, if $R_1 = R_3$ and $R_2 = R_4$, the output of the circuit with an op amp having an infinite CMRR is

$$V_{OUT} = \frac{R_2}{R_1}(V_{D+} - V_{D-}). \tag{5.42}$$

If the differential inputs V_{D+} and V_{D-} are connected together to a common input V_{IN}, any output voltage in this condition must be attributed to a nonzero common-mode gain and thus a CMRR that is less than perfect. However, because the CMRR of typical op amps can range from 80 to 120 dB or more, the resistors used must be extremely well matched to produce a meaningful measurement. This is illustrated by the following exercise.

Suppose R_1 and R_3 are nominally 1-kΩ 1% tolerance resistors and R_2 and R_4 are nominally 100-kΩ 1% resistors. For this exercise, assume the following exact values for these components, which are well within their tolerance ranges: $R_1 = 995\,\Omega$, $R_2 = 100.8\,\text{k}\Omega$, $R_3 = 1003\,\Omega$, and $R_4 = 99.7\,\text{k}\Omega$. The CMRR can be defined as

$$\text{CMRR(measured)} = \frac{(V_{OUT}(\Delta)/V_{IN}(\Delta))}{(V_{OUT}(\Sigma)/V_{IN}(\Sigma))} \tag{5.43}$$

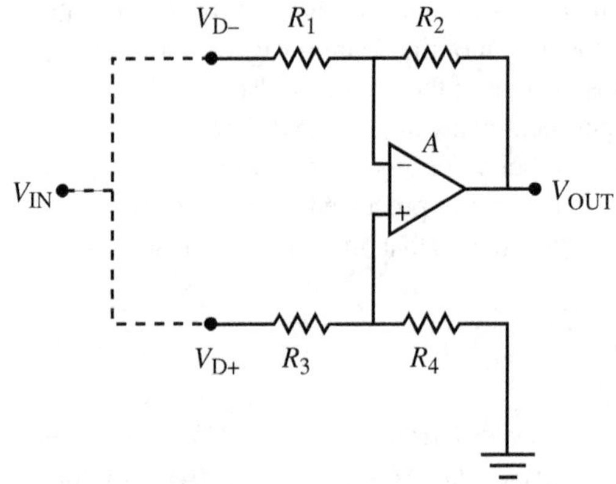

FIGURE 5.27 Differential amplifier using op amp, arranged with dashed-line connections to measure CMRR of op amp.

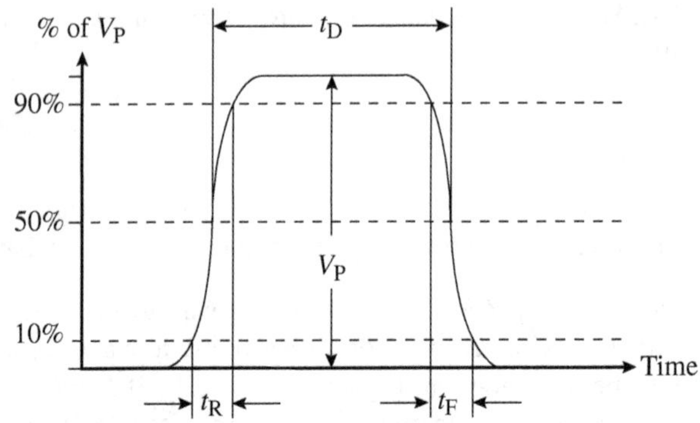

FIGURE 5.28 Definitions of rise time t_R, duration or dwell time t_D, and fall time t_F.

where Δ refers to the usual differential voltages ($2V_{IN}(\Delta)=V_{D+}-V_{D-}$) and Σ refers to the common-mode input and output. Assuming the op amp *itself* has infinite CMRR, calculate *CMRR* (measured) with the specific values of resistors given earlier. You should find that the ratio, when expressed in dB, is lower than even a low-grade op amp's actual specified CMRR.

An Analog Devices app note MT-042 (available at http://www.analog.com/static/imported-files/tutorials/MT-042.pdf) describes a CMRR measurement technique that involves shifts in the power-supply voltage and avoids the need for precision resistors, yet can measure CMRR up to 140 dB.

5.2. *Effect of slew-rate limit.* A digital pulse that starts at zero, rises to a fixed high level, and falls back again to zero can be characterized by three timing parameters called the **rise time** t_R, the **duration** (or **dwell time**) t_D, and the **fall time** t_F. The rise time and fall time are defined with respect to the times when the pulse crosses 10% and 90% of its ultimate full amplitude V_P, as shown in Figure 5.28. The duration or dwell time t_D is defined as the time interval between the 50% points.

Bearing these definitions in mind, suppose an op amp having a slew-rate limit of 1 V μs^{-1} is used in an amplifier circuit that delivers a voltage gain A (a numeric ratio, not dB). For each of the gains and input waveforms listed in the following, decide whether the output encounters the slew-rate limit or not. If it does, estimate the actual t_R, t_F, and t_D of the output pulse that results.

We assume that the output voltage proceeds linearly from 0 to V_P during the rise time and from V_P to 0 during the fall time. (This is not strictly true because these times are defined between the 10 and 90% points, but the resulting error is small for typical waveforms.)

(a) $V_P = 200$ mV, $t_R = t_F = 0.5$ μs, $t_D = 6$ μs, $A = 2$.
(b) $V_P = 200$ mV, $t_R = t_F = 0.5$ μs, $t_D = 6$ μs, $A = 20$.
(c) $V_P = 3$ V, $t_R = t_F = 0.5$ μs, $t_D = 6$ μs, $A = 1$.

5.3. *Bandwidth of op amp circuits.* Find the 3-dB-down bandwidth f_C (the cutoff frequency, at which the transfer-function magnitude is 3 dB below its DC value) for the following op amp circuits. Assume the op amp used has a DC gain $A_{V0} = 10^5$ and a gain–bandwidth product GBP = 1 MHz. (These specifications imply that the amplifier's bandwidth $f_0 = \omega_0/2\pi$ without feedback is $f_0 = \text{GBP}/A_{V0} = 10^6/10^5 = 10$ Hz.)

Equation 5.15 (rewritten to express Hz instead of radian frequency) is $f_{FB} = f_0(A_{V0}\beta + 1)$. Once f_0, A_{V0}, and β are known, f_{FB} can be determined.

(a) Voltage-follower circuit of Figure 5.6.
(b) Inverting amplifier of Figure 5.9 with $R_1 = 1$ kΩ, $R_2 = 10$ kΩ.
(c) Inverting amplifier of Figure 5.9 with $R_1 = 1$ kΩ, $R_2 = 470$ kΩ.
(d) Noninverting amplifier of Figure 5.13 with $R_1 = 1$ kΩ, $R_2 = 1$ kΩ.

5.4. *Test circuit for input offset voltage.* If the differential amplifier circuit of Figure 5.27 has the V_{IN} circuit connected and the V_{IN} terminal is grounded ($V_{IN} = 0$), it can be used to measure the input offset voltage V_{OS} of the op amp.

(a) Show that under these conditions, $|V_{OS}| = |V_{OUT}|\dfrac{R_1}{R_1 + R_2}$.
(b) If $R_1 = 1$ kΩ, $R_2 = 100$ kΩ, and $|V_{OUT}| = 52$ mV, what is $|V_{OS}|$?

5.5. *Plotting Bode plot and asymptotes for lowpass filter function.*

(a) Using a calculation and plotting application such as *Excel* or *MATLAB*™, plot the following function of frequency, which is the response of a single-pole lowpass filter with a cutoff frequency f_0 of 1 kHz: $|H(f)|_{dB} = 20\log_{10}\left(\left[1 + \left(\dfrac{f}{10^3}\right)^2\right]^{-1/2}\right)$. Use the Bode-plot format: the X-axis logarithmic in frequency from 1 Hz to 1 MHz and the Y-axis linear (in dB) with 0 dB at the top and −60 dB at the bottom. To produce a set of 20 frequency points per decade that are evenly spaced on a logarithmic axis, start with 1 Hz and derive each subsequent frequency point by multiplying the previous one by $10^{1/20} \sim 1.122018$.

(b) Once you have plotted the graph in 5.5. (a), using a straight edge, draw a straight line horizontally along the lower straight 0-dB part of the curve and extend it to the right beyond the point that the curve begins to fall. Then draw another straight line along the straight downward-sloping 20-dB-per-decade part of the curve and extend it to cross the first straight line. What is the vertical distance (in dB) between the point where the two straight lines intersect and the actual curve at 1 kHz? (It should be very close to 3 dB). This shows how you can sketch response curves with well-isolated poles and zeroes without doing any calculations at all!

5.6. *Gain and bandwidth improvement with voltage follower.* Suppose a low-level signal source has an output resistance of $R_S = 25\ \text{M}\Omega$ and is initially connected to a load that consists of a capacitor $C_L = 100\,\text{pF}$ in parallel with a resistor $R_L = 100\,\text{k}\Omega$.

(a) If the Thévenin equivalent circuit of the signal source has an RMS output voltage source $V_S = 1\,\text{mV}$ (with no load attached), what will the voltage V_L be across the load at DC? At what frequency f_C will the load voltage fall by 3 dB from its DC value?

(b) Now, suppose a voltage follower (Fig. 5.6) is connected between the signal source and the $R_L - C_L$ parallel-connected load. If the op amp used is ideal (infinite input resistance, zero output resistance), what is the DC load voltage V_L' once the voltage follower is added? What is the "improvement factor" V_L'/V_L (in dB) gained by adding the voltage follower? Is there a 3-dB cutoff frequency with the ideal voltage follower added? Why or why not?

***5.7.** *Current-to-voltage converter.* When the negative-feedback principle is used to calculate the transfer function of the current-to-voltage converter of Figure 5.8, the answer is that $V_{OUT}/I_{IN} = -R$. Perform a more exact calculation without using the negative-feedback principle. Instead, assume that $V_{OUT} = -\dfrac{A_{V0}}{1+\dfrac{s}{\omega_C}}(V_-)$, where V_- is the voltage at the inverting input, and find a frequency-dependent expression for V_{OUT}/I_{IN} that includes the effect of a finite DC voltage gain A_{V0} and the low-frequency unity-gain compensation pole at ω_C. What is the relationship between ω_C and the 3-dB-down bandwidth ω_{FB} of the current-to-voltage converter?

5.8. *Noninverting amplifier and input bias current.* Suppose there is a need to amplify a small AC signal $v_S = 100\,\mu\text{V}$ from a high-impedance source with an output resistance of $100\,\text{k}\Omega$. The voltage gain of the first amplifying stage must be at least 300, and the lowest frequency to be amplified is 10 Hz. Another student has designed the following circuit shown in Figure 5.29, but without the resistor R_B (in dashed lines).

In his design, $C = 0.1\,\mu\text{F}$, $R_1 = 1\,\text{k}\Omega$, and $R_2 = 299\,\text{k}\Omega$. Your fellow student tells you the circuit does not work.

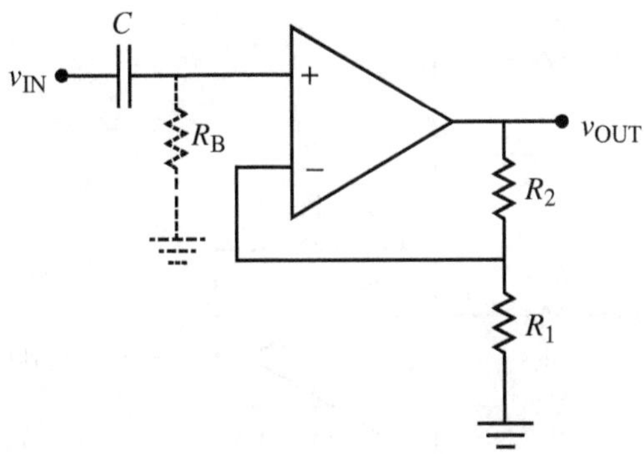

FIGURE 5.29 Noninverting amplifier with AC-coupled input.

(a) Looking at the specification sheet for the op amp used, you find that the op amp is a BJT type and requires a DC input bias current up to $I_{BIAS} = 100\,\text{nA}$. You find a 10-MΩ resistor on the bench and connect it as R_B. The circuit now amplifies the small signal properly. Why did adding R_B allow the circuit to work?

(b) Although the circuit now works for small signals, you find that there is a large DC offset of +9 V at the amplifier's output with no signal at the input. This severely limits the circuit's dynamic range, because the AC output waveform can have a peak amplitude of only 4 V before it begins to clip on the positive-going peak of the waveform at +13 VDC. The reason for this is that the input bias current flowing into the noninverting input terminal could develop a DC voltage of as much as $(100\,\text{nA})(10\,\text{M}\Omega) = 1\,\text{V}$, while the same current flowing (primarily) through R_1 could develop a DC voltage of only $(100\,\text{nA})(1\,\text{k}\Omega) = 100\,\mu\text{V}$. The difference in input voltages is amplified and appears at the output as the undesirable DC offset.

To fix this problem, the DC resistance to ground from each of the two input terminals must be the same, producing the same voltage drop at each input terminal and eliminating the *spurious* (false) DC offset. If $R_B = 1\,\text{k}\Omega$ to fix the DC offset problem, what value of C is required to make the 3-dB-down frequency of the highpass filter formed equal to $f_C = 10\,\text{Hz}$? (*Hint*: Use $f_C = 1/(2\pi R_B C)$.) Unfortunately, the input impedance of the circuit will now be only 1 kΩ, which will cause severe loading of the high-impedance source circuit. However, inserting a voltage follower between the signal source and the input of this amplifier will solve that problem.

5.9. *Summing amplifier problem.* The summing amplifier in Figure 5.30. needs to have the following voltage gains: $V_{OUT}/V_{IN1} = -10.0$, $V_{OUT}/V_{IN2} = -5.0$, and $V_{OUT}/V_{IN3} = -1.0$. If $R_2 = 22\,\text{k}\Omega$, find the required exact values for all other resistors to provide the required voltage gains.

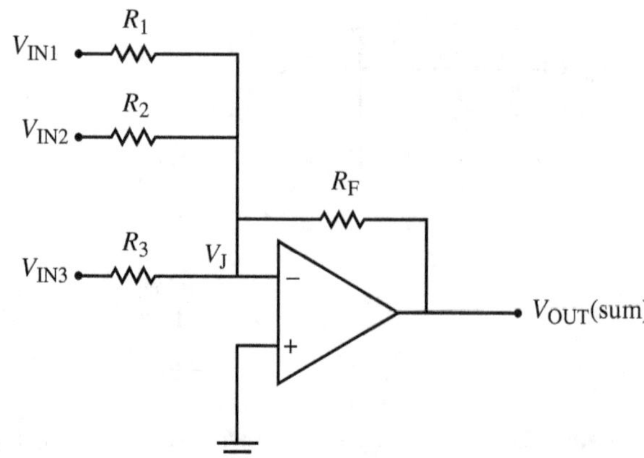

FIGURE 5.30 Summing amplifier with three inputs.

5.10. *Integrator with limited DC gain.* A detector circuit produces a DC output that also has AC components ranging from 100 Hz to 10 kHz. It is desired to average the detector output over a 1-second interval by using an integrator circuit whose output V_{OUT} represents the average of the input V_{IN}. The proposed circuit is shown in Figure 5.16, including resistor R_A to limit the low-frequency gain. Begin your design with $R = 1\,\text{M}\Omega$ and $C = 1\,\mu\text{F}$.

(a) What value of R_A should you choose so that the DC gain of the amplifier circuit is −1?

(b) With the R_A you chose in part (a), what is the 3-dB-down cutoff frequency f_C of the circuit?

(c) If an AC signal $v_{IN} = 100\,\text{mV}$ at 100 Hz appears at the input, what will the output voltage v_{OUT} be? This voltage represents a sort of worst-case error caused by the AC components of the detector output.

5.11. *Precision rectifier problem.* The detector for Problem 5.10 could be the following precision full-wave rectifier circuit shown in Figure 5.31, which uses an instrumentation amplifier circuit (see Fig. 5.14 and accompanying text for details).

The instrumentation amplifier is assumed to have infinite input impedances and a voltage gain of $A = V_{OUT}/(V_+ V_-)$.

(a) If the input voltage V_{IN} is a 1-V (peak) AC sine wave, will the DC component of the output voltage V_{OUT} be negative or positive? Explain.

(b) If $R_1 = R_2 = R_3 = 10\,\text{k}\Omega$, $A = 1$, D_1 and D_2 are silicon diodes with a forward-voltage drop of 0.6 V, and V_{IN} is a 1-V (peak) AC sine wave, sketch the waveforms of V_{IN}, V_{OUT}, and V_1, the output voltage of the op amp, for two complete cycles of the input waveform.

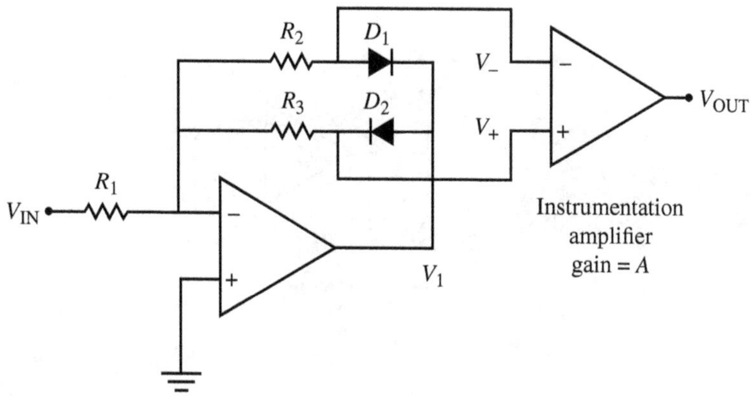

FIGURE 5.31 Full-wave precision rectifier circuit using op amp and instrumentation amplifier with voltage gain A.

5.12. *Symmetrical precision limiter design.* Limiters are often used to protect *analog-to-digital converters* (*ADC*s) from excessive voltages beyond their designed conversion range, because inputs beyond the acceptable range can cause spurious and erratic readings. For AC signals, it is desirable that the positive and negative voltage limits be equal. Design a precision limiter circuit that will limit the output voltage to a range of $-V_{REF}$ to $+V_{REF}$, given a *single* reference input voltage of $-V_{REF}$. Starting with the "floor" limiter of Figure 5.19, design a "ceiling" limiter by reversing the diode polarity. Cascade the output of the floor limiter to the input of the ceiling limiter, and derive the $+V_{REF}$ for the ceiling limiter from the $-V_{REF}$ voltage with an inverting amplifier having a gain of −1. By deriving both limiting voltages from a single source, it is possible to make the reference voltage source variable and provide variable limiting, for example.

5.13. *Temperature compensation of logarithmic amplifier.* In addition to BJTs, semiconductor diodes can be used as the circuit element that produces a logarithmic function for logarithmic amplifiers. As mentioned in Chapter 2, Problem 2.4, Equation 2.8, a good approximation to the current–voltage relationship of a semiconductor diode is $I(V) \approx I_S e^{\frac{qV}{k_B T}}$, where I_S is a current constant roughly proportional to the diode's area, $(k_B T/q)$ is 25.27 mV at room temperature (20°C or 293 K), and we have assumed the current is large enough to neglect the −1 term in Equation 2.8 compared to the exponential term. Solving for the voltage V across the diode in terms of the current I, we find $V(I) \approx \frac{kT}{q} \ln\left(\frac{I}{I_S}\right)$. While V is proportional to the natural logarithm of I, it is also directly proportional to the (absolute) temperature T, which varies about ±12% over the temperature range of 0 to 70°C. This much variation is unacceptable for many applications, and so some form of *temperature compensation* is usually needed.

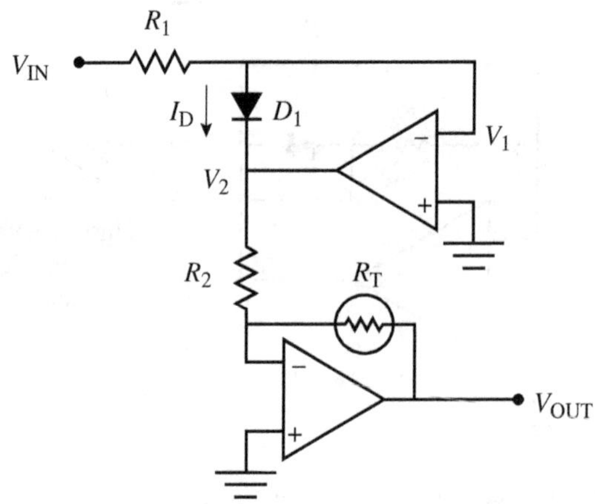

FIGURE 5.32 Temperature-compensated logarithmic amplifier using thermistor R_T.

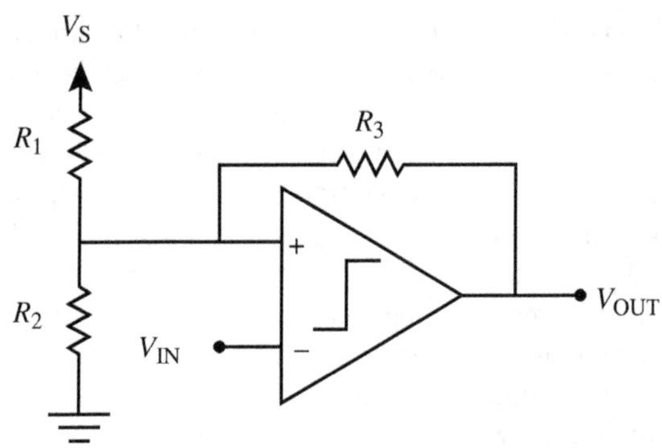

FIGURE 5.33 Schmitt trigger of Problem 5.14 with voltage-divider reference voltage.

One way temperature compensation can be achieved is through the use of a *thermistor* whose resistance R_T is inversely proportional to absolute temperature T (in degrees Kelvin or K): $R_T = K_T/T$, where K_T is a constant. The following circuit shown in Figure 5.32. is one form of temperature-compensated logarithmic amplifier:

(a) Find the current I_D for positive input voltages V_{IN}.

***(b)** Show that the output voltage V_{OUT} is proportional to the natural logarithm of the input voltage V_{IN} and is (theoretically) independent of temperature.

5.14. *Schmitt trigger circuit with prescribed thresholds.* In the design of a Schmitt trigger circuit using a fixed power-supply voltage as a reference, it is desirable to be able to select the positive-going threshold V_{TR+} and the negative-going threshold V_{TR-} independently. The circuit shown in Figure 5.33. will allow this.

Because the input signal is applied to the inverting input of the comparator, the output is HI when the signal goes low (past the negative-going threshold) and LO when the signal goes high (past the positive-going threshold). (This is not a significant complication for digital circuitry, because the digital output can easily be reinverted with an inverter gate.) Assume the power-supply voltage for the R_1–R_2 voltage divider is V_S. Also, assume the comparator output voltage V_{OUT} is V_{HI} when HI and V_{LO} when LO.

(a) If $V_S = V_{HI} = 5\,\text{V}$, $V_{LO} = 0\,\text{V}$, $R_1 = R_2 = 30\,\text{k}\Omega$, and $R_3 = 10\,\text{k}\Omega$, find values for the positive threshold V_{TR+} and the negative threshold V_{TR-}.

***(b)** Given values for V_S, V_{HI}, V_{LO}, V_{TR+}, V_{TR-}, and R_3, find general algebraic expressions for the values of R_1 and R_2 required. (*Hint*: The problem is easier if you solve first for the Thévenin equivalent voltage V_T and Thévenin equivalent resistance R_T of the voltage divider formed by R_1 and R_2. The expressions for V_T and R_T are fairly simple, and you can then solve for R_1 and R_2 in terms of V_T, R_T, and power-supply voltage V_S.)

For further resources for this chapter visit the companion website at **http://wiley.com/go/analogmixedsignalelectronics**

6

THE HIGH-GAIN ANALOG FILTER AMPLIFIER

6.1 APPLICATIONS OF HIGH-GAIN FILTER AMPLIFIERS

Because of advances in digital processing power and software, **digital signal processing** (**DSP**) and related techniques are now the best approach to many design problems that were formerly solved exclusively by means of analog circuits. But analog signals must first be converted to digital form before DSP can be used, and it is still relatively difficult and expensive to perform direct analog-to-digital conversion accurately on extremely low-level signals (below a few millivolts). This is why there is still a frequent need for high-gain analog amplifiers in situations where the input signal is at a very low level. The amplifier can also perform filtering operations if the bandwidth of the signal is known in advance. Such filtering reduces noise outside the spectrum covered by the signal and improves the output's **signal-to-noise ratio**.

This chapter treats topics related to such amplifiers, including special precautions one must take when designing high-gain amplifiers, analog filter circuits that are useful in the design of such systems, and a complete design example of an actual high-gain amplifier using op amps and active filters. In the following section, we will describe some applications that require high-gain audio-frequency amplifiers and some special techniques used to deal with problems that arise.

Analog and Mixed-Signal Electronics, First Edition. Karl D. Stephan.

Companion Website: http://wiley.com/go/analogmixedsignalelectronics

6.1.1 Audio-Frequency Applications

Acoustic sensors such as microphones and the pickup coils on electric guitars typically provide signals that are in the range of 10 mV or less. Depending on the dynamic range of the sounds that are being received by the microphones, the actual voltage can be much less than this. So that you can calculate the expected voltage levels and gains required, we will briefly discuss the way microphones are rated and specified with regard to their electrical output for a given acoustic input.

The standard way of referring to the intensity of a sound in air is the **sound pressure level** (**SPL**), which is typically expressed in units of dB SPL. Sound pressure, like other forms of pressure, is measured in units of **pascals** (equal to newtons per square meter and abbreviated as Pa). The usual reference pressure for 0 dB SPL is 20 μPa, which is approximately the quietest sound that can be heard by the normal human ear. There are several ways of specifying the voltage a microphone will deliver for a given acoustic sound pressure at the microphone element, but one of the most useful is the voltage output rating measured in **mV/Pa**. For example, the Electrovoice PL-37 condenser microphone has a voltage output of $6\,\mathrm{mV\,Pa^{-1}}$, while the Shure PE85 dynamic microphone has an output of $1.3\,\mathrm{mV\,Pa^{-1}}$.

Given these ratings, we have developed an imaginary but realistic situation in a recording studio. A drummer is making a sound that has a level of +80 dB SPL, which goes into the condenser microphone mentioned in the previous paragraph. In an isolated sound booth nearby, an artist is whispering into the dynamic microphone at a level of only +20 dB SPL. Both of these signals need to be fed to a mixer board that, for good signal-to-noise ratio and prevention of spurious noise such as power-line hum, requires that every input must be boosted to a level of at least 775 μV, which can be expressed as −60 dBm into a load resistance of 600 Ω. Why 600 Ω? This is a standard resistance that is used as an impedance reference for analog signals in audio communications and originates from the typical impedance shown by open-wire telephone lines. The **volume unit** or **VU** meters found in many pieces of audio gear are standardized so that a steady sine-wave signal of +4 dBm (1.228 Vrms) across 600 Ω produces a 0-dB VU reading. (The absolute unit of dBm expresses power in dB above or below 1 mW.)

Going back to the studio example, to provide a margin of safety, we will say that for various reasons having to do with the internal design of the mixer board, the drum signal should be at a level of about −28 dBm, and the voice signal at a level of −46 dBm at the mixer inputs. The mixer board has enough internal gain to boost both of these signals up to a combined level of −10 dBm, which is sufficient to drive either a power amplifier for monitoring headphones or the **analog-to-digital converter** (**ADC**) in a digital recorder.

The question now is, what do we need in terms of **microphone preamplifiers** in order to have the signals going into the mixer board at the desired levels? Figure 6.1 shows the situation. Every condenser microphone has an internal preamp (designated A_0) that is not accessible to the user. This amplifier is necessary because of the way condenser microphones work, but unless the amplifier is very carefully designed, it will add a noticeable amount of noise. The preamplifier often enables these types of

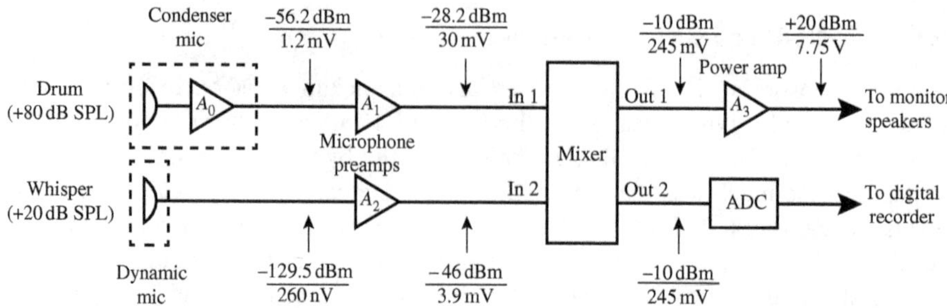

FIGURE 6.1 Hypothetical sound studio setup showing two microphones (one with internal preamp A_0), microphone preamps A_1 and A_2, mixer board, power amplifier A_3, and analog-to-digital converter (ADC). Sound levels are given both in dBm and volts (rms) into 600 Ω.

microphones to have a higher-voltage output (mV Pa^{-1}) for a given SPL than passive (dynamic) microphones can provide, but the noise level of the preamplifier may be unacceptably high for very soft sounds.

On the other hand, a dynamic microphone typically has no internal preamp, so its internal noise is very low. However, because there is no amplification internally, the dynamic-microphone voltage output may be lower than the condenser type. In the example, we assume a dynamic microphone is being used for the whispered voice signal.

The usefulness of dB units becomes evident when we calculate the gains needed for microphone preamps A_1 and A_2. To find the gain required in dB, we simply subtract the input voltage level (in dBm) from the required output voltage level (also in dBm). The difference is the gain that the preamps must provide:

$$A_1 \text{ gain (dB)} = -28.2\,\text{dBm} - (-56.2\,\text{dBm}) = 28\,\text{dB} \tag{6.1}$$

$$A_2 \text{ gain (dB)} = -46\,\text{dBm} - (-129.5\,\text{dBm}) = 83.5\,\text{dB} \tag{6.2}$$

The numeric-ratio gain of condenser-microphone preamp A_1 is a numeric ratio of only $10^{(28/20)} = 25.2$. But the gain needed for the dynamic-microphone preamp is quite large: $10^{(83.5/20)} = 14{,}962$! In practice, the studio engineer would probably just tell the whispering performer to get closer to the microphone. But sometimes, this is not possible, and extraordinary measures are needed to acquire certain sounds.

Typical microphone preamps provide gains of up to 40–60 dB or more and have optional features such as *equalization* (adjustable frequency response). This example shows why a microphone preamp is often desirable in many audio studio settings and why one with a gain of more than 60 dB might be needed for unusual situations.

6.1.2 Sensor Applications

The term "sensor" covers a tremendous variety of devices that deal with many types of inputs such as pressure, temperature, light intensity, radiation level, and even the concentration of specific chemicals in biological samples. Although some sensors provide direct digital outputs, many others operate by converting the quantity of interest into a continuously varying (analog) voltage or current. Typically, the signal of interest varies at a rate that can be anywhere from the radio-frequency range (1 MHz or more) down to almost DC, and the signal level can vary from volts down to the nanovolt range. Two types of amplifiers are especially useful for these kinds of sensor signals: the straight or conventional DC amplifier and the **chopper-stabilized** DC amplifier, which is really an AC amplifier used to amplify DC. Related to the chopper-stabilized amplifier is an instrument called a **lock-in amplifier**, which we will also describe briefly.

If a sensor signal is at least 100 mV or so in amplitude, a DC-coupled amplifier using a single op amp can boost it to a level that will exploit the full dynamic range of an ADC whose input is designed for signals in the range of −10 to +10 V, for example. But for signal levels much lower than 100 mV, problems of **offset**, **drift**, and **gain instability** arise.

The transfer function of a linear DC amplifier can be considered as a linear equation relating the input voltage x to the output voltage y through the relation

$$y = mx + b \tag{6.3}$$

in which m is the voltage gain and b is ideally zero in a perfect DC voltage amplifier. However, as we have seen in the discussion of op amps in Chapter 5, every op amp has a nonzero input offset voltage V_{OS}, which produces a spurious voltage at the output when no input is present. This offset voltage has the effect of making b in Equation 6.3 something other than zero, and at the output, the contribution of a nonzero offset voltage cannot be distinguished from the true signal, which is represented by x.

If b is a known constant value, a manual adjustment called an **offset adjustment** can be made. An offset adjustment artificially adds a constant compensating voltage to force b in Equation 6.3 to be zero. But manual adjustments are expensive and inconvenient and rely on the offset voltage to remain constant. Depending on what causes it, the offset voltage may change with temperature, power-supply voltage variations, and time due to aging of components. These slow and largely unpredictable changes in a nominally constant value are collectively called **drift** and are obviously undesirable. Although automatic offset-zeroing and drift-compensation circuits can be designed, they add complexity to a system and can themselves cause problems.

Another difficulty can arise when the gain factor m changes. While a well-designed op amp circuit's gain depends primarily on resistor values and not on the actual open-loop gain of the op amp A_0, changes in A_0 can produce small changes in gain m. And if the gain is large to begin with and the sensor voltage changes are a small fraction of its average value, these gain changes can be mistaken for changes in the signal as well.

For these reasons, it is not generally possible to design a "straight" DC amplifier using a single op amp to have a usable gain of much more than about 100 (40 dB). **Cascading** multiple DC-coupled op amp gain stages with lower gains does not always help with offset problems either, because a small offset voltage in the first

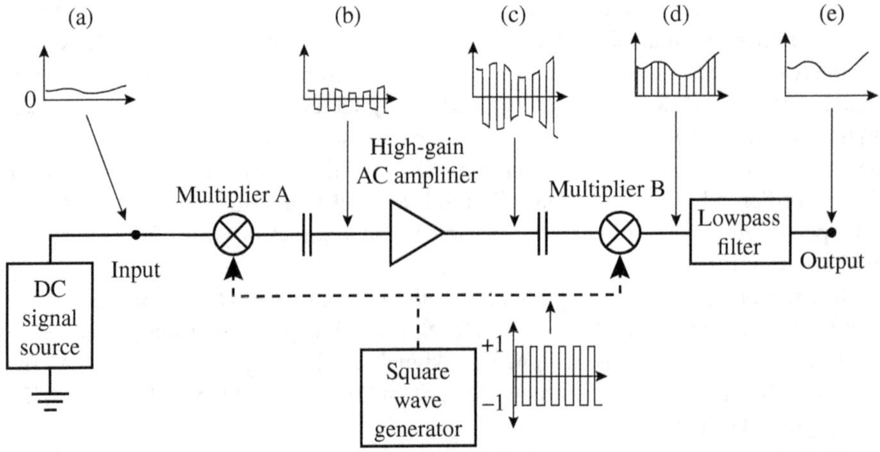

FIGURE 6.2 One form of chopper-stabilized DC amplifier. (a) DC input (b) input multiplied by square wave to produce AC input to amplifier (c) AC amplifier output (d) output multiplied by square wave to produce DC output (e) amplified DC output with switching transients filtered out.

stage is amplified by the second stage and can easily grow so large that it saturates the last stage in such a cascade.

When DC gains much in excess of 100 are required, one of the more commonly used solutions is to employ a chopper-stabilized amplifier. While there are many forms of chopper-stabilized amplifiers, one form that is easy to understand is shown in Figure 6.2.

The input signal is a slowly varying waveform with a large DC content, as the graph in Figure 6.2a indicates. Supposing that the highest significant frequency component in the input signal is f_{MAX}, a square-wave generator produces a square wave at a frequency of f_{CHOP}, which should be greater than $2f_{MAX}$ in order to preserve the highest-frequency components of the input signal. (This is a result of the **Nyquist sampling theorem**, which will be discussed more fully in Chapter 8 on analog-to-digital conversion.) We have labeled the high and low voltages of the square wave +1 and −1, respectively, because those are the factors by which two **analog multipliers** will multiply their input signals. An analog multiplier does just what its name implies: it multiplies the input voltage by a factor that depends on a second input voltage, in this case ±1. Such multipliers are readily available in the form of **electronic switches**. A well-designed electronic switch has very low offset and gain variations, much less than a typical op amp would show.

When the input signal reaches multiplier *A* in Figure 6.2, it has its polarity reversed for half of every square-wave cycle but is otherwise unchanged as to amplitude and DC content. At point (b) in the system, the input waveform with its DC component has been converted to a pure AC waveform but with all of its original information intact. As an AC waveform, it can now be amplified much more than is possible for a DC signal, because AC amplifiers can have high but stable gain without as much concern for drift or offset variations. Gains of 80–100 dB or more are possible, especially if the gain is limited to a fairly narrow bandwidth centered at f_{CHOP}.

The output (c) of the high-gain AC amplifier is a magnified version of the input with the same phase. Consequently, multiplying the output waveform with the same square-wave signal that multiplied the input waveform will "undo" the DC-to-AC transformation, because those parts of the waveform multiplied by +1 will pass through unchanged, while those that were multiplied by −1 will be multiplied by −1 again, restoring their original polarity. The result (d) is a good approximation of the original input waveform, only with a higher amplitude. However, high-frequency switching transients are present in the output (d) of multiplier B, because the switching operations take a non-zero time to occur and the multipliers are not perfect. These high-frequency components are eliminated in a lowpass filter whose bandwidth is greater than f_{MAX}, but lower than f_{CHOP}. The final output (e) is an amplified version of the input signal, including the DC component and free from virtually all drift and offset problems.

A **lock-in amplifier** is basically a packaged system containing everything from point (b) in Figure 6.2 onward to the right, including the square-wave generator but generally not including multiplier A. The user employs the square-wave generator output to produce an external sensor signal that is **modulated** at the frequency f_{CHOP}, which is amplified and **demodulated** by multiplier *B*. (Another name for the function performed by multiplier *B* is **synchronous detection**, and multiplier *B* is sometimes called a **synchronous detector** or **synchronous demodulator**.) More sophisticated lock-in amplifiers provide the user with variable chopping frequencies and a wide variety of filter options and gain selections. If the signal to be amplified is derived from a light beam, one can use a rotating **chopper wheel** (Fig. 6.3) to interrupt the beam periodically at a rate synchronized to f_{CHOP} so that the sensor receives a signal that is already converted to AC and suitable for amplification by a lock-in amplifier.

These are only two of many applications that high-gain amplifiers find in analog electronics. Others arise in systems that use high-frequency signals: radio receivers and **magnetic resonance imaging** (**MRI**) machines are just two examples. Many biomedical devices such as **ultrasound imaging sensors**, **electrocardiograph** (**EKG**), and **electroencephalograph** (**EEG**) machines and a wide variety of other applications in the medical and physical sciences need high-gain amplification of low-level signals.

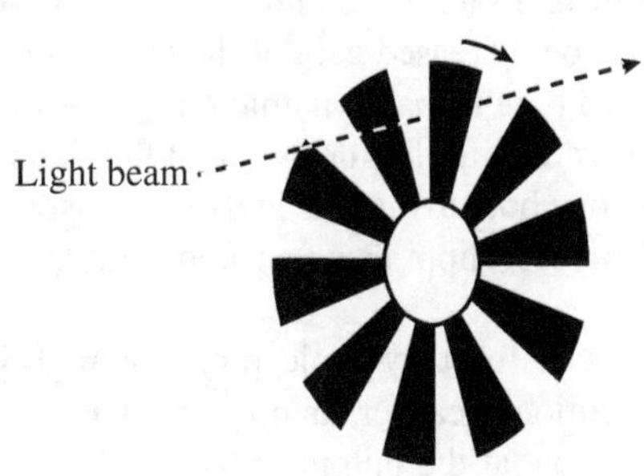

FIGURE 6.3 Optical chopper wheel used to interrupt light beams at a rate f_{CHOP}.

6.2 ISSUES IN HIGH-GAIN AMPLIFIER DESIGN

When a large amount of gain is desired from an amplifier circuit, certain problems can arise that generally do not affect lower-gain designs. The definition of "high gain" depends on the frequency range under discussion and the signal levels involved. At a microwave frequency above 1 GHz, a gain of 20 dB may be considered high, while at audio frequencies of 20 kHz or below, a gain of 40 dB presents relatively few problems. But in general, any attempt to achieve a gain of 60 dB or more at a single frequency or range of frequencies will often result in the types of problems discussed in the remainder of Section 6.2.

6.2.1 Dynamic-Range Problems

Other things being equal, a 10-dB increase in the gain of an amplifier system will reduce the system's dynamic range by 10 dB. In a well-designed amplifier, the noise floor will be determined by the input stage or stages. (We have taken the noise floor to be the input signal level at which the signal power equals the noise power at the output.) Adding gain in the amplifier cascade at any point past the input stage will simply amplify the noise power that is already present, which does not usually improve the signal-to-noise ratio. If the output-stage saturation limit level remains the same as the gain increases, there is less space remaining between the fixed upper saturation limit and the lower noise-floor limit, which has now moved higher due to the increase in gain. It is entirely possible for an amplifier system to be in hard saturation due solely to its own internal noise. While this situation is useful for certain special circumstances (e.g., analog FM radio receivers), it is undesirable for a linear amplifier to be anywhere close to saturated by either internal or external noise.

This problem is illustrated in Figure 6.4, which shows the transfer function of a hypothetical linear amplifier with an effective noise bandwidth of 1 kHz and a gain of 80 dB. The thermal noise in that bandwidth from a resistor at room temperature is about −143.8 dBm, which establishes the noise floor for this particular amplifier. Because the system has so much gain, the noise floor plotted on the vertical (output) scale is (−143.8 + 80) = −63.8 dBm. If the output saturates at about +20 dBm, that leaves a dynamic range of less than 80 dB. This is considerably smaller than the 137-dB dynamic range of the hypothetical 10-dB-gain amplifier illustrated in Figure 4.9, which has the same input and output conditions but only 10 dB of gain. And the difference is due solely to the increased gain of the higher-gain amplifier.

The lesson to be learned here is that dynamic range must be considered carefully when undertaking a high-gain amplifier design, and this includes an estimate of the noise present at the input (due both to internal and external sources) and the maximum acceptable output level before clipping or an undesirable amount of nonlinearity occurs.

Another problem that can limit dynamic range of high-gain amplifiers is DC offset at the output. As mentioned earlier, an op amp's small input offset voltage of a few millivolts can easily grow at the output to the level of volts in a high-gain DC amplifier, and this voltage limits the total AC voltage swing that the circuit can handle

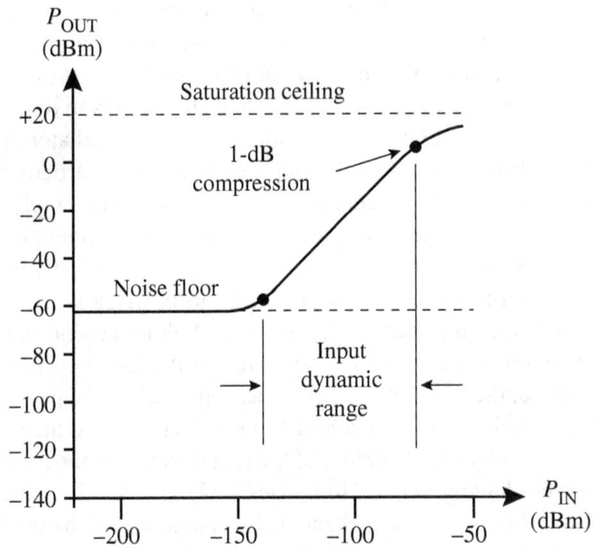

FIGURE 6.4 Dynamic-range graph of hypothetical amplifier with 80dB gain.

before clipping occurs. For example, suppose a particular amplifier's output saturation limits are at −11 and +12V. If the no-signal output is at 0V, the amplifier can deal with a signal level as high as 11V (peak) without first encountering clipping at the negative (−11V) peak. But if a DC offset voltage of, say, −3V is constantly present at the output, the amplifier will clip negative peaks when a peak output amplitude of only (−3−(−11))=8V appears. That is why AC amplifiers of all kinds (not just op amp circuits) should be designed so that the **DC operating point** is as close as possible to midway between the upper and lower saturation limits. In advanced designs, one can incorporate feedback circuits to set the DC output voltage at a desired point while leaving the AC gain largely unaffected.

6.2.2 Oscillation Problems

As we will see in more detail in Chapter 7 concerning signal generation, you can calculate the conditions under which a nominally linear circuit will spontaneously break into **oscillation**, which is defined as the generation of a periodic output waveform with no input to the system. While the term **stability** has a number of meanings in electronic engineering, it can refer in amplifier discussions to the tendency of an amplifier to oscillate. A stable amplifier, in this sense, does not oscillate, while instability in an amplifier means it tends to oscillate. (This has nothing directly to do with **gain stability**, which is a different criterion, although changes in gain can cause a stable amplifier to become unstable and oscillate.)

As you might expect, oscillation is an undesirable effect in amplifiers, but as the gain at a single frequency or range of frequencies increases, it becomes increasingly difficult to **isolate** the output of the amplifier from the input sufficiently to prevent oscillation.

To oversimplify a principle that will be stated later in more detail, oscillation can occur at a given frequency *f* if (i) the gain magnitude around a given closed signal path (the **loop gain**) is greater than one and (ii) the total phase shift around the loop at the frequency *f* is zero degrees (or a multiple of 360°, which amounts to the same thing at a given frequency). How can a high-gain amplifier become an oscillator in this way?

Consider an amplifier cascade with a gain of 80 dB at a particular frequency *f*. Portions of the output signal can appear in more places than simply at the output terminals. The output stage draws current from the power-supply circuits, and this current varies in proportion to the output signal. So the power-supply lines can potentially carry a smaller version of the output signal to other parts of the circuit. The return currents from the amplifier output also travel along ground planes and other pathways in the circuit board. It is quite possible that the level of one of these output currents or voltages at the input of the circuit, as conveyed by the power-supply lines or ground planes, may be no less than 80 dB lower than it was at the original output. The appearance of an attenuated version of the output signal at the input of an amplifier is called **reverse leakage** and is illustrated in Figure 6.5. It is clear that in this case, the 80 dB loss of the reverse leakage path is counteracted by the 80 dB gain of the amplifier, leading to a loop gain of 0 dB.

Eighty decibels is a large value of isolation, equal to a numerical ratio of 10^4, so the requirement here is for an output signal of 10 V to be no larger than 1 mV at the amplifier input. However, if the isolation between output and input is no better than that, and the phase shift at frequency *f* is 360° (or a multiple thereof), the circuit will quite possibly oscillate at frequency *f*. Even if the loop gain is only marginally less than 0 dB (say, −3 to −5 dB), the circuit will be only **marginally stable**, and if the gain increases by only 3–5 dB, it may oscillate. A marginally stable amplifier will often show severe distortions of the frequency response as the phase shift around the undesired feedback loop passes through multiples of 360° with changes in frequency. So even if the reverse leakage is not large enough to cause oscillation, it can still distort the circuit's frequency response.

For these reasons, the designer must pay special attention to isolating the output stages of a high-gain amplifier from the input stages. There are several techniques available for this. One of the most common is power-supply bypassing, mentioned briefly in Chapter 3.

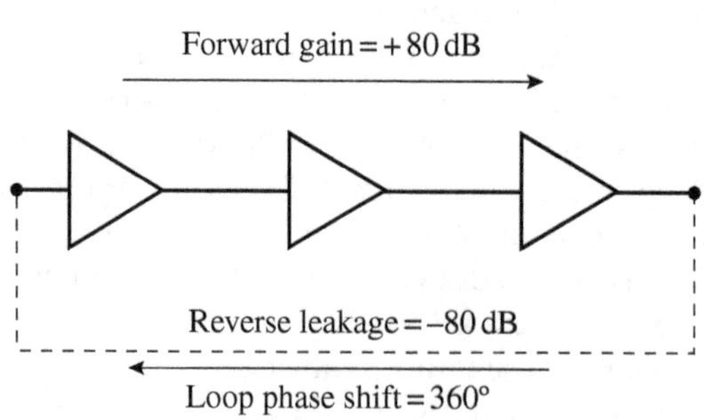

FIGURE 6.5 Undesirable feedback path around high-gain amplifier leading to potential oscillation.

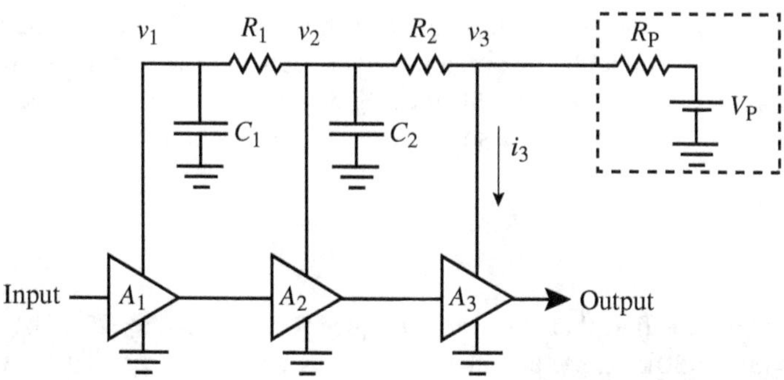

FIGURE 6.6 High-gain three-stage amplifier showing equivalent circuit of power supply and technique of power-supply bypassing.

Figure 6.6 shows a three-stage high-gain amplifier with a low-level input stage A_1, an intermediate-level stage A_2, and a high-level output stage A_3. Considerable AC current i_3 is drawn by the output stage when it produces the full linear output of which the amplifier is capable, and i_3 is proportional to the output signal.

If the power supply were an ideal voltage source, no amount of current drawn from it would affect its output voltage. But all real power supplies have some internal resistance, which is represented in Figure 6.6 by R_P. Changes in the current drawn by amplifier stage A_3, which we will represent by i_3, cause a small AC voltage $v_3 = i_3 R_P$ to appear on the power-supply line.

If this voltage v_3 goes directly to the power-supply connection of the input stage A_1, it may well complete a feedback loop with sufficient loop gain to cause oscillation. If this is the case, the reverse leakage loss can be increased substantially through the use of power-supply bypassing, examples of which are resistors R_1 and R_2 in combination with capacitors C_1 and C_2.

The goal of power-supply bypassing is to increase the attenuation of undesirable AC signals while minimizing the DC voltage drop through the circuit. In the example shown, the DC current drawn by amplifier stages A_1 and A_2 may be only a few mA. If this is the case, a 100-Ω resistor in series with a current of 10 mA results in a DC voltage drop of only (100 Ω)(10 mA) = 1 VDC. If the original power-supply voltage V_P is 9 V or larger, a 1-V drop can be accounted for in the design and need not impair circuit operation. Depending on the DC current, the series resistors can have values up to 1 kΩ or so for small current drains.

Capacitors C_1 and C_2 should be as large as possible, consistent with space requirements and the frequency range over which bypassing must be most effective. For audio-frequency bypassing, these capacitors are commonly electrolytics with values in the 1–100-μF range. For radio-frequency bypassing, the highest frequency to be bypassed should be below the capacitor's series-resonant frequency, and values below 1 μF are typically used.

The effectiveness of just one *R–C* section of power-supply bypassing can be estimated as follows. Treating R_2 and C_2 as a lowpass filter (discussed more fully in the following), suppose $R_2 = 100\,\Omega$ and $C_2 = 22\,\mu\text{F}$. If problems had been encountered with oscillation at $f = 30$ kHz, adding this *R–C* bypass circuit can potentially reduce the reverse leakage signal by an amount roughly equal to the voltage attenuation of the lowpass filter (this ignores the AC loading effects of the earlier stages, which are typically small). Using the lowpass filter attenuation equation

$$|H(\omega)| = \frac{1}{\sqrt{1 + (\omega / \omega_C)^2}}, \tag{6.4}$$

with $\omega_C = 1/RC$, we find that the cutoff frequency is $f_C = \omega_C/2\pi = 72.3$ Hz. At a frequency $f = \omega/2\pi = 30$ kHz, the attenuation factor is 2.41×10^{-3}, or 52 dB. Adding the second bypass circuit consisting of R_1 and C_1 adds roughly another 52 dB, for a total attenuation from the power supply to the first amplifier stage A_1's power-supply terminal of about 100 dB. If the amplifier gain is only 80 dB, this amount of reverse leakage loss should more than suffice to prevent an oscillation path from occurring through the power-supply lines.

Of course, the power-supply leads are not the only means by which a portion of the output signal can make its way back to the input. Poor ground-conductor layouts on printed circuit boards can cause return currents from the output stage to produce a voltage at the input. For this reason, it is best to keep connections to the power-supply ground leads close to the high-power output stages, rather than taking them from the area of the circuit near the sensitive input stages. (More information on this topic is provided in Chapter 12.)

Direct coupling from the output to the input via magnetic or electric fields is also possible. If inductors or transformers carry the output signal and are also used in the input circuit, mutual coupling of the components' magnetic fields can lead to a reverse leakage path and oscillation. This problem can be reduced by proper positioning of inductors so that their magnetic fields do not couple sufficiently to cause oscillation. High-impedance output circuits in which large voltages occur can produce electric fields that couple directly to the conductors in the input circuit. Such mutual-capacitance coupling can also produce oscillation if the forward gain is high enough. This type of problem is relatively easy to eliminate by means of grounded metallic shields that partly or completely enclose the input and/or output stages involved.

Once an amplifier oscillates, it is sometimes difficult to pinpoint the cause. These descriptions of the various ways that reverse leakage can occur around high-gain amplifiers should help you to determine the dominant reverse leakage pathway and which technique to use to eliminate the problem.

6.3 POLES, ZEROES, TRANSFER FUNCTIONS, AND ALL THAT

The purpose of an analog filter circuit is to provide a prescribed amplitude and phase response as a function of frequency. There are many reasons to use filters. In the amplification of low-level signals, a properly designed filter can reduce noise while leaving the signal largely unaffected, thus improving the signal-to-noise ratio at the output. When the output of an amplifier goes to an **ADC** for conversion to digital form, the signal must be **band limited** to prevent a problem called **aliasing**, which would otherwise introduce spurious signals in the converted waveform. And **equalizing filters** are used to compensate for frequency-response distortion due to room acoustics, transmission through distorting communications circuits, and other factors. So for these and other reasons, analog filters are often combined with high-gain analog amplifiers.

Like most other fields of analog electronics, the study of filters divides naturally into the complementary problems of **synthesis** and **analysis**. Filter synthesis is the process of starting with a desired frequency response (either amplitude, phase, or both), making sure it is physically realizable, and designing a circuit that will come reasonably close to delivering the desired response within the limits of available resources. Filter analysis is the process of figuring out what a given filter circuit's response actually is. Once you have synthesized a filter design to meet a given requirement, you use analysis to see if the design does what you want it to do.

Filter analysis is just the straightforward application of linear systems theory that was discussed in Chapter 3 and can be performed either manually or with circuit-analysis software. By far, the more challenging problem is filter synthesis. Not every response function you can imagine can actually be built. For example, **causality** is the very reasonable assumption that a signal has to go into the input of a filter some time before it emerges at the output after filtering. If a signal comes out before it goes in, the filter is doing something that amounts to predicting the future, and physics doesn't allow that. It turns out that certain ideal filter responses that would be nice to have actually *violate* causality! So unless you can build a time machine into your circuit, **noncausal** filters cannot be built.[1]

Because there are theoretically an infinite number of different kinds of filter responses, early filter designers developed a few general categories of response functions that were useful, fairly simple to express mathematically, and relatively straightforward to synthesize in circuit form. All these categories use the basic principle that any two-port circuit made with linear lumped components (not distributed, like a transmission line) has a transfer function that can be expressed as the ratio of two polynomials that are functions of the complex frequency variable $s=j\omega$. Such a function is called a **rational function**. In what follows, we will show you how to calculate the coefficients of a rational function by hand to yield a certain simple type of filter function. You should know, however, that tables and synthesis software are available for more complex filter designs, and once you master these basics, you will be able to use more advanced filter design tools intelligently.

Figure 6.7 shows such a two-port circuit driven at port 1 by a signal source v_s through a source resistance R_S and loaded by load resistance R_L at port 2, where the output voltage v_o appears. We will take the definition of the filter's complex transfer

[1]This restriction applies only to filters that operate in **real time** on a signal that is filtered as it arrives from the source. If the complete signal is *stored* in advance and the system has access to all significant portions of the signal at once, noncausal filters can be used, because the problem of predicting the future does not arise.

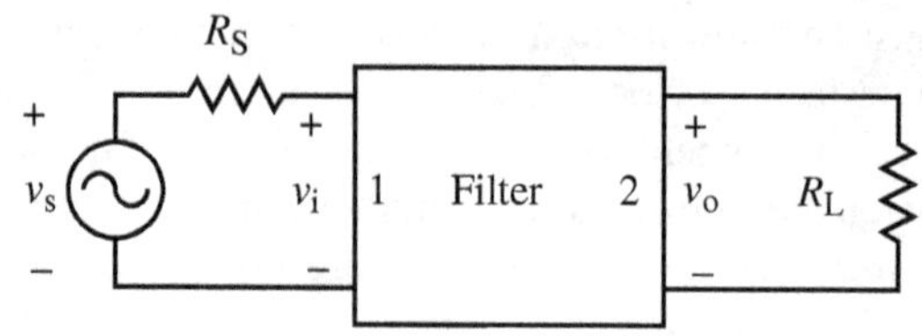

FIGURE 6.7 Generic two-port filter circuit driven by signal source v_s with source resistance R_S and load resistance R_L.

function $H(s)$ to be the ratio of the phasor *output voltage* v_o to the phasor *source voltage* v_s:

$$H(s) \equiv \frac{v_o(s)}{v_s(s)} = \frac{(s-z_1)(s-z_2)\cdots(s-z_m)}{(s-p_1)(s-p_2)\cdots(s-p_n)} \tag{6.5}$$

In Equation 6.5, the polynomials are written in *factored* form, as the product of terms like $(s-x)$ where x is either a *zero* or a *pole*. We will now explain what zeroes and poles are.

Let's examine the numerator polynomial first. There are m terms in the factored form. If we multiply it out so that it becomes the standard polynomial form in the variable s, the numerator polynomial Num(s) would look like this, in general:

$$\text{Num}(s) = s^m + a_{m-1}s^{m-1} + a_{m-2}s^{m-2} + \cdots + a_1 s + a_0 \tag{6.6}$$

where $a_0 = z_1 z_2 \ldots z_m$, and the other coefficients $a_1, a_2, \ldots a_{m-1}$ are various products of the z terms and constant integers. Equation 6.6 makes it clear that the *order* of the numerator polynomial (the highest power to which s is raised) is m. By the fundamental theorem of algebra, then, we know that the numerator polynomial has exactly m roots (a root is a value of s that makes the polynomial equal to zero). Those roots are in fact the zeroes $z_1, z_2, \ldots z_m$. That is why they are called zeroes: when s takes on the value of any zero, the numerator (and thus the entire function) becomes zero. Therefore, we can always write a polynomial like Equation 6.6 in factored form, which shows explicitly where all the zeroes are. Similarly, the denominator polynomial Den(s) can be written as

$$\text{Den}(s) = s^n + b_{n-1}s^{n-1} + b_{n-2}s^{n-2} + \cdots + b_1 s + b_0 \tag{6.7}$$

The quantities $p_1, p_2, \ldots p_n$ are called *poles*, because when s approaches one of these values, the function tends to go to infinity ("blows up"). (On a contour graph of the function, the graph near these values looks like, well, a pole, or at least a very steep mountain.) Expressing the denominator polynomial in factored form (as in Equation 6.5) makes it easy to tell where the poles are, and it is always possible to factor a polynomial such as Equation 6.7 in standard form, although some of the poles and zeroes may in general be complex.

The fact that lumped-element circuits have transfer functions that are rational functions is very helpful in circuit design, because circuit analysis generally provides a response function that gives the answer in terms of a ratio of two polynomials that are *not* factored, but instead are in the standard polynomial form, as Equations 6.6 and 6.7 are. It turns out that the behavior of any rational function (except for a constant multiplying factor) is entirely determined by the zeroes and poles in its numerator and denominator polynomials, respectively.

6.4 PASSIVE ANALOG FILTERS

To keep this discussion from getting too abstract, let's show how a very simple passive filter circuit fits into the pole-zero picture. Analog filter circuits can be either passive (consisting only of passive components) or active (including active devices such as transistors or op amps). Each type has advantages and limitations, but the passive type of filter is easier to understand, and the first filter circuit we will describe will be a passive filter.

Earlier in this chapter, we mentioned that a series-*R*, shunt-*C* circuit such as shown in Figure 6.8 acts as a **lowpass filter**. As the name implies, a lowpass filter passes most or all signals in a range called the **passband**, which for a lowpass filter extends from DC up to a **band limit**. Usually, the band limit is taken to be the frequency at which the response is 3 dB down from its maximum passband value (which usually occurs at DC for a lowpass filter of the type we are going to study). Above the band limit, there is a **stopband**, which extends (in principle) to infinite frequencies. Signals with frequencies in the stopband are attenuated by at least a specified amount, while those in the passband experience little to no attenuation.

6.4.1 One-Pole Lowpass Filter

Considering the lowpass filter of Figure 6.8 in the context of the general two-port shown in Figure 6.7, we take the source resistance R_S to be zero and the load resistance R_L to be infinity, so that $v_s = v_i$. For most low-level audio-frequency filters, these conditions are usually assumed, but at higher frequencies and power levels, you must often specify finite nonzero values for the source and load resistances.

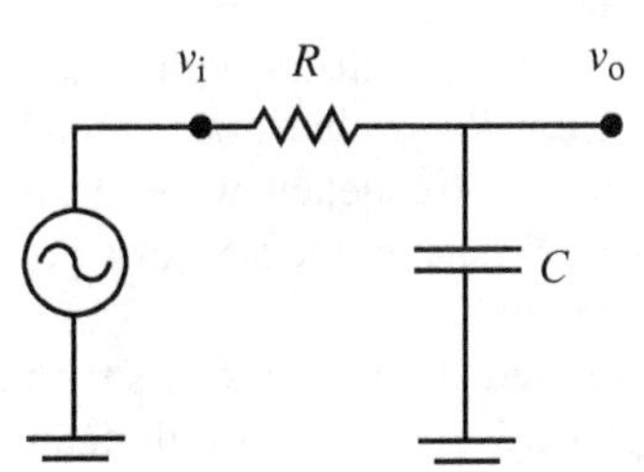

FIGURE 6.8 Single-pole passive *R–C* lowpass filter.

The transfer function $H_{LP}(s)$ of the R–C circuit in Figure 6.8 is easily found by using the voltage-divider formula with the impedance of C as the shunt impedance and the resistance of R as the series impedance:

$$H_{LP}(s) = \frac{v_o}{v_i} = \frac{1/sC}{R + 1/sC} = \frac{1}{sRC + 1} \tag{6.8}$$

In terms of the ratio of two polynomials, what can we say about this function? Clearly, the numerator polynomial is simply a constant, 1. So there are no zeroes in this function, because the numerator is not a function of s at all. However, the denominator polynomial has order $n=1$, because the highest power of s is s^1. And because the denominator is a function of s, we know there will be one pole. It is very easy to solve for the value of s that makes $sRC+1=0$. That value is $p_1=-1/RC$. So when s takes on the value $-1/RC$, the denominator of $H(s)$ goes to zero and we have a pole.

As we mentioned earlier, a **zero** is a value for s that makes the numerator (and therefore the entire function) equal to zero. A **pole** is a value for s that causes the denominator to go to zero, making the function approach infinity as s goes to the value of the pole. We will not go into too many details of complex mathematics in this text, but as you might expect, both zeroes and poles affect the value of the transfer function at various frequencies.

What is the magnitude of the frequency response $H_{LP}(s)$ as the real frequency ω is varied from zero to infinity? As ω goes from zero to infinity, the complex variable $s=j\omega$ travels from the origin of the complex plane to infinity along the positive imaginary (j) axis. This is a problem in complex math, but we are going to give it a geometrical interpretation that will prove handy later on with more complicated filter circuits.

In the complex plane, the quantity $s-p_1$ can be represented as a **vector**. This vector can tell us a lot, both qualitatively and quantitatively, about the behavior of the transfer function as the frequency variable $s=j\omega$ varies from zero up to high frequencies along the positive imaginary axis. But first, we must recast the transfer function so that the denominator is in factored form:

$$H_{LP}(s) = \frac{1}{sRC + 1} = \frac{1}{RC} \frac{1}{(s - (-1/RC))} = -p_1 \frac{1}{(s - p_1)} \tag{6.9}$$

That may seem to be a lot of trouble to go to, but the usefulness of having the denominator in factored form, and moving any constants that result to the front of the expression, will now be made clear.

The pole that the R–C lowpass circuit has is a *negative real* pole. It sits on the negative real axis at a distance $1/RC$ (radians per second) from the origin. Because this distance has the dimensions of frequency, we will call it the **pole frequency** and designate it as $1/RC=\omega_p=-p_1$. Whatever the sign of a pole itself is, pole frequencies are always positive by convention.

In Figure 6.9, we have plotted the vector $s-p_1$ for three different frequencies: $\omega_1=0$, $\omega_2=\omega_p$, and $\omega_3 \gg \omega_p$. There are two important features of this vector to notice: (1) its length relative to its length for $\omega_1=0$ and (2) its angle with respect to zero degrees (the direction of the positive real axis).

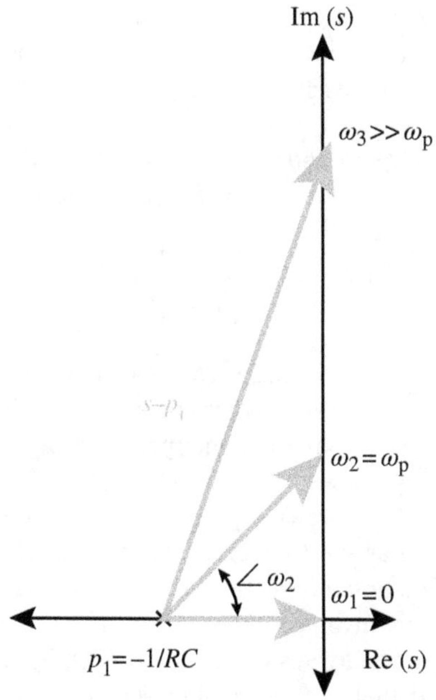

FIGURE 6.9 Complex s-plane with vector $s-p_1$ plotted for three values of $j\omega$, showing vector $s-p_1$.

At $\omega_1=0$, the vector is horizontal (0°) and its length is equal to p_1. The length of a vector is the magnitude of the complex number it represents, so the magnitude of the denominator at zero frequency is p_1. In the factored form of the transfer function $H_{LP}(s)$, the fraction is multiplied by p_1, which cancels out the p_1 in the denominator, so we conclude that at zero frequency, the magnitude of the overall transfer function is 1. That makes intuitive sense, because the impedance of a capacitor at DC is infinity, and so the input voltage passes through the series resistor R to the output without attenuation.

Next, observe what happens when the frequency rises along the imaginary axis to $\omega_2=\omega_p$. The vector $s-p_1$ forms the diagonal of a square, whose length is $\sqrt{2}$ times the side or $\sqrt{2}$ times longer than it was at zero frequency. Because the vector represents the denominator term, that means the magnitude of the response at $\omega_2=\omega_p$ is $1/\sqrt{2}$ times the response at zero. As you should know by now, the dB version of the amplitude ratio $1/\sqrt{2}$ is almost exactly −3 dB (−3.01 dB, to be precise). So we conclude that when the signal frequency $\omega=\omega_p$, the response magnitude is 3 dB down from what it is at DC.

For frequencies much greater than ω_p, the difference in length between the vector $s-p_1$ and the magnitude of ω will become smaller, and we conclude that the response magnitude will go approximately inversely with frequency. In terms of dB, this amounts to saying that the response falls at a rate of 20 dB per decade (a decade is a factor of 10 in frequency).

The angle that the vector makes with respect to the positive x-axis is also significant. Note that at zero frequency, the angle is zero, and as frequency increases to $\omega_2=\omega_p$, the angle $\angle\omega_2$ is exactly 45°, approaching 90° as the frequency rises toward infinity.

If you do the complex math and allow $\omega_p=\omega_C$, you will find that the response function $H_{LP}(\omega)$ is exactly that given in Equation 6.4, which we will repeat here:

$$\left|H_{LP}(\omega)\right| = \frac{1}{\sqrt{1+(\omega/\omega_p)^2}} \tag{6.10}$$

And the phase angle of $H_{LP}(\omega)$ is the negative of the angle that the vector $s-p_1$ makes with the horizontal axis, because the vector appears in the denominator. In addition to the math, you now have a geometrical interpretation of *why* the transfer-function magnitude behaves the way it does.

The same basic principle of drawing pole (or zero) vectors and seeing how their lengths and angles change as the frequency variable travels up the imaginary axis will apply to any filter function, no matter how complicated. Obviously, when more than two or three vectors are involved, the picture gets rather messy and it is difficult to interpret it for exact numerical results. Nevertheless, by knowing the location of the poles and zeroes and their corresponding vectors, you can gain a qualitative understanding of how a particular factored filter transfer function will behave without doing any additional math at all!

In Figure 6.10, we have done the exact calculations to plot the magnitude (in dB) and angle (in degrees) for the transfer function H_{LP}. The frequency axis is **normalized** (made dimensionless by dividing by a dimensioned constant) using the cutoff frequency $f_C=\omega_p/2\pi$ as the normalizing constant. This makes the graph universally useful to find the response of this circuit for any combination of R and C. For example, if $R=10\,\text{k}\Omega$ and $C=10\,\text{nF}$, $f_C=1/(2\pi RC)=1.59\,\text{kHz}$. To **denormalize** the x-axis for use with that particular combination, you simply multiply the normalized scale by f_C, so $f/f_C=1$ becomes 1.59 kHz, and similarly for any other cutoff frequency.

There are several important features to note about the curves in Figure 6.10. As we pointed out earlier in the discussion of Bode plots, the single-pole lowpass response asymptotically approaches 0 dB at DC and has a slope of −20 dB per decade at frequencies much higher than f_C. At exactly f_C, the response is almost exactly 3 dB below its DC value. Note that at 1/10th and 10 times f_C, the actual plot is 0.04 dB below the asymptotic values: −0.04 dB at $f_C/10$ and −20.04 dB at $10f_C$. So there is a symmetry of sorts about the cutoff frequency.

Something similar occurs with the phase response. As you would expect from the angle that the vector $s-p_1$ makes with the x-axis in Figure 6.9 for $\omega_2=\omega_p$, the phase shift (which is the phase angle of the response function $H_{LP}(f)$ expressed in polar form) reaches exactly −45° at $f=f_C$. At $f_C/10$, the phase shift is −5.7° (0.1 rad), and at $10f_C$, the phase shift is 90–5.7°, or 0.9 rad.

We emphasize these numbers because *any real pole* in a transfer function will have these same effects at the same respective frequencies. If the transfer function

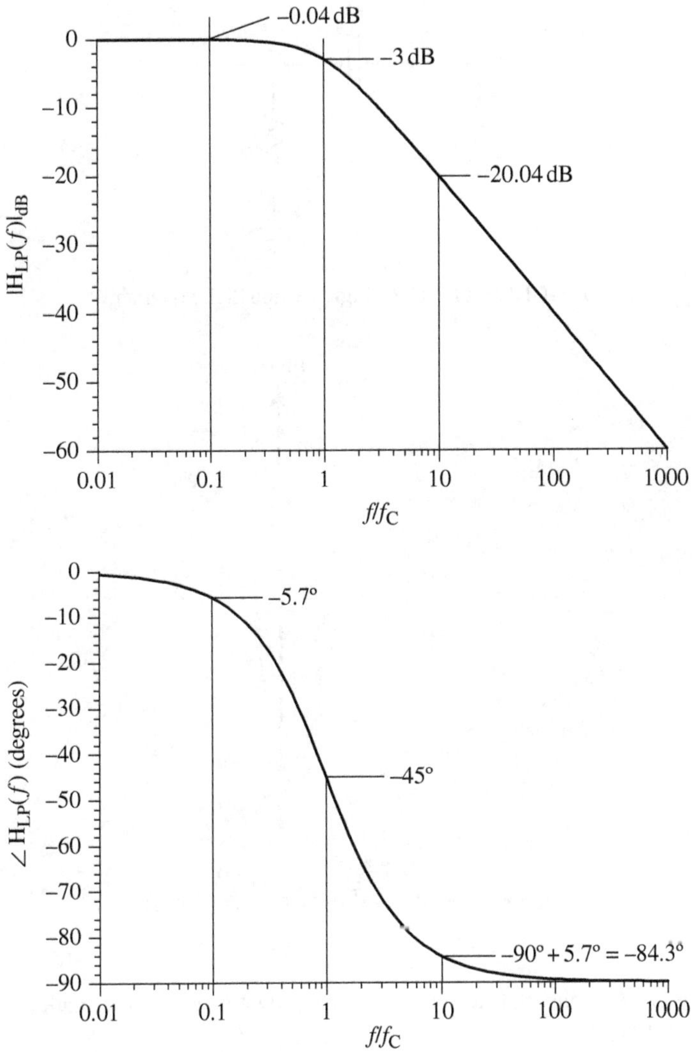

FIGURE 6.10 Magnitude and phase response of one-pole lowpass filter shown in Figure 6.8, with frequency axis normalized to f_C.

had 2, 6, or 10 real poles, all at different frequencies, they would each contribute the same dB loss and phase shift at their respective frequencies. While keeping track of more than two or three poles is best left to a calculator or a computer, you will still be able to estimate the effect of each pole based on its position on the real axis.

6.4.2 One-Pole, One-Zero Highpass Filter

If the positions of the *R* and the *C* are exchanged in the lowpass filter circuit of Figure 6.8, the highpass filter (hpf) of Figure 6.11 results.

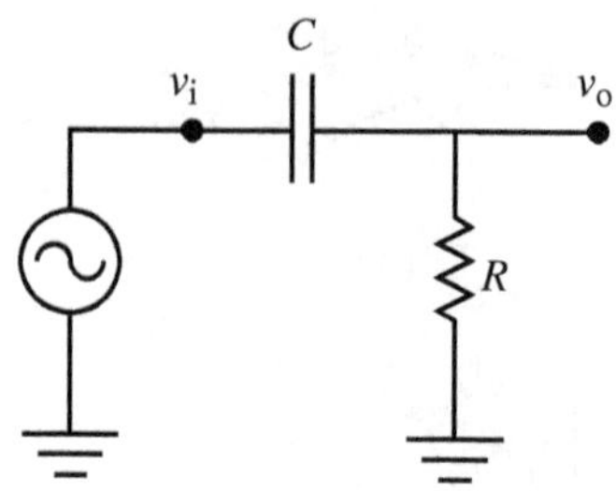

FIGURE 6.11 *C–R* single-section highpass filter.

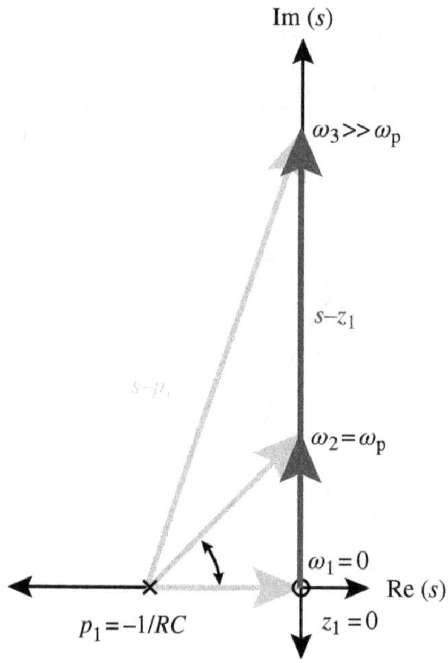

FIGURE 6.12 Complex *s*-plane with vectors $s-p_1$ (light gray) and $s-z_1$ (dark gray) plotted for three values of $j\omega$.

It is easy to show that the transfer function $H_{HP}(s)=v_o/v_i$ for this circuit is

$$H_{HP}(s)=\frac{sRC}{sRC+1}=\frac{s/\omega_p}{1+s/\omega_p}=\frac{s-z_1}{s-p_1} \tag{6.11}$$

Equation 6.11 differs from the lowpass function of Equation 6.8 in that besides the same pole in the denominator, there is now a zero in the numerator as well. The zero z_1 is at zero frequency: when $s=j\omega=z_1=0$, the numerator is zero. Before doing any math, we can draw the vectors s (which is $s-z_1$, the zero vector) and $s-p_1$, the same pole vector that we showed in Figure 6.9, to estimate what the magnitude and phase of the highpass transfer function will be. These vectors are plotted in Figure 6.12.

For frequencies much less than the pole frequency ω_p, the denominator vector is approximately constant and the function's behavior is determined by the numerator, which is proportional to ω. So for frequencies approaching zero, the transfer function will also approach zero. At the pole frequency rises to ω_p, the numerator vector's length is ω_p, but the denominator vector's length is $\omega_p\sqrt{2}$, which means that the overall response magnitude at the pole frequency is $1/\sqrt{2}$, or −3 dB. As the frequency rises far above ω_p, both the numerator and denominator vectors approach the same length, meaning that the response magnitude will approach 0 dB asymptotically. As for the phase shift, recall that the total phase is the numerator phase minus the denominator phase. Near DC, the numerator phase angle is +90° and the denominator is zero, giving +90° for the phase shift near zero. At $\omega = \omega_p$, the numerator still adds +90°, but the denominator subtracts 45°, giving a total of +45°. And for $\omega \gg \omega_p$, the denominator vector's angle approaches +90°, meaning that the phase shift approaches zero. These estimates are borne out by the exact calculated magnitude and phase for $H_{HP}(f)$ shown in Figure 6.13, again with the frequency scale normalized to $f_C = \omega_p/2\pi$. The addition of the zero at $f=0$ means that initially the magnitude of $H_{HP}(f)$ *rises* at 20 dB per decade until it encounters the pole at $f=f_C$. As always, a single pole causes the response to **break** downward at −20 dB per decade from that point onward, which cancels the zero's +20 dB per decade upward slope and causes the response to asymptotically approach 0 dB. If poles and zeroes are separated in frequency by at least a decade or so, one can trace out the response by just following the frequency upward and drawing asymptotes with appropriate slopes, breaking downward for poles and upward for zeroes.

6.4.3 Complex-Pole Bandpass Filter

What if a pole is *complex* rather than real? The same basic principles apply, except that now there are two variables for each pole: the real part and the imaginary part. Complex poles can arise from passive circuits having two or more reactive components. For example, a very simple **bandpass filter** is illustrated in Figure 6.14.

It consists of a series *L–C* combination in series with an output load resistor *R*. Whenever an inductor and a capacitor are connected together, a resonance is possible, and as we will show, this circuit tends to pass frequencies near the resonant frequency of the *L–C* combination and reject those far away from it.

If we write the transfer function $v_o/v_i = H_{BP}(s)$ by inserting the total series impedance of the *L–C* circuit into the voltage-divider formula, we obtain

$$H_{BP}(s) = \frac{R}{sL + 1/sC + R} \tag{6.12}$$

We will now introduce two new variables that will make it much easier to see what's going on with the poles of this function. The resonant frequency ω_0 (in radians per second) is

$$\omega_0 = \frac{1}{\sqrt{LC}}, \tag{6.13}$$

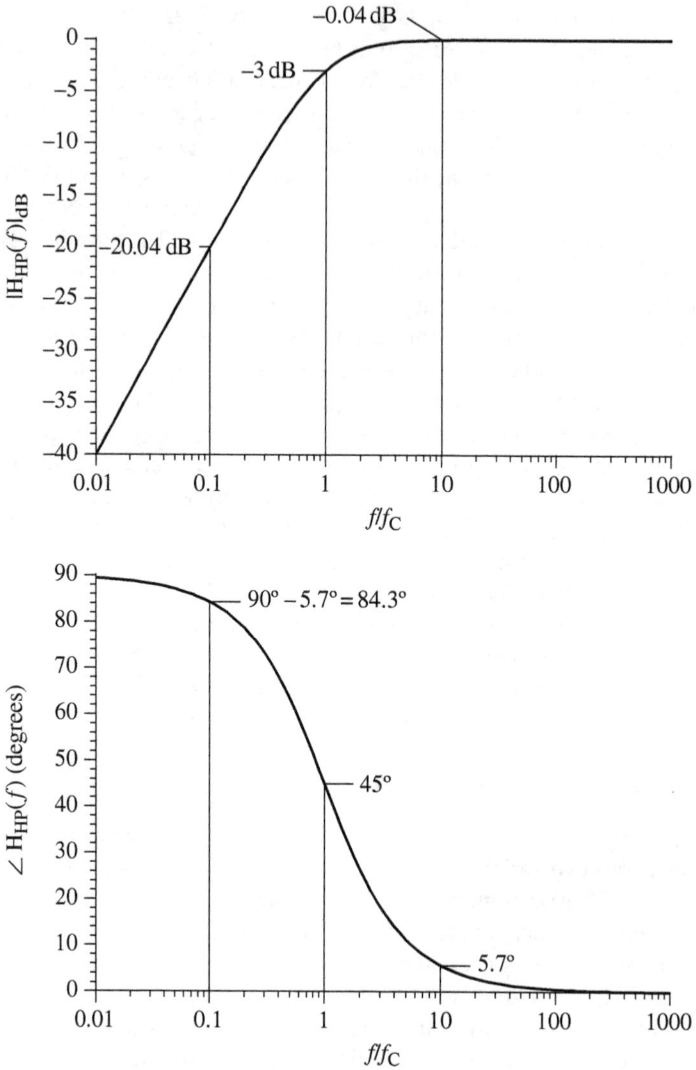

FIGURE 6.13 Magnitude and phase response of one-pole, one-zero highpass filter shown in Figure 6.11, with frequency axis normalized to f_C.

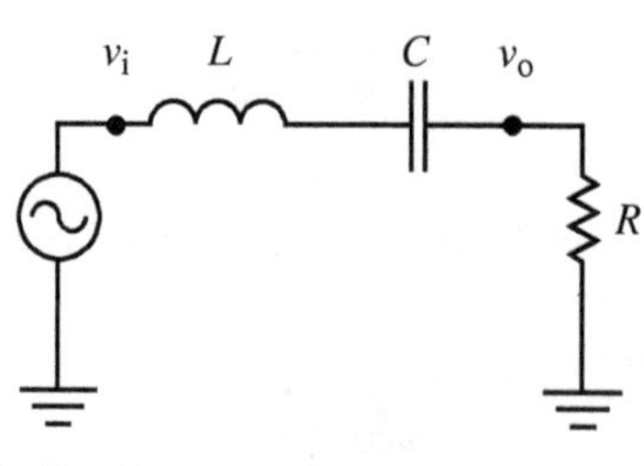

FIGURE 6.14 Idealized bandpass filter consisting of series L–C–R circuit.

which is derived from the familiar resonant-frequency formula

$$f_0 = \frac{1}{2\pi\sqrt{LC}} \tag{6.14}$$

that expresses the frequency f_0 (in Hz) at which the reactances of the capacitor and inductor are equal in magnitude and opposite in sign. A series L–C circuit with *ideal* reactances (no resistance or losses) has an impedance of zero at resonance.

But all real reactances have loss, which we have represented by the resistance R in the idealized circuit of Figure 6.14. Now that loss is included, we can introduce an important concept represented by the symbol Q. Q stands for **quality factor**, and its most general definition is the following ratio:

$$Q \equiv \frac{2\pi\,(\text{energy stored in resonant circuit})}{(\text{energy lost per cycle})} \tag{6.15}$$

It is easy to show that for this circuit, the Q is a function of the value L of the inductance, the value R of the resistance, and the resonant frequency ω_0:

$$Q = \frac{\omega_0 L}{R} \tag{6.16}$$

With the additional variables ω_0 and Q, if we make the substitutions

$$L = \frac{QR}{\omega_0} \tag{6.17}$$

and

$$C = \frac{1}{QR\omega_0}, \tag{6.18}$$

we can recast the response function of Equation 6.12 in a more convenient form:

$$H_{\text{BP}}(s) = \frac{s/\omega_0 Q}{s^2/\omega_0^2 + s/\omega_0 Q + 1} \tag{6.19}$$

The quadratic expression in the denominator has two roots, as the following solution for the quadratic equation shows. If we factor the denominator into the form $(s-p_+)(s-p_-)$, the roots are

$$p_+ = -\frac{\omega_0}{2Q} + \frac{\omega_0}{2}\sqrt{\frac{1}{Q^2} - 4} \tag{6.20}$$

$$p_- = -\frac{\omega_0}{2Q} - \frac{\omega_0}{2}\sqrt{\frac{1}{Q^2} - 4} \tag{6.21}$$

For $Q<2$, the roots are real and lie on the negative real axis. In that case, the circuit thus has two real poles widely separated in frequency, which is a behavior best achieved with two separate filter circuits.

For $Q>2$, the quantity under the square-root (radical) sign is negative, the imaginary constant j appears, and the poles become complex. Note that the poles have identical real parts and their imaginary parts are equal in magnitude but opposite in sign. For Q greater than 2, the poles will lie in the left half of the complex plane, equidistant from the negative real axis. This is another way of saying that the poles are **complex conjugates** of each other (they have equal real parts and imaginary parts of opposite signs).

This circuit is most useful when $Q \gg 2$ (say, 10 or more), because the response of the circuit becomes increasingly **selective** (a selective circuit passes a narrow range of frequencies and rejects all the others). Furthermore, the complex poles are now given by the following simplified expressions:

$$p_+ \,(\text{high } Q) \cong -\frac{\omega_0}{2Q} + j\omega_0 \tag{6.22}$$

$$p_- \,(\text{high } Q) \cong -\frac{\omega_0}{2Q} - j\omega_0 \tag{6.23}$$

There is always a zero at zero frequency for this function in any case, so the pole-zero plot in part of the complex plane for the high-Q case ($Q=10$, in this example) is shown to scale in Figure 6.15. The figure shows that the p_+ pole is very close to the imaginary axis (a distance of only $\omega_0/2Q$) compared to its distance from the real axis (ω_0). This proximity to the imaginary axis for high-Q cases has an important influence on the transfer-function behavior near the resonant frequency ω_0.

Near resonance ($\omega \sim \omega_0$), the vector $s-p_+$ dominates the behavior of the entire function, because its length is much shorter than the length of either the s vector from the zero or the $s-p_-$ vector from the p_- pole at $(-\omega_0/2Q, -j\omega_0)$. Figure 6.16 shows an enlarged view of what happens as the frequency ω goes through the resonant frequency ω_0.

If we define a 3-dB *bandwidth* $\delta\omega=\omega_0/Q$, we can look at the length of the $s-p_+$ vector for three values of ω: $\omega_1=\omega_0-\delta\omega/2$, $\omega_2=\omega_0$, and $\omega_3=\omega_0+\delta\omega/2$. Clearly, when $\omega=\omega_2=\omega_0$, the $s-p_+$ vector is at a minimum, and because it is in the denominator, the overall transfer function will have its maximum (peak) value very close to ω_0. At the *edges* of the passband defined by ω_1 and ω_3, the geometry works out so that the vector is a factor of $\sqrt{2}$ longer than at its minimum, meaning that the transfer-function magnitude will be about 3 dB down from its maximum at those frequencies. Far away from the resonant frequency, the transfer function will no longer be dominated by the p_+ pole and will instead follow a 20-dB-per-decade slope away from the peak. Figure 6.17 shows the magnitude and phase response of the bandpass filter shown in Figure 6.14, with the frequency scale normalized to the resonant frequency f_0. As the vector diagram in Figure 6.16 leads you to expect, the enlarged insert in Figure 6.17 of the frequency range near the peak shows that the points on the

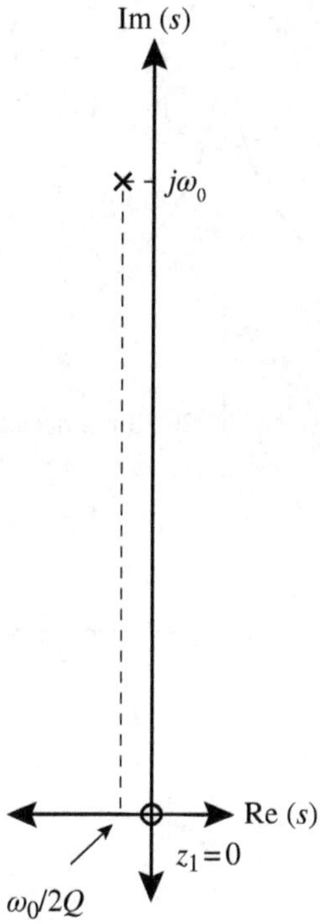

FIGURE 6.15 Complex plane showing Equation 6.19's zero and p_+ pole for $Q=10$.

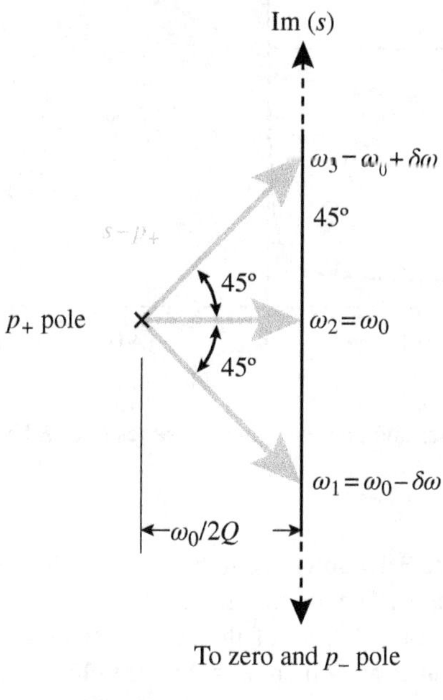

FIGURE 6.16 Enlarged view of neighborhood of p_+ pole as frequency ω passes through ω_0.

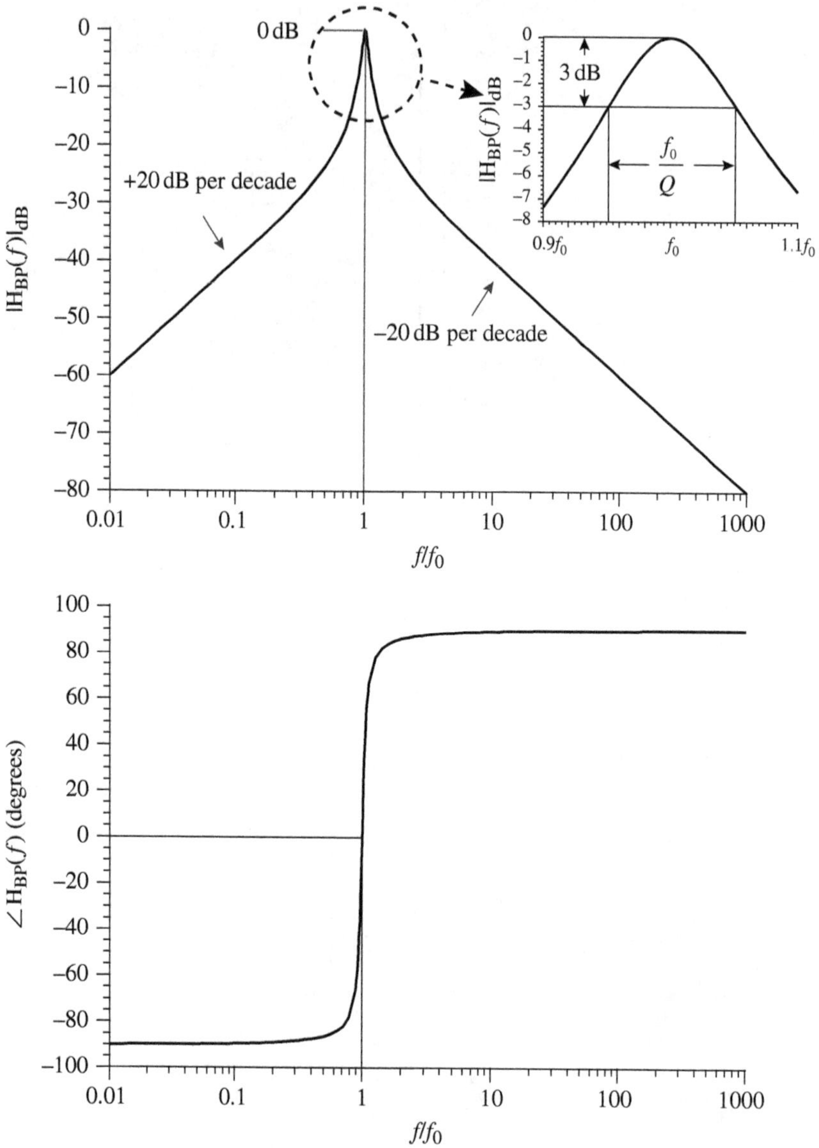

FIGURE 6.17 Magnitude and phase response of series *L–C–R* bandpass filter in Figure 6.14 for $Q = 10$.

magnitude curve that are 3 dB down from the peak lie almost exactly a distance f_0/Q apart on the frequency axis. The maximum occurs at a frequency slightly higher than f_0 itself because of the contribution of the zero's vector to the magnitude, but this difference becomes smaller as Q increases. You should also note that the phase shift encountered through the circuit undergoes a rapid change at the resonant frequency, from nearly –90° far below resonance, through 0° at resonance, to nearly +90° far above resonance. This rapid phase change with frequency will play an important role in the design of oscillators, as we will see in the chapter on signal generation.

The usefulness of a bandpass function such as that shown in Figure 6.17 is clear if you are faced with the problem of obtaining **selectivity** for a single narrow-band signal out of a range of signals at different frequencies. If the resonant frequency f_0 coincides with the desired signal's frequency, all other signals farther away than the 3-dB bandwidth of the bandpass filter will be more or less attenuated. Although digital filters are most commonly used currently for this type of problem, there are still applications where analog filters are the best choice. The main difficulty with passive bandpass filters is that their Q is limited by physical losses in the components. This limitation can be largely overcome through the use of op amps in **active filters**, which we will discuss in the following.

The main lesson to be drawn from the example of the passive *L–C–R* bandpass filter is that a complex pole near the imaginary axis causes a peak in the magnitude of the response function. The higher the pole's Q, the closer it is to the imaginary axis in comparison to its resonant frequency, and the peak becomes sharper (narrower) as Q increases.

6.4.4 Bandstop Filters

Besides bandpass filters that pass a range of frequencies around a center frequency f_0, there are circuits called **bandstop filters** that *reject* a relatively narrow range of frequencies. For example, if the series *L–C* circuit in Figure 6.14 is replaced by a *parallel L–C* pair, the peak (or maximum) in Figure 6.17 becomes a minimum called a **notch** because of its appearance. The frequency of the minimum can be arranged to coincide with a particular undesirable signal, for example. Although bandstop filters are not used as often as bandpass filters, there are situations such as the need to reject interference in an audio system at the power-line frequency where a bandstop filter comes in handy.

6.5 ACTIVE ANALOG FILTERS

As mentioned earlier, passive analog filters always have losses, which means the energy coming out of the filter circuit is always less than the energy that goes in. For simple filters with few components, these losses may be tolerable, but to realize complex or highly precise filter functions with passive components either involves intolerable losses or requires physically large and expensive components or both. For these reasons, **active filters** that incorporate amplifiers as an intrinsic part of the filter circuit are more commonly used for complex analog filter operations.

Whether a lumped-element filter circuit is active or passive, its transfer function can still be expressed as a rational function and is characterized by its poles and zeroes. The use of amplifiers in active filters means that within reason, quite complex functions with many poles and zeroes can be realized without concern that the circuit losses will become excessive, because the loss in passive components can be compensated for with amplification.

The active analog filter circuits we will describe all use op amps, and you should bear in mind that we make certain assumptions in these designs. The op amps are assumed to be ideal and operating well within their bandwidth limitations. If you try to design an analog filter to work at a frequency near the gain–bandwidth product limit of the op amps used, these assumptions break down and you will probably get poor results. So in any actual analog filter design using op amps, you should first make sure that the inherent bandwidth limitations of the op amps are not exceeded in the design.

With these precautions in mind, we will describe one type of active filter circuit that is easy to understand and design: an active lowpass filter using the Sallen–Key circuit.

6.5.1 Sallen–Key Lowpass Filter with Butterworth Response

The **Sallen–Key lowpass filter** is one of a class of active filter circuits first described by R. P. Sallen and E. L. Key in 1955.[2] Figure 6.18 shows a version of this circuit using an op amp, two resistors, and two capacitors. Despite its simplicity, this circuit produces a pair of complex poles that can be used in a variety of lowpass filter designs.

It can be shown that the transfer function of the circuit in Figure 6.18 is

$$\frac{v_o}{v_i} = H_{SK}(s) = \frac{1}{s^2 C_1 C_2 R_1 R_2 + sC_2(R_1 + R_2) + 1} \tag{6.24}$$

In general, this expression will allow you to put its two poles anywhere in the left half of the complex plane, but in order to show you how a particular lowpass filter design procedure works, we will choose values to make the Sallen–Key circuit have a **Butterworth lowpass response**.

A Butterworth lowpass filter (named after its inventor, Stephen Butterworth, who first described it in 1930) has the flattest response at DC for a given number of poles n. In this case, flatness is measured by the derivatives at zero frequency, and the Butterworth response has the most derivatives (first, second, etc.) equal to zero at DC. Recall that a common definition of the passband is the range of frequencies in which the response is within 3 dB of its maximum value. In the case of a Butterworth lowpass filter, the maximum response is at DC, so the cutoff frequency ω_C (in radians per second), which defines the upper edge of the passband, is the frequency at which the response is 3 dB lower than at DC (the lower edge of the passband).

The first step in any filter design is to decide what the requirements are. For many applications, the most important characteristics of a lowpass filter are:

1. *Flatness in passband.* As we mentioned, the Butterworth response has the flattest possible shape at DC for a given number of poles. As long as your design can tolerate a variation of 3 dB below the DC response, the Butterworth

[2]Sallen, R. P. and Key, E. L. "A practical method of designing RC active filters," *IRE Transactions on Circuit Theory* (March 1955), 2:74–85.

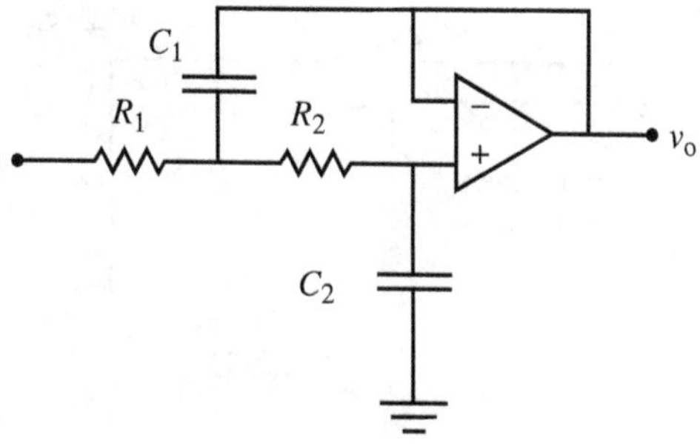

FIGURE 6.18 Sallen–Key lowpass active filter circuit.

response will do fine. The Butterworth response falls off smoothly with no ripples to the −3 dB passband edge. Other types of responses (notably one called the Chebyshev, named after the type of polynomial used in its design) allow a certain amount of **ripple** (up-and-down variations) in the passband. Allowing some ripple enables the filter's response to fall off faster in the initial part of the stopband than the Butterworth response's falloff rate in that region. Still other types of filter responses provide for minimum or linear phase shift with frequency in the passband, specific high-rejection frequencies in the stopband, or other special features. The references given at the end of this chapter have more information on these and other specialized types of filter responses.

2. *Steepness of cutoff.* The rate at which the filter response falls outside the passband is determined by the number of poles and their arrangement. Close to the passband edge, the Butterworth filter's rejection does not increase as rapidly as a Chebyshev filter's would. This is important if there are signals just above the passband edge that need to be rejected, for example.
3. *Rejection in stopband.* At frequencies far above the edge of the passband, the response of a poles-only lowpass filter (no zeroes) with n poles falls at a rate of $20n$ dB per decade, regardless of the type of response form. So if a wide range of signals in the stopband need a certain amount of rejection, the number of poles required is determined by the lowest frequency to be rejected and the amount of rejection. On the other hand, if the signal to be rejected is fairly narrow in bandwidth, it is possible to design a filter with a zero that coincides with the center frequency of the signal to be rejected.

Many filter requirements can be conveniently described graphically on a Bode-plot style of graph by drawing **specification zones** on the graph. For example, suppose that we have an audio-frequency application in which the passband must extend from 0 to 10 kHz with no more than 3-dB variation and must reject signals at 20 kHz or above by at least 20 dB. The way to express these requirements with specification zones is shown in Figure 6.19. The requirement that the response must fall by no more than 3 dB from 0 dB up to 10 kHz is expressed by the left-hand "forbidden

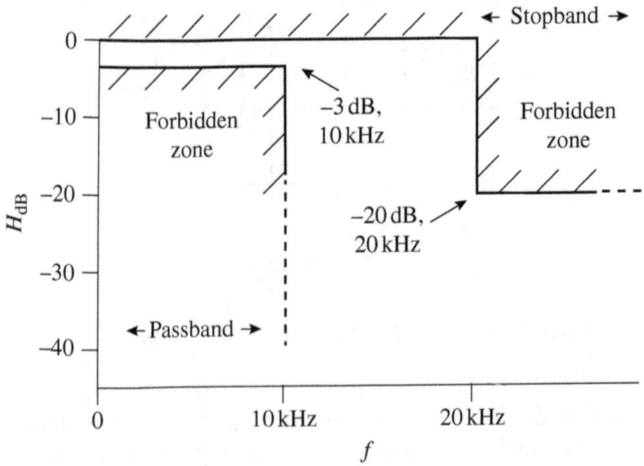

FIGURE 6.19 Specification zones for lowpass filter with 3-dB passband up to 10 kHz and stopband above 20 kHz.

zone," which means that the actual response magnitude cannot cross the line into the forbidden zone at any point. The requirement of at least 20-dB rejection above 20 kHz is expressed by the right-hand forbidden zone. In principle, any magnitude response function that avoids these forbidden zones will be acceptable.

You may recall how we plotted certain frequency responses using a dimensionless normalized frequency—that is, the actual frequency f divided by a normalizing frequency f_C. It turns out that the standardized types of filter responses are most convenient to design with a normalized cutoff frequency (in radians per second) of $\omega_C = 1$. The reason is that once such a **canonical filter design** is developed, it is easy to **frequency-scale** the design to fit the actual frequency requirements of the particular design you are doing. We will show how this works in the following design exercise.

Because the calculations for the Butterworth filter are fairly straightforward, we will design a Butterworth filter to meet the requirements shown in Figure 6.19. It turns out that because $\log_{10}(2) = 0.3$, the rate per *octave* (factor of 2 in frequency) at which an n-pole filter response falls in the stopband is $(0.3)(20) = 6$ dB. The ratio of 20 kHz to 10 kHz in Figure 6.19 is exactly one octave, so in order for the response to fall by 20 dB in that frequency range, we can estimate the number of poles needed by dividing 20 by 6: $20/6 = 3.3$. You can't have a fractional number of poles, so we choose the next larger integer, $n = 4$, for our design.

To design a canonical Butterworth lowpass filter with $\omega_C = 1$, the poles must be placed on a semicircle of radius 1 centered on the origin and must be spaced in a certain symmetrical way. This is a consequence of solving for the complex roots of the denominator polynomial that will give the maximally flat Butterworth magnitude response function

$$|H_{BW}| = \frac{1}{\sqrt{1 + (\omega)^{2n}}}, \tag{6.25}$$

With the additional conditions that the two resistors are equal to 1 Ω each, the two capacitor values in the first Sallen–Key circuit for the $k=1$ pair of poles are easily found to be

$$C_2(k=1) = -\cos\left(\frac{5}{8}\pi\right) = 0.38268\,\text{F} \tag{6.31}$$

$$C_1(k=1) = \frac{1}{C_2(k=1)} = 2.6131\,\text{F} \tag{6.32}$$

For the second Sallen–Key circuit, we simply advance the constant k to 2 in Equation 6.30 and find that

$$C_2\,(k=2) = -\cos\left(\frac{7}{8}\pi\right) = 0.92388\,\text{F} \tag{6.33}$$

$$C_1\,(k=2) = \frac{1}{C_2(k=2)} = 1.0824\,\text{F} \tag{6.34}$$

Because there are an even number of poles, no real pole is needed, and the entire filter circuit will consist of two Sallen–Key lowpass filters in cascade (one after the other).

This completes the **canonical** or **prototype** filter design using $\omega_C = 1$. The next step in the design is to **frequency-scale** the circuit. Frequency scaling is the operation of "denormalizing" the circuit in order for the cutoff frequency to be changed to the actual desired value, instead of leaving it at $\omega_C = 1$. Recalling that the 3-dB-down frequency of our specification graph in Figure 6.19 is $f_C = 10\,\text{kHz}$, we will denote the corresponding radian frequency with a prime (′) to distinguish it from the prototype's cutoff frequency of $\omega_C = 1$:

$$\omega_C' = 2\pi f_C = 62.832 \times 10^3\,\text{rad s}^{-1} \tag{6.35}$$

Presented below is the set of equations that constitute the frequency-scaling rule. If the original (canonical or prototype) filter circuit's cutoff frequency is ω_C and the design cutoff frequency is ω_C', the relationships between the prototype's original component values R, C, and L and the specific designed circuit's values R', C', and L' are

$$\frac{C'}{C} = \frac{L'}{L} = \frac{\omega_C}{\omega_C'} \tag{6.36}$$

$$\frac{R'}{R} = 1 \tag{6.37}$$

In words, the capacitors and inductors (if any) are scaled *down* by the ratio of prototype cutoff frequency to design cutoff frequency, and the resistor values are unchanged.

This means that unless your filter design has a cutoff frequency of less than 1 rad s^{-1}, your design values for C' and L' will be smaller than the prototype values.

So far, all resistors in the design circuit still have the same value as in the prototype circuit, namely, 1 Ω. This is an inconveniently small value for most practical designs, so the final step in our design procedure is an **impedance-scaling** step. Impedance scaling has no effect on the frequency response of an active filter design (though it will affect the response of a passive filter with losses). But it allows the designer to adjust the values of resistors and capacitors so that commonly available values can be used. Again, the choice of a scaling value is up to the designer, so we will choose to impedance-scale the values to be larger by a factor of $Z'/Z = 10{,}000$, which will transform all the 1-Ω resistors into 10-kΩ resistors. We must also impedance-scale the capacitor values (and those of inductors, if any) as well. We will denote the final design values for components with double primes (″), and so the impedance-scaling rule using this notation is

$$\frac{R''}{R'} = \frac{L''}{L'} = \frac{C'}{C''} = \frac{Z'}{Z} \tag{6.38}$$

Note that when impedance is scaled *up* as in this example, values of resistors and inductors become *larger* but capacitor values become *smaller*, because a smaller value of capacitance has a larger impedance at a given frequency. When this final scaling operation is performed, the resulting values for all components are given in Table 6.1.

The initial design circuit with all component values labeled is shown in Figure 6.21. You can see that the circuit is DC coupled, which means that at DC it will act as a unity-gain buffer amplifier. But as the input frequency increases, the circuit's gain should fall to −3 dB at 10 kHz and eventually roll off at (4)(20) = 80 dB per decade at higher frequencies.

You will note that the capacitor values in Figure 6.21 are rounded off to only two significant figures. Even this level of accuracy is optimistic, because precision capacitors are rated with tolerances of only ±5% or so unless they are very expensive, and cannot be obtained with arbitrary component values. This fact leaves the designer with two choices: either (1) accept the slight deviations from a perfect filter response that will result from using capacitor values that are not exactly the calculated ideal ones or (2) recalculate the component values by fixing the capacitor values to be ones

TABLE 6.1 Normalized (prototype) and Unnormalized Capacitor Values for 4-pole Sallen–Key Filter Design for $\omega'_C = 2\pi\,(10\,\text{kHz})$

Component	Prototype value	Frequency-scaled value	Impedance-scaled value
$C_1(k=1)$	2.6131 F	41.589 μF	4.1589 nF
$C_2(k=1)$	0.38268 F	6.0905 μF	609.05 pF
$C_1(k=2)$	1.0824 F	17.227 μF	1.7227 nF
$C_2(k=2)$	0.92388 F	14.704 μF	1.4704 nF
$R_{1,2}(k=1,2)$	1 Ω	1 Ω	10 kΩ

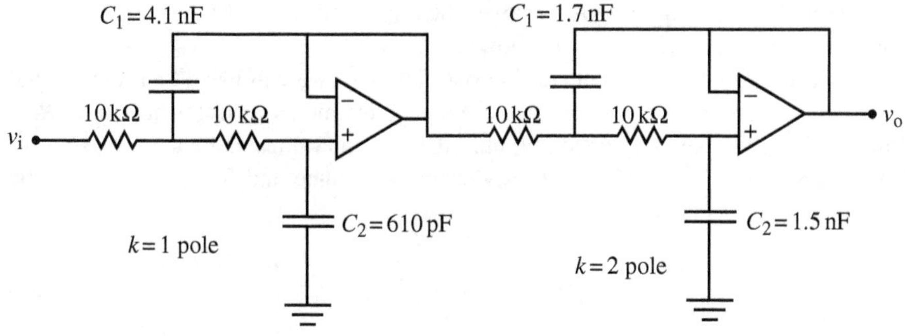

FIGURE 6.21 Initial $n=4$ Sallen–Key lowpass filter design for $f_C = 10$ kHz.

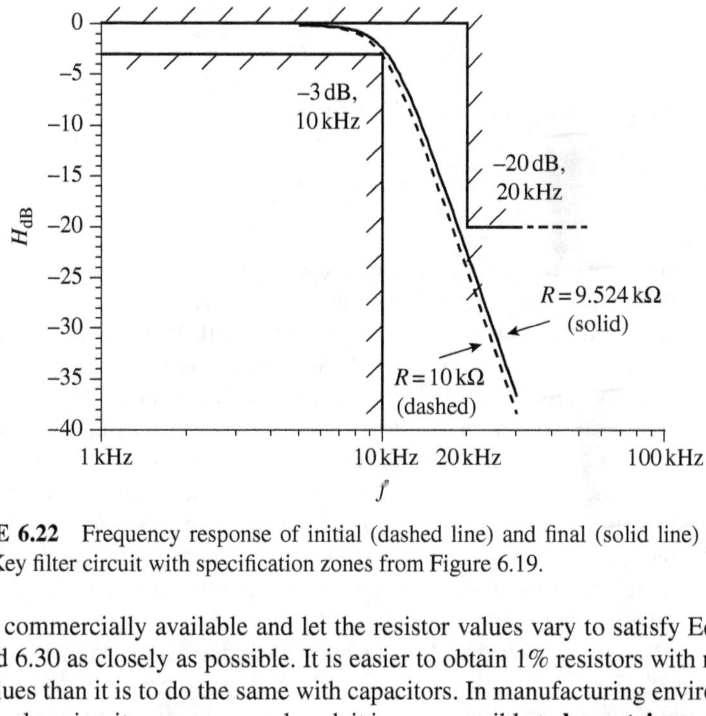

FIGURE 6.22 Frequency response of initial (dashed line) and final (solid line) cascaded Sallen–Key filter circuit with specification zones from Figure 6.19.

that are commercially available and let the resistor values vary to satisfy Equations 6.29 and 6.30 as closely as possible. It is easier to obtain 1% resistors with nonstandard values than it is to do the same with capacitors. In manufacturing environments where analog circuits are mass-produced, it is even possible to **laser-trim** resistors to arbitrary values by measuring the circuit's performance and adjusting the resistors with a laser to meet specifications more precisely. However, most filter applications are not so critical that unacceptable errors are encountered by the use of ±5% tolerance components for all resistors and capacitors.

In Figure 6.22, we have replotted the original specification zones from Figure 6.19 along with the actual calculated response of the circuit in Figure 6.21 as a dashed line. Because we specified the −3 dB cutoff frequency to be *exactly* 10 kHz, we should not be surprised that when we rounded off the capacitor values from their exact calculated quantities, the cutoff frequency spec was not quite met. As you can see, the dashed line slightly crosses over the corner of the −3 dB specification zone but easily avoids the −20 dB zone's limits.

To fix this problem without starting over entirely, we can use a handy trick that arises from the fact that you can make minor adjustments in frequency to an R–C filter (active or passive) by changing only the resistor values and leaving the capacitor values alone. If we simultaneously scale impedance and frequency so that the following equation is obeyed:

$$\frac{Z'}{Z} = \frac{\omega_C}{\omega_C'}, \tag{6.39}$$

the capacitor values will remain unchanged:

$$C'' = \frac{\omega_C}{\omega_C'}\frac{Z}{Z'}C = \frac{\omega_C}{\omega_C'}\frac{\omega_C'}{\omega_C}C = C, \tag{6.40}$$

but the resistors will scale inversely with frequency:

$$R' = \frac{\omega_C}{\omega_C'}R \tag{6.41}$$

A rough estimate of how far the initial (dashed) response curve in Figure 6.22 must move in order to clear both the passband and the stopband forbidden zones says that moving the cutoff frequency about 5% higher will do it. If we multiply the initial resistor values (10 kΩ) by 1/1.05 = 0.9524, we obtain resistor values of 9.524 kΩ. When these values are substituted for the initial ones in Figure 6.21 and the same capacitor values used, the response function shown as a solid line in Figure 6.22 results. As you can see, it meets both the passband and stopband specifications, though without much error margin. If components with larger tolerances than ±5% will be used and the specifications are critical, it might be best to use an $n = 5$ filter to ensure that the specifications are met for the worst-case combination of components. The subject of tolerances and specification errors is an important one but lies outside the scope of this chapter.

Although the Sallen–Key circuit can be reconfigured to act as a high-pass filter by exchanging R's for C's and vice versa, a different type of active filter circuit with only one additional op amp can act as either a lowpass, highpass, or bandpass filter. We will describe that circuit next.

6.5.2 Biquad Filter with Lowpass, Bandpass, or Highpass Response

Once a lowpass filter design is in place, there is a mathematical way to transform a lowpass filter function into either a highpass filter with a different cutoff frequency or a bandpass filter with an arbitrary center frequency and an adjustable **fractional bandwidth.** The fractional bandwidth is a dimensionless number that is the ratio of the bandpass filter's bandwidth $\delta\omega$ (typically measured between the −3 dB points) to the center frequency ω_0 of the filter. (If the filter in question has a response that goes unattenuated to DC, the lower limit of the bandwidth is taken to be 0 Hz.) Thus defined, the fractional bandwidth for the simple passive *L–C–R* filter in Figure 6.14 is the inverse of the *Q* factor we mentioned earlier:

$$\text{Fractional bandwidth} \equiv \frac{\delta\omega}{\omega_0} = \frac{1}{Q} \tag{6.42}$$

Suppose ω_C is the cutoff frequency of a given lowpass filter with an arbitrary number of poles arranged to produce a certain response. You can always write the transfer function in terms of the imaginary frequency variable $j\omega/\omega_C$, although it may take some algebraic manipulation to do so. Once the lowpass transfer function is in that form, you can convert the function to that of a highpass filter with a different cutoff frequency ω'_C by performing the substitution

$$\frac{j\omega}{\omega_C} \rightarrow \frac{\omega'_C}{j\omega} \tag{6.43}$$

Everywhere the term $j\omega/\omega_C$ appears in the lowpass transfer function, you simply write $\omega'_C/j\omega$ instead. This operation is called the **lowpass-to-highpass transformation**. The new function will be that of a highpass filter instead of a lowpass filter but will have the same response at ω'_C that the original lowpass filter did at ω_C.

Similarly, suppose you want a bandpass filter that behaves the same either side of a center frequency ω_0 that the lowpass filter behaves with respect to DC and has a fractional bandwidth corresponding to a certain value of *Q*, in accordance with Equation 6.42. The **lowpass-to-bandpass transformation**

$$\frac{j\omega}{\omega_C} \rightarrow Q\left(\frac{j\omega}{\omega_0} + \frac{\omega_0}{j\omega}\right) \tag{6.44}$$

will convert the lowpass filter function into a bandpass filter function. In terms of the poles of, say, a Butterworth lowpass filter, which all appear on a unit semicircle centered at $j\omega = 0$ for a normalized lowpass filter, the poles of the bandpass filter will now lie on two semicircles centered around $\pm j\omega_0$ on the imaginary axis.

Just having the transfer functions does not tell you how to design a filter circuit. But the following active filter circuit, which is called a **biquad filter**, will allow you to realize lowpass or bandpass functions with the same circuit by simply taking the output from different points.

Figure 6.23 shows one version of the biquad circuit. (The name **biquad** comes from the fact that both the numerator and denominator of the circuit's transfer function can be quadratics in *s*.) This particular version provides either a lowpass or a bandpass function, depending on which output is used (v_{BP} or v_{LP}). What follows is an analysis aimed at yielding a transfer function for the biquad in the same form we obtained for the passive *L–C–R* bandpass filter of Figure 6.14.

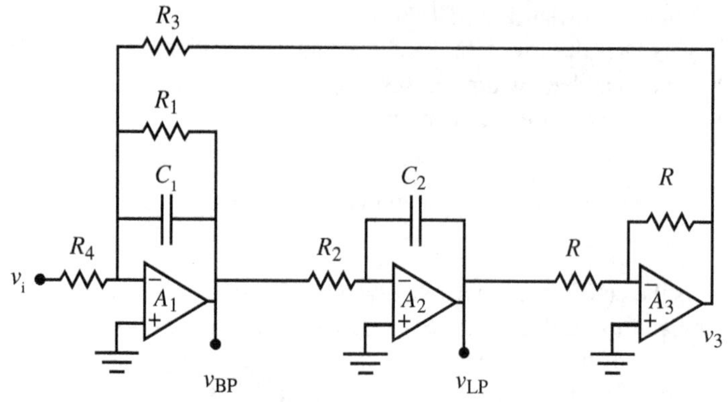

FIGURE 6.23 Biquad active filter with input v_i and outputs v_{BP} for bandpass function and v_{LP} for lowpass function.

First, we realize that with equal resistors R as the input and feedback elements of amplifier A_3, that op amp acts as an inverter with a gain of -1, meaning $v_3 = -v_{LP}$. Next, it is easy to see that A_2 acts as an ideal integrator, leading to the following relation between v_{BP} and v_{LP}:

$$v_{LP} = -\frac{v_{BP}}{sR_2C_2} \tag{6.45}$$

Next, we see that the negative-feedback principle allows us to assume the inverting input of A_1 is a virtual ground. Because we have variables for the voltages at the far ends of each component that join at that input, we can use Kirchhoff's current law to say that the sum of the voltages at the inverting input is zero (assuming the op amp is ideal, which we always do for a first-cut analysis). That gives us the following equation:

$$\frac{v_i}{R_4} + \frac{v_{BP}}{sR_2C_2}\frac{1}{R_3} + v_{BP}\left(\frac{1}{R_1} + sC_1\right) = 0 \tag{6.46}$$

Equation 6.46 can be solved for the ratio of v_{BP} to v_i, and in terms of the components in Figure 6.23, the expression is rather messy:

$$\frac{v_{BP}}{v_i} = \frac{-R_4}{sC_1 + (1/R_1) + (1/sR_2R_3C_2)} \tag{6.47}$$

What we would like to do is to get Equation 6.47 into a form similar to the passive *L–C–R* filter's transfer function (Eq. 6.19), which we could then modify in order to put the poles where we want them. This can be done if we put some constraints on the component values, namely, letting

$$C_1 = C_2 = C \tag{6.48}$$

$$R_4 = R_1 \tag{6.49}$$

$$R_3 = R_2 \tag{6.50}$$

and defining

$$Q = \frac{R_1}{R_2} \tag{6.51}$$

and

$$\omega_0 = \frac{1}{R_2 C} \tag{6.52}$$

With these constraints, it is relatively straightforward to show that the response from the input to the bandpass output becomes

$$\frac{v_{\mathrm{BP}}}{v_i} = -\frac{s/\omega_0 Q}{(s/\omega_0)^2 + s/\omega_0 Q + 1}, \tag{6.53}$$

which is exactly the same as Equation 6.19 for the passive *L–C–R* bandpass filter, except for the minus sign.

Situations may arise in which you start the design with fixed values for capacitors C_1 and C_2 (not necessarily the same) and wish to adjust the resistor values to satisfy requirements for gain, bandwidth, and Q. In that case, equations developed by one of the inventors of the biquad circuit, James Tow, will come in handy.[3] Suppose that you wish to synthesize the general bandpass response

$$\frac{v_{\mathrm{BP}}}{v_i} = \frac{cs}{s^2 + as + b}, \tag{6.54}$$

where a, b, and c are known from the requirements of pole location and overall gain. A parameter called k_1 can be freely adjusted to set the value of a particular resistor, for example, without affecting the overall response. Once C_1, C_2, and k_1 are determined, the remaining resistor values to obtain the response in Equation 6.54 are explicitly given by the following equations:

$$R_1 = \frac{1}{aC_1} \tag{6.55}$$

$$R_2 = \frac{k_1}{C_2\sqrt{b}} \tag{6.56}$$

[3]Tow, J. "A step-by-step active filter design," *IEEE Spectrum* (Dec. 1969), 6:64–68.

$$R_3 = \frac{1}{k_1 C_1 \sqrt{b}} \tag{6.57}$$

$$R_4(\text{BP}) = \frac{1}{cC_1} \tag{6.58}$$

Values for the inverting amplifier resistors R are noncritical as long as they are closely matched to give an accurate gain of −1.

For use as a two-pole lowpass filter, the biquad can be used in a way similar to the Sallen–Key lowpass filter circuit discussed earlier. However, the biquad has the advantage that fairly high values of Q—in excess of 100—can be obtained with good circuit stability. Such high-Q values can be difficult to achieve with the Sallen–Key approach. To obtain the following biquad lowpass response from the input v_i to the output v_{LP}

$$\frac{v_{LP}}{v_i} = \frac{d}{s^2 + as + b}, \tag{6.59}$$

the same Equations 6.56–6.58 should be used for R_1, R_2, and R_3. The only difference (besides taking the output from v_{LP} instead of v_{BP}) is that the value of R_4 is now

$$R_4\,(\text{LP}) = \frac{\sqrt{b}}{k_1 d C_1} \tag{6.60}$$

It is also possible to design a highpass filter in biquad form, although the circuit to do so is slightly more complicated than the one shown in Figure 6.23.

There are many other types of linear op amp circuits that perform filter functions. We have not addressed the question of phase response, although with some signals (especially video and digital signals), a filter's phase response can be just as important as its amplitude response. Poles and zeroes can be arranged to synthesize many types of phase responses as well as given types of amplitude responses, and the same general types of active circuits can be made to deliver the so-called "all-pass" responses whose response magnitude is nearly constant but whose phase varies in a prescribed way.

6.5.3 Switched-Capacitor Filters

A **switched-capacitor filter** is a type of mixed-signal circuit that involves both analog and digital techniques. Originally developed to deal with situations that were difficult to handle with purely analog active filters, switched-capacitor filters now have more competition, because DSP techniques can often perform the same functions better, cheaper, and faster. However, switched-capacitor circuits in packaged integrated-circuit (IC) form have continued to improve, and when faced with a complicated filtering problem, the analog filter designer should consider whether a switched-capacitor solution would help.

The basic idea of a switched-capacitor circuit is that by periodically switching a capacitor between two or more sources, the amount of charge transferred becomes a function of the switching frequency. Unlike fixed values of components such as resistors and capacitors, frequency is a variable that is precisely controllable to an absolute value of a few parts per million or less. If a switched-capacitor design's parameters are mainly dependent on the switching frequency, the system's response becomes as precise as the switching frequency, which can be extremely accurate. Switched-capacitor systems have the disadvantages that the highest usable signal frequency must be considerably less than half of the switching frequency, and the switching frequency itself can appear in the filter's output and cause problems. But depending on the application, these problems can be manageable and can lead to a sophisticated and precise filter design.

Figure 6.24 shows one of the simplest switched-capacitor circuits, one that acts as a lossless resistor. A voltage V_{IN} is sampled at a rate of f_S Hz by a clock waveform ϕ_1 that actuates its respective (electronic) switch whenever it goes high. Clock waveform ϕ_2 has the same frequency but is out of phase with ϕ_1, so that only one switch is on at a time. If the capacitor's voltage before ϕ_1 goes high is V_{OUT}, the amount of charge ΔQ transferred to V_{OUT} when ϕ_2 is high and ϕ_1 goes low is

$$\Delta Q = C_1(V_{IN} - V_{OUT}) \tag{6.61}$$

Because the inverse of the switching frequency f_S is $1/f_S = \Delta t_s$, the time per current transfer, the average current passing between V_{IN} and V_{OUT} is thus

$$\frac{\Delta Q}{\Delta t} = i = fC_1(V_{IN} - V_{OUT}) \tag{6.62}$$

A current whose magnitude is proportional to a voltage difference can be considered to be the result of an effective resistance R_S, where

$$R_S = \frac{1}{fC_1} \tag{6.63}$$

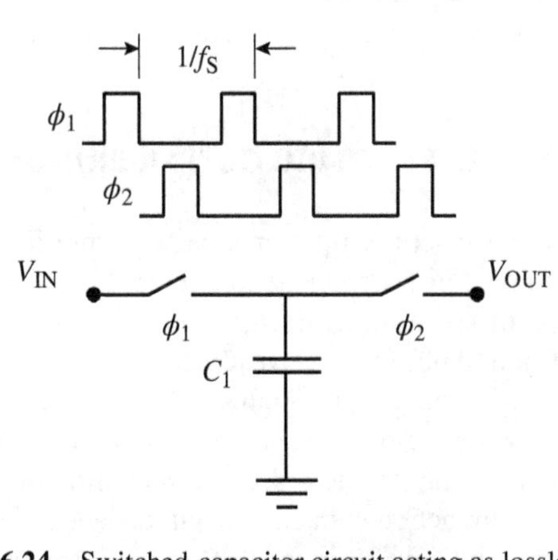

FIGURE 6.24 Switched-capacitor circuit acting as lossless resistor.

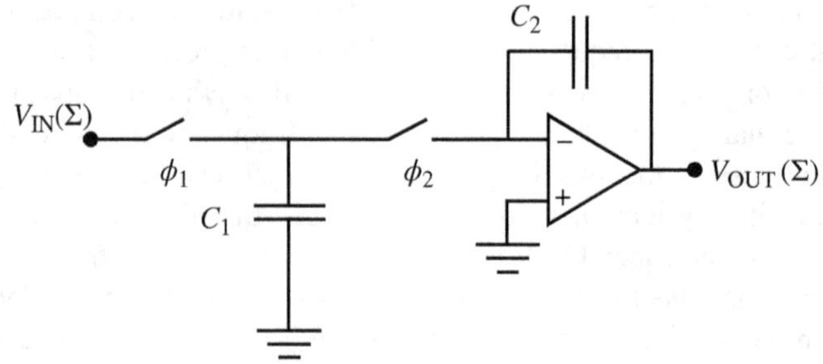

FIGURE 6.25 Switched-capacitor integrator using lossless switched-capacitor resistor of Figure 6.24.

The real advantage of switched-capacitor circuits over purely analog systems with physical resistors and capacitors shows up when we use the switched-capacitor circuit in Figure 6.24 to make a simple one-pole integrator.

Figure 6.25 shows the switched-capacitor "resistor" connected to the input of an op amp integrator using capacitor C_2 as the feedback element. Because the inverting input of the operational amplifier is a virtual ground, the amount of charge passed to C_2 for each cycle is exactly $\Delta Q = C_1 V_{IN}(\Sigma)$ and so the expression for the phasor relation between $v_{IN}(\Sigma)$ and $v_{OUT}(\Sigma)$ is

$$\frac{v_{OUT}(\Sigma)}{v_{IN}(\Sigma)} = -\frac{f_S C_1}{sC_2} \tag{6.64}$$

The corresponding relation for a conventional analog *R–C* integrator would be $-1/sC_2R$, which will vary directly with any variations in C_2 or R. While it is still difficult to achieve absolute precision for the value of any IC capacitor, making capacitor *ratios* precise on the same chip is easily done. Because the constant multiplying the complex variable s in Equation 6.64 is the product of a frequency f_S and a ratio C_1/C_2 of two capacitors, very precise complex filters with low-tolerance design constants can be executed with switched-capacitor circuit techniques.

6.6 DESIGN EXAMPLE: ELECTRIC GUITAR PREAMP

In audio equipment parlance, a **preamp** is an auxiliary amplifier that is used to boost a signal from an instrument or microphone to a level that is compatible with a **line input**, which is a type of signal input designed to deal with levels up to approximately 0 VU, or +4 dBm into 600 Ω. We have already discussed microphone preamps earlier in this chapter. Electric guitar pickups, which work typically by magnetic induction from vibrating steel guitar strings, provide an output signal level that is similar to that of a microphone, on the order of a few millivolts or less, and so preamps are often used in conjunction with electric guitars and similar instruments. It is useful to have a preamp with gain of up to 60 dB to boost the raw pickup output to a level that will be compatible with the line input of a mixer board or power amplifier. In this section, we will set out the specifications for such a preamp, describe the general approach in terms of a block diagram, and then proceed to a detailed design.

The specifications are as follows:

- Minimum voltage gain in passband: 60 dB.
- Low-frequency 3-dB-down cutoff: <10 Hz.
- High-frequency 3-dB-down cutoff: >20 kHz.
- Attenuation (relative to passband) at 35 kHz: >20 dB.
- No more than a fifth-order filter may be used.

Step 1: Block-Diagram Approach. The first step in any amplifier design that incorporates a filter characteristic is to decide on the type of filter response (lowpass, bandpass, or highpass), whether the response to be synthesized should be Butterworth (maximally flat) or a different type, and how many poles and zeroes will be required. Strictly speaking, the characteristic called for is a bandpass response, but the separation between the low-frequency and high-frequency cutoffs is so large (a ratio of 2000:1) that the best approach is probably to design separate lowpass and highpass filters independently and cascade them to obtain the desired response.

The lowpass response characteristic is the most critical, because it involves both a 3-dB-down cutoff frequency of at least $f_C = 20$ kHz and a requirement that the response at $f_A = 35$ kHz must be at least 20 dB below the maximum passband response. To find out whether this characteristic can be satisfied with a Butterworth response using n poles, we can calculate the theoretical responses at 35 kHz for a Butterworth lowpass filter of orders $n = 1$–5 using Equation 6.25, and substituting f_A/f_C for the normalized frequency variable ω, we can solve for $|H|_{dB}$ given n:

$$|H|_{dB} = 20\log_{10}\left(\frac{1}{\sqrt{1+(f_A/f_C)^{2n}}}\right) \tag{6.65}$$

The results of calculating $|H|_{dB}$ for $n = 1$–5 are given in Table 6.2. Because attenuation is the same as loss (in dB, a negative number represents a loss), the numbers listed in Table 6.2 are positive. (If we were listing gain instead of loss, the numbers would be negative.)

TABLE 6.2 Attenuation At 35 kHz of 20-kHz-bandwidth Butterworth Lowpass Filter with *n* Poles

n	Attenuation (dB)
1	6.09
2	10.16
3	14.73
4	19.49
5	24.32

Clearly, the minimum number of poles is 5, because $n=4$ does not quite provide 20 dB of attenuation at 35 kHz. Adding in tolerances and the fact that designing for a 3-dB cutoff of exactly 20 kHz will result in missing the specification half the time, the $n=5$ choice is clearly the only one that can work.

Allowing for a ±5% variation in actual cutoff frequency due to parts tolerances, let's find out if the $n=5$ design will still work even if f_C is purposely designed to be 21 kHz (20 kHz × 105%). Using $f_C=21$ kHz, $f_A=35$ kHz, and $n=5$, we find that Equation 6.65 predicts an attenuation of 22.21 dB, still within spec. If the specifications were any tighter, we might go to $n=6$, but $n=5$ will do for noncritical applications.

The next step is to design the filter. For an $n=5$ prototype filter ($\omega_C=1$), we will need two pairs of complex poles and one real pole (see the $n=5$ diagram in Fig. 6.20). The block diagram of the planned circuit is shown in Figure 6.26. Note that the first part of the circuit that the incoming signal encounters is the filter portion. In general, it is a good idea to put whatever filter functions are needed at or near the input of the system. In this way, undesirable out-of-band signals are reduced or eliminated by the filter before they can be amplified significantly and contribute to noise or distortion.

To arrive at actual component values for the Sallen–Key filter circuits, some iteration is needed between the ideal values called out for by Equations 6.29 and 6.30 and what is available in the form of standard-value capacitors and resistors.

The tolerance rating of a component's value represents the manufacturer's guarantee that the actual value of a given component lies within the tolerance range around the nominal value. For example, a 1 − μF capacitor with a tolerance of ±10% can in principle measure anywhere from 0.9 to 1.1 μF and still meet the manufacturer's specification. Because it would be pointless to manufacture many classes of nominal values that would overlap significantly, parts with larger tolerances are available in fewer nominal values. For example, resistors and capacitors with 20% tolerances are generally available in only the following values (and their multiples of 10): 10, 15, 22, 33, 47, 68, and 100. For 10% tolerance components, the preceding values plus the following are available: 12, 18, 27, 39, 56, and 82. For 1% tolerance components, which are rather costly and often unavailable except for resistors, almost any value with two significant figures can be obtained.

We will bear these guidelines in mind as we select capacitor and resistor values for the design. We will assume that available capacitors have 10% tolerance, while available resistors can be chosen for any decimal multiple of a two-digit integer (e.g., 22, 23, 24, …). Recall that the final value of a component in a filter circuit is obtained by both frequency scaling and impedance scaling. While the frequency-scaling

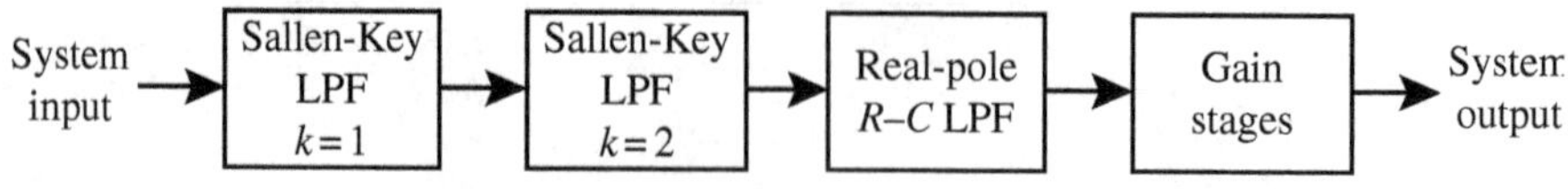

FIGURE 6.26 Block diagram of electric guitar preamp, showing filter stages followed by gain stages.

factor is determined by the frequency-response specifications and cannot be altered, the impedance-scaling factor is at the discretion of the designer. We will use the impedance-scaling factor to adjust the final capacitor values so that we can obtain standard-value capacitors that are reasonably close to the theoretically exact design values.

The frequency-scaling factor is determined by the 3-dB cutoff frequency f_C. For the tolerance reasons explained previously, we will set f_C at 5% above the exact specification, or 21 kHz. This gives a frequency-scaling factor

$$\frac{\omega_C}{\omega'_C} = \frac{1}{2\pi(21\times10^3)} = 7.5788\times10^{-6} \tag{6.66}$$

After some trial and error, we arrived at a value for the impedance-scaling factor of $Z'/Z = 2200$. The resulting theoretical capacitor values of the initial $n=5$ prototype filter, the frequency-scaled filter, and the impedance- and frequency-scaled filter are shown in Table 6.3. Capacitor C_3 is the C of the passive R–C real-pole lowpass filter, which has $R=1\,\Omega$ and $C_3=1\,\text{F}$ in the prototype design.

When we tried using the values in the "actual value selected" column in Multisim™, we found that the 20-dB attenuation frequency was about 5% too high. Accordingly, we left the capacitor values alone and raised the resistor values from 2.2 to 2.3 kΩ. With this slight change, the frequency response made both specifications for the lowpass characteristic, and we proceeded with the remainder of the design. The complete circuit design is shown in Figure 6.27, where amplifiers A_1 and A_2 are used in the two cascaded Sallen–Key lowpass filters.

Sixty decibels (a voltage factor of 1000) is too large to attempt in a single op amp stage, so we decided to split the total gain needed between two stages (amplifiers A_3 and A_4). Choosing standard-value resistors of 1 kΩ for the shunt element to ground and 47 kΩ and 22 kΩ, respectively, for the first and second gain stages provides a total voltage gain of (48)(23) = 1104, about 10% above the minimum required level. This will provide a reserve of gain in the case that the actual resistor values are unfavorable for gain.

In developing this type of circuit, the designer faces a choice in simulating the circuit. One can use a full-scale op amp model of the particular device chosen (e.g., a model of a specific brand of op amp such as the LM741) or else use a simplified

TABLE 6.3 Capacitor and Resistor Values for $n=5$ Lowpass Filter

Component	Prototype value	Frequency-scaled value	Impedance- and frequency-scaled value	Actual value selected
$C_1(k=1)$	3.2361 F	24.526 μF	11.148 nF	10 nF
$C_2(k=1)$	0.30902 F	2.342 μF	1.0645 nF	1 nF
$C_1(k=2)$	1.2361 F	9.3682 μF	4.258 nF	3.9 nF
$C_2(k=2)$	0.80902 F	6.1314 μF	2.787 nF	2.7 nF
R	1 Ω	1 Ω	2.2 kΩ	2.2 kΩ

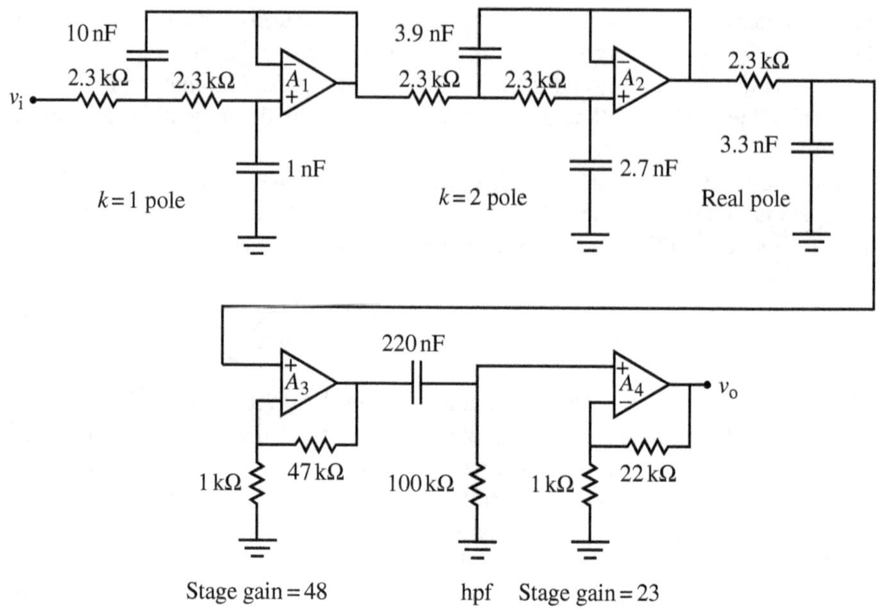

FIGURE 6.27 Complete electric guitar preamp design using block diagram of Figure 6.26.

generic **virtual op amp**. Unless a particular device's characteristics need to be modeled in the design for reasons of noise or nonlinear modeling, it is easier and faster to use virtual op amp models. The simplified equivalent circuits in virtual op amp models run faster in simulation and provide results that are usually very close to what the actual circuit will do.

Virtual op amps do not show effects of offset voltage, but because real ones do, we have placed a circuit labeled "hpf" between amplifiers A_3 and A_4. This circuit has two purposes. First, it provides a low-frequency −3-dB cutoff of about 7.2 Hz, safely below the specified value of 10 Hz or less. Second, it blocks any DC arising from the input offset voltage of A_3 so that it will not be further amplified in the second gain stage (A_4).

A Multisim™ calculation of the small-signal gain magnitude of the overall system of Figure 6.27 is shown in Figure 6.28, along with the specification zones setting out the requirements for the response to meet. As you can see, the design does meet the specifications but just barely. If this were a "one-off" special-purpose design for a unique application, such a close shaving of specifications might be acceptable. But for large-quantity production, the designer would need to take into account the possible variations of all components, observe their effects on the response, and design the circuit so that it would meet all specifications for any combination of component tolerances. While this is an important step, we have carried this design as far as we can without more information on the details of manufacturing variations and other factors affecting the specifications.

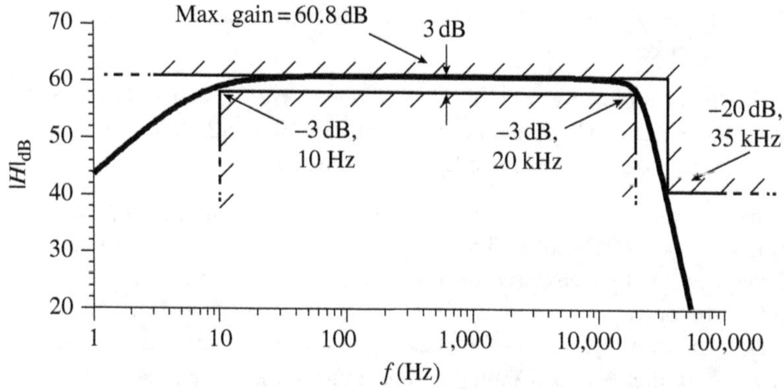

FIGURE 6.28 Gain magnitude response (in dB) from Multisim™ simulation of the circuit of Figure 6.27.

BIBLIOGRAPHY

Bianchi, G., and R. Sorrentino. *Electronic Filter Simulation & Design*. New York: McGraw-Hill, 2007.

Hollister, A. L. *Wideband Amplifier Design*. Raleigh, NC: SciTech Publishing, 2007.

PROBLEMS

Note: Problems of above-average difficulty are marked with an asterisk (*).

6.1. *Sound pressure level (SPL), noise floor, and saturation ceiling.* Table 6.4 gives typical SPL in dB relative to 0 dB = 20 μPa for various types of sounds. Assume all levels are rms.

Suppose a microphone having a sensitivity of 2.5 mV Pa^{-1} is used to sense each of these sounds. The microphone is connected to a preamp that has 40 dB gain and a maximum output level before unacceptable saturation of 2 V (rms). When the microphone is connected to the amplifier and placed in an absolutely silent place, the microphone–preamp combination produces a noise floor equivalent to an input noise voltage of 5.7 μV. For each of the six sounds in Table 6.4, determine if the signal produced by the sound is (a) above the noise

TABLE 6.4 Typical Sound Pressure Levels for Various Types of Sound[a]

Type	SPL (dB)
Breathing	10
Quiet conversation at 1 m	40
TV at 1 m	60
Road traffic at 30 m	85
Jackhammer at 1 m	100
Jet engine at 30 m	150

[a] Source: Wikipedia article "Sound Pressure Levels."

floor of the system (yes or no?) and (b) is below the saturation ceiling of the system (yes or no?).

6.2. *Noise floor, saturation ceiling, and dynamic range.* Solve Problem 6.1, but assume the preamp has a gain of 65 dB instead of 40 dB. Assume the increase in gain has no effect on the noise level of the system.

6.3. *Power-supply bypassing*. Assume a power amplifier and preamp are both powered by a +12 VDC source. The DC power source is connected to the amplifiers through wires that have a total resistance (supply and return leads) of 0.9 Ω. When the power amplifier supplies its peak output of 30 W, it draws a current waveform whose peak value is 3.2 A. The no-load current consumption of the power amplifier is 200 mA and the (constant) drain of the preamp is 15 mA. The preamp will operate satisfactorily with any voltage from 8 to 15 VDC.

(a) Calculate the peak-to-peak voltage V_P that will appear at the power amplifier's DC supply terminals when its current drain varies from the no-load value to the peak load value. Assume there is no filter or bypass capacitor at the power amplifier supply terminals.

(b) Assuming that the worst-case coupling through the power supply occurs at 100 Hz, design an *R–C* power-supply bypass circuit to be installed between the power amplifier supply terminals and the preamp supply terminals. Choose the value of *R* so that at least 8 V appears across the preamp supply terminals when 10 V (average) appears across the power amplifier supply terminals. Choose the value of *C* so that at least 40 dB of suppression at 100 Hz is provided by the bypass circuit.

6.4. *Poles and zeroes of rational function.* Find the values of the two poles p_1 and p_2 and two zeroes z_1 and z_2 in the following rational function: $H(s) = (s^2 + 4s + 3) / (s^2 + s + 2.5)$. (*Hint*: The poles are complex.)

6.5. *R–C circuit with zeroes on imaginary axis*. Figure 6.29 shows a circuit called a **Wien bridge** (the name Wien is pronounced "veen"), which is used in oscillator circuits because of its bandstop frequency response when measured from the input v_i to the differential output at v_o.

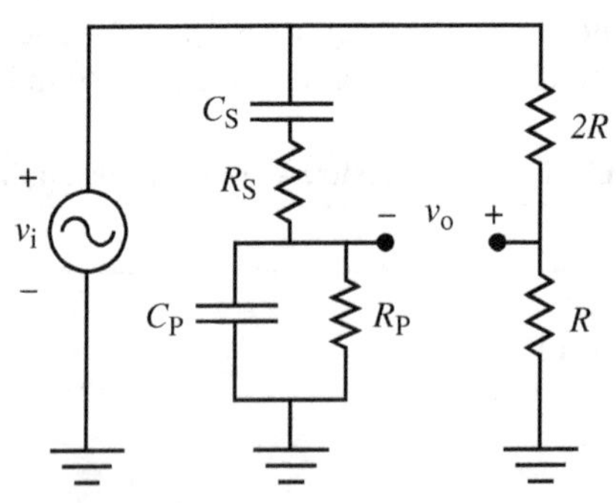

FIGURE 6.29 Wien-bridge circuit exhibiting a bandstop frequency response.

In common with other bridge circuits, the output v_0 will be zero when the ratio of impedances in each side of the bridge is balanced. In the particular case shown, this means that if Z_S is the impedance of the series circuit R_S–C_S and Z_P is the impedance of the parallel circuit R_P–C_P, the bridge is balanced when $Z_S/Z_P = 2R/R = 2$. If $R_P = R_S = R$ and $C_P = C_S = C$, find the frequency f_0 at which the transfer function $H(s) = v_o/v_i = 0$. (*Hint*: Solve for Z_S/Z_P, let $s = j(2\pi f_0)$, and find the value of f_0 that balances the bridge.)

6.6. *Resonant frequency and Q of passive L–C–R bandpass filter*. Figure 6.14 illustrates a passive *L–C–R* bandpass filter. In practice, *R* represents the internal losses of the inductor and cannot be reduced below a minimum value (although it can be increased with an external resistance). Suppose a certain circuit uses an inductor whose inductance $L = 88\,\text{mH}$ and whose effective resistance $R = 77\,\Omega$ and a capacitor $C = 22\,\text{nF}$. Calculate (a) the resonant frequency f_0 of the circuit (in Hz), (b) the *Q* of the circuit at its resonant frequency, and (c) the 3-dB-down full bandwidth of the circuit's bandpass response (in Hz).

***6.7.** *Sallen–Key transfer-function derivation.* Derive Equation 6.24 using the negative-feedback assumption and assuming an ideal op amp.

6.8. *Sallen–Key lowpass filter circuit design.* Using the prototype filter coefficients given in Table 6.1, design a 4-pole Sallen–Key lowpass filter circuit to have a 3-dB-down frequency of 500 Hz using 56-kΩ resistors. State the exact theoretical capacitor values for this exercise. To check your design, analyze its response using circuit-analysis software such as Multisim™, and use virtual op amps (not actual circuit models of a particular type of op amp). Using exact capacitance values, the response should agree with the theoretical response with almost no error.

***6.9.** *4-pole bandpass filter using biquad active filters.* The same frequency- and impedance-scaling operations that work for lowpass filters also work for bandpass filters. In this problem, you will design a prototype two-stage biquad bandpass filter with a center frequency of $\omega_0 = 1\,\text{rad s}^{-1}$ and a Q_{BPF} of 10 and then transform it to have a new center frequency of $\omega_0'/2\pi = f_0' = 10\,\text{kHz}$. Your final design should be a bandpass filter with a 3-dB-down bandwidth of $f_0'/Q_{BPF} = 1\,\text{kHz}$. Figure 6.30 shows the prototype 2-pole lowpass filter poles after they have been transformed from the lowpass to the bandpass configuration using Equation 6.44. The actual values of the poles are $p_{1\pm} = -0.0341 \pm j0.9646$ and $p_{2\pm} = -0.0366 \pm j1.0353$. The circle on which they lie is now centered at $\omega_0 = 1\,\text{rad s}^{-1}$ and the circle's radius is now $1/2Q_{BPF}$ instead of 1. But their angular relation to each other and the imaginary axis is approximately the same as for the lowpass poles. Your task is to first design a two-stage biquad amplifier cascade (six op amps) that realizes these poles and then to frequency- and impedance-scale it to make a practical bandpass filter.

(a) *Prototype design.* To make the initial design easier, we will give you some simplifying assumptions to make. Referring to the biquad filter schematic in Figure 6.23, for the prototype filter, use $C_1 = C_2 = 1\,\text{F}$ and $R = 1\,\text{k}\Omega$. If

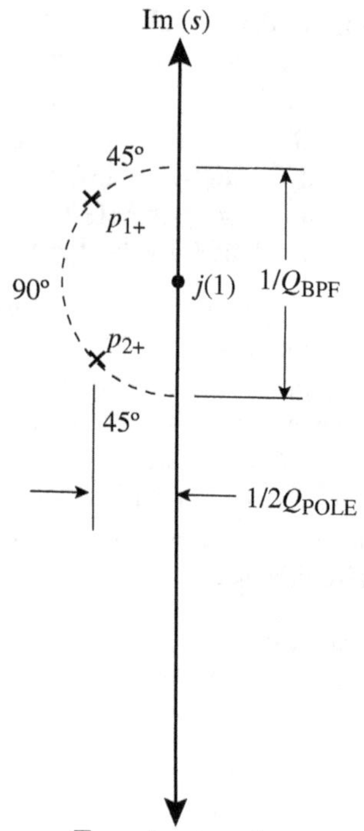

FIGURE 6.30 Location of poles for 2-pole bandpass prototype filter.

we take $k_1 = 1$ in Equations 6.56 and 6.57, the equations for the remaining resistors simplify to

$$R_1 = \frac{1}{a} \tag{6.67}$$

$$R_2 = R_3 = \frac{1}{\sqrt{b}} \tag{6.68}$$

$$R_4 = \frac{1}{c}, \tag{6.69}$$

where the coefficients a, b, and c appear in the bandpass transfer function $v_{BP}/v_i = cs/(s^2 + as + b)$ (Eq. 6.54). The value of c is an arbitrary gain factor and can initially be set equal to a. You may find the values for a and b by multiplying out the factored polynomial: $(s - p_+)(s - p_-) = s^2 + as + b$. You will do this twice: once for the $p_{1\pm}$ pair of poles (use these values in the first biquad circuit) and once for the $p_{2\pm}$ pair of poles (use these values for the second biquad circuit). Each biquad circuit provides one pair of conjugate poles. You can test the resulting circuit in circuit simulation software, but do not be disappointed if the results do not perfectly agree with theory, because you will need to adjust values manually after scaling. Notice that unlike the lowpass filter, the center-frequency loss of the bandpass filter is not 0 dB. You can compensate for the center-frequency loss by lowering the value of R_4 to make up for it, because R_4 influences only the overall gain.

(b) *Frequency and impedance scaling.* Once you have obtained the pair of biquad circuits that work at a center frequency of $\omega_0 = 1\,\text{rad}\,\text{s}^{-1}$, frequency-scale all the capacitors by a factor of $1/(2\pi(10\,\text{kHz}))$. This will bring them into the μF range. Next, select a value for the impedance-scaling factor Z'/Z that will make all the capacitors equal to 10 nF. This will bring the resistors into the kilohms range. Model your initial scaled circuit using a software package such as Multisim™ or equivalent. Use virtual (not real) op amps.

(c) Because this circuit is very sensitive to small errors in component values, the final design will require *tuning*, which means slight small adjustments in the component values to obtain the desired final response. If you have performed the scaling calculations correctly, the two biquad circuits should be tuned to different resonant frequencies, namely, (0.9646) (10 kHz) = 9.646 kHz and (1.0353)(10 kHz) = 10.353 kHz. The Q of each circuit (measured by finding the peak or resonant frequency f_0 and dividing by the 3-dB-down bandwidth δf) should be about 14. By connecting each biquad individually to a Bode plotter (be sure to take the output from the correct point for the bandpass function as shown in Fig. 6.23), you can measure the actual f_0 and Q of each circuit. If they are more than 1% or so in error, you can adjust the Q by means of varying resistor R_1 and the frequency f_0 by varying resistor R_2. If you tune these components to obtain the correct f_0 and Q for each biquad circuit individually, the overall response when connected in cascade should be very close to an ideal four-pole bandpass response, with a center frequency of 10 kHz and a 3-dB bandwidth of 1 kHz. Turn in your final schematic and a plot of the actual simulated response.

6.10. *Switched-capacitor integrator for long time constant.* The integrator circuit in Figure 5.16 (without auxiliary resistor R_A) produces the integral of the input voltage in accordance with Equation 5.33: $V_{OUT}(t) = -1/RC \int_{t'=0}^{t} V_{IN}(t')\,dt'$. In the frequency domain, the relation is $v_{OUT}(s)/v_{IN}(s) = -1/sRC$. Linear analog integrator circuits with long time constants present problems when the time constant $\tau = RC$ is much longer than 1 s, because the component values of R and C become impracticably large. But a switched-capacitor integrator can deliver similar performance with components of practical values.

(a) Assuming the largest resistor value that can be used without leakage and drift problems is 10 MΩ, find the value of capacitor C in a conventional (nonswitched) integrator circuit that will provide a time constant τ of

1000 s. Note that if your value of C is greater than 1 μF, only electrolytic capacitors are generally available, and they tend to have a leakage resistance that adds an unpredictable component (R_A in Fig. 5.33) to the circuit. Such leakage causes erratic and inaccurate outputs.

(b) Assuming the largest nonelectrolytic capacitor available has a value of 1 μF, design a switched-capacitor integrator (shown in Fig. 6.25) by choosing values for C_1 and C_2 and switching rate f_S so that the same effective time constant of 1000 s is obtained. Also assume that the signal must be sampled at a rate of at least 1 sample s^{-1}.

6.11. *Redesign of Sallen–Key lowpass filter for $n=6$.* As we have seen, the 5-pole lowpass filter design in the electric guitar preamp example barely meets the specifications given for the lowpass filter. Redesign the 60-dB-gain preamp to meet the same set of specifications with a 6-pole lowpass filter (using three Sallen–Key lowpass circuits) so that the redesigned filter exceeds the filter specifications by a larger margin than the 5-pole design. Use only 5% tolerance standard capacitor values, which are decimal multiples of the following values in pF: 10, 11, 12, 13, 15, 16, 18, 20, 22, 24, 27,30, 33,36, 39, 43, 47, 51, 56, 62, 68, 75, 82, 91, and 100. Verify the performance of your redesign with Multisim™ or a similar circuit-analysis simulator.

For further resources for this chapter visit the companion website at
http://wiley.com/go/analogmixedsignalelectronics

7

WAVEFORM GENERATION

7.1 INTRODUCTION

There are many uses for circuits that produce periodic waveforms with predictable characteristics. Test equipment, timekeeping devices, clock generators in digital systems and communications networks, and many other electronic systems all need circuits that produce periodic waveforms having a fixed or adjustable frequency. A circuit called an **oscillator** lies at the heart of most of these systems. Generally speaking, an oscillator is a circuit designed to produce a periodic output with no input (other than power-supply energy). The design of oscillators is a specialty, and many system designers now simply purchase inexpensive packaged oscillators instead of "rolling their own." But even if all you are going to do is to specify the performance of an oscillator or clock generator, an understanding of how oscillators work and what the design trade-offs are will enable you to set specifications intelligently to meet particular requirements.

We begin this chapter with a discussion of oscillator theory: what it takes for a circuit to begin oscillating and what happens once oscillation commences. Because this subject is intimately related to the question of amplifier stability, we treat that issue along with the question of what conditions are necessary for oscillation to occur. Next, we discuss several common types of frequency-determining components used in oscillators. These range from simple discrete components (such as resistors, capacitors, and inductors) to special devices such as quartz-crystal resonators and **microelectromechanical system (MEMS)** resonators. After descriptions

Analog and Mixed-Signal Electronics, First Edition. Karl D. Stephan.

Companion Website: http://wiley.com/go/analogmixedsignalelectronics

of the two main types of oscillators (sine-wave oscillators and two-state or **relaxation** oscillators), we conclude with a design example of an oscillator whose frequency is controlled by means of a quartz crystal.

Analog circuits are by no means the only way to generate waveforms. A **digital-to-analog converter (DAC)** can be used with digital circuitry to generate an arbitrary waveform whose frequency, harmonic content, and other characteristics are limited only by the capabilities of the digital system producing it. Such systems are called, appropriately enough, **arbitrary waveform generators** (**AWG**s) and represent an important class of waveform and signal-generation instruments. We will defer the description of such systems to the chapter on *analog-to-digital* and *digital-to-analog* conversion. But even AWGs rely for their frequency accuracy on a stable fixed-frequency clock generator whose internal circuitry is ultimately analog in nature.

7.2 "LINEAR" SINE-WAVE OSCILLATORS AND STABILITY ANALYSIS

The word "linear" is in quotation marks in the heading of this section because no functioning oscillator is truly linear, just as no amplifier is truly linear. All circuits that can provide power gain are (eventually) nonlinear, as we showed in Chapter 4. Nevertheless, linear analysis can provide information about the conditions under which oscillation can begin to occur, and these conditions are necessary (but not sufficient) for any oscillator circuit to work. This is a good place to discuss the general question of system stability, and we will begin with what is probably the simplest possible circuit that can produce sinusoidal oscillations: a series *L–C–R* circuit shown in Figure 7.1.

7.2.1 Stable and Unstable Circuits: An Example

To demonstrate the various kinds of stable and unstable behavior that linear systems can exhibit, we will analyze the circuit in Figure 7.1 by means of a differential equation. Once we have shown what happens in the time domain, we will apply the Laplace transform to the system and generalize its behavior to other linear systems.

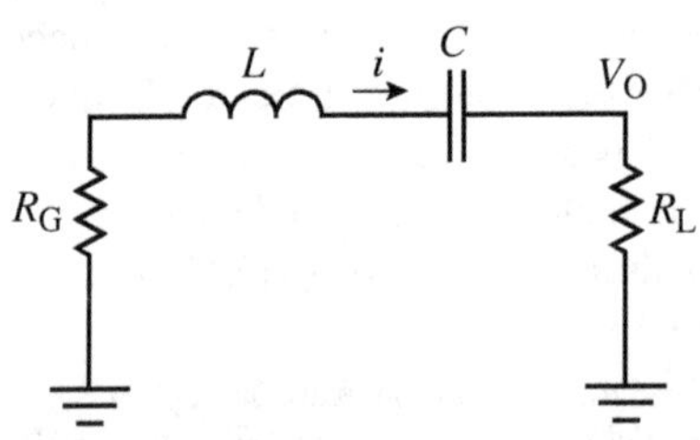

FIGURE 7.1 Series *L–C–R* circuit with (negative) resistor R_G and (positive) resistor R_L.

Suppose that while R_L in the system of Figure 7.1 is a positive resistance (representing circuit losses or a load resistance), resistor R_G's value is *negative*. Of course, negative resistors as independent components do not exist in reality, because they would violate the principle of conservation of energy. But when supplied with DC power, certain types of active devices can behave as a negative resistance over a limited range of voltage or current, and we will suppose that we have such a device and, furthermore, that it is linear over the entire range of current and voltage in question. We will also suppose that we can vary the value of R_G from 0 to $-\infty$ and that the load or loss resistance R_L can also vary from 0 to $+\infty$.

If we write the expression for each component's terminal voltage in terms of the current i, Kirchhoff's voltage law says that the sum of these voltages equals zero:

$$i(R_L + R_G) + L\frac{di}{dt} + \frac{1}{C}\int i\,dt = 0 \tag{7.1}$$

Differentiating once with respect to time gives a second-order differential equation for $i(t)$:

$$L\frac{d^2 i}{dt^2} + (R_G + R_L)\frac{di}{dt} + \frac{i}{C} = 0 \tag{7.2}$$

Let's try the general solution $i = Ae^{kt}$, where A is a constant whose dimensions are amps and k is a constant to be determined. Inserting this trial solution for i in Equation 7.2 gives the following characteristic equation for k:

$$Lk^2 + (R_G + R_L)k + \frac{1}{C} = 0 \tag{7.3}$$

This is a quadratic equation for k, which has solutions given by the quadratic formula

$$k = -\frac{R_G + R_L}{2L} \pm \frac{\sqrt{(R_G + R_L)^2 - 4L/C}}{2L} \tag{7.4}$$

Let

$$r = -\frac{R_G + R_L}{2L} \tag{7.5}$$

where r has the dimensions of s^{-1} (radian frequency). Then if we let the resonant frequency (in rad s^{-1}) be ω_0, so that

$$\frac{1}{LC} = \omega_0^2, \tag{7.6}$$

Equation 7.4 can be rewritten as

$$k = r \pm \sqrt{r^2 - \omega_0^2} \tag{7.7}$$

The sum $R_G + R_L$ can vary anywhere from $-\infty$ to $+\infty$, and thus so can r.

Let's examine the behavior of the current *i* for four cases: (1) $r_1<-\omega_0$, (2) $-\omega_0<r_2<0$, (3) $0<r_3<\omega_0$, and (4) $r_4>\omega_0$. As we will see, the behavior of the circuit is markedly different for each of these four conditions:

1. $r_1<-\omega_0$. In this range, the total resistance (R_G+R_L) is positive and so large that it exceeds the impedance at resonance of either the inductor *L* or the capacitor *C*. In other words, the dominant impedance in the circuit is resistive. In this case, *k* takes on one of two real negative values, because even when the plus sign in front of the square-root radical in Equation 7.7 is chosen, the square-root quantity is smaller in magnitude than *r*, leaving *k* negative. The function $i=Ae^{kt}$ is therefore an exponential that *decreases* with time. This is easily seen if we let $L=0$, which gives a first-order differential equation rather than a second-order one. The solution for the resulting first-order equation is the familiar exponentially decaying current of an *R–C* circuit. At any rate, for any initial condition (e.g., a given initial charge on the capacitor *C*), the circuit's excitation eventually dies away completely as *t* goes to infinity. In the terminology of control-system theory, the circuit is **stable**, in the sense that any initial voltage, charge, current, or other perturbation will eventually die away.
2. $-\omega_0<r_2<0$. In this range, the net resistance R_G+R_L is still positive, but small enough so that $|r_2|<\omega_0$. In this case, the quantity under the square-root sign is negative, making *k* complex. Rearranging Equation 7.7 makes this more obvious:

$$k = r_2 \pm j\sqrt{\omega_0^2 - r_2^2} \tag{7.8}$$

As *r* becomes smaller in magnitude (less negative), which happens if we make the negative resistance R_G only a little smaller in magnitude than the positive resistance R_L, the net resistance in the circuit becomes almost negligible compared to the reactances of the capacitor and inductor at the resonant frequency ω_0. Because the current must be a real function of time, we combine the two complex exponential solutions to obtain an exponentially decaying cosine wave:

$$i(t) = Ae^{-|r_2|t}\cos\left(\left(\sqrt{\omega_0^2 - r_2^2}\right)t\right), \tag{7.9}$$

which for $|r_2| \ll \omega_0$ can be approximated by $Ae^{-|r_2|t}\cos(\omega_0 t)$. These solutions represent sine or cosine waves whose amplitude decreases exponentially with time. The smaller the magnitude of r_2 is, the slower is the rate at which the amplitude decreases.

If $R_G=-R_L$ exactly, the net resistance in the circuit is exactly zero, and we reach the boundary point between cases (2) and (3). In this (theoretical) situation, any sine wave once initiated will continue forever, because the expression for the current is exactly $A\cos(\omega_0 t)$. Note, however, that this case is a mathematical fiction and cannot exist in reality.
3. $0<r_3<\omega_0$. In this case, the magnitude of the negative resistance R_G exceeds the magnitude of the positive resistance R_L, making the net circuit resistance *negative*. If the net resistance is allowed to be negative but is small enough

in total magnitude so that $|r_3| < \omega_0$, the quantity under the square-root radical in Equation 7.7 is still negative, leading to an imaginary number that produces an exponentially *growing* oscillation:

$$i(t) = Ae^{+|r_3|t}\cos\left(\left(\sqrt{\omega_0^2 - r_3^2}\right)t\right) \tag{7.10}$$

Any initial current, no matter how small, will now lead to increasingly large oscillations that (in theory) will grow without limit. In control-system theory, this type of behavior is termed **unstable**, because any perturbation or signal fed to the circuit will cause exponentially growing effects. Of course, in real circuits, nothing can grow without limit, and nonlinear effects will begin to limit the actual amplitudes. But the linear theoretical model does not take such nonlinear effects into account.

4. $r_4 > \omega_0$. This corresponds to a large net negative resistance whose magnitude exceeds the impedance of the inductor or capacitor at resonance. It is a kind of negative-resistance version of an *R–C* circuit. The quantity under the square-root bracket in Equation 7.7 becomes real, and the constant *k* is therefore real and positive, even when the – sign in front of the square-root radical is used in Equation 7.7. The result for this situation is that the current grows in a non-oscillatory exponential fashion, again theoretically without limit. This is also a type of **unstable** behavior. Figure 7.2 shows graphically how the current behaves in each of the four ranges of *r* considered, as well as at $r = 0$, where constant-amplitude oscillation occurs.

The takeaway from this entire discussion is that for this particular circuit, a net negative resistance leads to exponentially growing current and unstable behavior, while a net positive resistance leads to exponentially decaying current and stable behavior. The issue of circuit stability in the sense we have just analyzed is critical in a wide variety of situations, ranging from amplifier and control-system design to oscillator design. This is why we will use this example to demonstrate a general principle of linear circuit theory that you can use to tell whether a given design will be stable.

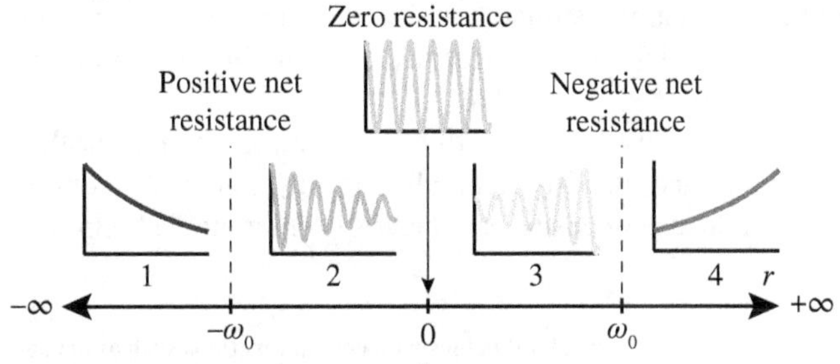

FIGURE 7.2 Four ranges of *r* and corresponding behavior of current *i* in circuit of Figure 7.1.

7.2.2 Poles and Stability

The Laplace transform of the time-domain current function discussed in the previous section is very simple to perform. The **Laplace transform** of a function $f(t)$ is defined as

$$F(s) = \int_{t=0+}^{t=\infty} f(t)e^{-st}dt, \tag{7.11}$$

where the lower limit is zero *approached from the positive side.*[1]

If $f(t) = i(t) = Ae^{kt}$, the Laplace transform of $f(t)$ is

$$\int_{0}^{\infty} Ae^{(k-s)t}dt = \frac{A}{k-s} \tag{7.12}$$

From the discussion of poles and zeroes in Chapter 6 (or from your general knowledge of Laplace transforms), you should know that the value $s = k$ is therefore a pole of the Laplace transform of the current $i(t)$ in the complex s-plane. Because the equation for k has two solutions, there are two poles, which are identical to the poles described in Equations 6.20 and 6.21 in Chapter 6's discussion of the simple bandpass filter:

$$k_+ = p_+ = -\frac{\omega_0}{2Q} + \frac{\omega_0}{2}\sqrt{\frac{1}{Q^2} - 4} \tag{7.13}$$

$$k_- = p_- = -\frac{\omega_0}{2Q} - \frac{\omega_0}{2}\sqrt{\frac{1}{Q^2} - 4} \tag{7.14}$$

where $Q = \omega_0 L/(R_G + R_L)$. In this discussion, Q is allowed to take on positive and negative values as well as infinity. Infinite Q corresponds to the case in which the poles lie exactly on the imaginary axis. With reference to Figure 7.2, when the poles have negative real parts and lie in the left half of the complex plane, any nonzero initial voltages or currents in the system will eventually die away, implying stability. When the poles have positive real parts and lie in the right half plane, any nonzero initial currents or voltages will produce exponentially increasing waveforms, implying that the system is unstable. When the poles lie exactly on the imaginary axis, any initial excitation will theoretically continue at the same amplitude for an infinite time, but this is a mathematical fiction that never occurs in reality. And in any case, such a situation would be regarded as unstable.

We have illustrated these effects with a very simple circuit, but the conclusions about pole locations and stability generalize to any linear, time-invariant system. Essentially, the whole task of ensuring a linear system is stable boils down to making

[1] Strictly speaking, to avoid mathematical difficulties with certain functions such as the unit step function, the Laplace transform is defined in terms of a limit: $F(s) = \lim_{\varepsilon \to 0+} \int_{\varepsilon}^{\infty} f(t)e^{-st}dt$, where the lower limit of the integral approaches zero from the positive side but never reaches it.

sure that the Laplace-transformed expression for its output voltage or current has poles only in the left half plane.

As of this writing, widely available mathematical analysis software such as MATLAB™ can be used to find poles and zeroes of arbitrarily complex functions, and circuit simulation software can deal with both linear and nonlinear systems. Investigating the stability of nonlinear systems is a much more challenging and less straightforward problem than solving for the poles of a linear system, but modern software is often up to the task. This is fortunate, because, strictly speaking, all oscillators are nonlinear and linear analysis can provide only general guidelines in the design of oscillator circuits. To see why this is so, we will next discuss two widely used criteria in oscillator design in turn: the **Nyquist stability criterion** and the **Barkhausen criterion**.

7.2.3 Nyquist Stability Criterion

To determine the poles of a system function exactly, one must know the explicit mathematical form of the function. In the case of a lumped-element linear system, that means knowing all the terms of the numerator and denominator polynomials that comprise the rational function in question. Sometimes, this function is known, but sometimes, a system is so complex that one can only make experimental measurements of its transfer function over a certain range of frequencies. The Nyquist criterion can deal with the latter case, in which the mathematical expression is not known but in which measurements are available for a quantity called the **loop gain**.

We mentioned loop gain briefly in Chapter 6 in connection with undesirable oscillations in an amplifier. Figure 7.3 illustrates the general problem that the Nyquist criterion addresses. Either as part of a feedback control system (which is designed to be stable) or an oscillator system (which is designed to oscillate, which means it is intentionally unstable), a gain block consisting of one or more amplifier stages having the overall transfer function $G(s)$ drives a feedback network with a transfer function $H(s)$. Usually (but not always), the feedback network is composed of passive components. The feedback network's output is connected to the input of the gain block through a **summing junction**, which is denoted by the Greek letter Σ (capital sigma). The summing junction has both a positive input and a negative input, and its output is the difference between the two.

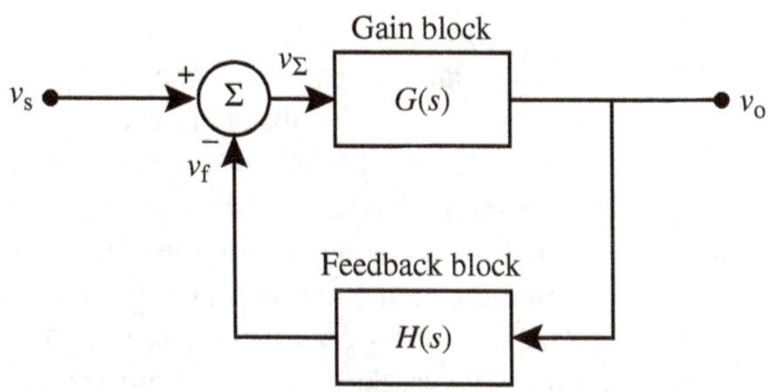

FIGURE 7.3 General feedback system with forward gain block transfer function $G(s)$ and feedback function $H(s)$.

The system input (if used) is the voltage v_s, which goes to the positive input of the summing junction. There, the negative input of the summing junction, namely, v_f, is subtracted from it, and the difference, namely,

$$v_\Sigma = v_s - v_f, \tag{7.15}$$

is sent to the gain block. The output of the gain block is

$$v_o = G(s)v_\Sigma \tag{7.16}$$

and becomes both the output of the overall system and the input for the feedback network. The transfer function $H(s)$ of the feedback network multiplies the output signal to form the feedback signal v_f:

$$v_f = H(s)v_o \tag{7.17}$$

and the loop is closed when the feedback voltage v_f reaches the summing junction.

Analysis of networks that include a closed signal path that meets up with one of its own inputs can be confusing. Up to now, the independent variable in our analyses has been an input voltage, for example, and each subsequent signal in a signal path has been a straightforward dependent function of the preceding one. Feedback circuits cannot be analyzed that way. Instead, one must view the entire circuit as a whole and write down the relations among the various voltages and then solve the equations for the desired variable.

When this is done, we find that the relation between the input voltage v_i and the output v_o is

$$\frac{v_o}{v_i} = T(s) = \frac{G(s)}{1 + G(s)H(s)} \tag{7.18}$$

This is the transfer function of a linear time-invariant system, and from our previous discussion of poles, we know that the system's stability depends on the location of its poles, which are the roots of the denominator, namely, the solutions of $1 + G(s)H(s) = 0$. But if all we have are *measurements* (experimental data) of the product $G(s)H(s)$ and no explicit knowledge of the mathematical form of the functions G or H, we can still answer the question, "Does the function $T(s)$ have poles in the right half plane of the complex variable s?" We find the answer by doing the measurement shown in Figure 7.4, which consists of opening the feedback loop at a suitable point and measuring the transfer function product $G(j\omega)H(j\omega) = v_{to}/v_{ti}$, and then plotting the resulting data on a **Nyquist diagram**. This can be done by varying the frequency $f = \omega/2\pi$ of a sine-wave test signal v_{ti} and measuring the resulting output v_{to}, in both amplitude and phase. (In performing this measurement, we assume that the system is **open-loop stable**, meaning that the loop-gain function $G(s)H(s)$ is itself a stable function. This is usually, but not always, the case, and there are ways to deal with unstable open-loop functions.)

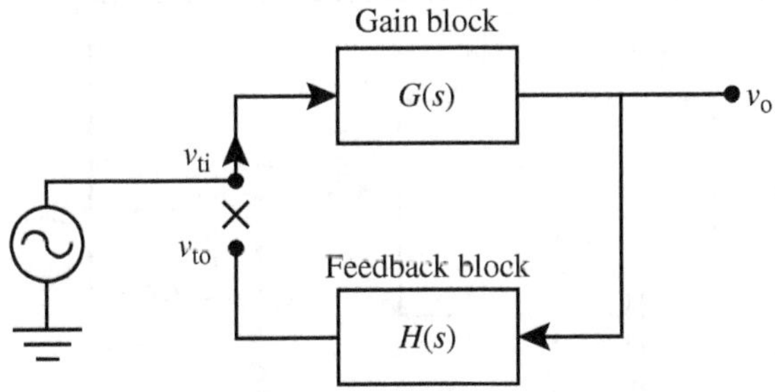

FIGURE 7.4 Experimental measurement of loop gain $G(s)H(s)$ in feedback system of Figure 7.3.

The Nyquist diagram of a feedback system consists of the **locus** (path) of the complex value of $G(j\omega)H(j\omega)$ as ω varies from positive infinity, to 0 (DC), and then to negative infinity, in principle.[2] Experimentally, it is impossible to produce negative frequencies or infinite frequencies. Practically speaking, the magnitudes of all loop-gain functions eventually go to approximately zero if the frequency is high enough, so one must take discrete-frequency measurements only up to the frequency at which the loop response is negligibly small. Plotting these measurements of $G(s)H(s)$ as points on the complex plane and connecting the points will produce the locus curve mentioned earlier. In doing these measurements, care must be taken not to take too large a frequency step between points, because significant parts of the locus might be missed that way. When we add the mirror image of the original curve reflected about the real axis, it will complete a closed loop and represents what one would obtain for negative frequencies as well as positive ones.

Harry Nyquist of the Bell Telephone Laboratories showed in 1934 that if the closed locus curve of points *enclosed* the point (−1, 0) on the negative real axis of the s-plane, then poles exist in the right half plane of the transfer function $T(s)$, and the system is *potentially* unstable.[3] That is, the stability of the linear system cannot be guaranteed.

If the curve does *not* enclose the (−1, 0) point, the linear system is guaranteed to be stable: that is, any disturbances or signals will eventually die away if the input is turned off. Use of the Nyquist criterion is best illustrated by an example.

Figure 7.5 shows a type of oscillator circuit known as the *twin-T*, for the reason that two T networks are connected as shown to produce a response that has a real zero at

[2]MATLAB generates a Nyquist plot by varying ω from negative infinity to positive infinity, which is the opposite of the conventional definition (see, e.g., B. C. Kuo, *Automatic Control Systems* (Prentice-Hall, 1962)). This merely has the effect of reversing the direction-indicating arrows on the locus, but does not change its shape.

[3]"Enclosed" is a more specific term than "encircled." A point is said to be *encircled* by a closed curve merely if it is inside the curve. A closed curve *encloses* a point if the point is in the region to the left of the path as the path is traced in a prescribed direction, which will be clear from the examples that follow.

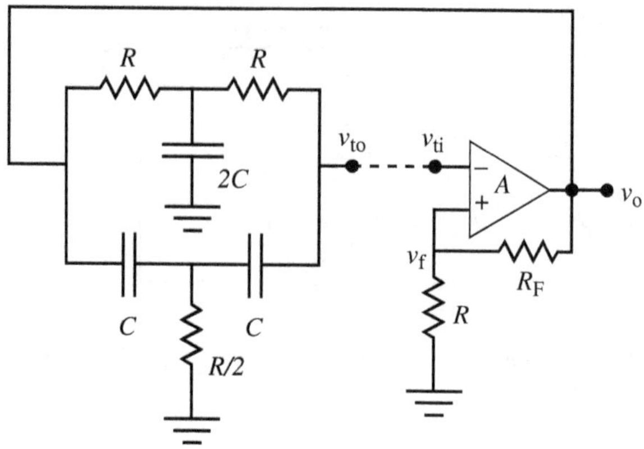

FIGURE 7.5 Twin-T oscillator circuit used for Nyquist-criterion example.

$$f_0 = \frac{\omega_0}{2\pi} = \frac{1}{2\pi RC} \tag{7.19}$$

To assess the stability of this circuit using the Nyquist criterion, we will proceed as follows. This circuit happens to be simple enough that its transfer function can be found analytically, but we will use the Nyquist criterion to show that stability of the closed-loop system can be investigated without knowledge of the actual transfer function, only its response product $G(j\omega)H(j\omega)$.

In the circuit of Figure 7.5, the gain block consists of the differential amplifier with gain A and the voltage divider formed by R_F and the shunt resistor R to ground. We have broken the closed feedback loop at the point indicated by the dashed line, and we can find the gain block's transfer function $-G(s)$:

$$-G(s) = \frac{v_o}{v_{ti}} \tag{7.20}$$

We call this ratio $-G(s)$ with a minus sign because by convention, $G(s)$ does not include the −1 factor contributed by the summing junction in Figures 7.3 and 7.4. Because v_{ti} goes into the inverting input of the amplifier, we must put a minus sign in front of $G(s)$ in Equation 7.20. (If we had drawn an external summing junction in the diagram and used only noninverting inputs for the amplifier, this problem would not arise.)

We can write the output voltage v_o in terms of the input voltages v_{ti} and v_f, the voltage fed back through the R_F–R network:

$$v_o = A(v_f - v_{ti}) \tag{7.21}$$

But v_f is itself a function of the output voltage, so we express it as a fraction β of the output:

$$v_f = v_o \frac{R}{R_F + R} = \beta v_o \tag{7.22}$$

We can now solve for the ratio v_o/v_{ti} by using Equation 7.22 to eliminate v_f in Equation 7.21 and obtain

$$v_o = A(\beta v_o - v_{ti}) \tag{7.23}$$

$$\frac{v_o}{v_{ti}} = -G(s) = \frac{-A}{1 - A\beta} \tag{7.24}$$

Assuming the product $A\beta$ is very large, which can be satisfied with typical op amp open-loop gains, we have

$$G(s)\big|_{A\beta\to\infty} \approx -\frac{1}{\beta} \tag{7.25}$$

Equation 7.25 says that the gain of the gain block is negative in sign (i.e., it inverts the phase of a sine wave) and inversely proportional to the feedback factor β. With the circuit as shown in Figure 7.5, the magnitude of $G(s)$ cannot be less than 1, but it can be made arbitrarily large.

In Figure 7.5, the feedback network described by $H(s)$ consists of the twin-T circuit. With the values of components in the proportions shown, the transfer function of the feedback network is

$$H(s) = \frac{v_{to}}{v_o} = \frac{(s/\omega_0)^2 + 1}{(s/\omega_0)^2 + 4(s/\omega_0) + 1} \tag{7.26}$$

You can easily see that for $s = j\omega_0$, the numerator has a real zero and so the twin-T circuit has a (theoretically) perfect *rejection* of any signal at that frequency. At all other frequencies, the twin-T feedback circuit will transmit some fraction of its input voltage to the inverting input of the amplifier. On the other hand, the R_F–R network provides *positive* feedback around the amplifier, so we can already see that at the frequency f_0, the net feedback from the amplifier output to either of its inputs is going to be positive. This qualitative fact is important to remember in the following Nyquist-criterion analysis.

To form the product $G(j\omega)H(j\omega)$ that can be experimentally measured, we simply multiply Equation 7.25 by Equation 7.26 to obtain

$$G(j\omega)H(j\omega) = -\frac{1}{\beta}\frac{1-(\omega/\omega_0)^2}{1-(\omega/\omega_0)^2 + 4j(\omega/\omega_0)} \tag{7.27}$$

It is the function in Equation 7.27 that is to be plotted on the complex plane as ω goes from zero to infinity and then its mirror image plotted as well.

Already, we can tell a few things about the Nyquist plot. Starting with very large positive frequencies, the value of $G(j\omega)H(j\omega)$ is approximately $-1/\beta$. As the frequency goes lower along the positive imaginary axis toward ω_0, the zero in the numerator will cause the value of $G(j\omega)H(j\omega)$ to approach zero. At $\omega_0 = 0$ (DC),

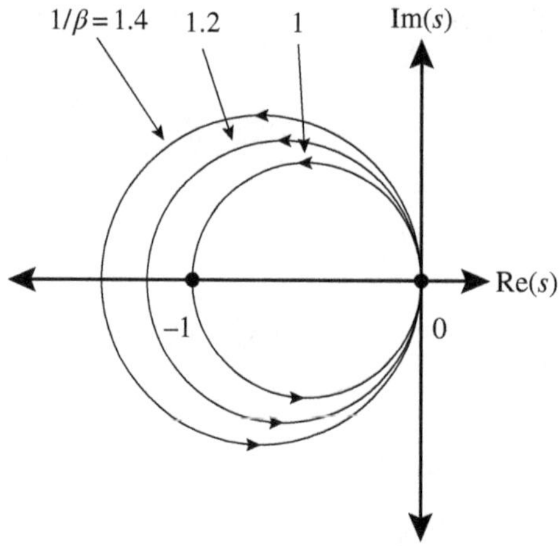

FIGURE 7.6 Nyquist plots for the circuit of Figure 7.5 using three values for $G(s) = -1/\beta$.

the function returns to $-1/\beta$ again. If the actual calculations are performed, it turns out that the locus of the Nyquist plot for this function is a perfect circle that intersects the real axis at two points: $(0, j0)$ and $(-1/\beta, j0)$. As the gain magnitude of the gain block increases from 1 to higher values, the circles traced by the Nyquist plot get larger and end up enclosing the critical point $(-1, j0)$. This behavior is illustrated in Figure 7.6.

You will note that when the gain block's gain magnitude is exactly 1, the Nyquist locus exactly crosses the critical $(-1, j0)$ point. Just as with the theoretical situation of a negative resistance exactly canceling out a positive resistance, this mathematical situation is impossible to realize in practice. Its practical importance is that it furnishes a threshold or minimum gain below which it will not be possible to obtain oscillations.

If one is designing a negative-feedback system to be stable, this critical point is to be avoided, and designs whose Nyquist locus strays too near the critical point may prove to be only marginally stable in practice, with small component or gain changes likely to "send it over the edge" into oscillation. On the other hand, if one is designing an oscillator to be *unstable*, the design should show a Nyquist locus that definitely encloses the critical point by a good margin as well.

We will discuss the Nyquist stability criterion in more detail when we take up the topic of phase-locked loops, which are a type of electronic negative-feedback loop. At this point, we will move on to the question of the Barkhausen criterion for oscillator design.

7.2.4 The Barkhausen Criterion

Heinrich Barkhausen (1881–1956) probably first described a form of what came to be known as the Barkhausen criterion in his 1907 doctoral thesis and later stated it in German-language editions of his textbook on vacuum tubes beginning in the early

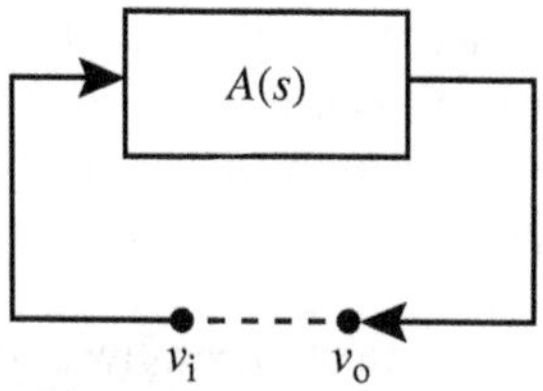

FIGURE 7.7 Feedback-loop setup for application of Barkhausen criterion.

1920s. Since these works were not translated into English, his criterion evidently made its way into English-language texts when English and American authors who could read German described it in their own publications and attributed it to Barkhausen. Textbooks written for undergraduates rarely cite extensive references for well-known concepts, and most references to the Barkhausen criterion merely cite other textbooks. But most sources agree on what the Barkhausen criterion is, whatever its origins might be.

Referring to Figure 7.7, a linear time-invariant system with a transfer function $A(s)$ has an output voltage v_o when provided with an input voltage v_i. The question to be answered is, what will happen when the feedback loop from output to input is closed and the dashed line in Figure 7.7 is replaced with a solid connection? The Barkhausen criterion states simply that in order for oscillation to occur, the loop gain, which is represented by $A(s)$, must be exactly unity:

$$A(s)=1 \tag{7.28}$$

Interpreting this condition in terms of amplitude and phase, Equation 7.28 means that the loop-gain magnitude is 1 and the loop-gain phase shift is zero degrees, or a multiple of 360°:

$$|A(s)|=1 \tag{7.29}$$

$$\angle A(s)=n(360°)\,(n=0,1,2,\ldots) \tag{7.30}$$

Imaginary values of s that satisfy Equations 7.29 and 7.30 supposedly satisfy the Barkhausen criterion.

The Barkhausen criterion has the advantage that it is easy to state and understand, and makes a certain intuitive sense. If you imagine a sine wave "going around the loop" represented by $A(s)$, it stands to reason that when it gets back to its starting place, it must have the same amplitude and phase as it started with in order to go around the loop again without change. Unfortunately, this intuitive notion has little or no basis in reality. In order for you to trace the path of such a wave in the time domain, it would have to change in amplitude or frequency with time, which means it is no longer a pure sine wave. But the Barkhausen criterion implicitly assumes a constant amplitude and frequency, so the whole picture breaks down under scrutiny.

Besides, it has been shown[4] that while the Barkhausen criterion (or something similar) is a *necessary* condition for oscillation to occur in a feedback-type oscillator, it is not a *sufficient* condition. That is, while every circuit that actually oscillates satisfies the Barkhausen criterion, not every circuit that satisfies the Barkhausen criterion will oscillate.

The initiation of oscillation is fundamentally a transient, not a steady-state, phenomenon. That is why the behavior of a linear system over a wide range of frequencies (theoretically infinite) must be known before anything definite mathematically can be stated about its stability, or lack thereof. The Nyquist criterion does this because it examines the entire right half of the complex plane for poles. But the Barkhausen criterion is inadequate as a model of how an oscillator design can be made to function.

There are reasons why the Barkhausen criterion is still used, however. Once an oscillator is running, the Barkhausen criterion is a fairly good description of conditions that one will actually observe in a sine-wave oscillator whose feedback loop is closed. If the phase of the output waveform is taken as the reference phase and an observer traces the feedback path in a type of oscillator that can be analyzed this way, you will indeed find that the measured loop gain approximately satisfies the Barkhausen criterion in both amplitude and phase. In a sense, it has to, because by definition the initial and final waveforms are the same, because they are measured at the same point. But the intermediate amplitudes around the loop will yield intermediate transfer functions whose overall product satisfies the Barkhausen criterion.

It turns out that the Barkhausen criterion, the Nyquist criterion, and the loop-gain transfer function are all related in the following way. The Barkhausen criterion, you will recall, states that the gain around the closed oscillator loop, which we have written as $-G(s)H(s)$, is exactly 1; that is, the gain magnitude is 1 and the phase shift is 0° (or an integer multiple of 360°). The Nyquist criterion is more general and requires that the locus of the loop-gain function $G(s)H(s)$ encircle the point $(-1, 0)$ on the complex plane. Note that when the Nyquist criterion is barely violated—that is, when the locus of $G(s)H(s)$ simply *intersects* $(-1, 0)$—then the Barkhausen criterion is satisfied. So the Barkhausen criterion is simply a special case of the Nyquist criterion. While the Barkhausen criterion is useful for initial design concepts, the Nyquist criterion is likely to model the linearized behavior of the circuit more accurately, because it theoretically examines the entire spectrum of frequencies at which a circuit can operate, not just the single frequency at which the circuit was designed to work.

As we will see, oscillator design is a combination of linear analysis, which is used to set up conditions favorable to oscillation, and nonlinear analysis, which is used to determine the amplitude and waveforms that actually exist in a design. Modern circuit simulation software is capable of modeling many of the nonlinear effects that establish the operating parameters for oscillators, although some oscillator designs

[4]See, for example, V. Singh, "Discussion on Barkhausen and Nyquist stability criteria," *Analog Integrated Circuits and Signal Processing* (2010), 62:327–332.

may tax the capacity of less sophisticated software packages. Nevertheless, a good understanding of the basic linear and nonlinear techniques that are useful in oscillator design is very helpful even when such software is available.

7.2.5 Noise in Oscillators

At this point, we will address the question of how noise affects the operation of oscillator circuits. Up to now, we have ignored noise in oscillator design, except to assume that either enough noise or a turn-on transient is sufficient to produce the exponentially growing oscillation needed to initiate operation. Even after an oscillator starts up and is producing a steady-state waveform, noise is still present, and it can affect the output waveform in several ways.

The ultimate effect of noise on the output of an oscillator can be treated as a type of modulation. An ideal feedback oscillator would produce a mathematically perfect sine wave whose spectrum would be a single infinitely narrow line. In actuality, the spectrum of real oscillators is widened by various processes, and the energy at frequencies other than the nominal frequency of operation can be considered as **modulation sidebands**. Although some oscillators can show changes in both amplitude and phase (or frequency), most of the noise in the output of typical oscillators appears as **phase modulation**.

When operating normally, most oscillators perform a **limiting** function whereby the amplitude of the output waveform is made essentially constant. This comes about either because the amplifier in the circuit encounters its clipping limits or because an auxiliary amplitude control system acts to control the output amplitude. However, the *phase* of the output waveform is not affected by the limiting action, in the sense that the times when the waveform crosses zero are the same regardless of what the amplitude is. Anything that influences the times when the waveform crosses zero, including the addition of noise, will change the instantaneous phase of the waveform. And this has consequences for the spectral purity and frequency stability of the output.

It turns out that the Q of the frequency-determining circuit in a sine-wave oscillator has a large influence on the noisiness of the oscillator's output. We can make a semiquantitative argument for the importance of using a resonator or frequency-determining circuit with as high a Q as possible in oscillators, with the aid of the hypothetical oscillator equivalent circuit shown in Figure 7.8.

For convenience in doing the analysis, we have isolated various functions and effects that in practice take place throughout a typical oscillator circuit. We assume the oscillator is operating normally and therefore satisfies the phase part of the Barkhausen criterion, namely, that the sum of phase shifts around the closed oscillator feedback loop is zero. To make things simple, we have lumped all the circuit phase shifts into two quantities: the phase shift ϕ_R associated with the frequency-selective part of the circuit (hereafter called the "resonator") and the phase shift ϕ_C encountered in the remainder of the circuit due to phase shifts in the amplifier, coupling capacitors, and any other part of the feedback loop outside the resonator circuit.

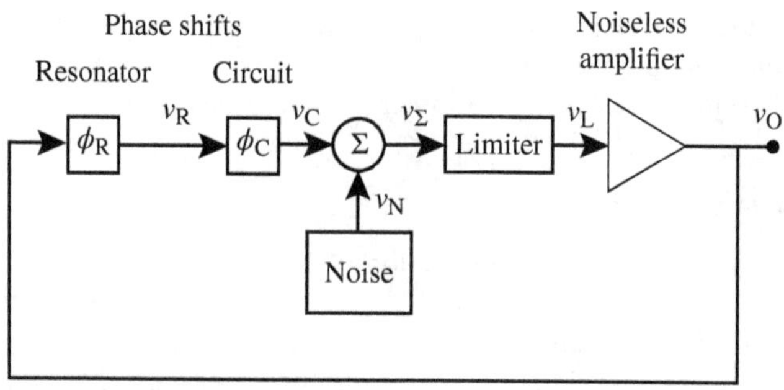

FIGURE 7.8 Hypothetical feedback oscillator block diagram used for oscillator noise analysis.

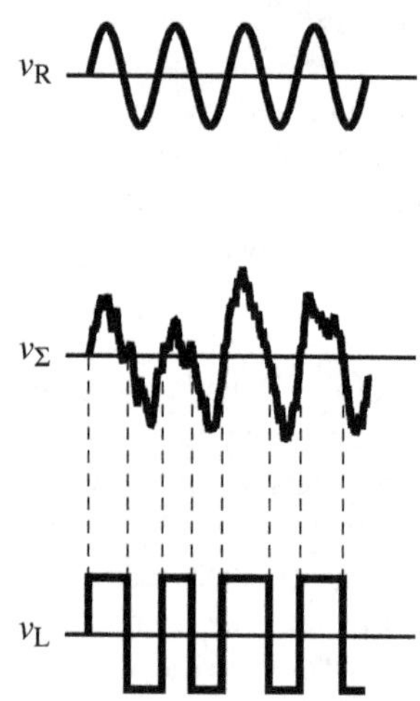

FIGURE 7.9 Example illustrations of voltage waveforms at v_R, the resonator output; v_Σ, the resonator output with noise added; and v_L, the limiter output. Note that noise causes the zero crossings of the limiter output to deviate from what they would be in the absence of noise.

Following the feedback loop from the point designated as v_C, a random noise voltage v_N is added to the circuit voltage v_C to form the sum voltage v_Σ. The sum voltage is then sent to an ideal limiter and amplifier whose output can be one of only two states: a constant positive voltage when the input voltage to the limiter is greater than zero or a constant negative voltage when the input is below zero. The resulting output is a rectangular wave at v_O, which passes to the resonator circuit. The resonator circuit eliminates the harmonics of the rectangular wave and passes a nearly pure sine wave v_R on to the remainder of the circuit, and the feedback loop is closed.

The waveforms at various points in the circuit are shown in Figure 7.9. The sine-wave output of the resonator at v_R has noise added to it and becomes v_Σ. It is important to note the effect of noise voltage on the points where the sum voltage v_Σ crosses zero. In general, the addition of noise will move the times of zero crossing away from the

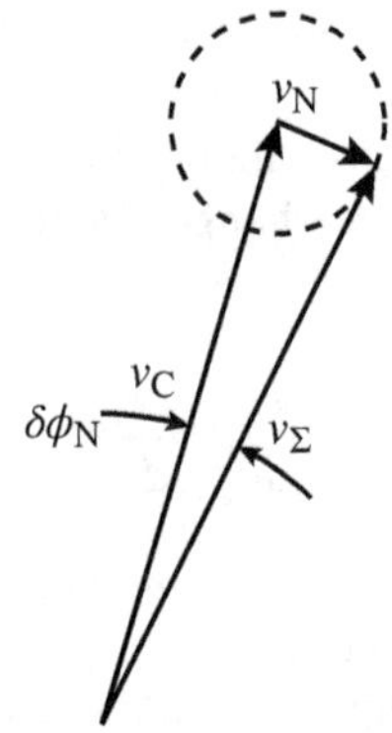

FIGURE 7.10 Addition of circuit voltage vector v_C and noise voltage vector v_N to produce sum voltage vector v_Σ, with resulting instantaneous phase shift $\delta\phi_N$.

equally spaced times when they would occur in the absence of noise. In other words, the addition of noise causes **jitter** (a random time shift from pulse to pulse) in the zero crossing times. Because the ideal limiter acts only on zero crossings, its rectangular output waveform v_L will reflect these deviations from the ideal sine wave's zero crossing times. The alteration of a wave's zero crossing time amounts to a change in phase, and we can estimate the RMS phase change due to the addition of a noise voltage by means of the vector diagram in Figure 7.10.

You should imagine both phasors v_C and v_N initially rotating at the oscillator's resonant frequency ω_0 and that you the observer are also rotating at the same speed, so that v_C appears stationary. In that case, if the noise phasor v_N moves slowly in a random walk with RMS amplitude V_N, it will represent that portion of the noise that has frequency components near the "carrier" represented by the resonant frequency ω_0. (In communications theory, a **carrier** is the steady single-frequency wave upon which information-carrying modulation is impressed.) As long as the amplitude V_C of the oscillation waveform is much greater than the RMS amplitude V_N(RMS) of the noise, we can approximate the instantaneous phase difference $\delta\phi_N$ (in radians) between the incoming sine wave v_C and the sum v_Σ of the sine wave and the noise wave by Equation 7.31:

$$\delta\phi_N(\text{RMS}) \approx \frac{V_N(\text{RMS})}{V_C} \tag{7.31}$$

This phase shift, if it persists during a few cycles of the oscillator waveform, will have to be included in the sum of phase shifts that satisfies the Barkhausen criterion. This addition of noise will have an effect on the short-term *frequency* of the oscillator output, as the following analysis will show.

Including the effects of noise on the instantaneous phase, we write the sum of phase shifts around the feedback loop, which must add to zero, as

$$\phi_R + \phi_C + \delta\phi_N = 0 \tag{7.32}$$

The phase shifts ϕ_R and ϕ_C depend on frequency. If the circuit has been well designed, the change in the resonator's phase ϕ_R for a given frequency deviation $\delta\omega$ will be much greater than the change in phase shift for the remainder of the circuit for the same frequency change. Mathematically, this is stated by the following inequality:

$$\frac{d\phi_R}{d\omega} \gg \frac{d\phi_C}{d\omega} \tag{7.33}$$

so that we can neglect the change of the circuit phase shift ϕ_C in analyzing the effects of noise on the oscillator output frequency.

The quantity $d\phi_R/d\omega$ is an important characteristic of any resonator circuit or device and is related to the Q of the resonator. For example, if we look at the phase response of the simple *L–C–R* bandpass filter in Figure 6.14, whose transfer function H_{BP} is given by Equation 6.19, the rate of change of the phase angle of the output with small frequency changes $\delta\omega$ near the resonant frequency ω_0 is

$$\left.\frac{d\angle H_{BP}(j\omega)}{d\omega}\right|_{\omega\approx\omega_0} = -\frac{2Q}{\omega_0} \tag{7.34}$$

In other words, the rate at which the phase changes per hertz of frequency shift is directly proportional to the resonator Q. This is an important general result that applies to all types of resonators, whether they are purely electrical (e.g., *L–C* circuits) or electromechanical types such as quartz crystals and MEMS resonators.

For a given noise-induced phase shift $\delta\phi_N$, how much frequency shift $\delta\omega_N$ must take place to satisfy the Barkhausen criterion, over a period of a few cycles? If we approximate the phase shift through the resonator as a linear function of frequency shift $\delta\omega = \omega - \omega_0$, we can use Equations 7.32 and 7.34 to solve for $\delta\omega_N$ in terms of $\delta\phi_N$:

$$\phi_R(\delta\omega) = \phi_R(0) + \delta\omega_N \frac{d\phi_R}{d\omega} = \phi_R(0) - \delta\omega_N \frac{2Q}{\omega_0} \tag{7.35}$$

$$\phi_R(0) - \delta\omega_N \frac{2Q}{\omega_0} + \phi_C + \delta\phi_N = 0 \tag{7.36}$$

In the absence of noise, the steady-state Barkhausen phase criterion requires that $\phi_R(0) + \phi_C = 0$, so we can write

$$\delta\phi_N = \delta\omega_N \frac{2Q}{\omega_0} \tag{7.37}$$

$$\delta\omega_N = \delta\phi_N \frac{\omega_0}{2Q} \tag{7.38}$$

Equation 7.38 shows that for a given amount of noise-induced phase shift $\delta\phi_N$, the resulting noise-induced frequency shift $\delta\omega_N$ is inversely proportional to the resonator Q: the better the Q, the less the frequency shift. Turning to a mechanical analogy, a high-Q resonator acts somewhat like a heavy flywheel on a motor, smoothing out short-term torque variations and supplying a nearly constant frequency (and phase) to the output.

A similar argument can show that a high-Q resonator reduces frequency changes due to changes in the circuit phase shift ϕ_C as well. Because of aging or temperature changes, the circuit phase shift can slowly vary with time over periods of hours or days. The exact same analysis that shows the noise-induced frequency shift is inversely proportional to Q will show that any frequency change due to circuit phase-shift changes will also be inversely proportional to Q. (This assumes, of course, that the resonator frequency itself does not change.)

The moral of this story is that raising the Q of the resonator in a feedback oscillator circuit will improve both the short-term phase noise spectrum and the long-term frequency stability with regard to changes in circuit components besides the resonator. This is why oscillators designed to have very stable outputs always use high-Q resonators. The *R–C* oscillators to be discussed in Section 7.3 are at the low end of the stability spectrum, because their effective Q, in terms of the rate of change of phase shift with frequency through the frequency-determining circuit, is rarely higher than about 3. Circuits with typical inductors and capacitors (*L–C* oscillators) have Q values in the range of about 30 to several hundred and therefore show noise performance intermediate between the relatively noisy *R–C* oscillators and the very stable crystal and MEMS oscillators, whose equivalent resonator Q values are much higher than one can obtain with practical *L–C* circuits made with discrete components.

We have seen that the effective Q of the frequency-determining circuit in a feedback circuit determines how rapidly the phase shift through it changes with frequency and in turn how well the circuit opposes changes in frequency due to noise and phase-shift changes elsewhere in the circuit. While resonators using passive electronic components are limited to Q values of a few hundred, electromechanical devices such as quartz crystals and MEMS components can reach effective Q values of 10,000–100,000 or higher. Although such devices are now routinely incorporated in integrated-circuit (IC) oscillator packages as complete units ready to use, a basic knowledge of how these systems work will help designers apply them intelligently in suitable system designs. For this reason, descriptions of a MEMS resonator and equivalent circuits for quartz-crystal resonators will be given later in this chapter.

7.3 TYPES OF FEEDBACK-LOOP QUASILINEAR OSCILLATORS

The phrase "quasilinear" means that an oscillator circuit can be designed with the aid of linear analysis to be unstable, in the sense that the output of the linearized circuit will be oscillatory and will grow without limit. In practice, nonlinear effects set in to limit the output amplitude, and we will discuss later in Section 7.3 how you can use intentionally nonlinear elements to minimize distortion in sine-wave oscillators.

All the oscillators we will discuss in this section take the form of a feedback loop, with a traceable signal path that closes on itself. There are some quasilinear oscillators that cannot be analyzed this way and instead use a component such as a *tunnel diode* that shows negative differential resistance over a part of its characteristic *I–V* curve. But such negative-resistance oscillators are rarely used today, and most quasilinear oscillator circuits can be analyzed with the feedback-loop approach.

The oscillator circuits we will examine have been classified according to the type of frequency-determining element used. Most electronic oscillators employ some type of resonator or energy-storage element as the component that establishes the frequency of oscillation. Because the frequency of an oscillator is one of its most important characteristics, we should briefly discuss the ways an oscillator's frequency is characterized.

One class of oscillators, sometimes referred to as **variable-frequency oscillators** (**VFOs**), is designed to produce a variable or adjustable output frequency, where the adjustment can be made either by varying the value of a component (e.g., a variable resistor or capacitor) or by varying a control voltage, which in turn changes the frequency. The latter type is termed a **voltage-controlled oscillator**(**VCO**) and is used in phase-locked loops and **frequency synthesizers**. A frequency synthesizer is a system designed to produce any one of a number of specific frequencies, usually in response to a digital command. However, most frequency synthesizers depend for the basic accuracy of their output frequencies on a single fixed-frequency oscillator called the **master clock**. And this brings us to the second main type of oscillator: the fixed-frequency **clock oscillator** or **reference oscillator**.

The **US National Institute of Standards and Technology** (**NIST**) maintains a national standard time based on the output frequency of an ultrastable oscillator, which is controlled by a **cesium-fountain** device, colloquially termed an **atomic clock**. As of this writing (2013), the state-of-the-art atomic clock used by NIST has a frequency uncertainty of about 3×10^{-16}, meaning that it loses or gains no more than about 26 ps (ps = 10^{-12} s) day^{-1}. This makes it one of the world's most stable oscillators.

The **stability** of an oscillator, in the sense we will use it in this section, is the degree to which the frequency is nearly constant, measured with respect to an (ideal) perfectly constant frequency. The **accuracy** of an oscillator's frequency is measured with respect to the frequency it was designed to produce. Accuracy and stability are related but independent concepts. Stability depends on the type of oscillator and the care with which it was designed, while accuracy depends on how close the average frequency is to an independent nominal or design frequency. You can tell how stable an oscillator is without knowing what frequency it is designed to produce, but to assess its accuracy, you must compare its frequency to the one specified by the maker.

The types of oscillator designs described in the following text are ranked roughly in order of increasing intrinsic stability. *R–C* oscillators are inexpensive and easy to make, but are not particularly stable. Oscillator circuits using discrete inductors and capacitors in an *L–C* circuit can show improved stability compared to the *R–C* type, but inductors tend to be bulky and somewhat unreliable. The best combination of stability and compactness in currently available oscillator designs is shown by oscillators that use an electromechanical element to stabilize the oscillation frequency. This class includes **quartz-crystal** oscillators, in which the frequency is controlled by the electrical characteristics of a mechanical quartz resonator.

7.3.1 *R–C* Oscillators

You have already encountered one type of *R–C* oscillator: the twin-T circuit shown in Figure 7.5. Two other common types of *R–C* oscillators are the **Wien-bridge** circuit and the **phase-shift** circuit.

The Wien-bridge oscillator, based on a type of measurement circuit devised by physicist Max Wien (pronounced "veen") in 1891, uses a series *R–C* circuit as a highpass element and a parallel *R–C* circuit as a lowpass element as shown in Figure 7.11.

At the frequency

$$f_0 = \frac{\omega_0}{2\pi} = \frac{1}{2\pi RC}, \tag{7.39}$$

the transmission loss through the series–parallel *R–C* circuit from v_o to v_b is a minimum, and it turns out that this is the frequency at which the circuit is most likely to begin oscillating. If we realize that the transfer function of the gain block in this circuit is

$$-G_W(s) = \frac{v_o}{v_a} = 1 + \frac{R_F}{R} \equiv \frac{1}{\beta} \tag{7.40}$$

and the transfer function of the feedback network is

$$H_W(s) = \frac{v_b}{v_o} = \frac{s/\omega_0}{(s/\omega_0)^2 + 3(s/\omega_0) + 1}, \tag{7.41}$$

the Nyquist plot of the overall loop gain $G_W(s)H_W(s)$ exactly intersects the point $(-1, j0)$ when $\omega = \omega_0$ and $1/\beta = 3$. This corresponds to the situation in which the gain block has a gain magnitude of 3. For a gain larger than 3, linear system theory predicts that the output will grow without limit, but one of two things will happen before that, depending on details of the design.

If the circuit is built just as shown and the gain block gain $1/\beta$ is made slightly larger than 3, any initial disturbance such as a power-supply transient at turn-on will

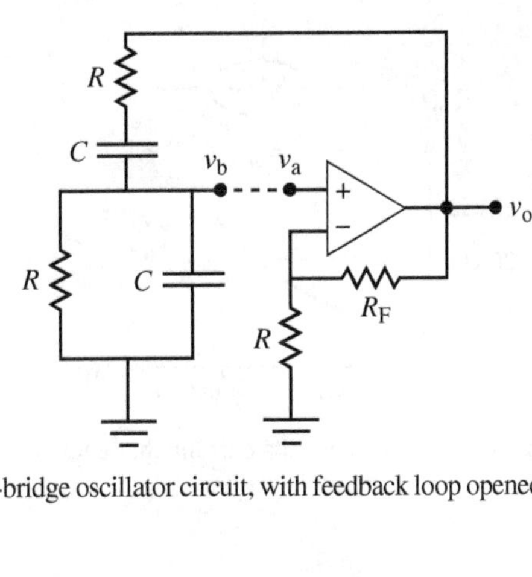

FIGURE 7.11 Wien-bridge oscillator circuit, with feedback loop opened at gap between v_b and v_a.

excite a growing sinusoidal wave, which will eventually cause the op amp to reach its *rails*—the maximum and minimum output voltage determined by the op amp's design and the power-supply voltages. When this happens, the waveform at v_0 will become clipped and distorted, and the circuit's steady-state output will depend on details of how the op amp clips and other factors. One thing is for sure, though: the output will not be a sine wave. If a sine-wave voltage is desired, this circuit will not deliver it as shown.

If, on the other hand, the designer implements a form of "gentle" gain control, the oscillator can be made to operate in a barely nonlinear fashion that greatly lowers the distortion level, making the output nearly sinusoidal. There are several ways to do this. One of the simplest ways is to use a nonlinear element for the resistor R_F, which affects the gain magnitude of the gain block. Assuming $R = 10\,\text{k}\Omega$, the value of R_F that will make the Nyquist diagram barely enclose the $(-1, j0)$ point is $R_F = 20\,\text{k}\Omega$. A **thermistor** is a type of resistor whose resistance varies with temperature. Its temperature depends on the power it dissipates, which in turn depends on the voltage across it. If R_F is made to be a thermistor with the resistance-current characteristic shown in Figure 7.12, an interesting thing can happen. Initially, the thermistor's resistance of less than $18\,\text{k}\Omega$ produces a gain magnitude of greater than 3 for the gain block, which leads to exponentially growing sine-wave oscillations in the circuit. When these oscillations at v_0 reach an amplitude of 5 V (rms), the rms current through R_F rises to $167\,\mu\text{A}$, which according to the graph in Figure 7.12 causes its resistance to rise above $20\,\text{k}\Omega$. This is the threshold resistance between growing and decreasing oscillations, and in practice, the oscillations will stabilize at a voltage of about 5 V (rms). This assumes that the feedback loop formed by the voltage amplitude–current amplitude–temperature–resistance causality chain is also stable, but this can be ensured by proper design of the thermistor's thermal mass and other factors. This particular design has the defect that ambient temperature will affect the stabilized voltage level of the output, but there are other less temperature-sensitive stabilization approaches involving detection of the output amplitude and controlled electronic resistances. The basic principle is the same, however: the designer uses the output voltage amplitude of the oscillator to control the gain of the gain block so as to maintain the loop-gain magnitude very close to 1. The advantage of such feedback

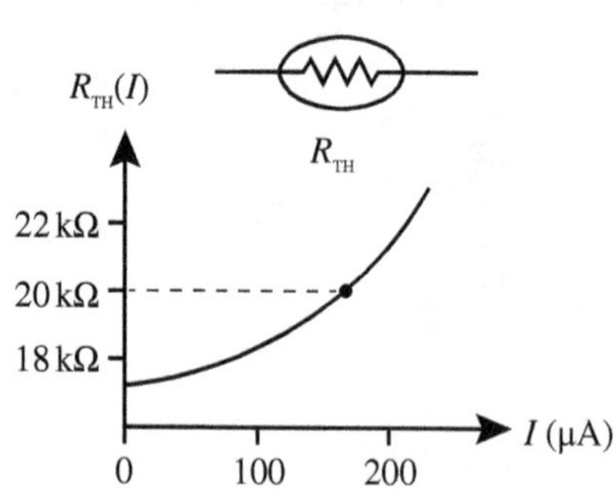

FIGURE 7.12 Hypothetical resistance versus current characteristic of thermistor used for R_F in Figure 7.11.

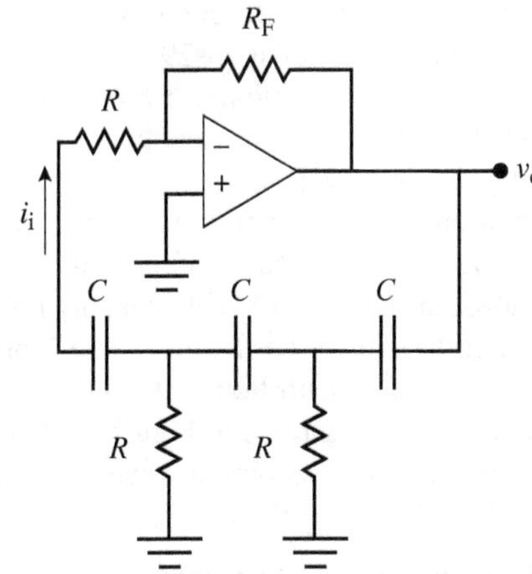

FIGURE 7.13 Phase-shift *R–C* oscillator circuit.

gain control is that the amplifier operates in its nominally linear range and the distortion of the output sine wave is held to a very low value.

Such linearized oscillator circuits are not as widely used as formerly, because digitally generated waves from **AWGs** can be produced that have a distortion level limited only by the accuracy of the *DAC* used and the algorithm used to produce the sine wave. But linearized sine-wave oscillators are still useful in less expensive equipment and for special purposes.

Another type of sine-wave *R–C* oscillator is the phase-shift oscillator, shown in Figure 7.13. This circuit consists of a phase-inverting gain stage with feedback resistor R_F followed by three highpass *R–C* filters connected directly to one another in cascade. In isolation, each such stage contributes a phase advance ranging from 90° at low frequencies to 0° at high frequencies, so obviously at some finite frequency, the phase shift of the transfer function of all three stages in cascade (represented by i_i/v_o) will pass through 180°. Adding another 180° phase shift contributed by the inverting gain stage yields 360°, which satisfies the phase requirement of the Barkhausen criterion. But to satisfy the amplitude part, one must know the exact transmission loss through the *R–C* network at the frequency where the phase shift is 180°, and this can be obtained from the transfer function *H*(*s*) of that network:

$$H(s) = \frac{i_i}{v_o} = \frac{sC(sRC)^2}{(sRC)^3 + 6(sRC)^2 + 5sRC + 1} \tag{7.42}$$

It is straightforward to show that the lowest frequency ω_0 at which the phase angle of $H(j\omega)$ reaches 180° is

$$\omega_0(\text{phase} - \text{shift}) = \frac{1}{RC\sqrt{6}} \tag{7.43}$$

At the frequency f_0(phase shift) = $\omega_0/2\pi$, the magnitude $|G(j\omega_0)H(j\omega_0)| = 1$ when $R_F/R = 29$, which means that a voltage gain of 29 from the left-hand *R–C* junction to v_o is needed to compensate for the loss through the passive *R–C* network at the frequency f_0. Any gain magnitude larger than 29 makes the Nyquist plot barely encircle the $(-1, j0)$ point, leading to a potential for oscillation. With the highpass configuration shown, high-frequency noise has an opportunity to initiate oscillation. In principle, a cascade of three lowpass *R–C* sections would work just as well (with series *R*s and shunt *C*s to ground), but that configuration may cause problems with initiation of oscillation. Without external means of limiting the oscillation amplitude, the phase-shift circuit shown in Figure 7.13 will probably not produce a very pure sine wave, because the high-frequency harmonics passed through the phase-shift network will contribute to harmonic distortion at the output. If a form of nonlinear gain control such as the thermistor described in Figure 7.12 is applied to the phase-shift oscillator, however, the quality of its output waveform will improve.

7.3.2 Quartz-Crystal Resonators and Oscillators

Crystalline quartz (silicon dioxide, or SiO_2) shows the **piezoelectric effect**, discovered by Jacques and Pierre Curie in 1880. They demonstrated that an electric field applied in certain directions to a quartz crystal caused its shape to change and also found that applying a mechanical stress would produce a voltage difference in the crystal. In 1921, physicist Walter G. Cady showed that this effect could be used to couple the mechanical resonance of a bar of quartz to an electronic circuit, forming an electromechanical resonator that allowed him to make an extremely stable oscillator. In every such oscillator, the energy stored in the mechanical vibrations of the crystal couples to the electronic circuit by means of the piezoelectric effect. It turns out that quartz's mechanical losses are very low when a good crystal is properly processed and mounted, and the equivalent electrical circuit that models a typical quartz crystal has an effective *Q* of several thousand or more.

The equivalent electrical circuit of a two-terminal quartz-crystal resonator is shown in Figure 7.14. At frequencies far removed from the designed resonant frequency of the crystal, the component behaves like a small capacitance C_0 that is called the **crystal shunt capacitance**. This capacitance results from the fact that physically, most quartz-crystal resonators consist of a thin slab of quartz sandwiched between two conducting metal films. This structure forms a capacitor with the value C_0, which is usually a few picofarads or less.

As the frequency of the applied AC voltage on the crystal nears the crystal's series-resonant frequency f_S, the mechanical vibrations in the crystal caused by the piezoelectric effect begin to excite the crystal's resonance, and the current through the crystal increases. This process is represented electrically by the series *C–R–L* circuit in parallel with the crystal capacitance C_0. The tuned circuit contains a **motional inductance** *L*, a **motional capacitance** *C*, and a **loss resistance** *R*. Some actual measured values from a crystal with a series-resonant frequency of $f_S = 427.4$ kHz are $C_0 = 5.8$ pF, $C = 42$ fF, $L = 3.3$ H, and $R = 385\ \Omega$. (The value of 42 fF—a *femtofarad* is 10^{-15} F—is not an error; equivalent circuits of crystal resonators typically have extremely small values of series-resonant capacitance and large values of inductance.)

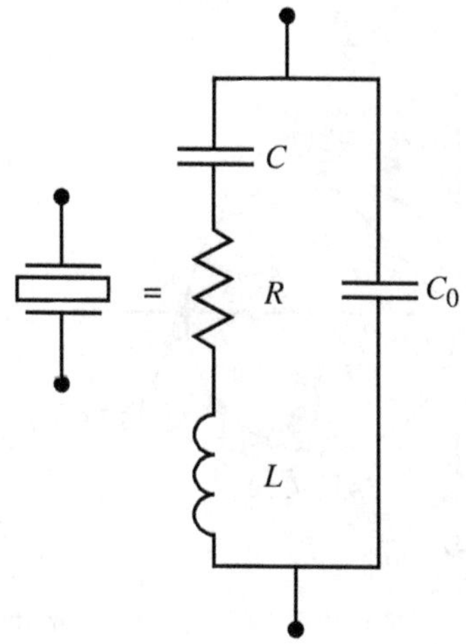

FIGURE 7.14 Schematic symbol for quartz-crystal resonator and equivalent circuit, showing resonant circuit C–R–L and crystal capacitance C_0.

Any two-terminal crystal has two characteristic frequencies: the series-resonant frequency f_S and the parallel-resonant frequency f_P. The series-resonant frequency f_S is simply

$$f_S = \frac{\omega_S}{2\pi} = \frac{1}{2\pi\sqrt{LC}} \tag{7.44}$$

The Q of the series resonance of the C–R–L circuit is the ratio of the inductive reactance of L to the series resistance R at the series-resonant frequency f_S:

$$Q = \frac{2\pi f_S L}{R} \tag{7.45}$$

In the case of the example equivalent-circuit values mentioned in the previous paragraph, its Q is about 23,000. Building a physical circuit with an equivalent Q out of actual passive components (inductors and capacitors) would require a 3.3-H inductor whose series resistance at 427.4 kHz was only 385 Ω. A physical inductor with these specifications is practically impossible to build, but a crystal contains in a small package an equivalent circuit that behaves like such an inductor was present. The reason is that mechanical losses of quartz are extremely low, allowing for such low-loss equivalent circuits.

At the exact series-resonant frequency of the crystal, the reactances of the equivalent components C and L cancel, leaving only the 385-Ω resistor in parallel with C_0, whose reactance is much higher than that at f_S and can be neglected at that frequency. However, at a frequency only slightly higher than f_S, the series C–R–L circuit behaves like an inductor that can resonate with the crystal capacitance C_0 to produce a *parallel* resonance at the frequency f_P.

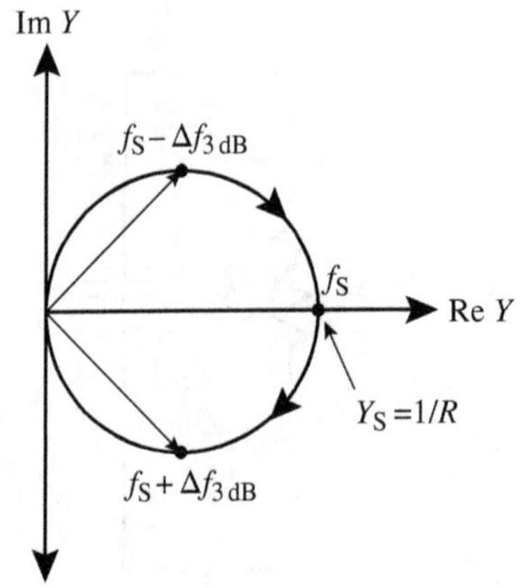

FIGURE 7.15 Plot of series-resonant C–R–L circuit's admittance Y_S near the series-resonant frequency f_S.

We can understand the parallel resonance better if we examine the **admittance** (inverse of impedance) versus frequency $Y_S(j\omega)$ of the series C–R–L circuit (only) on the complex admittance plane, as shown in Figure 7.15.

Near the series-resonant frequency, if we define $\delta\omega = 2\pi(f - f_S)$, it can be shown that the series-circuit admittance is

$$Y_S(f) \approx \frac{1}{R(1 + jQ(\delta\omega/\omega_S))} \tag{7.46}$$

At the series-resonant frequency, $\delta\omega = 0$, and the admittance is simply $1/R$, as it should be. Above f_S, the circuit behaves like an inductor, and this region is where parallel resonance is achieved. It turns out that many circuits that use crystal resonators employ the inductive region of the admittance-versus-frequency function as a large, high-Q inductor combined with external circuit capacitance to form a useful parallel-resonant circuit.

How large is the largest effective inductive admittance? If we define a "3-dB-down" admittance bandwidth as

$$\Delta f_{3dB} = \frac{f_S}{Q}, \tag{7.47}$$

at a frequency $f_S + \Delta f_{3dB}$, the admittance according to Equation 7.46 is

$$Y\Big|_{f_S + f_{3dB}} = \frac{1}{R(1+j)} = \frac{1}{2R} - \frac{j}{2R}, \tag{7.48}$$

and the value $-j/2R$ is the largest inductive admittance that the series C–R–L circuit ever shows. For the equivalent-circuit values given in the example of a 427.4-kHz crystal in Section 7.3, this value is about $-j1.3\,\text{mS}$. To make a parallel-resonant

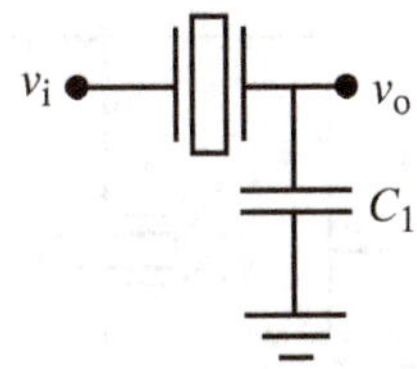

FIGURE 7.16 Feedback network for oscillator using series resonance of crystal resonator.

circuit with that value of inductive admittance near 427 kHz requires a capacitance of almost 2 nF, but it would not be wise to use that large a value, because small changes in Q might lower the maximum value of inductive admittance and the circuit might fail to resonate altogether. The total external capacitance connected across the crystal when it is used in a circuit is termed the **load capacitance**. Crystals designed for parallel-resonance use will have a specified load capacitance and are designed to show their parallel resonance at the specified frequency with that total load capacitance across the device. A typical value of load capacitance is in the range of 5–32 pF, depending on the C_0 capacitance of the device, and problems will arise if the external circuit applies much more capacitance than this across the crystal. In the worst case, excessive load capacitance can completely swamp the largest inductive admittance the crystal can show, and no parallel resonance appears at all, which means the circuit will not work as intended. So any design using crystal resonators should be carefully checked to ensure that the load capacitance is within requirements.

The type of feedback network in which a crystal is used depends on whether it is specified for use as a series-resonant or a parallel-resonant circuit. A simple feedback network that uses a series-resonant crystal is shown in Figure 7.16.

The crystal is connected between an input terminal v_i and an output terminal v_o, which is open-circuited except for capacitor C_1. If C_1's value is much greater than the crystal capacitance C_0, at frequencies far away from the series-resonant frequency f_S, the circuit forms a capacitive voltage divider with a large loss between input and output. However, as long as the crystal's series resistance R is much smaller than C_1's reactance at f_S, the circuit's transfer function v_o/v_i will show a sharp peak at f_S. If the circuit of Figure 7.16 is incorporated in a feedback loop with the proper phase shift and start-up conditions are met, the system will oscillate at a frequency near f_S that is controlled primarily by the crystal's characteristics.

The use of a crystal's parallel resonance can be illustrated by the circuit in Figure 7.17. In this circuit, the crystal is placed in parallel with a series combination of capacitors C_1 and C_2. Initially assuming an ideal voltage source drives the terminal labeled v_i, the circuit will present a real impedance to the output terminal labeled v_o at the parallel-resonant frequency $f_P = \omega_P/2\pi$ given by the solution to the equation

$$j\omega_P(C_0 + C) + \frac{1}{jQR(\omega_P/\omega_S)} = 0 \quad (7.49)$$

Because Q is typically so large for crystal resonators, the parallel-resonant frequency f_P needs to be only slightly higher than the series-resonant frequency f_S in order to

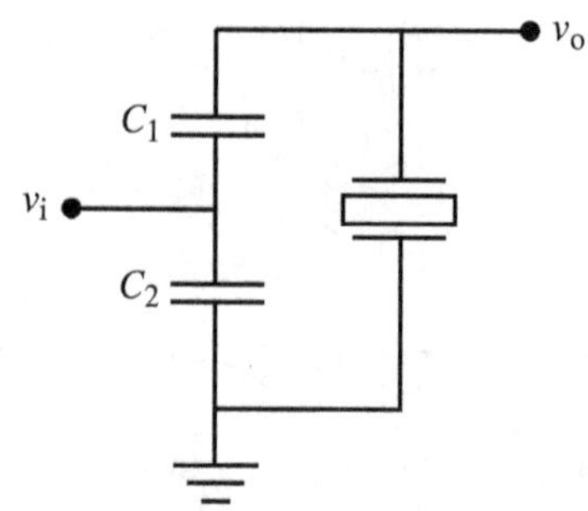

FIGURE 7.17 Feedback network for oscillator using parallel-resonant crystal resonator.

satisfy Equation 7.49. It is also evident from the same equation that changes in the value of capacitor C_1 will produce slight changes in the parallel-resonant frequency. This is a way to make slight adjustments or corrections in the frequency of a crystal oscillator without losing the advantage of having the crystal determine the frequency of oscillation.

In use, a circuit with a nonzero output resistance R_S would be connected to the terminal labeled v_i in Figure 7.17. It can be shown that at resonance, the circuit behaves like a step-up transformer whose turns ratio depends on the ratio of C_1–C_2. By adjusting this ratio and keeping the total effective capacitance across the crystal about the same, the designer can obtain a voltage step-up from the parallel-resonant circuit that can be helpful in certain types of feedback oscillator designs.

Besides their use in feedback oscillator circuits, crystals are useful as circuit elements in narrowband filters, and some types of crystals with three terminals are made for filtering purposes.

7.3.3 MEMS Resonators and Oscillators

MEMS use fabrication methods originally developed for semiconductor devices and ICs to make devices and systems with moving parts for various special purposes. One of the earliest applications of MEMS technology was in inkjet printers, which became popular along with the personal computer in the 1980s. A MEMS resonator is a mechanical structure designed to vibrate at a specific frequency, much like a quartz-crystal resonator. However, because the MEMS structure can be engineered to have a variety of useful characteristics, the designers of MEMS resonators can exploit standard CMOS IC manufacturing techniques to develop oscillators that are smaller than quartz-crystal oscillators, are much easier to make, and can outperform the older technology in some ways.

One of the most important specifications of a clock oscillator is its frequency stability with respect to temperature changes. Quartz-crystal oscillators can use specially cut crystals that have a fairly small sensitivity to temperature over a limited range, and there are ways to design **temperature-compensated crystal oscillators (TCXOs)** that lessen the system's temperature dependence even more. However, these systems are not integrated into one IC and typically require assembly of discrete components: the crystal itself (which is often housed in a separate container), the other oscillator components, and temperature-compensating circuitry.

As an example of recent MEMS oscillator developments, the firm Silicon Laboratories Inc. has developed a MEMS-based line of IC oscillators that eliminate most of the assembly problems associated with quartz-crystal oscillators. Their device uses a thin square resonator composed of a combination of pure silicon and silicon–germanium (SiGe) material. It turns out that the temperature coefficients (change in frequency with temperature) are opposite in sign for the two materials, so by choosing the right combination of silicon and SiGe, the intrinsic temperature sensitivity of the resonator can be made extremely small.

The resonator itself vibrates rather like a trampoline suspended on four springs, one at each corner, as shown in Figure 7.18. (The actual springs are not coiled wires, of course, but micromachined pieces of silicon with a somewhat different shape than our schematic picture shows.) Four electrodes separated from the resonator by sub-micron-wide gaps capacitively couple to the resonator, providing both electrostatic excitation forces and an output signal to use in an electronic feedback loop that keeps the resonator vibrating in the desired mode, which is in the low-megahertz region.

Instead of designing different resonators for different output frequencies requested by the customer, the same resonator frequency is used in all devices of a given type, and the desired output frequency is programmed into an on-chip system called a **frequency-locked loop**, a simplified version of which is shown in Figure 7.19.

A frequency-locked loop is a type of feedback system that stabilizes the frequency of a **VCO** by comparing its output with that of a stable reference oscillator. A frequency counter inside the IC uses the temperature-stabilized MEMS oscillator

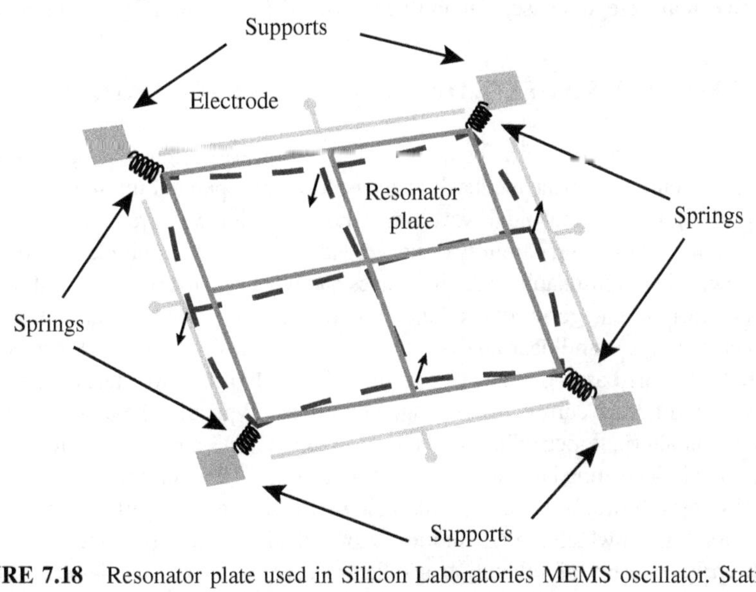

FIGURE 7.18 Resonator plate used in Silicon Laboratories MEMS oscillator. Stationary position shown in solid lines; vibration mode shown in dashed lines.

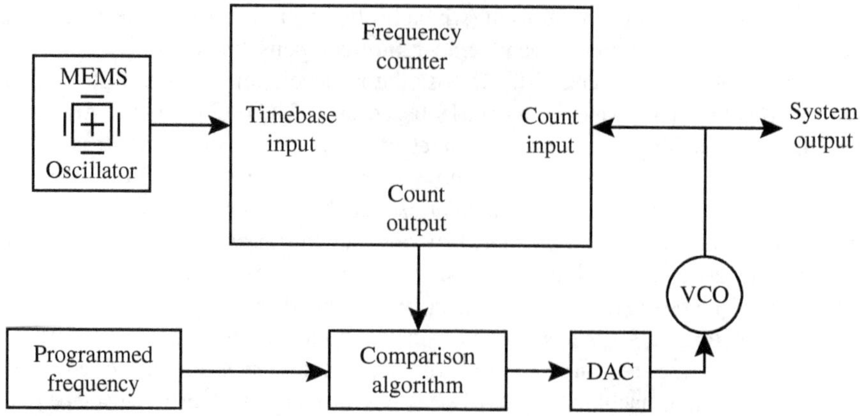

FIGURE 7.19 Frequency-locked loop used in one type of MEMS oscillator system.

output as its frequency reference and counts the VCO's frequency periodically. The counter's output is compared to the desired programmed-in frequency with a digital comparison algorithm, which produces a correction that is sent to a DAC that produces a control voltage sent to the VCO. Not shown in the diagram is a temperature sensor that applies further compensation to the output frequency as temperature varies, improving the system's frequency stability even more. By changing the programmed frequency on the chip with external inputs, the user can obtain any output frequency from 32 kHz up to 100 MHz with no change in hardware.

While quartz-crystal oscillators are currently still widely used, it is likely that MEMS-based oscillators will gradually supplant them, especially in applications for which the small size and ease of manufacture of MEMS-based ICs are important.

7.4 TYPES OF TWO-STATE OR RELAXATION OSCILLATORS

Up to now, all the oscillator circuits we have discussed were designed using a linearized equivalent-circuit and linear feedback analysis. They all operated through the gradual buildup of a growing sinusoidal waveform whose amplitude is eventually limited by circuit nonlinearities, but if carefully designed with appropriate amplitude controls, such circuits can deliver reasonably pure sine waves and work in a nearly linear fashion.

By contrast, the categories of oscillator circuits we will now discuss are intrinsically nonlinear in nature, and linear analysis cannot be used to study them. This is because they are best treated as two-state systems in which each of the two states is characterized by very different current and voltage conditions. Typically, the active devices in such circuits alternate periodically between extreme "on" and "off" states, and the cycle period is determined by passive component values. These circuits are also termed **relaxation oscillators**, because they alternate between a periodic influx of energy and a relaxation time in which the circuit moves toward a lower-energy state.

The earliest forms of relaxation oscillators used two-terminal devices such as neon-filled lamps, which exhibit a nonlinear current–voltage relationship characterized by a threshold or breakdown voltage. Solid-state devices that can be used in similar circuits include the two-terminal "*diode* for *a*lternating *c*urrent" (**DIAC**) and the **unijunction transistor**. However, most two-state oscillators in use today employ three-terminal devices, typically bipolar junction transistors (BJTs) or CMOS FETs, and we will describe a couple of such circuits to illustrate the uses and limitations of these oscillators.

7.4.1 Astable Multivibrator

The term **multivibrator** denotes a type of two-stage amplifier whose output is connected to its input, forming a positive feedback loop. The active devices in a multivibrator are either on or off and rapidly transition between on and off states. Multivibrators are classed into three types: **bistable**, **monostable**, and **astable**. Bistable multivibrators are usually known by the more familiar name of **flip-flops** and are used primarily in digital circuits. A monostable multivibrator stays in one state until triggered by an external input, whereupon it goes to a second state for a fixed interval of time before reverting to its original state. And an astable multivibrator has no stable state but alternates periodically between each of two states.

An astable multivibrator oscillator using NPN BJTs is shown in Figure 7.20. The two BJTs are connected in the common-emitter configuration as inverting amplifiers, so in the brief times when the devices transition through their linear regions, the phase shift around the loop is 360°, indicating there is positive feedback. This positive feedback in practice simply serves to cause the state transition to occur very rapidly, as fast as the high-frequency characteristics of the transistors will allow.

The operation of the astable multivibrator is best explained with the aid of waveforms measured at the base and collector terminals of both transistors. An actual circuit using 2N3904 BJTs was constructed and powered with V_{CC} = +5 V, using values for base resistors $R_1 = R_2 = 470\,\text{k}\Omega$, collector resistors $R_3 = R_4 = 2.2\,\text{k}\Omega$, and coupling capacitors $C_1 = C_2 = 1\,\text{nF}$. The resulting measured waveforms are shown in Figure 7.21.

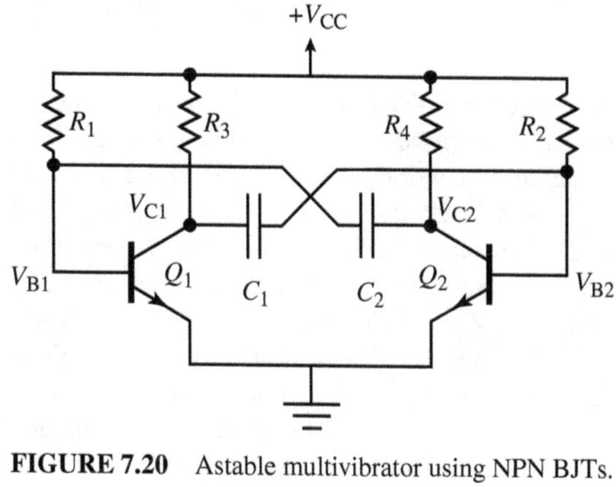

FIGURE 7.20 Astable multivibrator using NPN BJTs.

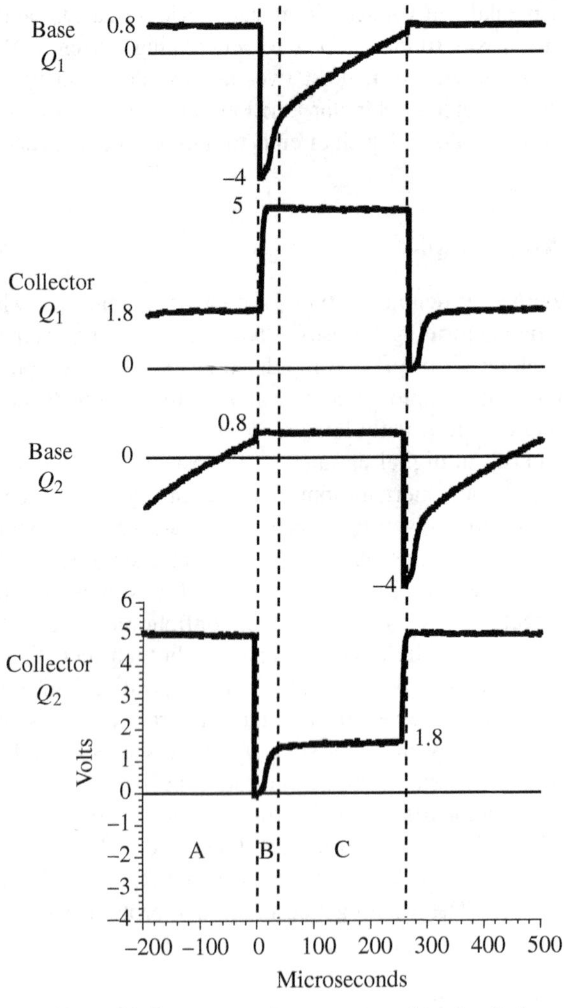

FIGURE 7.21 Astable multivibrator waveforms measured in circuit shown in Figure 7.20.

At the beginning of the time slice in Figure 7.21 shown as region *A* (from −200 to 0 μs), transistor Q_2 is cut off because its base voltage is less than +0.6 V. This is because the right-hand terminal of capacitor C_1 is being charged positively from an initial negative voltage by Q_2's base resistor R_2 toward V_{CC} = +5 V. The left-hand terminal of capacitor C_1 is held at a constant voltage of about 1.8 V, because transistor Q_1 is biased in its active region by its base resistor R_1. Capacitor C_2 has its right-hand terminal at Q_2's collector voltage of +5 V and its left-hand terminal at Q_1's base voltage, which in region *A* is about 0.8 V. In time slice *A*, essentially no current is flowing through C_2.

At a time determined primarily by the time constant $\tau = C_1R_3 = C_2R_1$, Q_2's base voltage rises through 0 to +0.6 V. This is the threshold voltage at which the circuit triggers or switches rapidly to a different state. (In general, all relaxation oscillators work by means of threshold or trigger voltages, which determine the time when the

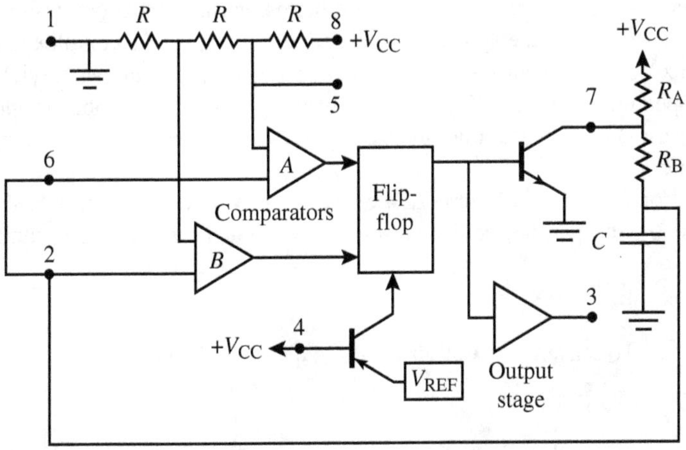

FIGURE 7.23 555 timer wired as an astable multivibrator with external components R_A, R_B, and C.

goes below 1/3 V_{CC} for a trigger signal, the 555 acts as a **monostable multivibrator**, producing a single pulse of known width at its output every time it is triggered. This function can be useful in event counters and other situations in which a reliable fixed-width pulse is needed, but the available trigger signal is erratic in amplitude or duration.

Many other relaxation-type oscillator circuits are used in a variety of analog electronics applications ranging from phase-locked loops to clock recovery circuits and others. Because they all operate by means of a threshold voltage, they are somewhat sensitive to circuit noise and interference and should be used with caution where highly precise timing waveforms are needed. But for many noncritical applications, analog timing and signal processing circuits using the relaxation-oscillator principle are often the best and cheapest solution.

7.5 DESIGN AID: SINGLE-FREQUENCY SERIES–PARALLEL AND PARALLEL–SERIES CONVERSION FORMULAS

Before we complete this chapter with an oscillator design example, it will be helpful to explain some formulas we will use that come in handy for many design situations involving resonant circuits in radio-frequency (RF) and high-frequency designs. For lack of a better name, we will call them the single-frequency series–parallel and parallel–series conversion formulas. They are useful in analyzing resonant circuits involving resistors, capacitors, and inductors, which covers a lot of ground.

We assume that the problem deals with a single sine-wave frequency f_0 with corresponding radian frequency $\omega_0 = 2\pi f_0$. The problem to be solved is this: given a series combination of a resistor and capacitor (or inductor), what is the equivalent parallel circuit that presents the same impedance (or admittance) at the frequency f_0? And given a parallel combination of R and C (or R and L), what is the corresponding equivalent series circuit?

There are several ways to solve these problems. Perhaps the most straightforward approach is to express the series (parallel) circuit in terms of a complex impedance (admittance) and use your calculator to invert the complex number to yield admittance (impedance) and then interpret the resulting complex number in terms of a parallel (series) circuit. But that process involves a number of steps, and errors can creep in along the way.

The set of formulas in Figure 7.24 expresses this process in terms of the radian frequency ω_0; the parallel equivalent-circuit component values R_P, L_P, and C_P; the

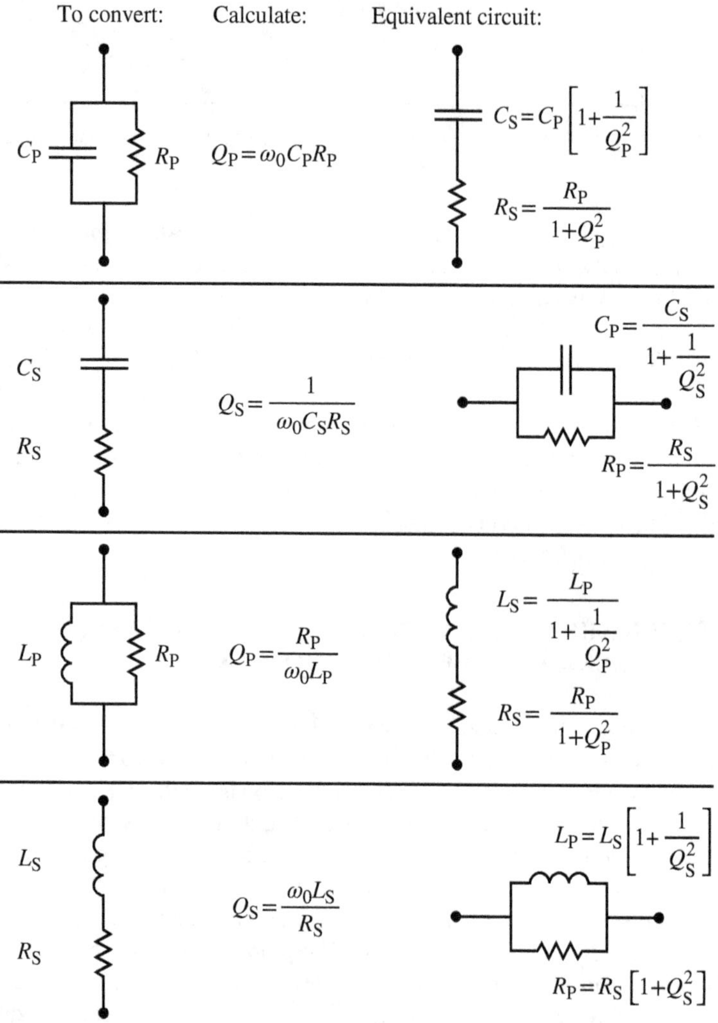

FIGURE 7.24 Formulas for conversion between series and parallel equivalent *R–C* and *R–L* circuits at a single frequency f_0.

series equivalent-circuit values R_S, L_S, and C_S; and two auxiliary quantities Q_P and Q_S. We will illustrate the use of these formulas with an example.

Suppose you have a parallel combination of a capacitor $C_P = 33\,\text{pF}$ and a resistor $R_P = 330\,\text{k}\Omega$ at a frequency of $f_0 = 60\,\text{kHz}$. Because this circuit happens to be in series with an inductor, you would like to know what its series equivalent circuit is, because then you can simply add the impedances in series to find the overall circuit impedance.

From the top row of circuits and equations in Figure 7.24, we find that we must first calculate a quantity $Q_P = \omega_0 C_P R_P$. This quantity is useful both for further calculations and for telling you whether the *R–C* circuit is mainly resistive or capacitive at the frequency in question. For $C_P = 33$ pF and $R_P = 330$ kΩ at $f_0 = 60$ kHz, $Q_P = 4.105$, which means that the circuit looks mainly like a capacitor, but with substantial losses. (A Q of <1 means the circuit is mainly resistive, while a Q of >10 means the circuit is mainly reactive and losses are low enough to be neglected in many cases.)

At the right side of the top row are two equations that give the series equivalent-circuit component values. Notice that the larger Q_P is, the closer C_S will be to C_P. For high-Q circuits, the value of the reactive component stays nearly the same for the parallel and series equivalent circuits. But note that the parallel equivalent resistor R_P gets divided by the quantity $(1 + Q_P^2)$, which in this case is 17.8, resulting in a series equivalent resistor $R_S = 18.48\,\text{k}\Omega$. Because Q_P includes the term R_P, for large $Q_P > 10$ we have the approximate relation

$$R_S \approx \frac{1}{(\omega_0 C_P)^2 R_P} \tag{7.51}$$

That is, the series equivalent resistor R_S is *inversely proportional* to the parallel equivalent resistor R_P. As R_P increases, R_S decreases. This effect will be significant in the design example we will now present.

7.6 DESIGN EXAMPLE: BJT QUARTZ-CRYSTAL OSCILLATOR

Oscillator design is made difficult by the fact that it always involves the nonlinear behavior of active devices. While linear analysis can be used to create conditions that are necessary for an oscillator to start oscillating, to determine whether these conditions are also sufficient generally requires numerical modeling with circuit simulation software. The following design example has been laboratory tested and works well even with the small crystals used in digital watches and other miniature equipment. The small physical size of these crystals means that the circuit using them must operate with very small currents at a high impedance level. The design begins with a given circuit topology shown in Figure 7.25. We will show how the discrete-component values are chosen for the particular crystal and frequency in use.

We will select the two most important components at the outset. The crystal X_1 is a quartz-watch-type unit with a parallel-resonant frequency of 60.00 kHz when connected in a circuit with a load capacitance of 33 pF. The transistor Q_1 is a general-purpose NPN BJT, the 2N3904.

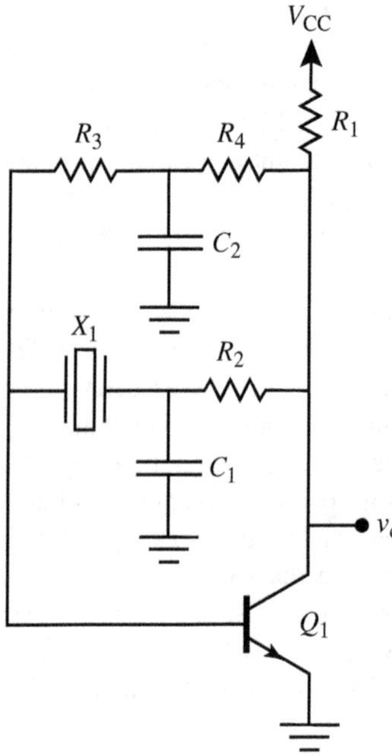

FIGURE 7.25 Quartz-crystal oscillator circuit design example using BJT.

The crystal oscillator circuit consists of an inverting amplifier, consisting of common-emitter BJT Q_1 and collector load resistor R_1, connected to two independent feedback circuits: a negative-feedback bias circuit (C_2, R_3, and R_4) and a positive-feedback oscillation path (C_1, R_2, and X_1).

The bias-circuit values will be chosen so that there is negligible transmission of the oscillation signal (in this case, 60 kHz) through the bias circuit, which passes only DC. The bias circuit operates in a negative-feedback mode to maintain the DC collector voltage at about 1/3 of the power-supply voltage V_{CC}. This biases the transistor near the middle of its active region and provides AC gain for the signal feedback path.

In operation, the crystal resonates at its parallel-resonant frequency f_p, and the crystal's load capacitance is primarily provided by C_1. A 180° phase shift is encountered by the oscillation signal as it goes from the transistor's base to its collector, and an additional 180° shift is provided by the combination of C_1, the crystal's effective inductance L_{EFF}, and the transistor's input capacitance, which we will determine with an AC equivalent circuit for the BJT. But first, we will design the DC bias circuit to establish a suitable DC collector current.

At DC, the base–emitter voltage V_{BE} is about 0.7 V for silicon transistors (including the general-purpose-type 2N3904, which we have selected for this design), and the only components of interest for the DC bias design in Figure 7.25 are R_1, R_3, and R_4. For convenience, we will choose $R_3 = R_4 = R$. If the transistor's DC current gain is β_{DC}, we can write an equation for the DC collector voltage V_C:

$$V_C = V_{CC} - I_C R_1 = V_{CC} - \beta_{DC} I_B R_1 \tag{7.52}$$

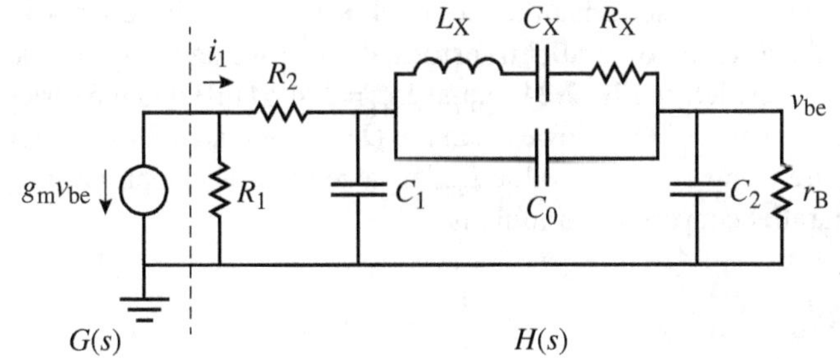

FIGURE 7.27 AC equivalent circuit of crystal oscillator: $R_1 = 56\,k\Omega$, $R_2 = 330\,k\Omega$, $C_1 = 33\,pF$, $L_x = 2549.76\,H$, $C_x = 2.76005\,fF$, $R_x = 9357\,\Omega$, $C_0 = 1.4\,pF$, $C_2 = 420\,pF$, and $r_b = 44\,k\Omega$. Value of g_m is unknown to be found. Gain block $G(s)$ and feedback block $H(s)$ are separated by dashed line.

C_{M1} and can be neglected, so for simplicity in the oscillator equivalent circuit of Figure 7.27, we have set $C_{M1} = C_2 = 420\,pF$.

Because a small crystal resonator is a high-impedance device, using it as a parallel-resonant circuit requires that if one terminal has a low impedance to ground, the other terminal should be at a fairly high impedance to ground. The large Miller capacitance of the BJT's base makes that terminal the low-impedance end, which is why we will place a large resistor R_2 (330 kΩ) in series between the collector and the crystal on the high-impedance side.

We can show why this is important if we use the series–parallel equivalents developed earlier to find the series equivalent circuit of the crystal, including its own internal losses. We measured an actual 60-kHz watch-type crystal in a simple test fixture, using an AWG to provide a synthesized signal source that can be set with 0.1-Hz accuracy and stability. The equivalent-circuit values of the particular crystal we measured are (with reference to the components in Fig. 7.14):

$L = 2549.76\,H$

$C = 2.76005\,fF$ (1 fF = 1 femtofarad, 10^{-15} F)

$R = 9357\,\Omega$

$C_0 = 1.4\,pF$

The resulting equivalent circuit has a Q of 102,700. When this circuit is placed in parallel resonance with a 33-pF capacitance, the effective inductance will be that which resonates at 60 kHz with (33 + 1.4) = 34.4 pF. That inductance works out to be $L_{EFF} = 204.54\,mH$. As long as the problems we are solving involve a single frequency, we can replace the crystal's four-component equivalent circuit with L_{EFF}, and the results will be unchanged. This is a convenient simplification for initial hand-calculation circuit designs, although it should not be used for detailed circuit simulation because it is inaccurate at frequencies other than f_p. L_{EFF} results from the inductive reactance that is the difference between the very large actual inductive reactance of L_x and the oppositely signed and nearly as large capacitive reactance of C_x.

In particular, we can see what the motional resistance R_x in series with L_{EFF} looks like if we make a series-to-parallel transformation. If we let the series inductance L_S (on the bottom row left in Fig. 7.24) equal $L_{EFF}=204.54$ mH and the series resistance $R_S=9357\,\Omega$, the value of Q_S is lower than the Q we calculated for the original circuit, because we are using the lower value L_{EFF} instead of L. The appropriate Q_S to use in the series–parallel conversion formula is

$$Q_S = \frac{\omega_0 L_{EFF}}{R} = 8.24 \tag{7.61}$$

which is considerably lower than 102,700. When we use this Q_S in the series–parallel conversion formulas shown in Figure 7.24 for the resistor–inductor circuit, we find that the parallel equivalent values are $L_P=207.55$ mH and $R_P=644.8\,\text{k}\Omega$. This is the same order of magnitude of resistance ($R_2=330\,\text{k}\Omega$) that we have placed across the high-impedance terminal of the crystal and further justifies our policy of keeping one terminal of the parallel-resonant crystal at a fairly high impedance.

This is a specific example of a general rule: the impedance levels of certain types of components require that the circuits they are used in approximately match those impedance levels. For example, if we tried to build an oscillator with the same crystal, but using a power BJT that required a base current of several amperes to operate, it would never work. Many analog designs use CMOS-based circuitry, which can present an extremely high impedance level to external components, and so circuit components used with such circuitry need to work effectively with very small currents.

The critical question to be answered next is this: does the transistor have enough gain to allow the circuit to oscillate? To answer this question, we will apply the Nyquist criterion to the equivalent circuit in Figure 7.27 and adjust the transconductance g_m so that the Nyquist locus crosses the (−1, 0) point. That transconductance will be the minimum theoretically necessary to initiate oscillation. We treat the equivalent circuit of Figure 7.27 as a gain block with transconductance $G(s)=g_m$ (the BJT equivalent-circuit voltage-dependent current source) connected to a feedback network whose *transfer impedance* is $H(s)=v_{be}/i_1$ (the crystal equivalent circuit and associated components). Note that the dimensions of $G(s)$ are Siemens and those of $H(s)$ are ohms, so that the product is dimensionless, as it must be for analysis with a Nyquist plot.

At frequencies far removed from the crystal's resonances near 60 kHz, the feedback network is a R–C network only, because the impedance of the L_x–C_x–R_x circuit is so high that it effectively disappears. At low frequencies, the transfer impedance $H(s)$ is zero because capacitor C_0's impedance goes to infinity. At high frequencies, the circuit capacitance means that for a fixed input current, the voltage at v_{be} again approaches zero, so we know that at zero and infinite frequencies, the product $G(s)H(s)$ will be zero. The interesting behavior happens in between, of course.

Figure 7.28 shows the Nyquist diagram of the equivalent circuit of Figure 7.27 with special emphasis on the frequencies very near the crystal's parallel resonance at 60 kHz. The small circle to the right of the imaginary axis is the locus of frequency points above and below 60 kHz. When the frequency is within a few hertz of 60 kHz, however, a large circle forms due to the resonant action of the motional components L_x, C_x, and R_x. The circuit will be unstable if the large circle encloses the (−1, 0) point on the negative

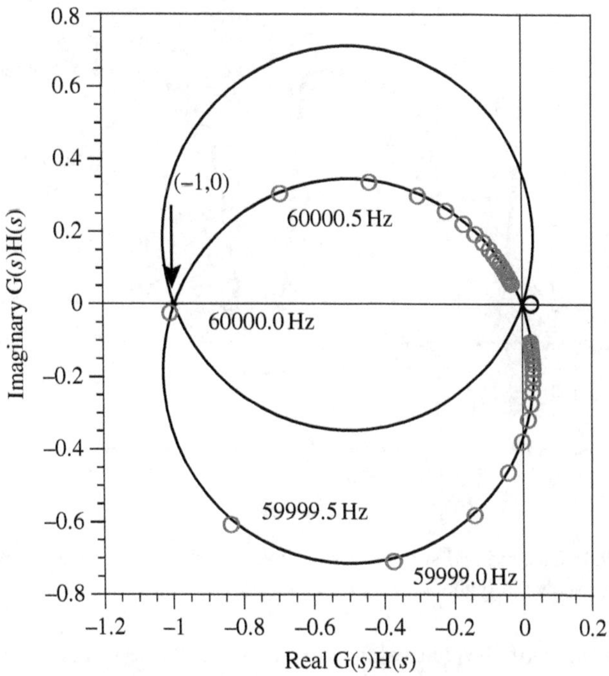

FIGURE 7.28 Nyquist diagram for equivalent circuit of Figure 7.27 with $g_m = 0.404$ mS. Small circles indicate 0.5-Hz frequency steps either side of 60 kHz.

real axis. As Figure 7.28 shows, this will occur for values of transconductance larger than about 0.4 mS. The aforementioned DC bias calculations showed that the actual transconductance with the bias conditions we established is considerably larger, about 2.27 mS. Because the entire Nyquist diagram scales with g_m, the diagram for $g_m = 2.27$ mS will definitely enclose the (−1, 0) point and will therefore have a pole in the right half plane, a necessary condition for oscillation. The fact that the feature of the diagram that encloses that point is restricted in frequency to a very narrow range around 60 kHz suggests that the resulting oscillation will occur near 60 kHz, and both simulation and laboratory experiments confirm that this is indeed the case.

The circuit in Figure 7.25 was simulated with Multisim™ circuit simulation software with one difference between the equivalent circuit of Figure 7.27 and what was actually implemented in Multisim. Circuit simulation software must solve for both the transient and the steady-state solutions for oscillator circuits. If one attempts to simulate a circuit with an extremely high-Q resonator, a difficulty arises. It can be shown that there is a time constant τ associated with the buildup of oscillation in an oscillator. Recall that for unstable oscillatory systems, the amplitude grows exponentially with time, which means as $e^{t/\tau}$. However, if τ is a large number, the exponential growth happens very slowly compared to the time scale of a single oscillation. For a passive *R–L–C* resonant circuit, the time constant can be shown to be

$$\tau = \frac{2Q}{\omega_0} \tag{7.62}$$

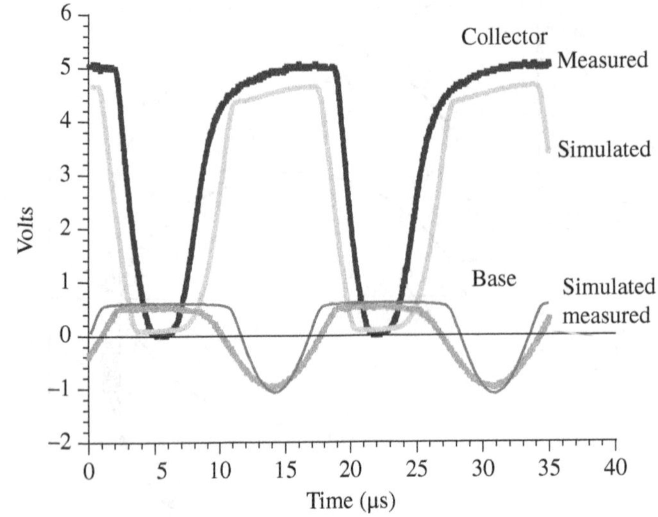

FIGURE 7.29 Measured and simulated base and collector voltage waveforms for crystal oscillator circuit shown in Figure 7.25.

The time constant of the equivalent circuit of the 60-kHz crystal used in this design example turns out to be over 0.5 s, or some 30,000 cycles of oscillation. This rate of increase is so slow that it can be observed by eye in the laboratory as a gradual rise in oscillator output voltage on an oscilloscope after the DC power is turned on. This long transient start-up delay means that a circuit simulation must run for a simulated time of several seconds (which may take many minutes on a computer) to have a chance of doing an accurate simulation. Often, simulation software runs into numerical difficulties when such long simulations are attempted, so we lowered the Q of the crystal's L_x–C_x–R_x equivalent circuit to 100 (by decreasing L_x and increasing C_x) before running the model on Multisim. The resulting simulation executed without a problem and gives a good indication of how the actual circuit operates, as the comparison of simulated and measured waveforms in Figure 7.29 shows.

These waveforms are taken after the transient start-up period when the circuit has reached its steady state. Further increase in output amplitude is limited by the fact that the transistor alternates between cutoff, lasting about 50% of the waveform's period, and saturation, which lasts about 20–30% of the period. The rest of the time the transistor is in its linear region as it transitions between cutoff and saturation. Note that the base–emitter junction acts as a diode clamp circuit that prevents the AC waveform at the base from rising any higher than about 0.6 V. Once the base–emitter diode turns on, the collector voltage begins to fall, though with a finite delay and fall time due to the Miller capacitance between base and collector. The transistor remains saturated until the base waveform goes into its negative half cycle, cutting off base current and moving the transistor's state from saturation through the linear region to cutoff. While the measured and simulated waveforms are not identical, the general shapes and amplitudes

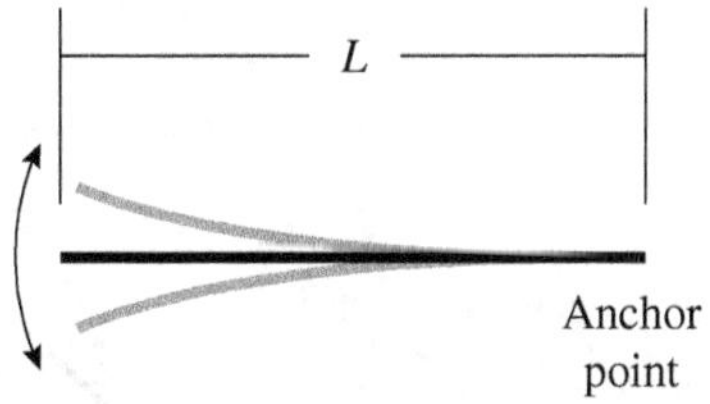

FIGURE 7.31 Simple silicon vibrating-beam MEMS resonator element.

forth in the direction the wave is propagating and is called the *longitudinal wave*. The other pertains to side-to-side motion perpendicular to the direction of wave travel, and it is called the *transverse wave*. In silicon, the longitudinal wave velocity is about 9000 m s^{-1}, and the transverse wave velocity is about 5500 m s^{-1}. One simple type of MEMS resonator element consists of a thin bar of silicon of length L that is held stationary at one end at an *anchor point* and is free to vibrate at the other end (see Fig. 7.31). The vibration is both excited and sensed by capacitive coupling to the free end. Although the detailed mechanical analysis of this structure is complicated, one can obtain an estimate of the dimension L required by assuming $L \sim 0.25\lambda$, where λ is the sound wave's length at the resonant frequency. If the desired resonant frequency is 8.0 MHz, use the appropriate speed of sound to calculate the length L required.

7.6. *Astable multivibrator design.* Redesign the astable multivibrator circuit shown in Figure 7.20 to operate with a power-supply voltage V_{CC}=+15 V and with collector load resistors $R_3 = R_4 = 6.8\,\text{k}\Omega$. Assume the silicon transistors' DC $\beta = 120$, and choose the base bias resistor values $R_1 = R_2$ so that with no current through the capacitors, the DC collector voltage is close to 7.5 V. Then choose values for capacitors $C_1 = C_2$ so that the oscillation frequency should be close to 2 kHz. Choose standard 5% tolerance values for the components. (*Optional*) Simulate your design with a circuit simulation package such as Multisim™ and see how close the actual oscillation frequency is to your design value. In order to obtain good simulation results, you may need to manually restrict the maximum time step the program takes to 1 μs. In Multisim™, this is done under the simulation menu.

7.7. *555 timer analysis.* In operation as an astable multivibrator, the voltage V_C across the capacitor at pins 2 and 6 in Figure 7.23 varies between 1/3 V_{CC} and 2/3 V_{CC} as shown in Figure 7.32. The figure uses as an example a circuit in which the time constant $\tau_1 = R_A C = 0.5$ s and $\tau_2 = (R_A + R_B)C = 1$ s. Based on what you know about R–C time constants and the fact that the trigger and threshold points are 1/3 V_{CC} and 2/3 V_{CC}, respectively, show that the oscillation frequency of the 555 in this configuration is $f = \dfrac{1}{T} = \dfrac{1.44}{(R_A + 2R_B)C}$

7.8. *L–C oscillator design.* Oscillator circuits that use discrete inductors and capacitors are especially useful at high frequencies in the RF range (~3–300 MHz). One popular form of *L–C* oscillator uses a FET in a circuit called a *Colpitts oscillator*, which is characterized by its resonant circuit consisting of a single

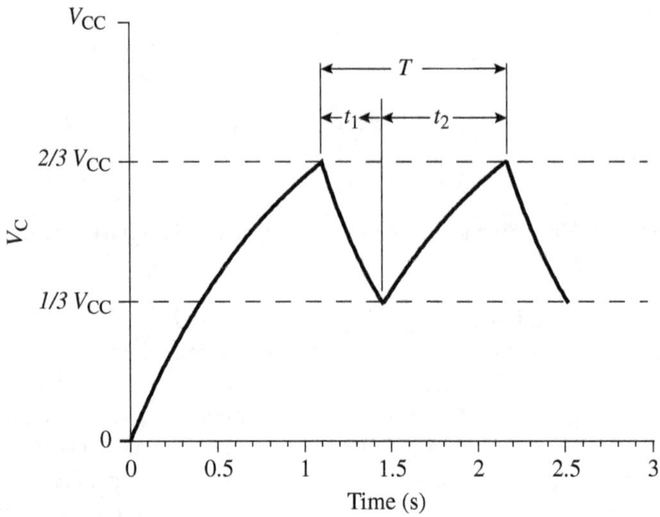

FIGURE 7.32 Capacitor voltage waveform in 555 timer circuit wired as astable multivibrator in Figure 7.23.

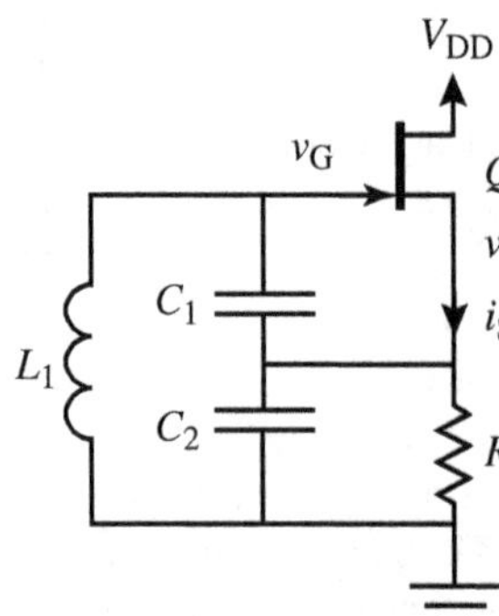

FIGURE 7.33 Colpitts oscillator analyzed in Problem 7.8.

inductor in parallel with two capacitors in series. The Colpitts *L–C* oscillator in Figure 7.33 uses a FET source follower (Q_1) to provide positive feedback from gate to source. As you know, source followers have a voltage gain of less than 1, so the feedback circuit (consisting of capacitors C_1 and C_2 in parallel with inductor L_1) performs a voltage step-up, somewhat like a transformer, at the circuit's resonant frequency f_0, which is given by

$$f_0 = \frac{1}{2\pi\sqrt{L(C_1C_2/(C_1+C_2))}}$$

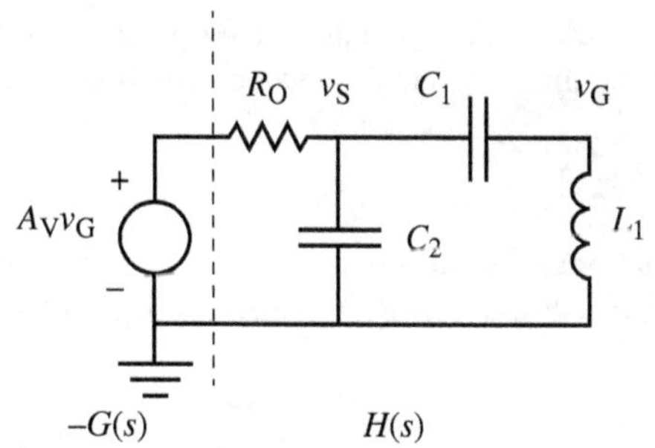

FIGURE 7.34 Simplified equivalent circuit of Colpitts oscillator in Figure 7.33.

(a) Assume an AC voltage v_S at the resonant frequency f_0 is present at the source terminal. Assuming the gate terminal is an open circuit (infinite impedance), which is not a bad approximation for lower frequencies, show that the resulting AC voltage on the gate is given by

$$v_G = v_S \frac{C_2 + C_1}{C_1}$$

This voltage gain is enough to more than compensate for the voltage loss through the FET source follower, which means that the amplitude part of the Barkhausen criterion can be satisfied. A more detailed analysis would include losses in the inductor and the finite input impedance of the gate terminal, but this simplified analysis is enough to guide an initial design.

***(b)** A simplified equivalent circuit of the Colpitts oscillator in Figure 7.33 is shown in Figure 7.34.

The source follower can be modeled as a voltage-dependent voltage source with voltage gain A_V and output resistance R_O, where $A_V = \frac{g_m R_S}{g_m R_S + 1}$ and $R_O = \frac{R_S}{g_m R_S + 1}$. The transconductance g_m of the FET can be calculated from its pinchoff voltage, saturation current, and DC bias conditions. Show that the Nyquist loop gain is

$$G(s)H(s) = -A_V \frac{s^2 L_1 C_1}{s^3 L_1 R_O C_1 C_2 + s^2 L_1 C_1 + s R_O (C_1 + C_2) + 1}$$

***(c)** If $L_1 = 4\,\mu\text{H}$, $C_1 = C_2 = 120\,\text{pF}$, and $R_1 = 1.5\,\text{k}\Omega$, find the frequency of oscillation f_0 and the minimum transconductance g_m needed to initiate oscillation. If you have access to MATLAB, you can use the tf and sys commands to draw a Nyquist plot ("nyquist(sys)") of the values you find to confirm that the circuit will oscillate.

For further resources for this chapter visit the companion website at
http://wiley.com/go/analogmixedsignalelectronics

8

ANALOG-TO-DIGITAL AND DIGITAL-TO-ANALOG CONVERSION

8.1 INTRODUCTION

Nearly all electronic systems deal with forms of energy (voltage, current, power, light intensity, etc.) that change with time. These changes represent information about significant quantities or events that the system is designed to interact with. For a simple example, consider the pilot lamp on the front of an electric stove top that shows if a burner is turned on. The current to the lamp depends on the state of the stove's controls and links that state to the eyes of the cook, who can tell at a glance whether any of the burners are turned on. As simple as this system is, it shares three main features with more complex systems such as thermostats, cell phones, and computers. There is an *input* where the outside world (in the form of the cook's hand) produces a change in the state of the system. There is a *processing* stage, which includes everything from the stove-top controls to the lamp itself. And there is an *output*, which is the light from the lamp that meets the cook's eyes (not to mention the heat produced by the burner).

A voltage, current, or other quantity that conveys information is called a *signal*. Signals are ultimately derived from the world outside the electronic system but may come from either *real-time* information sources or *information-storage* sources. Real-time sources include all types of sensors such as switches, microphones, and digital camera chips, as well as devices designed specifically for information input by humans, such as a keyboard or a mouse. Information-storage media are things like books, computer hard drives, and memory chips. You can think of a storage medium

Analog and Mixed-Signal Electronics, First Edition. Karl D. Stephan.

Companion Website: http://wiley.com/go/analogmixedsignalelectronics

as a place where "frozen" information can be thawed out by performing the operation of reading or playing back. Regardless of the source of signals, the task of most electronic devices is to process signals to perform some useful function.

The way signals are produced and processed in electronic systems has changed over time, although not as much as you might think. The first widespread use of electricity in commerce came with the development of the electromagnetic telegraph, beginning in the 1840s. Information in the form of a coded sequence of long and short current pulses was generated by a sending operator who opened and closed a hand-operated switch called a key. The recciving operator listened to clicks generated by a type of electromagnet called a sounder and decoded and wrote down the message as it emerged at the other end of the telegraph line. We would term this sort of system primarily digital today, although such terminology was not in use at the time.

The second major application of electricity to come along was the telephone, which was developed in the 1870s. The sound waves produced by the human voice are continuous variations in air pressure, which were most conveniently translated into continuous variations in voltage by a diaphragm-operated device called a transmitter. A similar device at the receiving end translated the voltage variations back into air-pressure variations to be heard by the ear. We would term the early telephone a primarily analog system, though it soon acquired digital features such as pulse-actuated automatic dialing.

As a matter of statistics, the vast majority of electronic signals today are in digital form, meaning that their interpretation involves the ones and zeroes of digital technology. This is mainly because **digital signal processing** and related digital technologies are vastly more powerful, flexible, and better performing than their analog-signal-processing counterparts, if such counterparts exist at all. Many routine digital processes today could not be performed at all with analog-only systems, no matter how elaborate or expensive.

However, many signals of interest exist in analog form—light waves, sound waves, motions of objects, positions of obstacles on a road, and so on. For digital processing to occur, all these analog signals must be transformed into a form that is easily dealt with by digital systems. And at the other end of the process, digital information and commands must often be converted back into an analog quantity or signal. So as long as there are real-world inputs and outputs that are not digital, there will be a need to convert between the analog and the digital domains. The conversion process is the subject of this chapter.

We will now turn to more specific definitions and discussions of the two main ways one can represent information in electronics.

8.2 ANALOG AND DIGITAL SIGNALS

8.2.1 Analog Signals and Measurements

An **analog** signal is a form of electrical energy (voltage, current, or electromagnetic power) for which there is (ideally) a linear relationship between the electrical quantity and the value that the signal represents. As an example, in the case of a telephone or microphone signal as discussed in Chapter 6, there is a direct proportion between the instantaneous sound pressure change p (in pascals) at the microphone and the microphone's output voltage v in volts:

$$v = kp, \tag{8.1}$$

where k is a proportionality constant (with units of mV Pa^{-1}) that is characteristic of the particular microphone in use. In general, the variable p can represent any numerically measurable quantity of interest that can be converted into a voltage, and the proportionality constant k will have the dimensions of volts per unit of p.

Real analog systems can show **distortion** and **noise**, both of which were discussed in Chapter 4. These effects can be considered as adding or subtracting a spurious quantity v_S to the ideal voltage v that a perfect linear and noiseless system would produce (*spurious* means "false"). So the actual voltage v_A produced by a real analog system can be expressed as

$$v_A = kp + v_S \tag{8.2}$$

The spurious voltage v_S may be either positive or negative, and so the resulting **absolute error** ε between the ideal output v and the actual output v_A is

$$\varepsilon = v_A - kp = v_S \tag{8.3}$$

In some disciplines, it is customary to take the absolute value $|\varepsilon|$ to be the error, which is therefore always a positive number. But in discussions of electronic systems, it is customary to retain the sign of the error, so that errors in electronic systems may be either positive or negative. In this example, the absolute error has the dimensions of volts, although it can also be expressed in terms of the quantity being measured.

Relative error ε_r is a dimensionless number obtained by dividing the absolute error by the value of the true error-free quantity:

$$\varepsilon_r = \frac{\text{absolute error } \varepsilon}{\text{true value}} = \frac{v_A - kp}{kp} = \frac{v_A}{kp} - 1 \tag{8.4}$$

Multiplying ε_r by 100 produces the **percent relative error**, which is often quoted in discussions of instrument accuracy and precision.

8.2.2 Accuracy, Precision, and Resolution

Accuracy and precision are related but distinct concepts. **Accuracy** refers to how close a measured quantity is to its true value, which is usually defined in terms of a **standard** quantity. For example, the standard of length in SI units is the meter, which is defined as the distance light travels in a vacuum in 1/299,792,458 s. So if repeated measurements of the thickness of a sheet of metal, for example, produce results that always lie between 0.9 and 1.1 mm, and the true thickness is 1.0 mm, the measurements fall within an accuracy range (or briefly, simply have an *accuracy*) of ±0.1 mm. In terms of

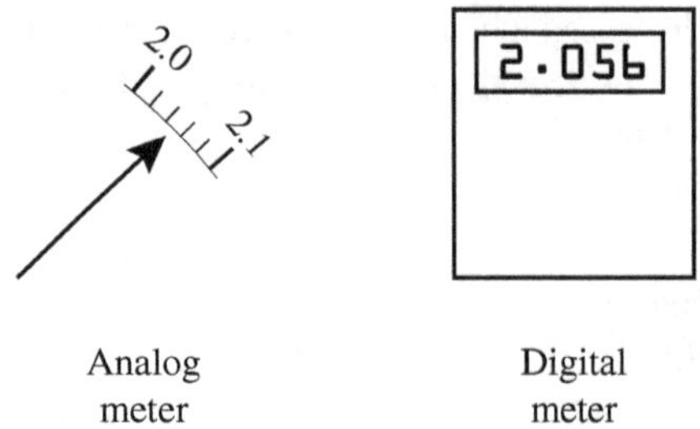

FIGURE 8.1 Example of voltage measurement with analog and digital voltmeters.

percentage accuracy, the accuracy would be expressed as ±10%, because the percent relative error as expressed by Equation 8.4 would be ±10%.

Precision is independent of the true value of a measured quantity. Instead, precision refers to how **repeatable** a measurement is; that is, how much *variation* occurs from one measurement to the next, usually considered over a sequence of several measurements. Precision is also related to the **resolution** of a measurement. Resolution is the smallest physical change in the measured quantity that will produce a corresponding change in the measurement. These concepts can be illustrated by an example of the measurement of a battery's voltage with both an analog and a digital voltmeter (DVM).

Figure 8.1 illustrates the typical form of readouts that each type of instrument has.

An analog meter has a mechanical pointer that moves across a curved scale that resembles a ruler. The reading is taken from the location on the scale indicated by the pointer. In the example shown, there are five minor divisions (spaces) in the major division between 2.0 and 2.1, so each minor division represents (2.1−2.0)/5 = 0.02 V. Unless one is prepared to estimate distances between minor divisions, the analog meter scale can be read only with a resolution of 0.02 V = 20 mV, meaning that any voltage change smaller than 20 mV will not necessarily result in a changed reading. In the example shown in Figure 8.1, the voltage as read by the analog meter is 2.04 V, although someone accustomed to reading analog meters might *interpolate* by eye between the minor divisions and read the meter as indicating 2.05 V.

Turning to the digital meter's display, you can see that it displays a total of four decimal digits. Digital meters are often characterized by the number of digits displayed. It is necessary for a very accurate digital meter to have a certain minimum number of displayed digits, but simply having lots of digits is not sufficient to make the meter either accurate or precise. The digits must also be meaningful. The resolution of the digital display in Figure 8.1, meaning the smallest voltage that will produce a change in the readout, is 0.001 V = 1 mV. This resolution is considerably smaller than the 20-mV resolution of the analog meter. But does this mean that the digital meter is inherently more accurate or precise? Not necessarily.

Suppose that the same (unchanging) battery voltage is measured by both the analog meter and the digital meter a total of five times. Supposing that the lab's temperature varied a good deal during these measurements, and the digital meter is a cheap one that is sensitive to temperature changes, the results could be something like what is shown in Table 8.1:

TABLE 8.1 Voltage Measurements with Analog and Digital Meters

Measurement no.	Analog meter	Digital meter
1	2.04	2.054
2	2.04	2.136
3	2.04	2.007
4	2.04	1.985
5	2.04	2.103
Average	2.04	2.057

The digital meter has better resolution (1 mV) than the analog meter (20 mV). But due to temperature changes or other influences, the five measurements taken with the digital meter varied from a low of 1.985 V to a high of 2.136 V. The usual way of expressing this variation in readings is with a statistical function called the **standard deviation**, denoted as σ. For a finite number N of readings x_i ($i = 1$ to N) whose mean (average) value is μ, the standard deviation is

$$\sigma = \sqrt{\frac{1}{N}\sum_{i=1}^{N}(x_i - \mu)^2} \tag{8.5}$$

The standard deviation has the same units as the quantity measured. In the case of the digital readings, $N_D = 5$, $\mu_D = 2.057$ V, and the digital data's standard deviation is $\sigma_D = 63.34$ mV. Because all the readings with the analog meter were the same, the average for the analog meter $\mu_A = 2.04$ V, and the analog data's standard deviation $\sigma_A = 0$.

Odd as it may seem, the fact that the digital meter data's standard deviation of 63.34 mV is larger than the zero standard deviation of the data from the analog meter shows that the analog readings are *more precise* than those of the digital meter, even though the analog instrument cannot be read to as many decimal places as the digital instrument can. This may go against an intuitive sense that the more decimal places an instrument has, the more precise it is, but this is not necessarily the case. Electronic calculators can provide results having seven or more decimal places, but if the input quantities are not known to better than two-decimal-place accuracy, the decimal places beyond the second in the result are meaningless. Large numbers of decimal places in an instrument's readout (or, as we will see, large numbers of bits in a binary representation of a quantity) allow only for the *potential* of good precision and accuracy. They do not guarantee that the measurement is either precise or accurate.

Suppose it happens that both the digital and analog meters were calibrated with an incorrect calibration voltage and that the true battery voltage, as measured by comparison to a standard voltage, is 2.000 V. In that case, the analog meter happens to be slightly more *accurate* than the digital meter as well, because the average of its readings (2.04 V) is closer to the true value of 2.000 V than the average reading (2.057 V) of the digital meter.

Now suppose the digital meter is properly calibrated and held at a constant temperature. If five measurements of the battery voltage are taken now, the results might be the value 2.003 V, repeated five times. The precision (using the concept of standard deviation) of such a result is equal to the precision of the earlier analog readings, namely, zero, because as far as we can tell with five readings, the results are perfectly precise. And the new measurement has an error of only +3 mV, which in terms of relative error is (+3 mV)/(2 V) = +0.0015 or +0.15%. Such accuracy in reading a 2-V quantity is not possible in general with the analog meter, because its resolution is limited to ±20 mV, which in the case of a 2-V reading amounts to 1% of the reading. In general, the accuracy of an instrument for a *single* reading cannot be better than the instrument's resolution. This example shows how the greater number of digits available from the digital meter allows for its having greater accuracy than the analog meter, but does not necessarily guarantee that it does.

Unless a design involves a measurement instrument or a critical input for which a specified absolute accuracy is required, most measurements performed by electronic systems need be accurate only in a relative sense. The absolute sound pressure sensed by a microphone is not usually of concern, but the relative proportions among the different pressure samples taken at different times need to be preserved if noise and distortion are not to be added to the signal. Nevertheless, it is important to distinguish among the concepts of resolution, precision, and accuracy, because different applications may require different levels of performance in all three of these areas.

8.2.3 Digital Signals and Concepts: The Sampling Theorem

In contrast to an analog signal, which can be represented mathematically as a continuous function of time $v(t)$, digital signals are invariably expressed as discrete numbers called **samples** associated with discrete times separated by intervals called **sampling intervals**. For simplicity, we will assume the sampling times are evenly spaced a sampling-interval time Δt_S apart, although this does not have to be true in general.

To illustrate the sampling concept, suppose a sine-wave voltage with an amplitude of 1 V (peak) and a frequency of $f = 500$ Hz is to be sampled for conversion into digital form. If we choose a sample interval $\Delta t_S = 500\,\mu$s, four samples for each cycle of the sine wave will be obtained, because the ratio of the sine wave's period $T = 1/f = 2$ mS to the sampling interval $\Delta t_S = 500\,\mu$s is (2 ms/500 μs) = 4.

The result of the first four sampling operations is a set of ordered pairs of numbers. The first number in each pair represents the time at which the sample was taken, and the second number represents the voltage at that time. So, for example, if the sample times are exactly synchronized to the starting point of the sine wave, the samples will be as shown in Figure 8.2. The first sample is (0.5 ms, +1 V) and so on through the other samples.

The reason digital systems must perform periodic sampling, rather than dealing with a continuous voltage directly, is that the nearly all digital systems of any complexity incorporate a clock that determines the maximum rate at which a continuous input signal may be translated into discrete numbers. While it is possible to perform

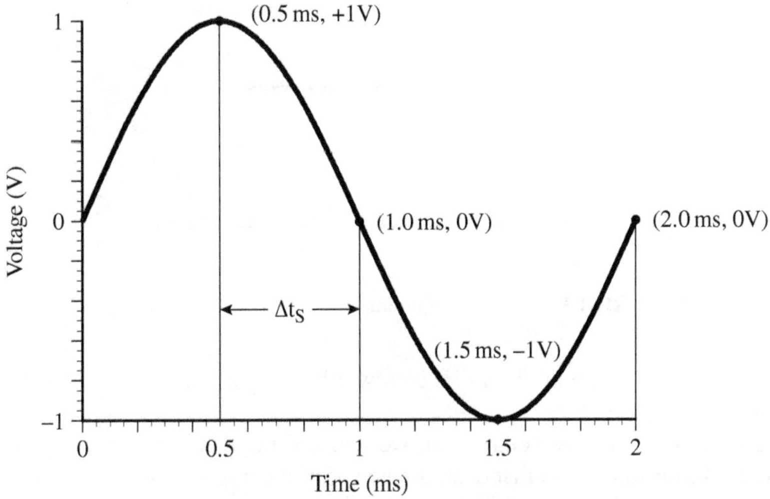

FIGURE 8.2 1-V (peak) 500-Hz sine wave sampled at intervals $\Delta t_S = 500\,\mu s$.

conversion of an analog input signal to digital form in a way that reflects changes in the input almost instantaneously, representing these changes in a numerical way presents problems to a system whose clock frequency is slower than the maximum rate of change of the input signal.

For example, suppose a digital system can register and transfer binary numbers at a rate of 10^8 numbers (also called **digital words**) per second; in other words, it can produce discrete numbers, one per sample, at a frequency of 100 MHz. This means its minimum time between samples is 1/(100 MHz) = 10 ns. If an input signal has significant frequency components as high as 500 MHz, however, it is clear that the digital system will not be able to keep up with all the changes that the analog signal undergoes. In the 10-ns period between sampling operations, the input signal may execute as many as five oscillations. But because the system's eyes are closed, so to speak, during this time, it will miss these high-frequency oscillations, and they will not appear directly in the digital system's record of the input signal. This is why digital systems must generally sample analog signals at a rate that is considerably less than the highest-frequency clock signal used in the system.

If you wish to convert a digitally stored signal back into an analog function of time, something has to happen between the specific times at which the digital signal is defined. Mathematically speaking, the digital table of values and times can be converted to what is called a **discontinuous function** of time. A simple example of a discontinuous voltage function is shown in Figure 8.3. For any time between 0 and 1 s, (including 0), the voltage is 0. But when the time reaches exactly 1 s, the function jumps instantaneously to 1 V and stays there. Such mathematical fictions cannot exist in reality, because no voltage or current can change absolutely instantaneously. But we can approximate the actual analog form of a digitized voltage with functions that show discontinuities such as the one in Figure 8.3 and then

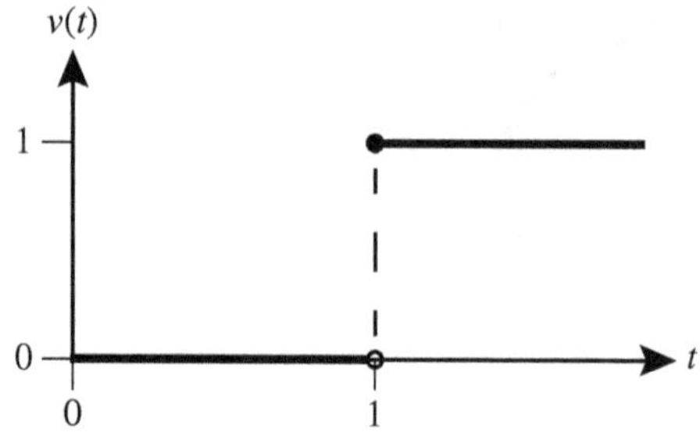

FIGURE 8.3 A voltage function with a discontinuity at $t = 1$ s.

apply what we know about the finite bandwidth of a particular system to discover what its real output will be.

In preparation for what comes next, we will assume that the analog signal to be sampled is **band limited** so that there are no spectral components (frequencies in the signal's Fourier analysis) above a maximum frequency f_{MAX}. This idealization can only be approximated in practice, of course, but it simplifies the following discussion.

There is an exact mathematical relationship between f_{MAX} (the highest-frequency spectral component of the band-limited signal to be sampled) and the sampling frequency $f_S = 1/\Delta t_S$ required to sample the signal so that no part of it in the frequency range from 0 to f_{MAX} is lost. The same Nyquist who devised the Nyquist stability criterion discovered the essence of what is now called the **Nyquist sampling theorem** in 1928, although information theorist Claude Shannon was the first to state the theorem in its usual form. For the sampling theorem to be strictly true, the signal to be sampled can have *no* energy above the frequency f_{MAX}. If that is the case, then the theorem states that as long as the sampling frequency f_S has the following relation to f_{MAX}:

$$f_S > 2f_{MAX}, \tag{8.6}$$

then no information is lost by converting the continuous input signal into an ordered list of sample times and corresponding signal values. For example, sampling a signal having a highest-frequency component of $f_{MAX} = 500$ MHz requires a sampling frequency (also called **sample rate**) greater than 2 (500 MHz) = 1 GHz.

If that is true, then it seems we were wasting time by taking more than two samples per cycle of the 500-Hz waveform in Figure 8.2. Actually, there are some advantages to doing what is called **oversampling**; that is, sampling at a higher rate than is strictly required by the sampling theorem. We will see what some of those advantages are in Section 8.3. For the present, however, it is enough to know that sampling of a band-limited signal can preserve all the information in the signal, as long as the sampling rate exceeds the minimum rate prescribed by the sampling theorem.

What happens if one violates the sampling theorem's requirement that $f_S > 2f_{MAX}$, either by **undersampling** (sampling a band-limited signal at a rate $< 2f_{MAX}$) or by allowing some frequency components to be present at frequencies above f_{MAX}? In either case, the result is the same: an undesirable effect called

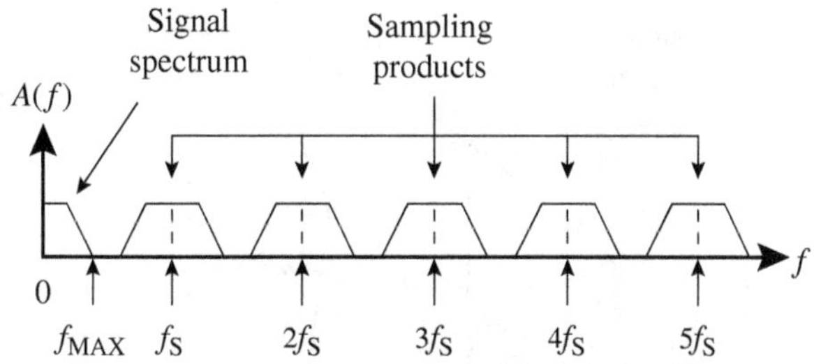

FIGURE 8.4 Spectrum of analog signal extending from 0 to f_{MAX} after sampling at a rate $f_S > 2f_{MAX}$.

aliasing. To understand how aliasing occurs, we will show what happens in the frequency domain when a signal is sampled.

Sampling can be considered mathematically as a multiplication by a periodic pulse at the sampling frequency f_S whose integral (area under the curve) is 1 and which is so narrow that its frequency components extend practically to infinity.[1] Such a pulse has an infinite number of harmonics at frequencies $2f_S$, $3f_S$, …, and its spectrum is called a **comb spectrum**, because the equally spaced equal-amplitude harmonics in its spectrum graph look like the teeth of a comb. When such a pulse is multiplied by a band-limited analog signal, this operation is a type of **modulation** that produces **sidebands** on either side of each pulse harmonic. In effect, the **baseband** signal spectrum, which extends over the frequency range 0–f_{MAX}, is copied and appears on either side of each pulse harmonic. The copy lying immediately above each sampling-frequency harmonic is called the **upper sideband** and is in the same order (low to high) as the original spectrum. The sideband immediately below each harmonic is called the **lower sideband**, and it appears as a mirror image of the original spectrum, with the original high frequencies showing up at a lower frequency than the original lower frequencies. This situation is shown in Figure 8.4, in which the sidebands on either side of the sampling-pulse harmonics are labeled as sampling products. The sampling theorem says that sending the sampled signal through a filter that passes only those frequencies up to f_{MAX} will eliminate all the sampling-product components, leaving only the original signal spectrum behind.

But that can happen *only* if $f_S > 2f_{MAX}$, which is true in Figure 8.4, but not in Figure 8.5. Figure 8.5 shows what happens if we violate the sampling theorem and undersample a signal so that $f_S < 2f_{MAX}$. The figure clearly shows a region below f_{MAX} (in gray) in which frequency components from the original baseband signal *and* from the lower sideband of the sampling-frequency fundamental f_S appear in the same frequency range. The sideband components are appearing "under an alias," so to speak, and are *spurious* in the sense that they are products of the sampling process and were not present in the original signal. When the lower sideband of the sampling frequency overlaps with the baseband spectrum in this way, the spurious aliasing products cannot

[1]This is a description in words of the **Dirac delta function** $\delta(t)$.

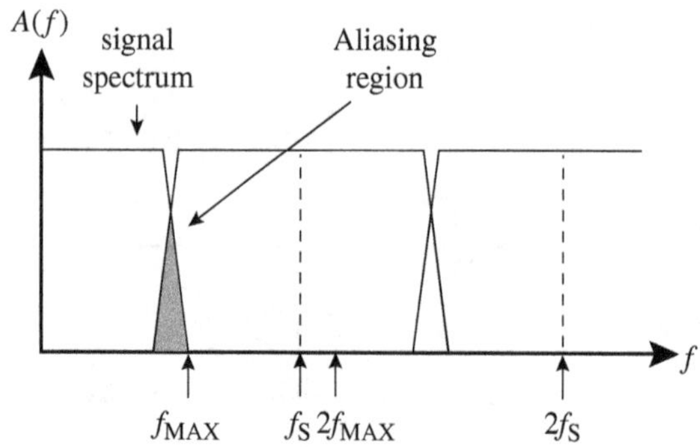

FIGURE 8.5 Spectrum of undersampled signal showing *aliasing* region (gray) where baseband and lower sampling sideband spectrum overlap.

be eliminated from the sampled signal by filtering. Consequently, an undersampled signal will contain aliased signals that appear as erratic noise and distortion, depending on the spectral content of the original signal at the time. This aliasing effect is almost always undesirable and is the main reason not to undersample signals.

The sampling theorem does for analog-to-digital and digital-to-analog conversion (DAC) what the principle of energy conservation does for energy-conversion technologies: it sets fundamental limits on what it is possible to do, regardless of the details of a design. Working within these limits, the intelligent designer chooses a solution to a design problem that meets the specifications in an efficient and economical way. Before we move on to describing types of conversion circuits, we will examine one more fundamental limit on the performance of such circuits: the quantum or discrete character of all electronic signals.

8.2.4 Signal Measurements and Quantum Limits

In a naive picture of analog signals, it is possible in principle to measure voltage, current, or power to any arbitrary level of precision. In this picture, you could zoom in on the sine wave in Figure 8.2, for example, and no matter how fine a voltage or time scale you used, you would still see a smooth, continuous curve. But in reality, there is a limit to the resolution an analog system can have—a limit imposed by the discrete or quantum nature of certain aspects of reality.

Suppose the quantity being measured is a current. Current is composed of electrons, which are discrete charges that come in multiples of 1.6×10^{-19} C. In fact, the type of noise discussed in Chapter 3 called **shot noise** arises from the discrete nature of electrons. So if you examine an analog record of current flow closely enough, which amounts to asking what the noise floor of the measurement is, you will find that it consists of a series of pulses, each of which represents a single electron. This fact is usually obscured by the huge number of electrons that comprise most measurable currents, but it is always true. So at a low enough level, current measurement becomes more of an electron-counting or digital type of operation, rather than the measurement of a smooth, continuous quantity.

What about voltage? An analog measurement of an unchanging voltage can in principle result in any one of an infinite number of answers, so in that sense it would be a truly continuous measurement. But if the voltage is AC with a minimum frequency component f_{MIN}, the system measuring the voltage can be regarded in a quantum-mechanical sense as having a series of discrete energy levels, with the distance between energy levels being given by

$$E = hf_{MIN}, \tag{8.7}$$

where h = Planck's constant (6.626×10^{-34} J-s). This energy spacing is extremely small. For a frequency of $f_{MIN} = 10$ Hz, for example, $E = 6.6 \times 10^{-33}$ J, which is the amount of energy that a single hydrogen atom acquires in falling by gravity a distance of only 400 nm! At room temperature, the energy associated with thermal noise is much larger than these energy-level steps, which can therefore be neglected except at very low temperatures encountered only in certain physics experiments. In fact, for frequencies below the microwave region, the quantum nature of signals can be neglected for nearly all practical purposes, but in principle, all changing voltages have a fundamentally discrete or quantum nature as well.

This is also true of a power or energy measurement, because the same quantum principles apply. In the case of light signals carried on fiber-optic cables, it is particularly easy to realize that light consists of discrete packets of energy called photons. So a measurement of a very feeble light signal is a matter of counting discrete photons, not a matter of measuring a continuously varying power level.

The point of this short detour into physics is to show that the traditional distinction between analog and digital signals is more apparent than real. Nevertheless, many signals are most easily acquired or transmitted in analog form, and it is therefore necessary to convert between its analog and digital representations. We will begin with a discussion of the basics of analog-to-digital conversion, followed by examples of commonly used circuit types.

8.3 BASICS OF ANALOG-TO-DIGITAL CONVERSION

We have already stated that a band-limited analog signal can be completely reconstructed from samples taken at a sufficiently frequent rate in accordance with the sampling theorem. Sampling by itself is not an inherently digital operation, however. The samples can be preserved and transmitted in their original analog form without ever converting them into digital numbers. However, the customary procedure is to first sample an analog signal at a fixed rate f_S and then convert each sample into its binary representation.

8.3.1 Quantization Error

The resolution of an analog-to-digital converter (henceforth abbreviated as **ADC**) is related to the number of meaningful bits N that it can produce. Not including the sign bit (which, if used, indicates whether a voltage is positive or negative), an N-bit binary number can represent any of 2^N discrete voltage levels. So a 1-bit converter (there are actually uses for such a rudimentary device) can distinguish between $2^1 = 2$ levels, while a 16-bit converter can handle $2^{16} = 65{,}536$ levels, and each additional bit doubles the number of levels available. But even a 16-bit converter cannot represent arbitrary voltage levels perfectly, and this unavoidable error in ADCs is called **quantization error** or **quantization noise**.

A simple graphical example of quantization error is illustrated in Figure 8.6, which shows what would happen if we digitized the range of voltages from 0 to 1V with both a 1-bit and a 4-bit ADC. Let V_A designate the digitized approximation to the input voltage V_{IN}. The error voltage V_E is thus

$$V_E = V_A - V_{IN} \tag{8.8}$$

To distribute the error evenly, the 1-bit converter produces 0.25V when its input V_{IN} is between 0 and 0.4999V, and the output goes to 0.75V for V_{IN}=0.5–1V. Consequently, there are only two input voltages (0.25 and 0.75V) for which the digital representation is exactly correct. For all other voltage levels, the error varies in the sawtooth manner shown. If we let ΔV represent the resolution (in volts) of the ADC, then the peak-to-peak amplitude of the error is exactly ΔV. Dividing up

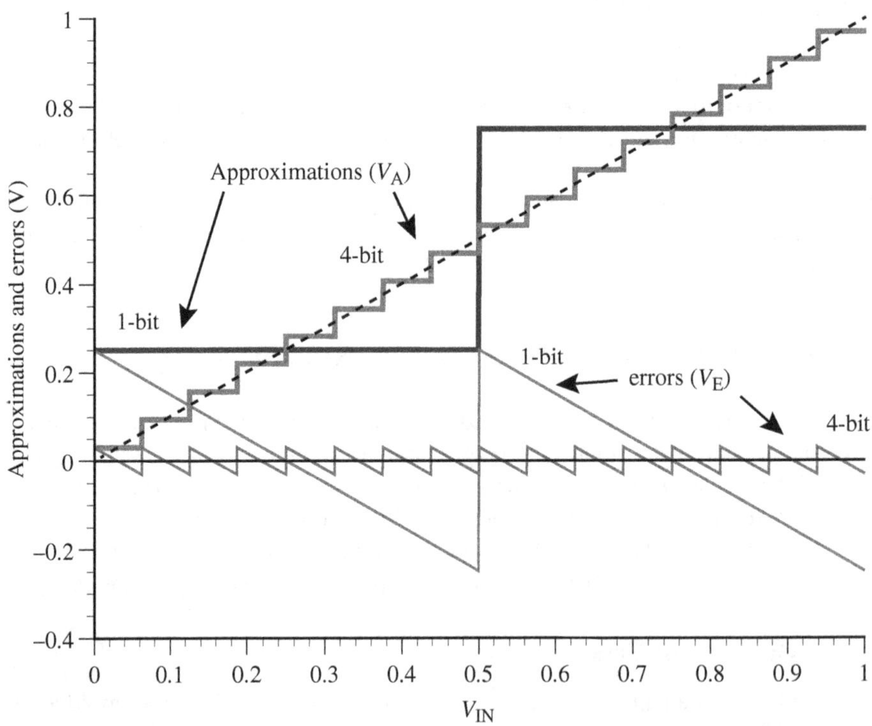

FIGURE 8.6 Analog voltage range of 0–1V digitized by 1-bit and 4-bit ADCs, showing digital approximations and resulting errors.

the entire 1-V input range in the manner illustrated means that the resolution is related to the maximum (single-polarity) input voltage V_{MAX} by

$$\Delta V = \frac{V_{MAX}}{2^N} \tag{8.9}$$

This resolution voltage is also the voltage change that results when the **least significant bit** (abbreviated as **LSB**) changes value. For example, the 1-bit converter ($N=1$) has a resolution of (1 V)/2 = 0.5 V, and so on. This example shows why more bits allows better resolution for a given V_{MAX}, but it also shows that some error is inevitable in digitizing signals.

We can calculate the RMS error encountered in digitizing a signal with an ADC having a resolution of ΔV. We simply integrate the squared error voltage over one resolution interval (one "tooth" of the sawtooth error function), divide by the interval, and then take the square root:

$$\left\langle \Delta V^2 \right\rangle = \frac{1}{\Delta V}\int_{-\Delta V/2}^{+\Delta V/2} (\Delta V)^2 \, d\Delta V = \frac{1}{\Delta V}\left[\frac{(\Delta V)^3}{3}\right]_{-\Delta V/2}^{+\Delta V/2} = \frac{(\Delta V)^2}{12} \tag{8.10}$$

$$\Delta V_{RMS} = \frac{\Delta V}{\sqrt{12}} = 0.288\ \Delta V \tag{8.11}$$

If we assume that the "signal" into an N-bit ADC occupies the full amplitude available from 0 to V_{MAX}, we can express the signal-to-quantization-noise ratio conveniently in dB as follows:

$$\frac{S}{N_{dB}} = (6.02N + 1.76)\ \text{dB} \tag{8.12}$$

So, for example, the 1-bit ADC would show an RMS signal-to-noise (S/N) ratio of 7.78 dB, a 4-bit unit would be (6.02×4+1.76) = 25.8 dB, and a 16-bit device with 65,536 accessible levels would show an S/N ratio of 98.08 dB, all with a signal that occupied the entire 0–1-V input range of the converter.

Nothing has been said yet concerning the time-domain behavior of the ADC or how it deals with the frequency spectrum of the input signal. The aforementioned analysis of S/N ratio assumes an input that is randomly distributed over the entire input range from 0 to V_{MAX} but otherwise assumes nothing about its frequency spectrum. It turns out that by **oversampling** (i.e., taking more samples per second than is strictly required by the Nyquist sampling theorem), we can move the spectrum of the quantization noise around so as to minimize its effect on the final output signal. Here is how.

8.3.2 Output Filtering and Oversampling

Although this section is focused on ADCs, we must look at the opposite process of digital-to-analog conversion briefly in order to show how oversampling and other techniques can be used to improve the S/N ratio of a digitized signal.

From the viewpoint of the sampling theorem, quantization noise is like any other part of the signal. It is an artifact introduced by the digitization process, but once it is introduced, the sampling theorem applies to quantization noise as well as to the signal. That is, the quantization noise can be reproduced exactly as long as its frequency components extend from zero up to half the sampling frequency f_S. Because there is no reason to believe otherwise, we will assume that the **power spectral density** of the quantization noise is evenly distributed in a range from 0 to $f_S/2$. This has important implications for how much of the quantization noise actually ends up in the useful output spectrum.

Suppose that the actual signal of interest has frequency components that extend only up to f_{MAX}. For a given signal, the quantization noise power integrated over the entire frequency range from 0 to $f_S/2$ is constant. If we denote the amplitude of the noise power as $A_N(f)$, then it can be shown that the power spectral density of the quantization noise power is given by

$$A_N^2(f) = \frac{v_N^2}{R}\left(\frac{2}{f_S}\right), \tag{8.13}$$

where v_N is the RMS noise voltage across a reference load resistance R. The dimensions of A_N are W Hz^{-1}, as with all power spectral densities.

Figure 8.7 shows a hypothetical signal spectrum that falls off to zero amplitude at f_{MAX}. With a sampling rate of f_{S1}, the signal is oversampled, but only moderately so because the frequency $f_{S1}/2$ is only slightly higher than f_{MAX}. The quantization noise spectrum forms a rectangle whose area is a constant, namely, $\left(v_N^2/R\right)$. After the digitized signal is converted back to analog form, it can be filtered with a filter that cuts off at f_{MAX} without losing any significant features of the signal. If this is done, only that part of the quantization noise that lies below f_{MAX} will appear in the output at a level A_{N1}.

Suppose now that we double the sampling rate from f_{S1} to $f_{S2}=2f_{S1}$. Because the area of the noise rectangle remains constant, the rectangle's height is cut in half, and its level $A_{N2}=0.5A_{N1}$. Although there is just as much quantization noise power as before, it is spread over a wider frequency range. Consequently, when

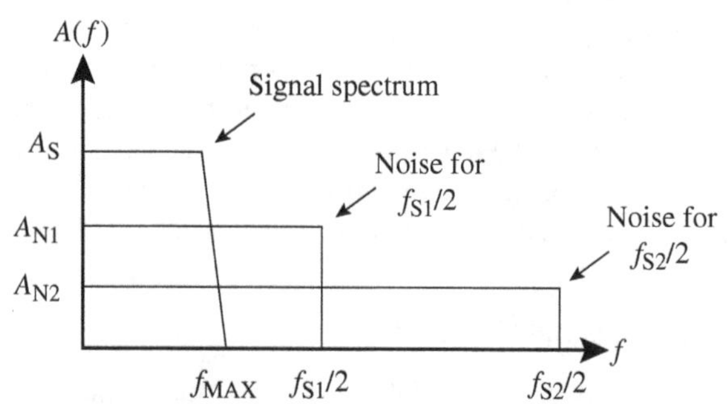

FIGURE 8.7 Hypothetical signal spectrum extending from 0 to f_{MAX} plotted with quantization noise spectra resulting from sampling rates of f_{S1} and $f_{S2}=2f_{S1}$.

postconversion filtering is applied to the resulting analog signal, the quantization noise that appears in the passband below f_{MAX} is reduced by half. As we will see in the next section, increasing the sampling rate for a given input bandwidth is costly in terms of circuit resources, so there is a trade-off between the advantages of oversampling and the costs associated with it. But the sampling theorem requires some degree of oversampling for all ADCs, and so the designer must decide what amount of oversampling will achieve the desired S/N ratio without using excessive resources in terms of circuit cost, complexity, and power consumption.

8.3.3 Resolution and Speed of ADCs

By now, it should be obvious that the bandwidth of the signal to be converted determines how fast an ADC must operate, with the sampling theorem setting a theoretical lower bound on the minimum sampling rate required for a given f_{MAX} of the input signal. And we have seen how the number of meaningful bits that the ADC can deliver—its resolution—is related to the quantization error or noise introduced by the conversion process. Different types of signals present widely differing requirements in terms of resolution and speed. A noncritical measurement of a DC voltage, such as whether a battery voltage has fallen below a specified minimum, might require a resolution of 0.2 V and a speed of one sample per minute. A very simple ADC circuit consisting of a discrete-component comparator might suffice for this application. On the other hand, applications involving high-quality optical images or wide-bandwidth analog signals may require a resolution of 24 bits or more and bandwidths exceeding 1 GHz. Generally speaking, the complexity, size, cost, and power consumption of ADCs increase as both the speed and resolution increase, with speed being the more influential of the two. That is, a low-speed, high-resolution ADC is likely to be simpler than a high-speed, low-resolution ADC simply because high-speed circuits require more complex and exacting designs and infrastructure support in terms of clocks, amplifiers, shielding, and other supporting electronics and hardware. This is why the specifications for an ADC design should be carefully considered to make sure that only the resolution and speed actually required by the application are implemented.

There are many different types of ADC circuits, but most of them have a few ingredients in common. At the heart of many ADC circuits, you will find one or more **comparators**. A comparator is a hybrid analog–digital circuit, which can be thought of as an overdriven differential amplifier whose output is compatible with the digital circuitry employed. A conceptual block diagram of a comparator is shown in Figure 8.8.

The inputs of the comparator resemble those of a differential amplifier or op amp. Typically, they have high input impedances, and the inverting input is often connected to a locally generated voltage termed the reference voltage V_{REF}. The analog input voltage V_{IN} is compared to the reference voltage, and if $V_{IN} - V_{REF} > 0$, the comparator output voltage V_{OUT} goes to a logical HI voltage to be interpreted by digital circuitry connected to the output. On the other hand, if $V_{IN} - V_{REF} < 0$, the output becomes LO. The transfer function of an ideal comparator is a step function, as Figure 8.6 shows.

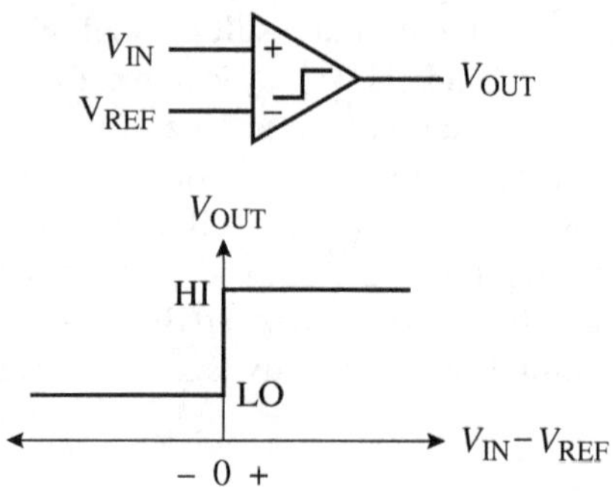

FIGURE 8.8 Block diagram of comparator circuit and ideal transfer function.

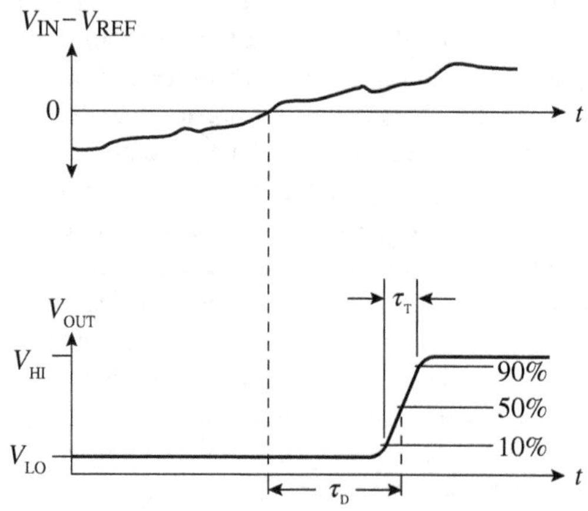

FIGURE 8.9 Propagation delay time τ_D and transition time τ_T of comparator.

Note the step-function symbol on the amplifier-symbol triangle that indicates it is a comparator.

Real comparators have limitations that restrict them to certain applications. The **offset voltage** of a comparator is the input voltage difference at which the comparator output actually switches, which is generally different from exactly zero. A nonzero offset voltage can reduce the accuracy of the ADC in which it is used, because it introduces a spurious voltage in the conversion process.

Two time-domain limitations of real comparators include the **transition time** τ_T and the **propagation delay time** τ_D. Figure 8.9 shows what these values mean. From the time that the input difference voltage crosses zero going positive (in the example shown), the comparator requires a propagation delay time τ_D to respond. By convention, the time at which the output rises to 50% of its ultimate voltage change ($V_{HI} - V_{LO}$, in this case) is regarded as the time at which the output change is registered. The speed at which the output rises is measured by the transition time τ_T, which, again by convention, is measured between the 10 and 90% points of the change in output voltage between the logic levels V_{LO} and V_{HI}. Obviously, if the input is changing so fast that $V_{IN} - V_{REF}$ changes sign more than once in a time τ_D, the comparator will have difficulty following these changes and will deliver an erratic output or no output at all. In order to ensure that the output change is properly recorded by the digital circuitry connected to the output, one must wait a time $t_W = \tau_D + \tau_T/2$ to be sure that the output is valid. This delay determines the maximum frequency at which the comparator can reliably monitor an input signal. If we assume two transitions per cycle (one up and one down), then the limiting frequency f_{LIM} of the comparator is roughly

$$f_{LIM} \approx \frac{1}{2\tau_D + \tau_T} \tag{8.14}$$

For example, if $\tau_D = 45$ ns and $\tau_T = 10$ ns, the highest analog input frequency it would be advisable to process with the comparator would be about 10 MHz.

Because erratic results can occur when a "raw" unprocessed signal is fed directly to the input of a comparator, the comparator's signal input is often preceded by a **sample-and-hold** circuit, sometimes abbreviated as **S/H**. The function of an S/H circuit is to **sample** or track the input signal at a definite time (which is usually synchronized to the other operations of the digital system) and then **hold** the sample at a constant voltage while the comparator examines it, possibly with respect to a reference voltage that may change during the conversion operation, depending on the type of ADC used. When the comparator's input voltage is held constant in this way for at least as long as the comparator output needs to settle down to its correct value, its output can be relied upon by the digital circuitry connected to the output.

One simple form of S/H circuit is illustrated in Figure 8.10. The signal voltage V_{SIG} to be sampled is periodically connected for a brief time to storage capacitor C by the closing of electronic switch S, operated by digital timing circuitry. When S opens and the voltage-follower op amp has enough time to propagate the sampled voltage to its output, V_{SAMPLE} remains at a constant potential and represents V_{SIG}'s level at the time the sampling switch operated. The sample can be held as long as desired, because the high input impedance of the op amp means that the charge and voltage on the capacitor remain practically constant for the short time between samples. The value of C is determined by the output impedance of the

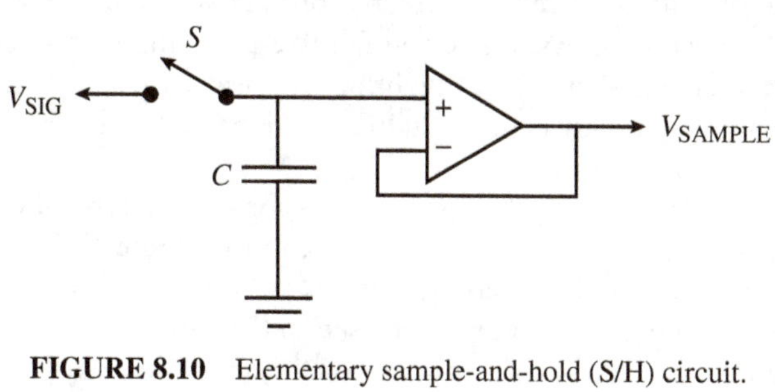

FIGURE 8.10 Elementary sample-and-hold (S/H) circuit.

signal source: lower-impedance sources allow for larger capacitance. Electronic switches can be implemented in a number of ways but typically consist of FETs operated in either the resistive mode or beyond cutoff by a digitally generated gate control voltage.

As we will find in the following discussions of ADCs, comparators and S/H circuits are two of the most important building blocks in ADC architecture. The other components used are mostly familiar elements such as op amps, integrators, and active filters, plus various digital functions as needed.

8.4 EXAMPLES OF ADC CIRCUITS

We will describe four types of ADC circuits in detail: the flash converter, the successive-approximation converter, the delta-sigma converter, and the dual-slope integrator. There are many variations and elaborations of each of these systems, but these four basic systems provide a good understanding of many of the engineering issues involved in the more complex systems. They are ranked roughly in order of decreasing complexity and speed for a given resolution. That is, an 8-bit flash converter will be more expensive and complex than an 8-bit dual-slope integrator but can operate at a much higher speed for a given class of technology. There is no one best ADC solution for a given engineering problem, but some choices are usually better than others for a particular application.

8.4.1 Flash Converter

One of the simplest and fastest forms of ADC is the **flash converter**, so called because it operates essentially instantaneously on the input signal V_{SIG}. A diagram of a basic flash converter circuit is shown in Figure 8.11. An N-bit converter of this type will use 2^N+1 comparators, each of which uses a different reference voltage derived from a voltage divider having 2^N identical resistors R. In this way, a voltage corresponding to each minimum resolvable voltage step is produced and sent to each comparator. The same input signal is provided to all the noninverting inputs of each comparator, so the corresponding **Boolean-logic** output variables A_0, A_1, A_2, ... indicate whether the input signal is greater or less than the ith comparator's reference voltage. The digital output of such a circuit is termed a **thermometer code**, because if the outputs were fed to a column of indicator lamps, for instance, the lighted part of the column would rise and fall like the column of an old-fashioned mercury thermometer. There are straightforward ways to decode thermometer-coded data to ordinary binary code, including the **sign bit** that indicates whether the input is positive or negative.

Flash converters are among the fastest ADCs for a given type of circuit family, such as CMOS or the less common **emitter-coupled logic** (**ECL**), which uses BJTs in differential amplifier configurations. However, because the number of comparators doubles for every increased bit of resolution, they are typically designed for a maximum resolution of 8 bits, which requires $2^8+1=257$

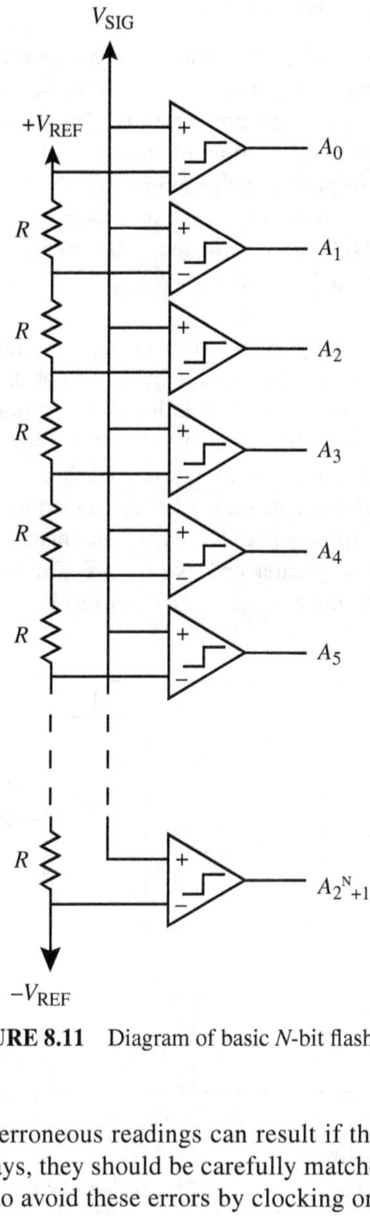

FIGURE 8.11 Diagram of basic *N*-bit flash ADC.

comparators. Because erroneous readings can result if the comparators have different propagation delays, they should be carefully matched for similar characteristics. There are ways to avoid these errors by clocking or gating the comparators or by preceding the entire system with an S/H circuit, but the addition of an S/H would reduce the high-speed limit of the system. High-speed comparators with short propagation delays tend to use more power than slower ones, so a large flash ADC often consumes a large amount of power and will require careful design for the high frequencies involved. These very fast converters are often used for video and other wide-bandwidth applications where the ultimate in speed is more important than high resolution or low cost.

8.4.2 Successive-Approximation Converter

Many ADCs incorporate a DAC or portions thereof in a feedback loop, and the **successive-approximation converter** is one of these. The block diagram of a basic successive-approximation converter is simple, as Figure 8.12 shows. A single comparator monitors the output of an S/H connected to the signal voltage V_{SIG}, and the logic system periodically commands the S/H to sample the input. Once a sample is obtained, it is held at the noninverting input of the comparator while the DAC successively approximates it by means of a decision tree. An example of this tree for a 3-bit successive-approximation ADC is shown in Figure 8.13.

Suppose the system is designed to cover an input voltage range of 0–7 V. The internal DAC is designed to produce 0–7 V by 1-V steps as its digital input ranges from $000_2 = 0_{10}$ to $111_2 = 7_{10}$ (the subscripts indicate the **numerical base**, binary being base-2 and decimal being base-10). Once the logic has told the S/H circuit to sample and hold an input voltage, the system goes through a sequence of comparisons between the sampled input voltage and voltages provided by the internal DAC. First, the DAC is commanded to produce 4 V. This amounts to asking the comparator whether the input voltage is greater or less than 4 V. The output of the comparator then answers the question either *yes* ($V_{SIG} > 4$ V) or *no* ($V_{SIG} < 4$ V). Depending on the

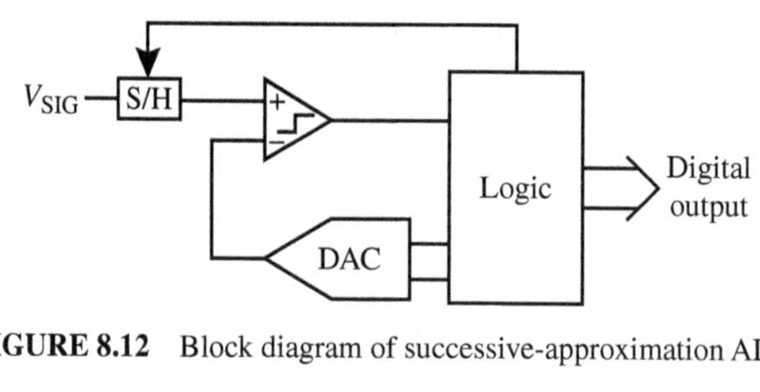

FIGURE 8.12 Block diagram of successive-approximation ADC.

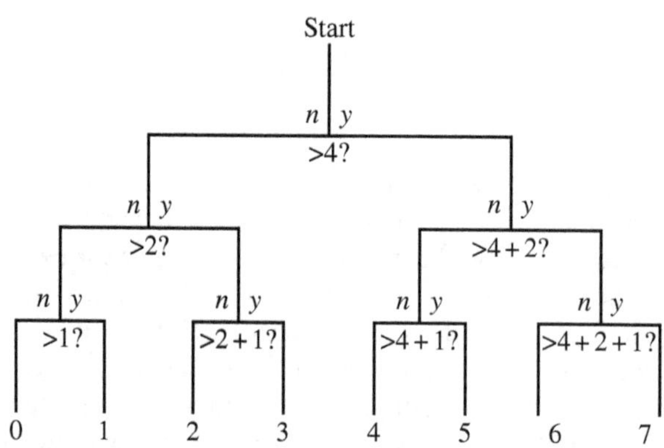

FIGURE 8.13 Decision tree for 3-bit successive-approximation ADC.

answer, the system will command two more voltages, and by the third query, it will be able to decide within ±0.5 V what V_{SIG} is. For example, if $V_{SIG} = 5.5$ V, the question–answer sequence will go like this:

$V_{SIG} > 4\text{V}?\ y$
$V_{SIG} > (4+2) = 6\text{V}?\ n$
$V_{SIG} > (4+1) = 5\text{V}?\ y$

If $y = 1$ and $n = 0$, the answers to the three questions taken in order become the binary representation of the input voltage: *yny* becomes $101_2 = 5_{10}$, which is as close as a 3-bit converter covering 0–7 V can get to the answer. Once the logic system has answered the last question, it indicates that its output is valid and the process begins with the next input sample.

As you can see, the successive-approximation ADC is potentially much simpler than the flash ADC, depending on the complexity of the DAC used. It takes longer to process a single sample, however, because the time taken for the logic to make decisions and for the DAC's output to change is proportional to the number of bits used: one measurement-decision cycle for each bit of resolution. But this delay grows only linearly with the number of bits, which makes the successive-approximation converter relatively easy to expand in comparison to the flash converter. The DAC determines the successive-approximation converter's accuracy, which obviously cannot be any better than the DAC's accuracy. For higher speeds at resolution of more than 8 bits, a modified form of the successive-approximation ADC called the **pipeline converter** combines features of both the successive-approximation and flash approaches to achieve higher speeds without the extreme complexity required by a "pure" flash ADC.

8.4.3 Delta-Sigma ADC

The name **delta-sigma** comes from the meanings of the Greek letters Δ (for *difference*) and Σ (for *sum*), implying that the system operates on differences and sums. This is true as far as it goes, although the details are somewhat more involved. In common with the successive-approximation DAC, the delta-sigma DAC (also sometimes referred to as **sigma-delta**) operates by means of a feedback loop that attempts to approximate or track the changes in the signal being converted.

A simplified form of delta-sigma ADC is shown in Figure 8.14. (This particular circuit is too simple to be very useful, but it demonstrates the basic principles clearly and is the basis for more sophisticated circuits that are more commonly used.) The entire ADC consists of two parts: a **delta-sigma modulator** that converts the analog input into a digital waveform, and a digital logic and control subsystem that converts the digital modulated output into a form suitable for further digital processing.

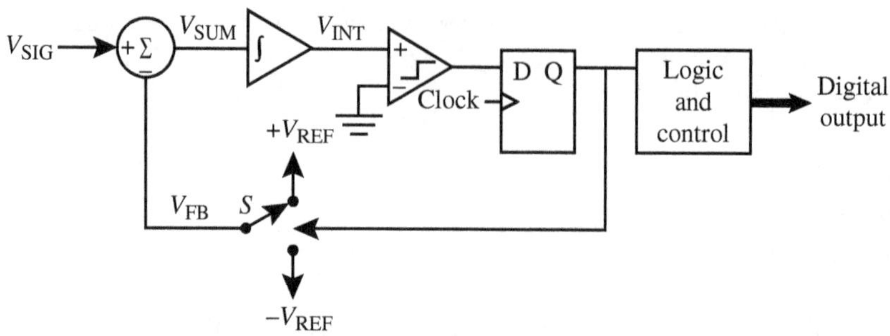

FIGURE 8.14 Block diagram of basic delta-sigma ADC.

Let's consider the delta-sigma modulator first. The input signal V_{SIG} goes to a *summing junction*, which in practice can be an op amp connected as a differential amplifier. The summing junction's output goes to an integrator circuit, again typically implemented with an op amp. Finally, the analog output of the integrator goes to the noninverting input of a comparator whose inverting input is grounded, so that the comparator simply indicates whether the integrator's output is greater or less than zero.

The comparator's (digital) output is periodically sampled by a *D*-type flip-flop. This type of flip-flop transfers whatever digital state is present on its *D* input to its *Q* output upon the arrival of the rising edge of the clock pulse, which of course happens once per clock-pulse cycle. (The clock-pulse frequency f_C equals the highest effective sample rate of the system, incidentally.) When the comparator's output is transferred to the *Q* output of the flip-flop by a clock pulse, it is sent both to the logic and control subsystem and to the input of an analog switch *S*.

The analog switch is controlled by the digital *Q* output so that when *Q* is HI, the switch connects its output to a positive reference voltage $+V_{REF}$. When *Q* is LO, the switch's output becomes $-V_{REF}$. The output of the switch is therefore a series of rectangular pulses alternating between $+V_{REF}$ and $-V_{REF}$ in a way that depends on the input signal.

Note that the output of switch *S* goes to the *inverting* input of the summing junction. This is a negative-feedback connection, in the sense that a positive output of the summing junction becomes a positive slope from the integrator, which feeds the noninverting input of the comparator and is not inverted by the flip-flop or switch until it reaches the summing junction. The negative-feedback connection means that the system will be constantly trying to cancel out the integrated effects of the input signal by producing sequences of pulses from the analog switch that are the opposite of the input signal, in an average sense.

A complete mathematical treatment of a delta-sigma modulator even as simple as this one is beyond the scope of this text. However, we can show how the system could operate in a stable state for a number of different constant input voltages. (In general, however, this simple system is *not* stable for arbitrary values of input voltage, which is one reason it is not used in practice.) For the following example, we will assume that $V_{REF} = 10\,V$, and in order to stay within the system's allowed input range, the signal voltage V_{SIG} cannot be allowed to go outside the range $-V_{REF} < V_{SIG} < +V_{REF}$. Suppose that $V_{SIG} = +6.66\,V$, and the system has settled down to a steady state. What do the waveforms at various points in the system look like under these conditions?

A key concept in analyzing the delta-sigma modulator is to understand that if it is working properly, the integrator output V_{INT} never saturates and ideally hovers around zero as the feedback system operates to produce a digitized approximation of what the actual signal is doing. In the case of a steady-state output for a constant input voltage, this means that the average feedback waveform must be the negative of the constant input voltage V_{SIG}. The averaging, of course, is performed by the integrator circuit. If the circuit is an *R–C* integrator of the type shown in Figure 5.16, its output is given by the following equation (ignoring any constant of integration and the DC-gain-limiting resistor R_A):

$$V_{INT} = \frac{1}{RC}\int (V_{SIG} - V_{FB})\,dt \tag{8.15}$$

The integrator's time constant $\tau = RC$ should be much greater than the inverse of the clock frequency f_C:

$$RC \gg \frac{1}{f_C}, \tag{8.16}$$

or else the system is liable to saturate on large-signal swings of the input voltage. The integrator time constant must not be too large, however, or else the integrator's output voltage changes will be too small to be reliably sensed by the comparator for small input signal changes.

At any rate, with a signal voltage V_{SIG} of 6.66 V, the system can be stable if its digital output is the sequence 100100100 …, because the average voltage produced by the analog switch under those conditions is

$$\langle V_{FB} \rangle = \frac{+10 + (-10) + (-10)}{3} = -6.66\text{ V}, \tag{8.17}$$

which will exactly cancel out the positive V_{SIG} in the long run. The waveforms $-V_{FB}$ and V_{INT} under these conditions are shown in Figure 8.15. As you can see, the integrated waveform is greater than zero for two out of every three sample times, and less than zero for the third. This example evades the question of how the initial condition of the integrator affects matters and other practical issues. In practice, a real delta-sigma modulator's output is not usually as periodic as the example shown, but there is a definite relationship between the number of 1's in the output and the V_{SIG} level.

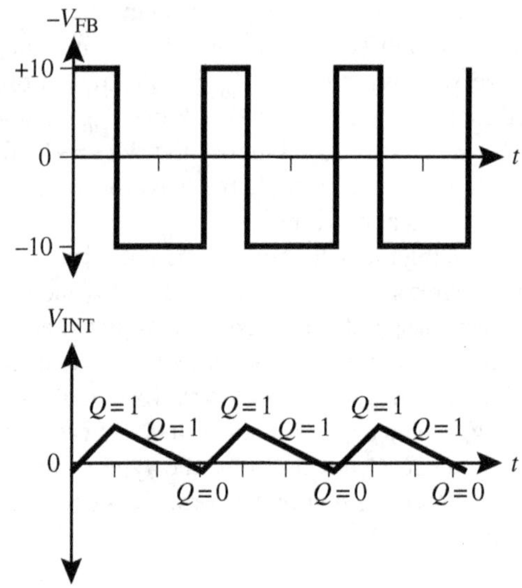

FIGURE 8.15 Waveforms $-V_{FB}$ and V_{INT} in the delta-sigma modulator of Figure 8.14 if V_{REF} = 10V and V_{SIG} = 6.66V. Tick marks on horizontal t-axis indicate positive clock-pulse transitions.

For this first-order circuit shown, this relationship is based on the average number of times n_{HI} the output state is HI during a given time interval T, compared to the total number of clock cycles n_T in T, where

$$T = \frac{n_T}{f_C} \tag{8.18}$$

The digital representation V_{DIG} of the input voltage V_{SIG} observed during an interval T is simply

$$V_{DIG} = 2V_{REF}\left(\frac{n_{HI}}{n_T}\right) - V_{REF} \tag{8.19}$$

So as the number of HI pulses during the interval T goes from zero to its maximum $n_{HI} = n_T$, the digital representation goes from $-V_{REF}$ to $+V_{REF}$, as it should.

As with other types of ADC circuits, there is a resolution/speed trade-off with the delta-sigma ADC, and in this case, the trade-off is particularly clear. For a given clock frequency f_C, the longer the sampling interval T is, the more total clock pulses n_T occur during that interval, and the finer the resolution is. If it isn't clear by now, the theoretical minimum resolution ΔV of this type of ADC is the digital voltage change represented by one bit out of the total n_T, or

$$\Delta V = \frac{2V_{REF}}{n_T} = \frac{2V_{REF}}{f_C T} \tag{8.20}$$

Equation 8.20 shows plainly that for a given reference voltage and clock frequency, the resolution is inversely proportional to the sample time T. The inverse of the sample time is of course the effective sampling frequency f_S

$$f_S = \frac{1}{T} \tag{8.21}$$

and so the resolution for a given reference voltage, the clock frequency, and the sampling frequency are all related by Equations 8.18, 8.20, and 8.21.

In principle, the sample time could be made equal to the clock frequency ($f_S = f_C$), which would yield 1-bit resolution. In voltage terms, one could tell whether the input was either positive or negative, but that is all. Although such a coarse ADC is actually useful for some purposes, most delta-sigma modulators are designed for resolutions better than this. For convenience, one can set the sampling time T so that the total number of clock pulses n_T in the interval is a power of 2. This will provide a "clean" number of bits that can be directly converted from a digital count into a binary number.

To give an example of how these parameters might be chosen for a practical design, let us suppose we wish to sample an audio signal at a rate of $f_S = 48$ kHz (a standard sampling rate for professional audio gear) with a resolution of 12 bits. The number of clock pulses per sample $n_T = 2^{12} = 4096$ so the minimum clock frequency $f_C = n_T f_S = (4096)(48\,\text{kHz}) = 196.6\,\text{MHz}$. This is rather fast for an analog circuit to operate and reveals the shortcomings in this basic first-order circuit. A resolution of 8 bits would lower the minimum clock frequency to $(2^8)(48\,\text{kHz}) = 12.28\,\text{MHz}$, a much more reasonable rate.

A significant advantage that the delta-sigma ADC possesses is that it can shape the quantization noise spectrum introduced by digitization so that most of it lands outside the bandwidth of the digitized signal. The linearized analysis that follows is not exact, but it approximates what actual circuits produce.

Figure 8.16 shows a conceptual model of the modulator portion of the delta-sigma ADC shown in Figure 8.15. The quantities shown in Figure 8.15 are Laplace transforms of the actual voltages in Figure 8.16, and the quantization noise N never appears as an independent voltage in Figure 8.15. Instead, the operations of quantization, the comparator, and the digital circuitry that produces the feedback voltage to the input summing junction are subsumed into the right-hand summing junction in Figure 8.16. With these assumptions in place, we proceed to find the relation between the input signal S and the output signal F.

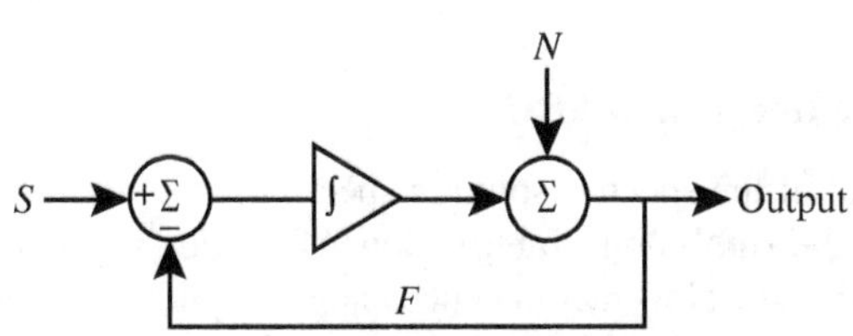

FIGURE 8.16 Laplace-transform model of delta-sigma modulator shown in Figure 8.15.

The operation of integration amounts to the multiplication of a Laplace transform by $1/f$, where f is the frequency variable. So the quantities summed at the right-hand summing junction are

$$F = \frac{S - F}{f} + N \tag{8.22}$$

Rearranging Equation 8.22 to find F in terms of f, N, and S, we obtain

$$F = \frac{S + fN}{f + 1} \tag{8.23}$$

The quantity of interest in the output is the output S/N ratio r, which works out to be

$$r = \frac{S}{fN} \tag{8.24}$$

The result in Equation 8.24 shows that if we assume that the input spectrum S of the signal is flat with regard to frequency and if the noise spectrum N is also flat, then the S/N ratio of the output is $r = S/fN$, which is inversely proportional to frequency. With the proper choices of sampling and clock frequencies, the shape of the S/N spectrum can be markedly suppressed in the signal's passband and made worse than otherwise in a high-frequency portion of the spectrum, which will be filtered out once the digital-to-analog conversion is performed. The actual noise performance of real delta-sigma ADC circuits is not as simple as this analysis portrays, but noise suppression in the passband is a real and measurable effect.

Certain limitations of the first-order delta-sigma ADC can be overcome in the design of a second-order delta-sigma DAC. The only significant difference between the first- and second-order circuits is that the second-order circuit has two summing-junction-integrator circuits in cascade, instead of a single input summing junction and a single integrator. As you might expect, the second-order delta-sigma converter shows even greater quantization noise suppression in the oversampled output signal's passband than the first-order circuit, but this improvement in performance is paid for with a cost that may or may not be justifiable: the expense of a second integrator and summing junction. Third- and higher-order delta-sigma ADCs are used, as well as more complex feedback circuits, which feed back one of three or five voltage levels instead of simply $+V_{REF}$ or $-V_{REF}$. In these and other ways, enhanced performance of delta-sigma converters with regard to linearity, distortion, speed, and resolution can be obtained in ways other than the brute-force approach of increasing clock speed.

8.4.4 Dual-Slope Integration ADC

At the opposite end of the speed-resolution spectrum from the fast but rather costly flash converter is the **dual-slope integration** ADC. For a given technology base, dual-slope integration is a slow method but capable of quite respectable accuracy. For this reason, it is frequently used in *DVMs* and similar applications in which sample rates on the order of 1 s^{-1} are acceptable.

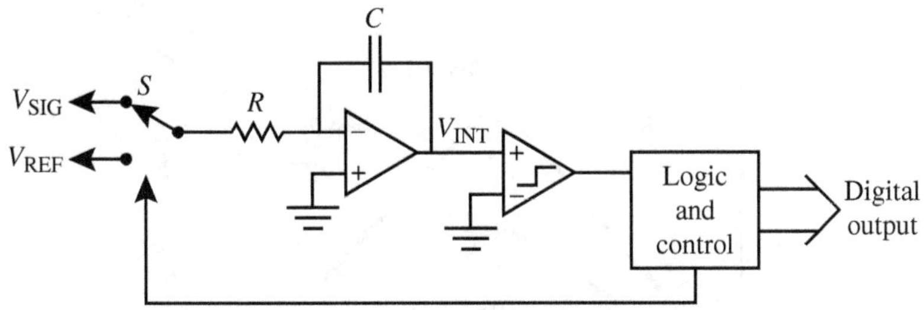

FIGURE 8.17 Dual-slope integration ADC block diagram.

The block diagram of a dual-slope integration ADC is shown in Figure 8.17. The input voltage V_{SIG} is connected to one terminal of a single-pole double-throw (*SPDT*) electronic switch *S*. The second terminal of the switch goes to a constant reference voltage $+V_{REF}$. The common terminal of the switch goes to an integrator consisting of resistor *R*, capacitor *C*, and an op amp. The integrator's output goes to the noninverting input of a comparator, whose inverting input is connected to ground. The digital output of the comparator goes to logic and timing circuits, which control the switch and measure time intervals with reference to an internal clock signal (not shown).

The integrator's processing cycle involves two steps, which we will refer to as 1 and 2. At the beginning of step 1, the integrator's capacitor has no charge, and the switch *S* connects the integrator input to the signal voltage V_{SIG}. So that the waveforms come out positive without cluttering up the diagram with inverting stages, we will assume that the signal voltage V_{SIG} takes on only negative values ($V_{SIG} < 0$). (This could be reversed easily but would not be as straightforward to explain!) The integrator's output voltage V_{INT} rises upward at a slope given by

$$\left.\frac{dV_{INT}}{dt}\right|_{UP} = -\frac{V_{SIG}}{RC} > 0 \tag{8.25}$$

(This slope is positive because we have assumed a negative value for V_{SIG}.) As illustrated in the graph of V_{INT} versus time in Figure 8.18, the integration of the signal voltage is carried out during a fixed, predetermined interval of time t_U established by the timing circuits of the logic system. Consequently, the voltage V_{INT} at the end of the period t_U is proportional to the signal voltage V_{SIG} but is still in analog form.

At the end of step 1, the logic system switches the electronic switch to the (positive) reference voltage V_{REF}, which is held constant. The integrator output now slopes downward at a constant rate given by

$$\left.\frac{dV_{INT}}{dt}\right|_{DOWN} = -\frac{V_{REF}}{RC} < 0 \tag{8.26}$$

The logic system times the duration of the period t_D with a counter until V_{INT} reaches zero and stops step 2 when the comparator indicates that V_{INT} has crossed zero going negative. The system then reverts to step 1 and the cycle repeats.

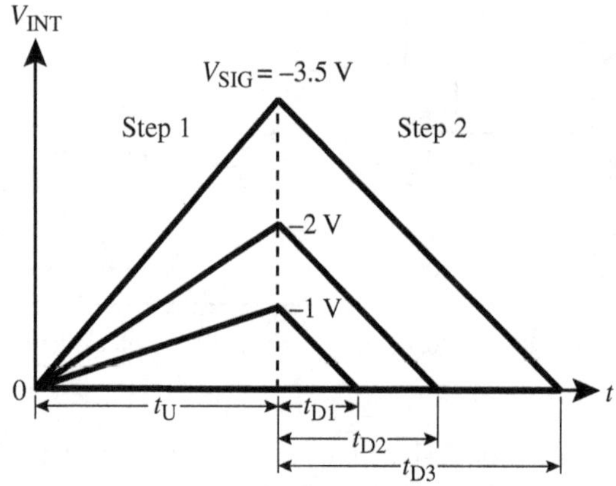

FIGURE 8.18 Dual-slope integrator output V_{INT} for three values of V_{SIG}: −1, −2, and −3.5 V.

Because the heights of the two triangles formed by the upward- and downward-sloping voltage curves are equal, we can equate the products of the slopes of the hypotenuses and the bases, taking proper account of signs:

$$-\frac{V_{SIG}}{RC}t_U = \frac{V_{REF}}{RC}t_D \tag{8.27}$$

The product RC cancels, yielding the following expression for the signal voltage:

$$V_{SIG} = -V_{REF}\frac{t_D}{t_U} \tag{8.28}$$

Because a clock-pulse counter can easily measure the ratio t_D/t_U with a precision that is limited only by the clock frequency and the size of the counter, the accuracy of the dual-slope integration ADC is usually limited by the accuracy with which the reference voltage V_{REF} is known. This explains why this type of ADC is often found in low- to medium-priced measurement equipment. It also shows why this type of conversion process is relatively slow: a single measurement can take as long as $2t_U$, which involves running the internal counter through its entire range twice. As with other ADC systems, the designer of a dual-slope integrator ADC will encounter a resolution/speed trade-off, because higher resolution is obtained for a given clock speed by increasing the maximum count of the counter. But a higher maximum count means that the maximum sample time $2t_U$ increases as well. As always, it is best if clear specifications for the application are established at the outset so that the proper compromise can be made between speed and resolution.

8.4.5 Other ADC Approaches

There is a bewildering variety of ADC methods in current use, but most of them combine various concepts already outlined in the descriptions of the four example methods earlier. For example, a circuit called a **voltage-to-frequency converter** or

V/F ADC does just what its name implies: it converts its input voltage to a frequency related to the input voltage by a constant K having dimensions of $\mathrm{Hz\,V^{-1}}$. Because frequency can be measured with extremely high precision and accuracy, the system's overall accuracy will be limited by that of the voltage-to-frequency converter itself and not usually by the precision of the frequency measurement. Counting pulses during a defined interval is a function employed in the V/F ADC, in the dual-slope integration ADC, and in the delta-sigma ADC.

For extremely fast ADC conversion, one can resort to hybrid optoelectronic methods. For example, a **time-stretch ADC** converts a wide-bandwidth electronic input signal to optical form, performs an operation with fiber-optic components that amounts to stretching the signal out in time and lowering its bandwidth, and reconverts it back to electronic form where it can be handled by conventional fast electronic ADC circuits. It cannot be emphasized too strongly that the overall system's requirements must be first specified clearly enough to inform the designer about the true maximum speed and resolution required, because only such information will allow the designer to make the most fitting choice among the many ADC approaches available.

8.5 EXAMPLES OF DAC CIRCUITS

Because many ADC circuits incorporate a DAC as an essential part of their operation, it is not possible to make a sharp division between the two types of circuits. However, one can always distinguish the overall purpose of a system, so in this section, we will focus on the problem of transforming a sequence of digital words, representing a voltage in time, into an actual output voltage that is a continuous function of real time.

Before we proceed with the description of DAC circuits, we will discuss certain specifications and performance measures for DACs. These include speed, accuracy, **monotonicity**, and resolution.

The speed of a DAC indicates how many digitally encoded voltage samples it can successfully convert into analog form per second. Regarded as partly digital systems, most DACs require several digital clock cycles to produce one analog voltage sample, so normally, the maximum speed (often quoted as a frequency limit) is much lower than the maximum system clock frequency. There are complications to this issue such as oversampling of the original signal and a process called **decimation** (also called **downsampling**) that reduces the sampling rate of the original signal without necessarily affecting the bandwidth. We limit this discussion to what happens to the digital signal after all previous digital signal processing has been carried out, and the only step left is to convert a sequence of binary numbers into an analog voltage or current.

Accuracy refers to the percentage error between the analog output that is intended for a given binary input and the output that is actually produced. Accuracy can be in absolute terms, as when an output voltage is compared to a voltage standard whose calibration can be traced ultimately to a national standards laboratory, or relative, in which case the accuracy is more properly described in terms of nonlinearity, which

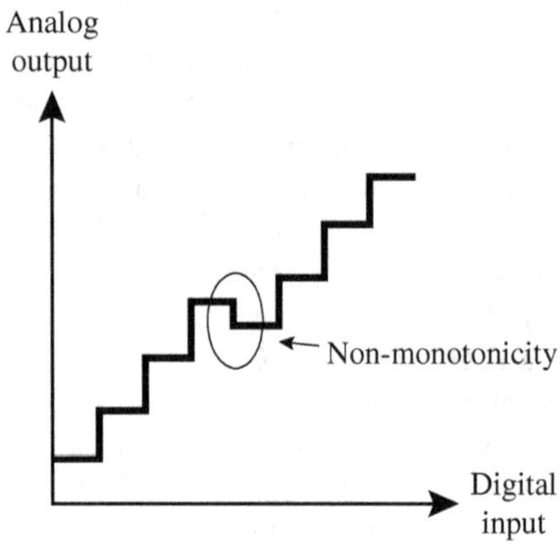

FIGURE 8.19 Illustrating the problem of nonmonotonicity in the input–output characteristic of a DAC.

was discussed extensively in Chapter 4. Nonlinearity can be measured several ways and is often characterized either in terms of maximum deviation from a true straight line or in terms of maximum levels of spurious products in the frequency-domain spectrum of a DAC output when it produces a standard single-frequency tone.

Monotonicity is a special characteristic that applies to DACs. The analog output of a typical DAC consists of a sum of a number of contributions from differently weighted bit values. For example, suppose a 12-bit DAC is designed to cover an output range of 0–10V. The **LSB** in the input digital word will then represent a voltage change of $10/(2^{12}) = 2.44\,\text{mV}$. The **most significant bit** (**MSB**), on the other hand, represents $10/2 = 5\,\text{V}$. If there is an error of only $(2.44\,\text{mV})/(5\,\text{V}) = 0.048\%$ in the accuracy with which the 5V representing the MSB is produced, that small error in the MSB will overwhelm the LSB change whenever the MSB is called for. Any time a problem like this happens, the result is that the digital-input analog-output function, which ideally is a series of smooth upward steps as the digital word's value increases by one bit per step, will in fact step *downward* at one or more points, as in the example input–output plot shown in Figure 8.19. A function that only increases as its argument increases and never decreases is called **monotonic**, and every DAC should have that characteristic. Nonmonotonicity was more of a problem in early DACs that often used discrete components and is not so much an issue today, but most manufacturers of DAC circuits currently guarantee the monotonicity of their product's output.

Resolution has the same definition for a DAC as for an ADC, namely, the voltage change ΔV produced when the LSB changes. Alternatively, the resolution can be specified in terms of the width (in bits) of the largest digital word that the DAC can convert. Most DACs (and ADCs too) rely on a **reference voltage**, which can often be supplied either internally or externally. The output of the DAC is typically expressed in terms of the reference voltage multiplied by the number represented by the digital word, so the voltage resolution is a function of the reference voltage. While the relative accuracy (or distortion) of a DAC is an intrinsic property of the device, the absolute accuracy of its output depends both on the linearity of the device and the accuracy of the voltage reference used.

With these preliminaries defined, we proceed to consider a few examples of DAC circuits, beginning with one of the oldest: the *R–2R* ladder circuit.

8.5.1 *R–2R* Ladder DAC

The **R–2R ladder** is still used to some extent, although it is not as efficient with regard to power consumption as some newer CMOS-based designs. A 4-bit version of this DAC approach is shown in Figure 8.20.

The circuit uses a **periodic ladder circuit**, which has the useful property of dividing the input voltage at each period (or stage) by exactly 2, assuming the resistor values are exact. This division ratio is required to produce currents I_0, I_1, I_2, and I_3 in the ratios of 1:2:4:8, as needed for a selectable binary-weighted output voltage. Here is how it works.

The op amp and its associated feedback resistor R_O form a **current-to-voltage converter** with an (ideal) input resistance of zero and a conversion constant $V_O/I_{IN}=R_O$. That is, if the input current $I_{IN}=1$ mA and the feedback resistor $R_O=1$ kΩ, the resulting output voltage $V_O=(1\text{ mA})(1\text{ k}\Omega)=1$ V. The output circuit's zero input resistance guarantees that the current bus connected to each electronic switch will always be at ground, because of the virtual ground at the inverting input of the op amp.

Because each of the switches goes either to a true ground or a virtual ground, we can analyze the *R–2R* ladder as though the shunt resistors are all connected to ground.

Starting at the right-hand end of the ladder, there are two 2*R* resistors connected in parallel, one going permanently to ground and the other going to the I_0 switch. Together, these paralleled resistors have a value of $2R/2=R$. Because the series

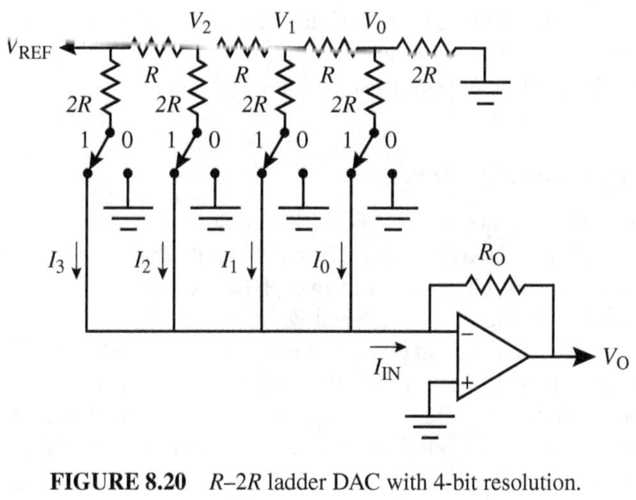

FIGURE 8.20 *R–2R* ladder DAC with 4-bit resolution.

resistor connecting the V_1 junction to the V_0 junction is also R, there is a voltage divider set up between V_1 and V_0 whose division ratio is exactly 2. Therefore, $V_1 = 2V_0$, and the current $I_0 = V_0/2R$ is also exactly one-half of I_1. Inspection of the circuit reveals that the exact same conditions apply to the V_1 junction with respect to V_2 and so on. The result is that this periodic network produces currents that scale as powers of 2: $I_1 = 2I_0$, $I_2 = 4I_0$, and $I_3 = 8I_0$, just as is required for the 1:2:4:8 ratio needed for a binary DAC. If A_3, A_2, A_1, and A_0 are the binary digits of the digital word to be converted (e.g., $A_3A_2A_1A_0 = 0110_2$ for a value of six units in decimal notation), then the output voltage of the R–$2R$ circuit can be expressed as

$$V_O = -V_{REF}\frac{R_O}{2R}\left(\frac{A_3}{2^0}+\frac{A_2}{2^1}+\frac{A_1}{2^2}+\frac{A_0}{2^3}\right) \tag{8.29}$$

The circuit can easily be extended to incorporate more bits but is limited by the precision with which the resistors can be made equal and the exact R–$2R$ ratio can be maintained. Note that the absolute values of the resistors are not critical as long as the ratios can be fabricated accurately. IC fabrication methods can provide very precise resistance ratios, although it is also possible to obtain discrete resistors with better than 1% tolerance as well.

We have not shown auxiliary circuits such as digital latches to hold the digital word constant during conversion, delay circuits to indicate how long it takes before the electronic switches have settled and the output is a valid voltage, and other features needed to make such a circuit practical. The resolution of the R–$2R$ circuit is limited mainly by the relative accuracy of the resistors used, and its speed is limited by the speed with which the electronic switches can perform their functions and the bandwidth of the output op amp. This type of DAC combines reasonably good resolution with the capability of high speed without excessive complexity, but at the price of fairly high power consumption compared to other types of circuits to be described. To avoid stray capacitance effects, the value of R cannot be arbitrarily large, which means that the circuit constantly consumes a fair amount of power in maintaining the currents through its various branches. This is an unavoidable difficulty with the circuit, but its other advantages may outweigh this problem for applications where efficient power use is not a high priority.

8.5.2 Switched-Capacitor DAC

A type of DAC that is more compatible with analog CMOS circuitry than the R–$2R$ approach is the class of **switched-capacitor** DACs, of which one example is shown in Figure 8.21. This circuit operates in two phases, denoted as ϕ_1 and ϕ_2. All the electronic switches in Figure 8.21 labeled ϕ_1 are closed during the first phase, and those labeled ϕ_2 are open. And in phase 2, some of the ϕ_2 switches are closed and the ϕ_1 switches are open. Capacitors C_1 through C_N are designed to have **binary-weighted values** so that $C_2 = 2C_1$, $C_2 = 2C_1$, and so on up to the largest capacitor C_N. The feedback capacitor C_F around the op amp has a value of $(2^N - 1)C_1$. The number of bits equals N.

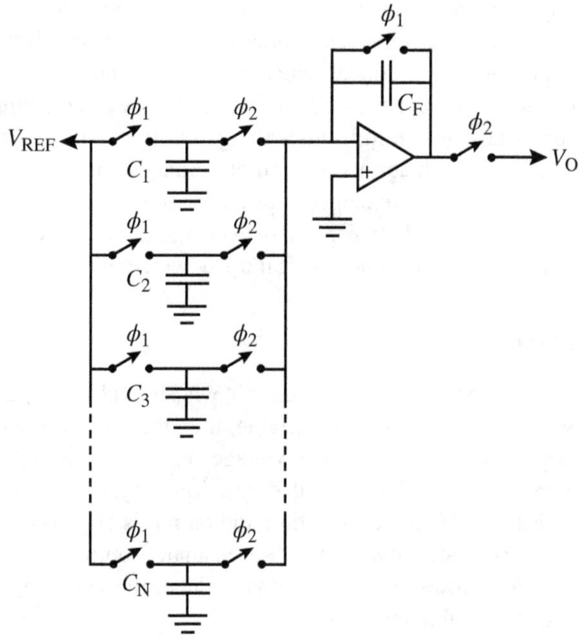

FIGURE 8.21 One type of switched-capacitor DAC circuit.

In the first phase, the circuit is prepared for a conversion by opening all the ϕ_2 switches and closing all the ϕ_1 switches. This charges capacitors C_1 through C_N to the reference voltage value V_{REF}. It also resets the feedback capacitor C_F to have zero charge.

In the second phase, only those ϕ_2 switches are closed, which connect to capacitors whose bit values are 1. For example, if the binary number to be converted is 0 … 011, only the ϕ_2 switches connected to capacitors C_1 and C_2 are closed. If we consider the charge held by C_1 when connected to V_{REF} to be one unit $Q_1 = C_1 V_{REF}$ of charge, then closing the two lowest bit value ϕ_2 switches sends $(2+1)=3$ units of charge to the inverting input of the op amp. Bearing in mind the principle that an op amp does whatever it has to in order to keep its inputs at the same voltage, the op amp proceeds to charge the feedback capacitor C_F with the same amount of charge dumped into its inverting input from the other capacitors. In a short time limited only by the speed of the op amp and the size of the capacitors, the output voltage becomes

$$V_O = \frac{3Q_1}{C_F} = -V_{REF} \frac{3}{2^N - 1} \tag{8.30}$$

Once the op amp output settles to a stable value, the output switch (also labeled ϕ_2) closes, sending the valid output voltage to the outside world. If all the other capacitors had been discharged into the op amp's input, the resulting output voltage would be $-V_{REF}$, so in this way the entire range from 0 to $-V_{REF}$ can be covered with the proper selection of switches.

There are many elaborations on this basic circuit: **thermometer-code** converters that have 2^N identical capacitors instead of binary-weighted ones, circuits with active feedback to compensate for nonideal capacitor values, and others. But the main advantage of any switched-capacitor circuit is that its power consumption is much less than a resistive-network circuit, other things being equal. The only time that current flows is when the capacitors are initially charged and reset and when the output is generated. Otherwise (except for power to the op amp), the circuit draws no power between transitions between phases. This is a large advantage in power-critical applications such as portable and battery-powered equipment.

8.5.3 One-Bit DAC

The **one-bit DAC** is a primarily digital circuit whose analog output is extremely simple. If a single-polarity output is acceptable, then the circuit consists of a single switch alternating between a reference voltage V_{REF} and ground, as shown in Figure 8.22. Electronic switch S is controlled by a digital system designed to produce an output whose **duty cycle** (fraction of time the output is HI), when averaged over some interval $\tau = RC$, is proportional to the desired analog-output voltage. The simple passive R–C filter that follows the switch output reduces the switching frequency and its harmonics to an acceptable level.

A variety of digital waveforms can be used to drive the one-bit DAC. The simplest is a form of signal called **pulse-width modulation** (**PWM**), which is illustrated in Figure 8.23.

For any form of one-bit DAC, the switching frequency of the raw digital output must be well above the highest desired analog-output frequency. In other words, if the switching frequency is equated to the sampling frequency, the signal must be greatly **oversampled**, sometimes by a factor of 100 or more. In PWM, the frequency of the digital output waveform is fixed at the (oversampled) sampling rate, but the duty cycle varies in proportion to the desired analog-output voltage. Mathematically, for any given cycle of the output waveform at time t, its duty cycle D is

$$D(t) = \frac{V_O(t)}{V_{REF}} \tag{8.31}$$

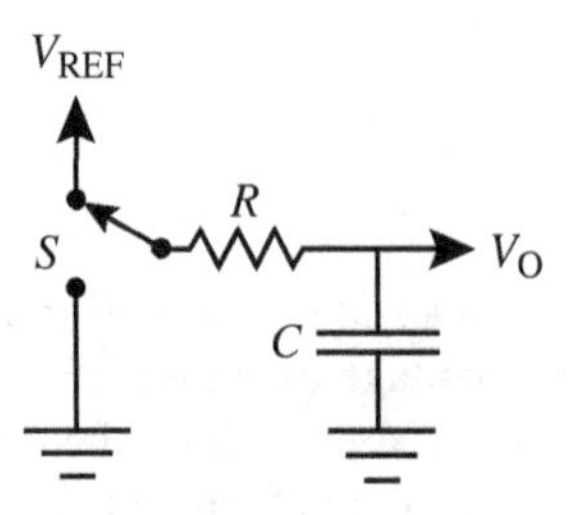

FIGURE 8.22 Analog portion of one-bit DAC.

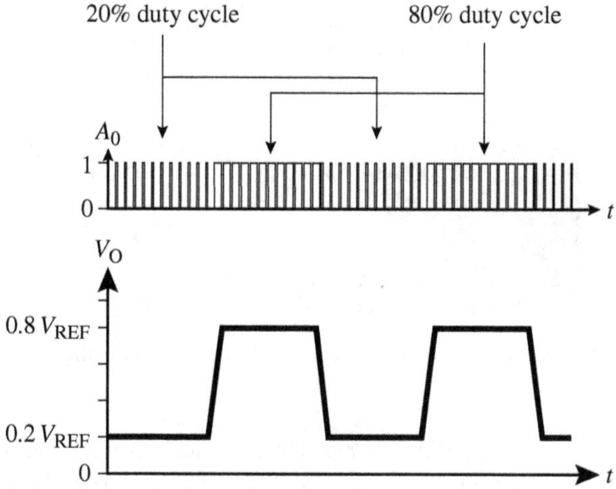

FIGURE 8.23 Example of digital drive signal A_0 to electronic switch in Figure 8.20 and resulting output waveform.

Because the duty cycle varies between 0 and 100%, the voltage range available for the output signal $V_O(t)$ is therefore limited to $0 < V_O < V_{REF}$. The reason for the large factor of oversampling is that it allows one to use a very simple 1-pole *R–C* lowpass filter at the output to eliminate most of the spurious sampling-frequency components from the signal. For example, if the time constant τ is chosen so that the desired analog output is attenuated by only 3 dB at the upper limit of its bandwidth, and the signal is oversampled by a factor of 100, the sampling frequency will be attenuated by about 40 dB below the level V_{REF} despite the simplicity of the passive *R–C* filter. The simplicity of the analog filter needed for a highly oversampled one-bit DAC makes it attractive for low-cost consumer applications in audio equipment, where a large oversampling factor is possible with relatively low clock speeds in the megahertz range.

8.6 SYSTEM-LEVEL ADC AND DAC OPERATIONS

Today's complex electronic systems are usually designed by teams of engineers, with different parts of the system being designed by individuals who work with specifications given to them by other engineers. It is a hazard of such a procedure that sometimes a simple means of achieving an overall goal gets lost in what might be called "divide-and-conquer" thinking. If a manager's staff includes one analog specialist and one digital specialist, for example, the tendency is to divide a project into purely analog and purely digital parts and assign the two parts to the appropriate designers. But sometimes, the optimum design is not something that is so easily divided into analog and digital pieces. Here is an example of how this sort of problem can be avoided.

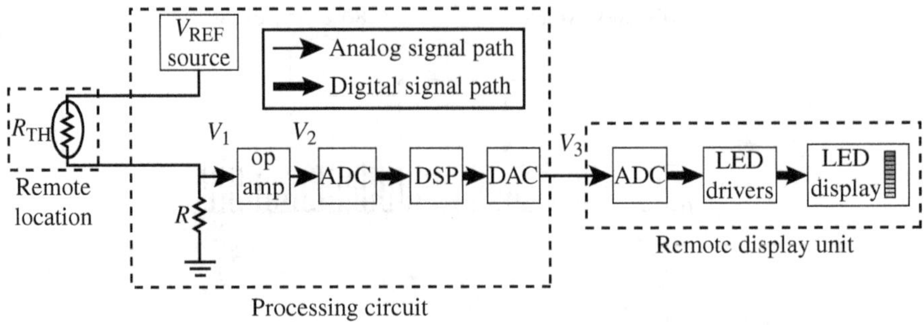

FIGURE 8.24 "Divide-and-conquer" approach to temperature-measurement design.

Suppose a team is assigned the task of sensing a temperature and transforming it into an analog display of a temperature "bar" on an LED array: the hotter the temperature, the taller the column of illuminated LEDs. The sensor used is a device called a *thermistor*, whose resistance changes nonlinearly with temperature over a fairly wide range, say 3:1. One "divide-and-conquer" way to achieve this design is shown in Figure 8.24. The thermistor is placed into one leg of a voltage divider driven by a reference voltage V_{REF}. The voltage divider's output V_1 is scaled and offset by an analog op amp circuit to become V_2 and is then fed to an ADC. The digital output is processed with custom software and then used to produce a linearized analog-output voltage V_3. This linearized analog voltage is sent to a packaged display circuit that incorporates another ADC and circuitry to drive the LED bar display.

This approach would certainly work. The analog designer would need to concentrate on the reliability and noise level of the reference voltage, possibly facing shielding issues if the thermistor happens to be in an electrically noisy environment such as an automobile, and the question would arise of what optimum voltage range is required for the ADCs and the LED display. The ADC itself might be a unit that comes as part of a microprocessor board, so its resolution and input voltage range may not be under the control of the analog designer, who has to adapt the analog design to the ADC's requirements. The digital designer takes the range of digital words produced by the ADC, categorizes them with some sort of lookup table that translates temperature into another voltage, and sends that analog voltage to a prepackaged LED-display voltmeter, which has an internal ADC that translates the output voltage (now proportional to temperature) into a bar display.

This solution has some problems and vulnerabilities. There are three physically separate parts: the thermistor sensor, the processing circuit, and the display circuit. The signal links between these three parts are all in the form of analog voltages, which are easily corrupted by noise, poor connections, and other types of interference. Supplying a voltage reference in a noisy environment such as an automobile chassis or a larger digital system is not a trivial thing either. And there are six or seven different parts of the design, all of which must work together for a reliable outcome.

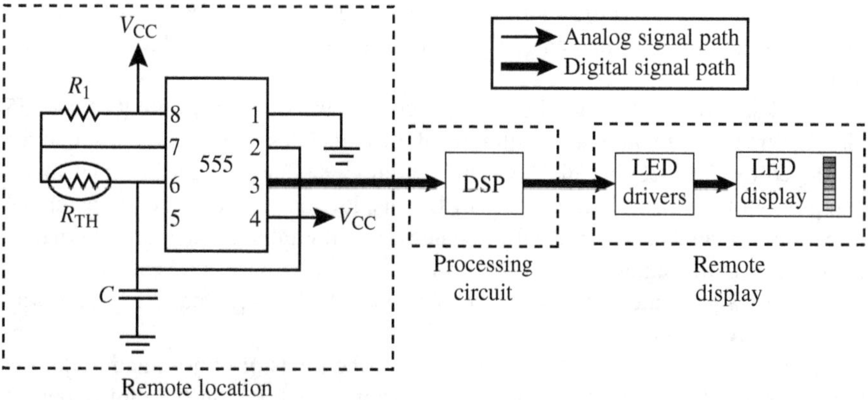

FIGURE 8.25 Circuit with same design goal of temperature display as in Figure 8.24, using 555 timer in more efficient approach that is less vulnerable to interference.

If one keeps in mind the overall goal of a design, even if it must use a purely analog input, it is sometimes possible to avoid ADC and DAC circuits altogether. Figure 8.25 shows an alternative approach to the same design problem. The second approach achieves the same goal with considerably less hardware and in a way that is less vulnerable to interference. At the remote location, this design places the temperature-sensitive thermistor with a 555 timer (see Chapter 7 for more information on 555 timers).

The timer is wired as an astable multivibrator whose digital-compatible square-wave output's frequency is given approximately by

$$f = \frac{1.44}{(R_1 + 2R_{TH})C} \tag{8.32}$$

(see Eq. 7.50 and accompanying text for an explanation). Note that neither the power-supply voltage V_{CC} nor the ambient temperature of the 555 circuit appears in this equation. While the temperature sensitivity of this circuit is not zero, the change due to the thermistor's variation with temperature is by far the largest contribution, and other temperature dependences are merely minor adjustments to it. The point is that one can derive a frequency-to-temperature function either experimentally or with data provided by manufacturers, and this function will in all likelihood be more than accurate enough to drive an LED temperature display having no more than 10 or so divisions.

Once the 555 produces its square-wave output, proper choice of the power-supply voltage V_{CC} ensures that this square wave is directly compatible with a digital input and can be treated as a digital signal. This means that the signal between the sensor (which now includes both the thermistor and the 555 timer) and the digital processing circuit is a digital one, which is much less vulnerable to corruption by interference. Even if interference does contribute a few extra counts, they will not have a significant effect on the overall frequency count as calculated in the digital circuit, which can be part of a processor primarily devoted to other tasks.

A lookup table in the digital circuit can convert the frequency output of the 555 circuit directly into another digital word that encodes the desired level to be displayed by the LED bar display. This digital signal can be sent in **serial form**, using only two wires, to an all-digital LED display that converts the digital signal into a bar display that is periodically updated by the digital processor. The alternative approach gets the signal into digital form at the location where it is sensed, which is generally a good idea, other things being equal. And it stays that way all the way to the display output.

These two designs are extreme examples of the kinds of options and trade-offs that designers face frequently. While this is not a textbook on the philosophy of engineering design, it is important to emphasize that the early stages of any design when the requirements are specified is a phase that designers often move through too quickly in order to get to the "fun stuff" that uses their more technical smarts. But all engineering is carried out under a variety of constraints: time, money, resources, abilities, space, weight, availability of power, and many others. Good designs that advance the practice of engineering take all these constraints into account and deliver a creative solution that achieves the most with the least amount of resources, not with unthinking "crank-turning" procedures. While ADC and DAC circuits are getting cheaper and better all the time, they still consume a fairly substantial amount of resources, and a holistic approach to a problem taking the design context into account can often yield a design solution that minimizes complexity and works better, too.

BIBLIOGRAPHY

Kester, W., editor. *Data Conversion Handbook.* Amsterdam: Elsevier, 2005.

Schreier, R., and G. C. Temes. *Understanding Delta-Sigma Data Converters.* New York: IEEE Press, 2005.

PROBLEMS

Note: Problems of above-average difficulty are marked with an asterisk (*).

8.1. *Accuracy and precision of individual measurements.* Suppose a sequence of five voltage measurements V_1, V_2, V_3, V_4, and V_5 are made of a reference standard $V_{REF} = 1.2280$ V(rms) with a 5-digit digital AC voltmeter undergoing calibration. The results are $V_1 = 1.2284$, $V_2 = 1.2349$, $V_3 = 1.2351$, $V_4 = 1.2277$, and $V_5 = 1.2366$.

Assume both the precision and accuracy of the reference standard are essentially perfect (0% error and standard deviation = 0 V). For each of the five individual independent measurements produced by the voltmeter, calculate the absolute error ε (allowing for either positive or negative sign) and the relative error ε_r compared to the reference standard (also either positive or negative). Also, calculate the standard deviation σ of all five measurements. (Standard deviation is a measure of the precision or repeatability of a measurement system.) If the meter is stated to have an accuracy of ±1%, is the meter within its specifications? Is it really useful to write down all five digits, or would four digits produce just as meaningful a measurement? Why or why not?

8.2. *Accuracy and precision of averaged measurements*. Suppose an analog sample-and-hold system takes a large number of samples of a voltage that consists of a constant DC term $V_{SIG} = 1.000$ V (the quantity of interest) plus a random noise voltage V_{NOISE} whose probability distribution is given by

$$P(V_{NOISE})dv = \frac{1}{\sigma\sqrt{2\pi}} e^{-\frac{V_{NOISE}^2}{2\sigma^2}} dv \tag{8.33}$$

where $\sigma = 100$ mV. This is called **Gaussian-distributed** noise and is the usual way to model random noise when nothing is known about it other than its RMS value. It turns out that the standard deviation of the noise voltage in this case (which equals the RMS value) is exactly σ. Using a function called the **probability integral**, one can calculate the probability that the noise voltage magnitude for any given sample is less than the following values shown in Table 8.2.

The chances are more than even that a particular sample will contain a noise voltage of less than 100 mV but are less than 10% that the noise voltage will be less than 10 mV.

A theorem from statistics states the following: if a large set of samples of data has a mean value μ and standard deviation σ, any sample will, on average, have the same *variance* σ^2 as any other sample. However, it can be shown that the variance of the *mean* of N-independent samples is

$$\sigma_N^2 = \frac{\sigma^2}{N} \tag{8.34}$$

and so the **standard deviation of the mean** of N-independent samples is

TABLE 8.2 Probabilities for V_{NOISE} Occurring in Various Ranges

Magnitude range (mV)	Probability that V_{NOISE} lies within range
$\lvert V_{NOISE}\rvert < 10$	0.079
$\lvert V_{NOISE}\rvert < 30$	0.235
$\lvert V_{NOISE}\rvert < 50$	0.383
$\lvert V_{NOISE}\rvert < 100$	0.683
$\lvert V_{NOISE}\rvert < 200$	0.954
$\lvert V_{NOISE}\rvert < 500$	0.999

$$\sigma_N = \frac{\sigma}{\sqrt{N}} \tag{8.35}$$

This shows that taking the *average* (mean) of N samples of a noisy signal can reduce the standard deviation due to random noise by a factor of the square root of N. (Of course, it takes longer to obtain N samples than to obtain one sample, so there is a noise–speed trade-off to be made.) How many independent samples N must you take and average together in order for the resulting mean (average) of the samples of the signal above to have a standard deviation of less than

(a) 50 mV?

(b) 5 mV?

(c) 500 μV?

8.3. *Sampling theorem.* The Nyquist sampling theorem says that to reconstruct a sampled signal without losing any information, one must (perfectly) sample it at a frequency $f_S > 2f_{MAX}$, where f_{MAX} is the highest-frequency component of the band-limited signal. In order to see what problems arise when a sampled signal's frequency is *equal* to $f_S/2$, consider the following problem. Assume the signal to be sampled is a sine wave $V_{SIG}(t) = V_0\sin(\omega t)$. Assume the sampling circuit takes instantaneous samples at times that satisfy the following equation: $\omega t_S = n\pi + \theta$, where n is a positive integer (0, 1, 2, ...). The sampled value is held constant until the following sample. This equation implies that the signal is sampled twice each cycle, with the sampling phase set by θ. Sketch two cycles of the original signal sine wave and of the sampled waveform for

(a) $\theta = 0$

(b) $\theta = \pi/2$

What is the average value of the sampled waveform in each case?

8.4. *Shot noise, thermal noise, and quantization noise.* Unless extremely weak signals are being dealt with, most electronic signals are far stronger than the **shot-noise limit**, which arises from the discrete natures of electrons and photons. Depending on the **effective noise temperature** of a signal source, the dominant source of noise in a signal processed through an ADC may be quantization noise, shot noise, or thermal noise. The circuit shown in Figure 8.26 has a signal source producing V_{SIG} (RMS).

The equivalent circuit of the signal source also includes a source resistance at an effective temperature T_{EFF} and a shot-noise source I_S. We have modeled the noisy resistor as a combination of a noiseless resistor R_S and a separate noise voltage source v_{NT}. Suppose the bandwidth of the system is limited by filtering or other means (not shown) to a range B Hz wide. The effective temperature T_{EFF} produces an RMS noise voltage

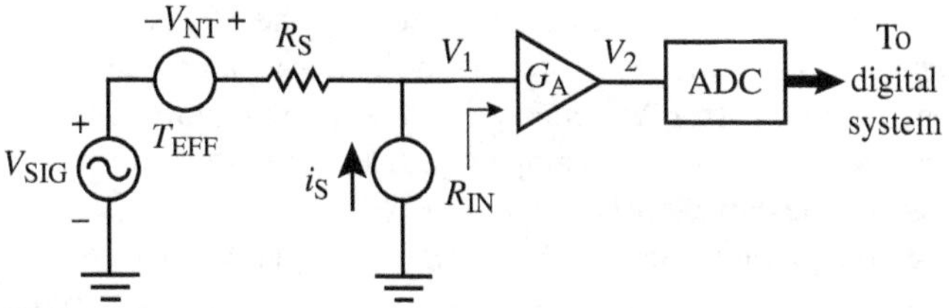

FIGURE 8.26 Hypothetical signal source connected to amplifier and ADC for Problem 8.3.

$$v_{NT} = \sqrt{4k_B T_{EFF} R_S B} \tag{8.36}$$

(see Equation 3.46 and accompanying text for an explanation of thermal noise). In addition, the signal source has a DC current I_{DC} flowing in it, which produces an AC noise current whose RMS value is (see Eq. 3.48)

$$i_S = \sqrt{2qBI_{DC}} \tag{8.37}$$

The resulting total voltage V_1 goes to an ideal noiseless amplifier with infinite input resistance $R_{IN} = \infty$ and voltage gain G_A. The amplifier output $V_2 = G_A V_1$ goes to an ADC whose total input voltage range is $0 < V_{IN} < +V_{MAX}$ and whose LSB voltage (resolution) is (see Eq. 8.9)

$$\Delta V = \frac{V_{MAX}}{2^N} \tag{8.38}$$

(a) Derive a mathematical expression for the total digitized voltage V_{OUT} (RMS), treating each voltage source (the signal voltage, the thermal noise, the shot noise, and the quantization noise) as *uncorrelated* with all the rest. Use Equation 8.11 for the value of quantization noise in terms of V_{MAX} and N. Because powers of uncorrelated voltages add, you should square each voltage, add the sums of the squares, and take the square root to find the total voltage V_{OUT}.

(b) Suppose you want the quantization noise voltage to be 20 dB lower than the dominant (largest) external noise voltage (either shot noise or thermal noise). If $B = 1$ kHz, $I_{DC} = 2$ mA, $R_S = 50\,\Omega$, $T_{EFF} = 1000$ K, $N = 8$, and $V_{MAX} = 5$ V, how much voltage gain G_A is required for the amplifier in order to ensure that the quantization noise is 20 dB below the dominant level of shot or thermal noise? This exercise shows why it is difficult to connect low-level signals directly to most ADCs without encountering severe problems with quantization noise.

8.5. *Design of sample-and-hold circuit.* In the elementary sample-and-hold (S/H) circuit shown in Figure 8.10, assume the signal source connected to the point labeled V_{SIG} can be modeled as an ideal voltage source V_0 in series with an output resistance R_0. In such a situation, an R–C time-constant circuit is formed when the switch S closes, so the assumption that the capacitor instantaneously charges to the sampled input voltage is no longer true.

(a) If the maximum voltage range covered by the *S/H* circuit is V_{MAX} and the sampling switch is closed for a time t_S, derive an equation that expresses the maximum error ΔV_S encountered when two samples separated by V_{MAX} are taken. Assume the op amp is ideal and draws no current from *C*. (*Hint*: Assume the capacitor is charged to V_{MAX} initially and then samples $V_0=0$. The voltage on the capacitor at the end of t_S is ΔV_S.)

(b) Suppose $R=50\,\Omega$, $C=1\,\text{pF}$, and the maximum allowable error $\Delta V_S=0.2V_{MAX}/2^N$, where $N=16$ is the number of bits of resolution in the ADC that follows the S/H circuit. (This is a reasonable condition, because if the sampling error exceeds the LSB resolution, it renders the LSB of the output meaningless.) What is the shortest sample time t_S that can be used under these conditions?

8.6. *Conversion times of flash and successive-approximation converters*. Suppose the same type of comparator circuit is used both in a flash converter and in a successive-approximation converter. The comparator circuit takes a maximum of $t_{PC}=50$ ns to produce a valid digital output once the voltages stabilize on its inputs. Both converters accept positive input voltages only and have 8-bit resolution.

(a) How many comparators will the flash converter use? If the digital circuitry that decodes the raw comparator outputs to standard binary has a propagation time $t_{PD}=200$ ns, what is the approximate maximum frequency f_{MAX}(flash) that the flash converter can successfully digitize?

(b) Suppose the successive-approximation converter takes $t_C=1\,\mu\text{s}$ to complete one approximate-test cycle (consisting of instructing the DAC to send a certain voltage to the comparator, reading the comparator's output, and deciding what the next voltage command should be). About how long will it take the successive-approximation converter to produce the digital word corresponding to one voltage sample? What is the approximate maximum input frequency f_{MAX}(SA) that the successive-approximation converter can handle accurately?

***8.7.** *Limitations of delta-sigma converter*. Although the delta-sigma ADC is very flexible and used in a wide variety of forms, it has certain limitations with regard to minimum voltage resolution and speed. It turns out that these limitations are not unrelated to each other, as the following exercise will show.

Consider the elementary delta-sigma ADC circuit shown in Figure 8.14. If V_{SIG} consists of a periodic AC waveform at the clock frequency f_{CLOCK}, it is possible that the circuit will produce an output at the D flip-flop's *Q* terminal that is a digital square wave at f_{CLOCK}. This square wave will have a 50% duty cycle and will (correctly) indicate that the average value of the input signal is zero. But the digital output will not change at all if the input voltage amplitude changes, indicating that the voltage resolution of the circuit has degraded to zero. This shows that the resolution of the delta-sigma circuit is not independent of the input frequency and is one reason why such circuits are usually operated at a clock frequency that represents a large factor of oversampling.

(a) If a designer wishes to achieve a resolution of N bits with the delta-sigma ADC shown in Figure 8.14, what is the maximum input-signal frequency f_{MAX} that can be converted, bearing in mind the Nyquist sampling criterion? Express f_{MAX} in terms of N and the clock frequency f_{CLOCK}.

(b) Suppose the comparator has a maximum input offset voltage error V_{OS}. This means that the input voltage difference may need to be as large as $\pm V_{OS}$ before the comparator output changes state, rather than 0 as with an ideal comparator. If the integrator operates during one period of the clock frequency f_{CLOCK} with a time constant RC such that its input–output relation is $V_{INT} = \frac{1}{RC}\int V_{SUM}(t)dt$, find the minimum input voltage change δV necessary to ensure that the comparator will change state for the whole range of possible comparator offset voltages. Express δV in terms of R, C, f_{CLOCK}, and V_{OS}. For voltage changes smaller than δV, the input signal is liable to be "swamped" by offset voltage changes, which contributes another noise source to the output.

8.8. *Accuracy and speed of dual-slope integrator converter*. Because the dual-slope integrator ADC does not depend for its accuracy on a precise rate of integration, its intrinsic accuracy is almost entirely dependent on the accuracy of the voltage reference used. However, there is a definite speed-resolution trade-off, as the following example shows.

(a) Suppose a digital counter is available that can count to $2^{16} = 65{,}536$ and a reference voltage of −2.0V is available. Find the design parameters requested below for a dual-slope integrator ADC that will have a total input range of 0 to +2V, can convert 10 samples/second at full-scale input (+2V), and produces a count of 2^{16} for +2-V input.

The design parameters are (1) the minimum time constant RC of the op amp integrator used in Figure 8.15 (assume the op amp integrates to negative voltages from a starting voltage of 0V and will saturate for an output voltage lower than -10 V) and (2) the frequency f_{COUNT} that the counter counts in order to transform a time delay into a digital version of the input voltage. You can use the fact that $\frac{dV_{OUT}}{dt} = \frac{V_{IN}}{RC}$ in solving part (1).

(b) Using the same counter, suppose that the number of samples must be increased to 100 per second. What is the approximate maximum number of bits of resolution that the ADC can now deliver, assuming f_{COUNT} cannot be adjusted?

8.9. *Comparison of R–2R and switched-capacitor converters*. Power consumption is an increasingly important characteristic of many designs for portable battery-operated equipment. This exercise will provide a comparison between the R–$2R$ and switched-capacitor DAC approaches.

(a) Design an R–$2R$ network for a 12-bit DAC, using the circuit shown in Figure 8.20 extended to 12 bits. Suppose the smallest current that can be reliably switched is 1 μA and the highest available reference voltage is 2.7 V. What is the total DC power consumption of the R–$2R$ network?

(b) Now suppose a 12-bit switched-capacitor DAC of the type shown in Figure 8.19 uses binary-weighted capacitors (i.e., $C_2=2C_1$, etc.). The available V_{REF} is 2.7 V for this circuit as well. Also suppose that the smallest capacitor that can be reliably made with a certain process is 20 fF (1 fF = 1 *femtofarad* = 10^{-15} F). If one cycle of operation to produce the analog equivalent of one digital word takes a time t_C and the cycle frequency is $f_C=1/t_C$, find the cycle frequency that will cause the switched-capacitor circuit's average power consumption to equal that of the R–$2R$ circuit. For any speed slower than this, the switched-capacitor circuit's power consumption will be lower.

8.10. *Oversampling and one-bit DAC*. Suppose a one-bit DAC of the form shown in Figure 8.22 is driven by a pulse-width-modulated waveform at a frequency f_{CLOCK}. The appearance of f_{CLOCK} or its harmonics in the analog output is undesirable, and the aim of this exercise is to choose the R–C filter's time constant $\tau=RC$ so that the worst-case RMS amplitude of the f_{CLOCK} component of the output is suppressed by 40 dB below the reference voltage V_{REF}. It is easy to show that the worst-case situation in which the largest fundamental-frequency signal is apt to appear at the output is when the duty cycle of the output is $D=50\%$. Under those conditions, the fundamental-frequency component at f_{CLOCK} going into the R–C filter has an RMS amplitude of $\left(\sqrt{2}/\pi\right)V_{REF}$. Using what you know about single-pole RC filters from Chapter 6, find the value of $\tau=RC$ that will suppress the RMS value of the f_{CLOCK} frequency component by 40 dB below V_{REF} if $f_{CLOCK}=4$ MHz.

For further resources for this chapter visit the companion website at
http://wiley.com/go/analogmixedsignalelectronics

9

PHASE-LOCKED LOOPS

9.1 INTRODUCTION

Phase-locked loops (**PLLs**) are used in a wide variety of applications ranging from communications devices such as cell phones, radios, and TVs to computers and motor controllers. Many on-chip computer clock systems use **PLLs**, and digital communications systems requiring the acquisition of synchronization signals from a remote source often use a PLL or equivalent to recover the clock signal. PLLs usually incorporate one or more analog oscillator circuits, although entirely digital PLLs can also be designed as well. However, other necessary functions in a PLL are best done with digital circuits, so a PLL is typically a mixed-signal system.

PLLs can be designed with a number of different approaches. The simpler PLLs can be analyzed with linear systems analysis, specifically **control theory**, and a brief summary of the control theory necessary to design simple PLLs will be included in Section 9.3. More advanced PLLs must be analyzed with more sophisticated mathematical techniques or by circuit simulation software, which can always be used to verify a paper design before it is built. The material in this chapter is sufficient to guide the designer in tackling the more straightforward PLL problems and will help those who simply deal with them in systems to specify and use them intelligently.

Analog and Mixed-Signal Electronics, First Edition. Karl D. Stephan.

Companion Website: http://wiley.com/go/analogmixedsignalelectronics

9.2 BASICS OF PLLS

The basic components of a simple PLL are shown in Figure 9.1. They consist of a **voltage-controlled oscillator (VCO)**, a **phase detector** (sometimes called a **phase discriminator**), and a loop amplifier. Every PLL must have a reference input frequency to which its output is phase locked. This can be generated internally (as with a crystal-controlled oscillator), or it can be derived from an external signal such as a digital communications channel. Whatever the source of the external reference frequency, the whole point of the circuit is to produce an output signal whose phase is locked to the reference frequency's phase.

On the face of it, that does not look too impressive. Why not just send the reference frequency straight to the output? There are a number of reasons why using a PLL driven by a reference signal can be useful. Here are a few:

1. *The output frequency can be a multiple or submultiple of the input frequency.* As we will show later, by placing a digital frequency divider between the VCO output and the phase-detector input, the output frequency f_{OUT} can be forced to be an exact multiple of the reference frequency f_{IN}. This is how cell-phone transmitters operating at frequencies of up to 2 GHz or more are controlled in frequency by crystal oscillators operating at much lower frequencies.
2. *A noisy or drifting reference signal can be "cleaned up" by a PLL.* One of the first important uses of PLL circuits was to recover very weak radio signals

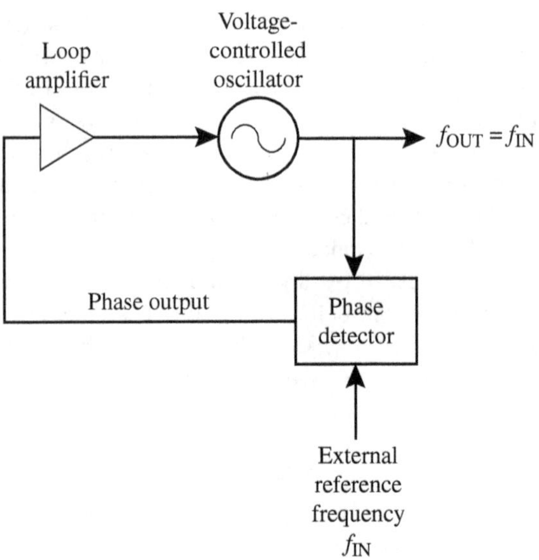

FIGURE 9.1 Basic phase-locked loop (PLL), showing voltage-controlled oscillator (VCO), phase detector, and loop amplifier.

from space probes. Even if a reference signal disappears entirely for a short time due to fading or other causes, a properly designed PLL can "flywheel" during the time the fade lasts and still maintain synchronization until the reference signal shows up again.

3. *PLLs can demodulate some types of digital signals.* A type of modulation known as **frequency-shift keying (FSK)** modulates the carrier by shifting its frequency a certain amount. Since the PLL's VCO frequency tracks the input frequency, the VCO's control voltage will be directly proportional to the input frequency and follows its frequency shifts, which means that the demodulated signal appears on the VCO control line. More complex forms of digital modulation are often demodulated with the help of a PLL to obtain and maintain a precise phase reference for the demodulator circuits.

Now that you know some uses for PLLs, we will outline the basic linear analytical approach to them, which is derived from control theory.

9.3 CONTROL THEORY FOR PLLS

In Figure 9.2, we have labeled the various voltages in the PLL for our analysis. Since phase is the controlled variable, we will not be too concerned about the amplitude of the oscillator output. Instead, we will show how the system acts with respect to the oscillator phase θ_0 and its relation to the reference phase θ_R.

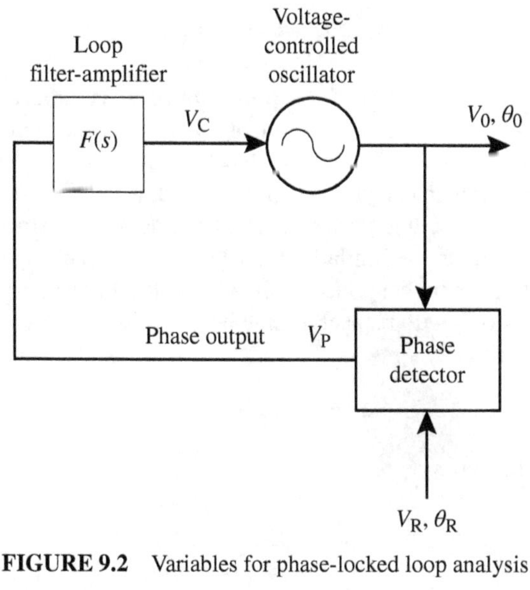

FIGURE 9.2 Variables for phase-locked loop analysis.

The VCO is characterized by a **VCO constant** K_C whose dimensions are (rad s^{-1}) V^{-1}. This constant expresses how the oscillator's radian frequency ω varies with the control voltage V_C:

$$\omega(V_C) = K_C V_C(t) \tag{9.1}$$

The oscillator output itself looks like this:

$$V_0(t) = V_A \cos(\omega t) = V_A \cos([K_C V_C(t)]t) \tag{9.2}$$

To find the instantaneous phase, we must integrate the argument of the cosine, since radian frequency is the derivative of phase:

$$\theta_0(t) = K_C \int V_C(t)\, dt \tag{9.3}$$

Next, consider the reference signal V_R (we will take the amplitude equal to the oscillator's amplitude for simplicity):

$$V_R(t) = V_A \cos(\omega_R t) \tag{9.4}$$

The phase of the reference signal is θ_R:

$$\theta_R(t) = \omega_R t \tag{9.5}$$

Now consider what the phase detector does. While there are many types of phase detectors, most of them can be modeled (at least for some conditions) as a device that produces a voltage V_P that is a linear function of the *phase difference* between the reference voltage and the oscillator voltage:

$$V_P = K_P(\theta_R(t) - \theta_0(t)) \tag{9.6}$$

The constant K_P (V rad^{-1}) is called the **phase-detector constant**. The sign of the right-hand side of Equation 10.6 is important—if you get the sign wrong, you will get positive feedback instead of a negative-feedback system!

At this point, rather than carrying integrals around, we will introduce the **Laplace transform** $p_0(s)$ of the oscillator phase $\theta_0(t)$. (The Laplace transform was discussed in more detail in Chapter 7.) Similarly, we will designate the Laplace transform of any voltage function of time $V(t)$ as small $v(s)$. In the Laplace-transform domain, integration of a function with respect to time becomes division by the complex variable s, so we can write

$$v_P(s) = K_P(p_R(s) - p_0(s)) \tag{9.7}$$

and Equation 9.3 becomes

$$p_0(s) = \frac{K_C v_C(s)}{s}, \tag{9.8}$$

so we end up with the following expression for the Laplace-transformed phase-detector output:

$$v_P(s) = K_P p_R(s) - \frac{K_P K_C v_C(s)}{s} \tag{9.9}$$

The phase-detector output goes through the loop filter, whose Laplace-transformed response we will call $F(s)$, and its output becomes v_C. So we finally close the loop to obtain this expression:

$$v_C(s) = F(s)\left[K_P p_R(s) - \frac{K_P K_C v_C(s)}{s}\right] \tag{9.10}$$

After some algebra, we can obtain the following expression that gives the ratio of the oscillator phase p_0 to the reference phase p_R:

$$\frac{p_0(s)}{p_R(s)} = \frac{F(s)K_C K_P}{s + F(s)K_C K_P} = H(s) \tag{9.11}$$

The ratio of the (Laplace-transformed) output phase p_0 to input phase p_R is the **closed-loop transfer function** $H(s)$. This transfer function contains important information about the behavior of the loop when it is **locked** (i.e., when the input phase is properly controlling the output phase.) Now that we've got a basic relation between the input and output phases, we will explore the consequences of using various types of loop filter characteristics for $F(s)$.

9.3.1 First-Order PLL

The **order** of a PLL (or any linear control system, for that matter) is the highest exponent of the Laplace-transform variable s in the denominator polynomial of the closed-loop transfer function $H(s)$. For example, if the denominator of $H(s)$ is $s+5$, it represents a first-order loop. If the denominator is s^2+7s+3, it represents a second-order loop, and so on.

The simplest possible loop filter function is a direct connection, meaning $F(s)=1$. Using such a connection produces a **first-order PLL** whose closed-loop transfer function is

$$H(s) = \frac{K_C K_P}{s + K_C K_P} \tag{9.12}$$

The product $K_C K_P$ has the dimensions of

$$\frac{\text{rad}}{\text{s}-\text{V}} \cdot \frac{\text{V}}{\text{rad}} = \text{s}^{-1} \tag{9.13}$$

which is in units of frequency. When you recall that a single-pole passive *R–C* lowpass filter has the transfer function

$$\frac{1}{1+(s/\omega_C)} = \frac{\omega_C}{s+\omega_C}, \tag{9.14}$$

where $(1/\omega_C) = RC$, you can understand that the first-order loop's $H(s)$ has a frequency response equal to that of a single-pole lowpass filter whose cutoff frequency f_C (in Hz) is

$$f_C = \frac{K_P K_C}{2\pi} \tag{9.15}$$

While a first-order loop can work, it has serious limitations. For example, if the reference (input) frequency varies, the steady-state phase error has to change in order to track the input frequency variation. This is because the phase-detector output V_P is zero for zero phase error. But to produce a specific output frequency, we must feed the VCO input a nonzero voltage, which can only result from a nonzero phase error that changes with input frequency.

The first-order loop is also fairly inflexible, because there is only one "knob" to adjust. Even if we replace the direct connection with a DC-coupled amplifier with gain G, all that changing the gain does is to change the loop bandwidth f_C. A larger gain means a greater loop bandwidth, but as we will see in the following text, a loop bandwidth that is too wide can cause problems. If you happen to want a small loop bandwidth and try to make G less than 1, you can run into a different obstacle. If the input frequency changes over a certain range, you cannot make G too small or else there will not be enough voltage variation at the VCO control-voltage input to tune the VCO over the desired range.

For these and other reasons, most designs use at least a **second-order PLL**, which we will now describe.

9.3.2 Second-Order PLL

While there are various types of second-order loop filters using both passive and active circuits, for simplicity, we will describe a type of filter response obtainable with only three passive components: a capacitor and two resistors. This will nevertheless allow you to vary the loop characteristics in a way that is sufficiently flexible for many applications.

Consider the filter circuit shown in Figure 9.3:

This filter's transfer function $F(s) = v_c/v_p$ is easy to calculate and results in the following expression:

$$F(s) = \frac{R_2}{R_1+R_2}\frac{s+\omega_2}{s+\omega_1} \tag{9.16}$$

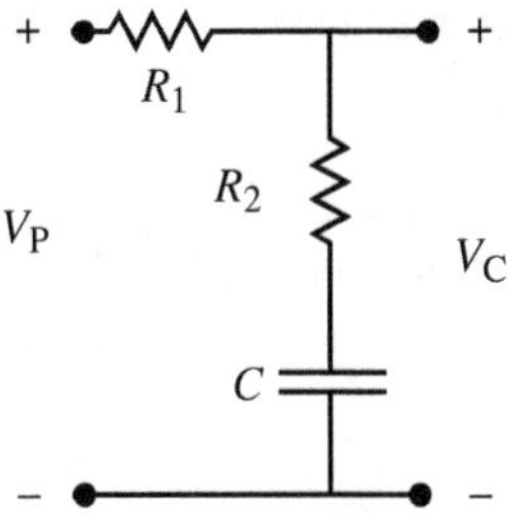

FIGURE 9.3 Passive loop filter for second-order PLL.

in which

$$\omega_1 = \frac{1}{(R_1 + R_2)C} \tag{9.17}$$

and

$$\omega_2 = \frac{1}{R_2 C} \tag{9.18}$$

When this filter function $F(s)$ is inserted into the closed-loop transfer function $H(s)$, you can show after some algebra that a *second-order* transfer function results. To make this easier to follow, let's define a constant G_0 that has the dimensions of frequency:

$$G_0 \equiv \mathrm{K_C K_P} \frac{R_2}{R_1 + R_2} \tag{9.19}$$

After some algebra, we can write $H(s)$ as

$$H(s) = \frac{G_0(s + \omega_2)}{s^2 + (\omega_1 + G_0)s + \omega_2 G_0} \tag{9.20}$$

Next, we resort to some control-theory conventions to recast Equation 9.20 into a standard form that is easier to work with, because the standard form has two parameters that signify distinct and significant things. These parameters are ω_n, the **natural frequency**, and ζ (the Greek letter "small zeta"), which is called the **damping factor**. It is straightforward to show that the transfer function $H(s)$ in Equation 9.20 can be written in terms of these two parameters plus the factor G_0:

$$H(s) = \frac{G_0 s + \omega_n^2}{s^2 + 2\zeta\omega_n s + \omega_n^2} \tag{9.21}$$

If we define the new parameters in terms of the filter frequencies ω_1 and ω_2 and constant G_0,

$$\omega_n = \sqrt{\omega_2 G_0} = \sqrt{\frac{K_C K_P}{(R_1 + R_2)C}} \tag{9.22}$$

and

$$\zeta = \frac{\omega_1 + G_0}{2\sqrt{\omega_2 G_0}} = \frac{(1/((R_1 + R_2)C)) + (K_C K_P R_2 /(R_1 + R_2))}{2\omega_n} \tag{9.23}$$

In Equations 9.22 and 9.23, we have reinserted the original component values and VCO and phase-detector constants for design purposes.

Ordinarily, the designer will begin with a choice of ω_n and ζ and will wish to derive component values for C, R_1, and R_2 in the passive loop filter circuit shown in Figure 9.3. (We will discuss how to choose ω_n and ζ shortly.) With only two parameters and three component values, the system of equations is underdetermined, so we can arbitrarily choose a value for one of the components and work from there. Initially choosing a value for C will determine the impedance level of the circuit and allows us to solve for the values of R_1 and R_2, given the circuit constants $K_C K_P$, C, and the design choices for ω_n and ζ.

There is a restriction on how small a damping factor can be chosen for a given set of the preceding parameters. The minimum possible damping factor ζ_{MIN} depends on ω_n and the product of the phase-detector constant K_P and the VCO constant K_C:

$$\zeta_{MIN} = \frac{\omega_n}{2K_C K_P} \tag{9.24}$$

For reasonable values of the circuit constants, this restriction is not inconvenient because an excessively low damping factor is not desirable, as we shall see.

As long as the damping factor selected is greater than ζ_{MIN}, you can use the following equations to find values for R_1 and R_2:

$$R_1 = \frac{1}{C}\left(\frac{K_C K_P}{\omega_n^2} - \frac{2\varsigma}{\omega_n} + \frac{1}{K_C K_P}\right) \tag{9.25}$$

$$R_2 = \frac{1}{C}\left(\frac{2\zeta}{\omega_n} - \frac{1}{K_C K_P}\right) \tag{9.26}$$

Equation 9.26 shows why there is a minimum limit for ζ, which occurs when $R_2 = 0$. Choosing a damping factor lower than this would require a negative resistor for R_2!

The terms "natural frequency" and "damping factor" originate from the fact that the transfer function $H(s)$ in Equation 9.21 has exactly the same form as that of a

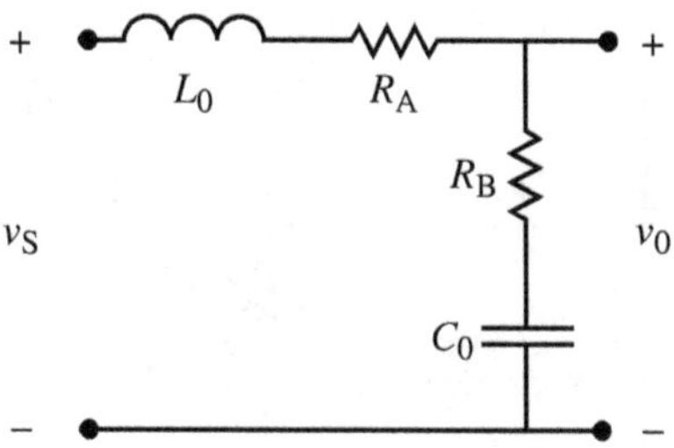

FIGURE 9.4 *R–L–C* circuit with transfer function similar to transfer function $H(s)$ of second-order PLL.

series-resonant *R–L–C* circuit of the general type we studied in Chapters 6 and 7. As you may recall, the series *R–L–C* circuit was characterized by a resonant frequency ω_0 and a quality factor Q. For values of Q higher than 0.5, the transient behavior of the circuit is **oscillatory**—that is, its reaction to an impulse will be the production of one or more cycles of a damped sinusoidal wave, rather than an exponential decrease to zero. It turns out that the damping factor $\zeta = 1/(2Q)$, so as Q increases, the damping factor decreases. Less damping means more oscillation cycles for a given transient input.

The second-order PLL circuit exhibits the same basic types of behavior as the passive *R–L–C* circuit because it is governed by similar equations. If we take the output voltage V_0 across the series combination of R_B and C in the series *R–L–C* circuit shown in Figure 9.4, it is straightforward to show that the transfer function of the circuit of is:

$$\frac{v_0(s)}{v_s(s)} = M(s) = \frac{s(R_B/L_0) + (1/L_0C_0)}{s^2 + s((R_A + R_B)/L_0) + (1/L_0C_0)} \tag{9.27}$$

This means we can simulate the PLL response $H(s)$ with an *R–L–C* circuit having the proper component values. More importantly, we can transfer all the understanding we gained about the passive *R–L–C* circuit's responses in the frequency and time domains to the behavior of the second-order loop.

In the frequency domain, if frequency or phase modulation is applied to the reference signal, you can think of the loop response as "filtering" that modulation to a greater or lesser degree. For example, suppose you phase-modulate a 10-kHz carrier with sine-wave modulation having a frequency of 1 kHz. That means the phase of the 10-kHz carrier varies periodically at a 1-kHz rate (1000 times a second). If the natural frequency of a second-order loop is only 100 Hz, the modulation frequency of 1 kHz is well beyond the loop's natural frequency, which acts as a kind of filter cutoff frequency for modulation. In other words, a loop with a natural frequency of 100 Hz is going to have a lot of trouble following phase variations at a 1-kHz rate. In practice, either the modulation will barely appear at all on the VCO control voltage, or if the **deviation** (amount of phase or frequency change) is large enough, the loop will fall out of lock and generate a lot of nonlinear and chaotic noise-like waveforms. Either eventuality is undesirable.

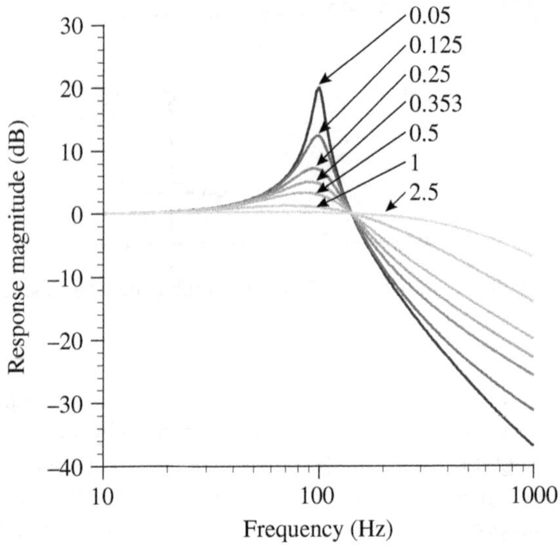

FIGURE 9.5 Frequency response of second-order PLL with natural frequency $\omega_n = 100\,\text{Hz}$ and damping factor ζ varying from 2.5 to 0.05.

With that in mind, let's examine the frequency responses of a second-order loop with a natural frequency $\omega_n/2\pi$ of 100 Hz, as we vary the damping factor from 0.05 to 2.5. (In order to do this without running into the ζ_{MIN} restriction, we had to vary the constants $K_P K_C$ as well, but this does not affect the outcome significantly.) We show this range of damping factors in Figure 9.5. The most highly damped response ($\zeta = 2.5$) is flat well beyond the natural frequency of 100 Hz and is only 3 dB down at about $2\omega_n$. This indicates a trend that continues for higher damping factors, in which the response degenerates into that of a single-pole *R–L* filter, because the term representing capacitive impedance in the response equation is dwarfed by the resistive impedance at frequencies near and above ω_n. As the damping factor decreases through 1 to smaller values, the frequency response begins to show a peak that gets sharper and higher as ζ decreases (or equivalently, as Q increases). While a small amount of peaking (2–3 dB for ζ in the range of 0.5–1) is acceptable in terms of distortion for demodulating a frequency- or phase-modulated carrier, peaks higher than that range will produce unacceptable frequency and phase distortion. That is why damping factors less than 0.2 or so are not recommended.

To summarize, too low a damping factor gives a sharp peak in the frequency response and results in frequency distortion, while too large a damping factor increases the loop bandwidth so that it exceeds the natural frequency. Consequently, one should generally design a second-order PLL circuit to have a damping factor between 0.5 and 1 (0.707 is the theoretical ideal, sometimes termed **critical damping**).

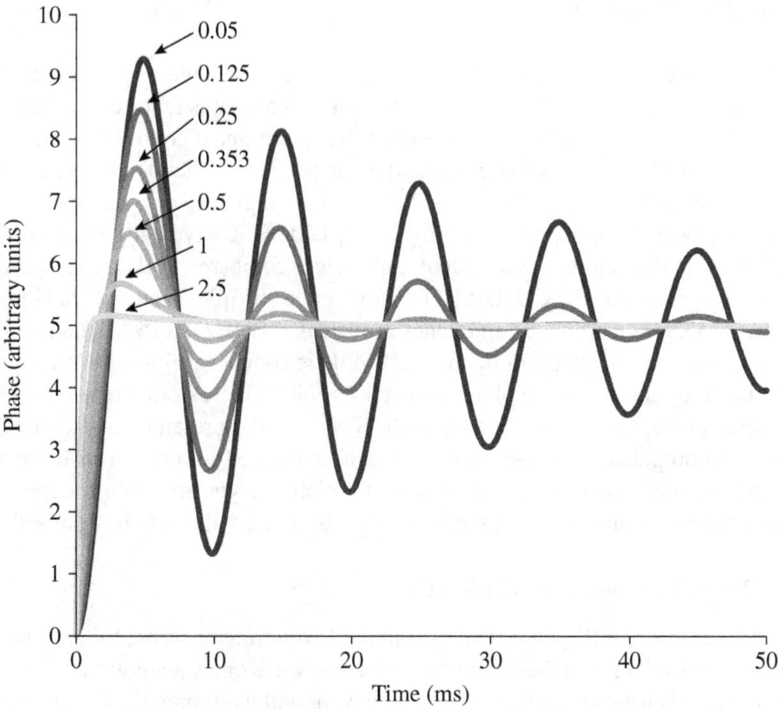

FIGURE 9.6 Time-domain response of PLL phase to arbitrary change in input reference phase of 5 units, for various damping coefficients ζ from 0.05 to 2.5.

The effect of varying the damping coefficient on the time-domain response is even more vivid. Figure 9.6 shows the response to an arbitrary step function in the reference phase of 5 units (the units could be degrees or radians, although to keep things linear, the step-function phase shift should be limited to less than 20° or so). As you can see, the response to the step function for the overdamped case ($\zeta = 2.5$) is very fast and settles almost at once to the correct value of 5, because of its relatively wide bandwidth. As damping decreases through critical damping ($\zeta = 0.707$), the response becomes slightly slower and shows a moderate amount of **overshoot**, meaning that the phase passes through the correct value to larger values before settling down. For very low values of damping factors, the response executes one or more complete oscillations around the correct value of 5 units, and these oscillations get bigger and last longer the smaller the damping is. This phenomenon is known as **ringing**, from its resemblance to what a bell does when you strike it. Ringing is the time-domain behavior that goes with a peaked frequency response higher than 2–3 dB above the low-frequency value and is also undesirable because it is the time-domain consequence of such frequency distortion.

In the next section, we will examine several specific types of phase detectors in the context of designing a specific PLL circuit with a particular PLL IC, the CD4046B.

9.4 THE CD4046B PLL IC

While there are many types of PLL ICs and systems available, we have chosen to describe the workings of one specific IC because it embodies features that are often found in PLLs, and it is simple enough to analyze with linear control theory.

The CD4046B IC is a CMOS device, which is one reason that it can operate with a wide range of supply voltages from 5 to 15 V. Because digital CMOS circuits consume almost no power except when changing states, CMOS ICs typically have very low power consumption and very high input impedances compared to BJT-based circuits. In a typical application, the CD4046B consumes under 100 μW in operation. However, the circuit's VCO can produce a frequency as high as 1.4 MHz if a 15-V supply is used.

A simplified block diagram of the CD4046B is shown in Figure 9.7. The upper half of the diagram shows the phase-detector circuitry. The external reference signal (represented by θ_R) goes into pin 14, and the local signal represented by θ_O (either the actual VCO output or a divided-down version) is applied to pin 3. The IC features two different phase detectors. The choice of which one you should use depends on the application and the nature of the input signals, as we will show in what follows.

9.4.1 Phase Detector 1: Exclusive-OR

Pin 2 is the exclusive-OR phase-detector output. If your reference and VCO signals are rather low in level, or take the form of either a sine wave or a square wave with a duty cycle of 50%, then the exclusive-OR phase detector will work fine. For the purposes of

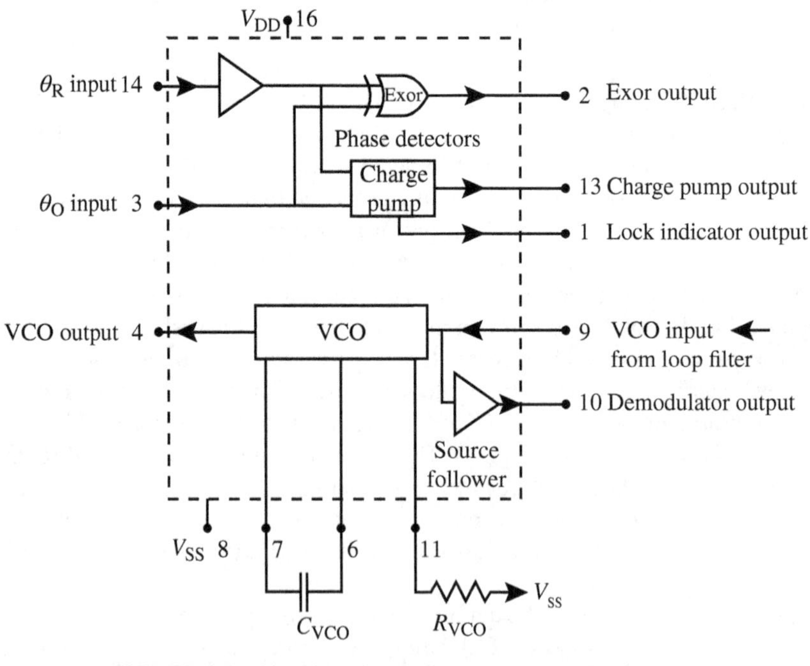

FIGURE 9.7 Simplified block diagram of CD4046B PLL IC.

the following explanation, we will assume both inputs are square waves, however, and vary between $-V_{max}$ and $+V_{max}$. Because the operation of the exclusive-OR phase detector is essentially the same as that of a radio-frequency (RF) **mixer** circuit, we will call the reference input (θ_R) the RF signal, and the VCO input (θ_O) will be called the **LO** signal, which stands for **local oscillator**. These terms derive from their original use in radio receivers, which receive a RF signal and multiply it by a locally generated LO waveform. (Do not confuse this term with the logical LO (0) of a logic gate.) Both an RF mixer and the exclusive-OR phase detector perform multiplication operations on their two inputs, and the product appears at the output. The phase-detector output will be called the **intermediate-frequency (IF)** signal. We will assign the logical state HI (1) to the voltage $+V_{max}$, and the logical state LO (0) to the voltage $-V_{max}$ in what follows.

Observe what happens as the relative phase $\Delta\phi$ of the two signals goes from 0° through 90° to 180° in the sequence shown in Figures 9.8a–c.

As you can see, the exclusive-OR function produces a $+V_{max}$ output when the two inputs are the same level (both HI or both LO) and a $-V_{max}$ output when the two inputs are different. Therefore, if the LO and RF inputs are in phase, they are always the same level and the output will be at $+V_{max}$ all the time. Conversely, if the LO and

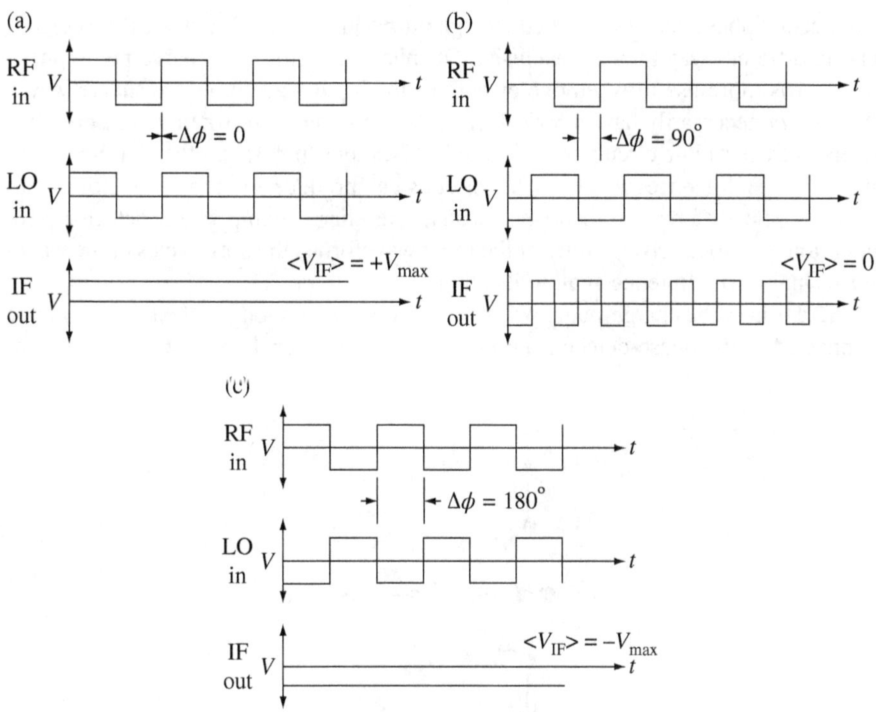

FIGURE 9.8 Inputs and output of exclusive-OR phase detector for input phase differences of 0°, 90°, and 180°. (a) Phase difference $\Delta\phi=0°$: average $<V_{IF}>=+V_{max}$. (b) Phase difference $\Delta\phi=90°$: average $<V_{IF}>=0$. (c) Phase difference $\Delta\phi=180°$: average $<V_{IF}>=-V_{max}$.

RF inputs are exactly 180° out of phase, the exclusive-OR output will be at $-V_{max}$ all the time. Any intermediate phase condition (whether leading or lagging doesn't matter) will lead to an average phase-detector output level that goes linearly from the $-V_{max}$ value to the $+V_{max}$ value as the phase shifts from perfectly out of phase (180° phase difference) to perfectly in phase (0° phase difference).

Assuming the phase-detector output is a fairly "stiff" voltage source, we can estimate the phase constant to use in terms of volts per radian. Ignoring any DC offset, it is clear that the phase constant K_p is given by

$$K_P = \frac{V_{max} - V_{min}}{\pi \text{ rad}} \tag{9.28}$$

For example, in the event that $V_{max} = +5\text{V}$ and $V_{min} = 0\text{V}$ (which are typical TTL levels), $K_p \sim 1.6\,\text{V rad}^{-1}$. If the phase-detector output goes **rail to rail** (meaning from the most positive power-supply voltage to the most negative power-supply voltage) and the power supply provides +15 V and ground, the phase constant would increase to about $4.8\,\text{V rad}^{-1}$.

9.4.2 Phase Detector 2: Charge Pump

The second phase detector, labeled **charge pump** in Figure 9.7, operates in a considerably different way than the exclusive-OR phase detector. The charge-pump phase detector is more useful when one or both of the input waveforms are square waves that do *not* necessarily have a 50% duty cycle. This can occur when you insert a frequency-divider logic circuit between the VCO output (pin 4) and the phase-detector input (pin 3). Such frequency-divider circuits can produce an output waveform with a duty cycle different from 50%. Because the charge-pump phase detector pays attention only to the rising edges of the input waveforms, the duty cycles of the waveforms make no difference at all to its output.

Here is how the charge-pump phase detector works. As long as the load resistance connected to the phase-detector output (pin 13) is greater than 5 kΩ, the equivalent

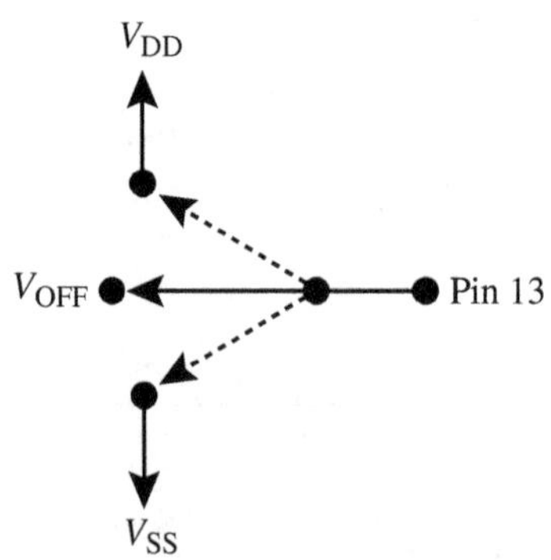

FIGURE 9.9 Equivalent "tristate" circuit of charge-pump phase-detector output.

circuit shown in Figure 9.9 is accurate to within 5% of the true supply voltages V_{DD} and V_{SS}. Typically in a TTL-compatible circuit, you would use $V_{DD}=+5\text{V}$ and $V_{SS}=0\text{V}$ (ground), although the CD4046B will operate with $V_{DD}-V_{SS}$ as high as 15V (and some specifications improve at higher supply voltages). The circuit in Figure 9.9 is called a **tristate output**, meaning it can be in any one of three possible states. In the V_{HI} state, the output is connected to V_{DD} through a p-channel MOSFET. In the V_{LO} state, the output is connected to V_{SS} through an n-channel MOSFET. And in the V_{OFF} state, the output is essentially open circuited and, if not connected to anything, will assume a voltage about halfway between V_{DD} and V_{SS}.

All the circuitry in the charge-pump phase detector is **positive-edge triggered**, meaning that the only parts of either input waveform (θ_R or θ_O) that do anything are the positive-going edges of the square waves. (This is why the duty cycles of the inputs don't matter to this circuit.) One part of the circuit detects which input (θ_R or θ_O) is *higher in frequency* than the other. It does this by waiting for a period of one input (the time between two positive-going edges) in which the other input has two positive-going edges that "fit inside" the first input's period. Once this happens, the circuit goes definitely into one of two modes, determined by which input has the higher frequency and illustrated in Figure 9.10 by case 1 ($f_R>f_O$) and case 2 ($f_R<f_O$). In Figure 9.10, the positive-going edges of the reference and oscillator signals are represented by vertical arrows.

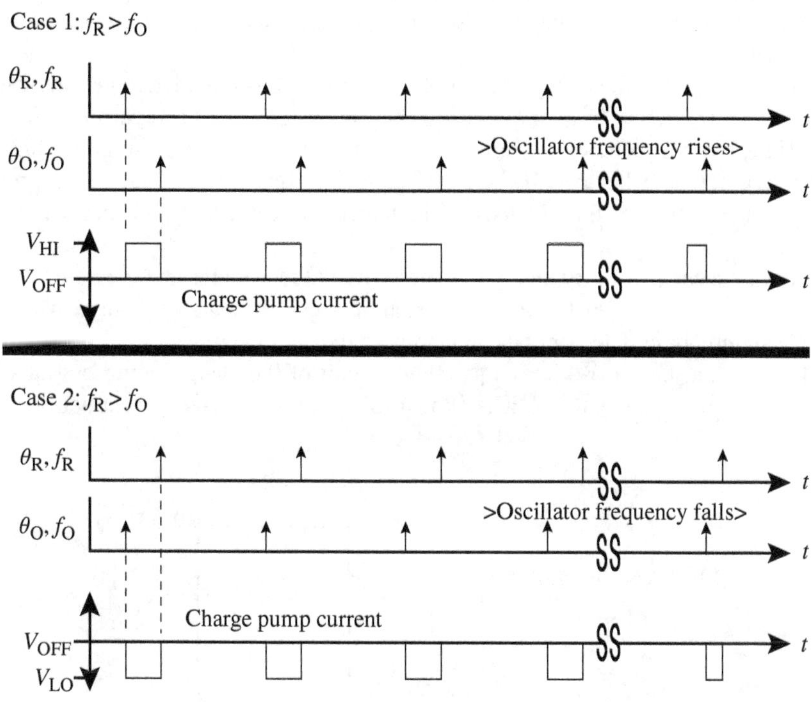

FIGURE 9.10 Waveforms produced by charge-pump phase detector.

If we consider case 1 first, the reference (external input) signal (f_R) is *higher* in frequency than the VCO frequency f_O (assuming no division circuits for the moment). In case 1, the circuit arranges itself so that when a *reference-signal* positive edge comes along, the tristate switch goes to the positive (V_{DD}) supply. And when the *oscillator-signal* positive edge comes along, the switch goes back to V_{OFF} (essentially open). The net effect of this is that a positive current flows out of the pump whose average value is proportional to the phase difference between the reference and the oscillator signals. The larger the phase difference (error), the larger the average positive current. When this current is applied to a capacitor, it will tend to charge *up* (higher voltage), which will *raise* the VCO's frequency closer to the reference frequency.

Now, consider case 2, in which the reference frequency is *lower* than the oscillator frequency. In this case, we must *lower* the VCO frequency to approach a lock condition. In case 2, when a positive edge from the oscillator arrives that sets the tri-state switch to V_{SS} (ground or negative voltage), and when a positive edge from the reference arrives, it sets the switch back to open. In this way, a *negative* current is pumped *out* of the capacitor, *lowering* its voltage and *lowering* the VCO frequency so it approaches the reference frequency.

When the reference and oscillator frequencies are essentially phase locked, it is mathematically impossible for them to be at *exactly* the same frequency and phase (because of noise, etc.), so there will always be very slight phase errors that produce short positive- or negative-going current pulses. These short pulses will provide small corrections to keep the oscillator phase locked to the reference over the long term.

You can see that the charge-pump phase detector has the advantage that it will lock to any reference frequency within the range of the VCO, regardless of how far off the two frequencies are initially. This is not true in general of the exclusive-OR phase detector, because if the two frequencies are too far apart to start with, the average output of the exclusive-OR detector is zero, and lock-in may never occur.

Because a capacitor is typically used at the output of a charge-pump circuit, a pole appears automatically in the transfer function. Here is how to calculate what the charge-pump phase-detector constant and pole frequency will be.

The Norton (current-source) equivalent circuit of the charge-pump output with an *R*–*C* network connected to it is shown in Figure 9.11. The current source $I_p(t)$

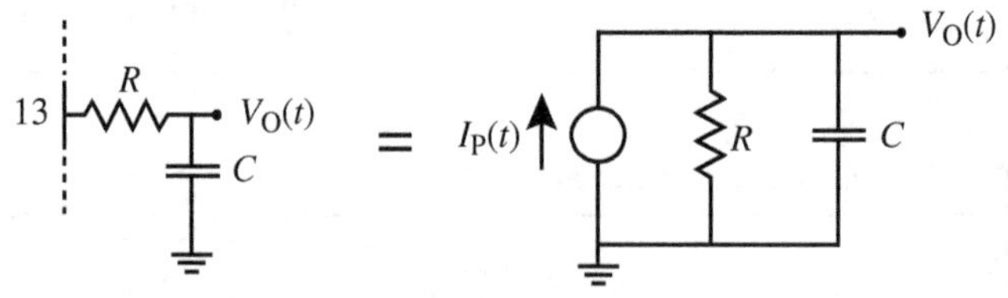

FIGURE 9.11 Typical charge-pump circuit at pin 13 (output) and Norton equivalent circuit.

produces either a positive or a negative current, corresponding to the V_{HI} or V_{LO} waveforms in Figure 9.10, respectively. If we define

$$\Delta V \equiv \frac{V_{DD} - V_{SS}}{2}, \tag{9.29}$$

then the magnitude of this current when the tristate switch is on is

$$|I_P| = \frac{\Delta V}{R} \tag{9.30}$$

If the phases θ_R and θ_O are in radians, the waveforms in Figure 9.10 tell us that the average current from the Norton equivalent current source is

$$\langle I_P \rangle = \frac{(\theta_R - \theta_O)}{\pi}\frac{\Delta V}{R} \tag{9.31}$$

When this current is applied to the parallel-*RC* circuit in Figure 9.11 and we go to the Laplace-transform domain, the result is that the output voltage $v_O(s)$ is

$$v_O(s) = \frac{K_P(p_R - p_O)}{1 + sRC} \tag{9.32}$$

in which the phase-detector constant for the charge-pump circuit is $K_P = \Delta V/\pi$. So the charge-pump circuit with an associated resistor and capacitor already has a pole in it, which sometimes eliminates the need for additional loop filtering altogether. Generally speaking, getting a circuit to work with the charge-pump circuit is easier than with the exclusive-OR circuit, although the performance in terms of phase noise and modulation sidebands may be inferior to what the exclusive-OR phase detector can deliver.

9.4.3 VCO Circuit

The VCO control-voltage input in the CD4046A has a high impedance, which means you can use relatively small capacitors and large resistors in the loop filter circuit without worrying about loading by the VCO input. The smallest capacitor you can use and expect reliable operation, however, is 100 pF (V_{dd} = 5 V) or 50 pF (V_{dd} = 15 V). And the resistors must be greater than 5 kΩ. These mild constraints nevertheless allow a wide range of frequencies with reasonable component values, from less than 1 Hz to over 1 MHz.

To find the VCO constant K_C (in units of rad s^{-1}V^{-1}), you need to know the values of R_{VCO} (ohms) and C_{VCO} (farads). Shown in Figure 9.7, these two components determine the relationship between the VCO control-voltage input and the frequency of the VCO output waveform. The following empirical expression is based on experimental work and gives a reasonably good estimate for the VCO constant:

$$K_C \approx 8.8(R_{VCO}C_{VCO})^{-0.8} \text{ rad s}^{-1}\text{ V}^{-1} \tag{9.33}$$

For example, if $R_{VCO} = 100\ \mathrm{k\Omega}$ and $C_{VCO} = 100\,\mathrm{pF}$, the VCO constant will be about

$$8.8[(10^5)(100\times10^{-12})]^{-0.8} = 8.8(10^{-5})^{-0.8} = 8.8(10^4) = 88,000 \qquad (9.34)$$

There are also datasheet graphs available that give the VCO's center frequency as a function of C_{VCO} for various fixed values of R_{VCO}, but Equation 9.33 is expressed directly in terms of the VCO constant and is accurate enough for initial designs.

Once the VCO constant has been determined, the actual frequency ω_{VCO} (in rad s^{-1}) of the VCO is given by

$$\omega_{VCO} = K_C(V_9 - 1\,\mathrm{V}) \qquad (9.35)$$

where V_9 is the voltage applied to pin 9, the VCO input terminal. When Equation 9.35 is Laplace transformed, the −1-V term goes away. However, you should be aware of the 1-V offset when you are checking out your circuit for proper operation.

Because the control voltage to the VCO is at a high impedance, if your circuit needs to use the VCO control voltage as an output (in a demodulator, for instance), it would cause problems to connect a low-impedance load directly to pin 9. For the designer's convenience, therefore, the CD4046B has an internal buffer amplifier (a source follower) that provides a low-impedance output voltage at pin 10 to drive a demodulator output amplifier or other load. The connections for R_{VCO} and C_{VCO} are shown in Figure 9.7, as is the output of the VCO at pin 4, which is TTL-compatible with a 5-V supply if you wish to drive digital frequency-divider circuits, for example.

9.5 LOOP LOCKING, TUNING, AND RELATED ISSUES

Up to now, we have assumed in our analysis that the loop is already *locked*: that is, somehow, the phase of the VCO output has already become synchronized to the reference signal's phase, and we are just looking at small changes to that phase relationship. We have said nothing about how this locking occurs, but it must happen somehow for PLLs to work at all.

For the charge-pump type of circuit that detects whether the VCO frequency is lower or higher than the reference frequency, the VCO's frequency will be automatically steered toward synchronism with the reference frequency, and locking is not usually a problem. However, the charge-pump circuit is difficult to analyze with the simplified linear analysis we have presented so far, so we will discuss locking with the exclusive-OR phase detector as well.

One-way locking occurs by simply making sure the VCO's **free-running frequency** (its frequency when there is no signal applied to the reference input of the PLL) is within the **capture range** of the reference (input) frequency.[1] If a reference frequency of a given amplitude shows up within the capture range, the PLL will spontaneously lock without any special measures taken to ensure lock.

[1]PLL terminology varies, and the capture range is sometimes called the pull-in range.

What happens is the following. When a signal initially appears at the reference input at a frequency ω_{REF}, and the VCO is running at a different frequency ω_{VCO}, the phase detector will produce a difference-frequency output waveform at the frequency $\omega_p = |\omega_{REF} - \omega_{VCO}|$. As long as the loop bandwidth is wide enough to pass ω_p without too much attenuation, the difference frequency will **frequency-modulate** the VCO. Frequency modulation is a nonlinear process. The nonlinearity produces a DC term in the phase-detector output that begins to push the VCO's frequency toward the reference frequency. This change reduces the difference frequency, and in a relatively short time that depends on the loop constants, but is typically only a few milliseconds, the difference frequency falls to zero, and the phase of the VCO locks to the reference frequency phase. Once lock is achieved, the reference frequency can shift within a wider range than the capture range called the **hold range** (also sometimes called the **lock range**), and the VCO will track it, maintaining the phase-locked condition. But if the reference frequency deviates too far from the VCO's free-running frequency, the system cannot push the VCO far enough fast enough, and the system **loses lock**, meaning that the VCO output is no longer phase locked to the reference input.

How do you tell whether your PLL is locked? By examining the phase relationship between the VCO and the reference frequency. This can be done in several ways. One is by looking at the phase-detector output directly. In a locked loop, the output has a large DC component with an AC component at twice the VCO frequency (at least for the XOR type of phase detector, as we explained earlier). But when the loop is unlocked, there will sometimes be an irregular waveform at the frequency difference f_p between the reference signal and the (unlocked) VCO signal.

Another way to determine if the PLL is locked is to display the reference input signal and the VCO waveform on two channels of a dual-channel oscilloscope, and compare their phases directly. If you **sync** the scope on one signal (say, the VCO), and if the loop is locked, both signals should be displayed as stable waveforms. If the reference signal doesn't stabilize but the VCO signal is stable, the PLL is probably not locked. This is also a good way to compare the phases of the reference and VCO signals once the loop is locked.

Still another way to determine phase lock is to feed the VCO signal to the X-input of a scope and the reference to the Y-input (most scopes can be made to show an X–Y display). The result is a type of mathematical figure called a **Lissajous** pattern, which can be interpreted for phase data. Regardless of the shape of the pattern, if it is stable and doesn't move with time, the loop is locked.

Finally, some PLL ICs (including the CD4046B) have a lock-indicator output that can be used to determine whether the loop is locked.

As you might expect, a PLL with a small loop bandwidth cannot acquire or maintain lock over as wide a frequency range as one with a wider loop bandwidth. This is because the relatively high difference frequency f_p that results when the VCO and reference frequencies are widely separated is filtered out by a small-bandwidth loop filter, while a wider-bandwidth loop filter might pass it enough to allow the loop to lock. This may or may not be a problem, depending on the application. If the reference frequency is known and constant, a narrow loop bandwidth can be used, as long as the VCO is **tuned** manually using an adjustable resistor in the timing circuit of the VCO (R_{VCO} in Fig. 9.7). The VCO is designed to be fairly stable in the event of variations of power-supply voltage and temperature, so this adjustment is not usually necessary to repeat often once it is set. On the other hand, if the reference signal's frequency varies considerably in normal conditions (as in FM or FSK modulation), you will need a wider loop bandwidth to allow the VCO to keep up with the input signal frequency's deviations. The loop bandwidth has to be higher than the highest significant frequency component of the demodulated signal, or else you will be in danger of losing lock during part of the signal and encountering severe distortion.

9.6 PLLS IN FREQUENCY SYNTHESIZERS

PLLs are very useful in the design of certain types of **frequency synthesizers**. A frequency synthesizer is a system that produces any of a set of predetermined frequencies on command (usually a digital command). They are often used in digital radio receivers to establish what is called the **local oscillator** frequency, which determines what reception frequency the radio is tuned to. Frequency synthesizers are also useful in certain types of spread-spectrum communications, in which a variety of different frequencies are transmitted in a rapid sequence.

One of the most general types of frequency synthesizers using a PLL is shown in Figure 9.12. In operation, the VCO's output f_{VCO} is divided by an integer N using a digital frequency divider to form one input of the phase detector. The other input of the phase detector is derived by starting with the reference frequency f_{REF} and dividing it by M with another digital divider circuit. When the PLL is locked, the two inputs to the phase detector must be at the same frequency, so we have $f_{VCO}/N = f_{REF}/M$ or

$$f_{VCO} = f_{REF} \frac{N}{M} \tag{9.36}$$

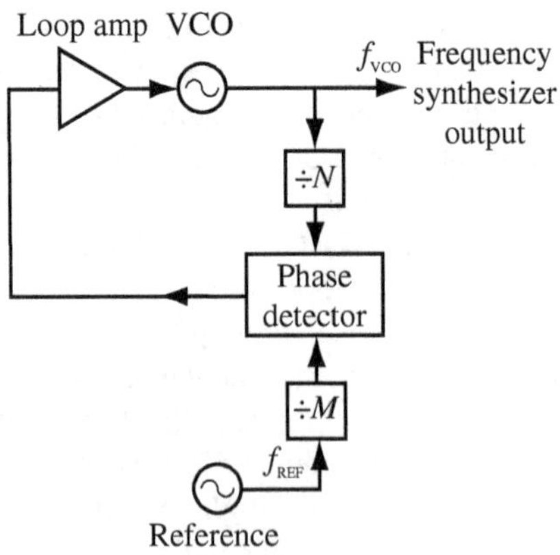

FIGURE 9.12 One type of PLL frequency synthesizer circuit.

By choosing different values for N and M, you can start with a constant reference frequency (typically from a crystal-controlled oscillator) and multiply it by any rational fraction such as 7/9 or 57/111! In this way, almost any desired frequency can be produced from a single constant reference frequency. By making N very large, you can start at a microwave frequency such as 2.4 GHz and divide down to the MHz range in order to lock a microwave transmitter (e.g., for a cell phone) to a lower-frequency crystal-controlled reference oscillator. The subject of variable-divide-ratio digital divider circuits is beyond the scope of this chapter, but such circuits are available and form the basis of this type of frequency synthesizer.

Of course, the digital frequency division must be taken into account when designing the PLL. The effective frequency of the VCO is f_{VCO}/N, and so the tuning constant K_C must be also divided by N. Since both inputs to the phase detector are in digital form, it is often convenient to use the XOR phase detector. Just be sure that both waveforms going into the XOR gate have a 50% duty cycle, or you may get unexpected results. (Some digital dividers provide waveforms like this, but others do not.)

9.7 DESIGN EXAMPLE USING CD4046B PLL IC

The following design for a PLL circuit using the CD4046B involves a simple synthesizer-type circuit that divides the VCO frequency by 2. The VCO output will therefore be exactly twice the reference frequency and will have a known phase relationship to it. This type of signal is useful in various types of digital signal demodulators.

The specifications are as follows:

1. Reference (input) frequency f_{REF} shall be 50 ± 1 kHz.
2. VCO shall operate at twice the reference frequency (e.g., $\omega_{VCO}/2\pi = f_{VCO} = 100$ kHz when $\omega_{REF}/2\pi = f_{REF} - 50$ kHz).
3. Damping factor $\zeta = 0.6$.

The design can proceed as follows.

The first design decision concerns the overall circuit. We will use the basic synthesizer block diagram shown in Figure 9.12 and let $N = 2$ and $M = 1$. So there will be no divider between the reference frequency and the phase-detector input, but we will use a divide-by-2 (e.g., a **J–K flip-flop**) to divide the VCO frequency by 2 to be equal to the reference frequency.

The next choice concerns the type of phase detector: XOR or charge pump? Because the lock-in range is fairly small (±1 kHz out of 50 kHz), we will probably not have problems with lock-in if we use the XOR phase detector. A J–K flip-flop automatically produces a duty cycle at its output of 50%, so this will work fine with the XOR circuit.

Next, we will choose a loop bandwidth. This is roughly equivalent to the natural frequency ω_n of a second-order loop, and as we showed in Section 9.3, if the transfer function is

$$H(s) = \frac{Gs + \omega_n^2}{s^2 + 2\zeta\omega_n s + \omega_n^2}, \tag{9.37}$$

there are specific relationships between the components of the passive loop filter shown in Figure 9.3 and the natural frequency ω_n and damping factor ζ. But first, we should choose a natural frequency, since that is going to determine our loop bandwidth.

For a damping factor between 0.5 and 1, the loop bandwidth (3 dB down) is comparable to the natural frequency ω_n. To be safe, the loop bandwidth should be a little larger than the one-sided locking range of 1 kHz (recall that the specified input frequency is 50 ± 1 kHz). Let's try for a natural frequency of $\omega_n = 2\pi(1.5\,\text{kHz}) = 9425\,\text{rad s}^{-1}$.

Now that we have both ω_n and ζ, we can determine the loop filter component values. We will use the loop filter circuit of Figure 9.3. Because there are three components (R_1, R_2, and C) and only two variables, we need to make an arbitrary choice of one component, and then the other two are determined. We should choose a value for C that gives resistor values in the range of 100 Ω to 1 MΩ, if possible. That is high enough to take advantage of the high input impedance of the VCO, but not so high that parts are hard to find or insulate. Generally, resistor values above 1 MΩ in discrete circuits are to be avoided unless there is a good reason to use them.

The next step is to select the VCO components R_{VCO} and C_{VCO}. Again, many combinations of R and C will suffice for a given frequency, but we need to choose particular values in order to determine the VCO control constant K_{C}. Let's choose $C_{\text{VCO}} = 100\,\text{pF}$ (a standard value) and $R_{\text{VCO}} = 25\,\text{k}\Omega$. According to Equation 9.33, this combination will give us a VCO control constant of

$$K_{\text{C}} \approx \frac{1}{2}[8.8(10^{-10} \cdot 25{,}000)^{-0.8}] = 133.3 \times 10^3\ \text{rad s}^{-1} \tag{9.38}$$

The reason for the leading factor of 1/2 is that we are dividing the VCO output frequency by two before it enters the phase detector, and this factor enters into the effective value of K_{C}.

Let's use a 15-V supply with the XOR phase detector, so according to Equation 9.28, $K_{\text{P}} = 4.8\,\text{V rad}^{-1}$. The phase-detector constant and the VCO control constant always appear together, so we find that the product $K_{\text{C}}K_{\text{P}} = 640 \times 10^3\,\text{s}^{-1}$.

Now that we have R_1, K_{C}, and K_{P}, we can pick a value for C, the capacitor in the passive loop filter of Figure 9.3, and see what resistor values result for R_1 and R_2. We will start with $C = 10\,\text{nF}$. Equation 9.25 for R_1 gives

$$\begin{aligned} R_1 &= \frac{1}{C}\left(\frac{K_{\text{C}}K_{\text{P}}}{\omega_n^2} - \frac{2\varsigma}{\omega_n} + \frac{1}{K_{\text{C}}K_{\text{P}}}\right) = \frac{1}{10^{-8}}(7.205 \times 10^{-3} + 127 \times 10^{-6} + 1.56 \times 10^{-6}) \\ &= 707.6\ \text{k}\Omega \end{aligned} \tag{9.39}$$

TABLE 9.1 Damping Factors for Various Values of R_2

R_2 (kΩ)	ζ
1	0.057
6.2	0.315
12	0.59
15	0.741

We will choose the closest 10%-tolerance value, namely, $R_1 = 680\,\text{k}\Omega$. We expect R_2 to be smaller, and using Equation 9.26, we find

$$R_2 = \frac{1}{C}\left(\frac{2\zeta}{\omega_n} - \frac{1}{K_C K_P}\right) = \frac{1}{10^{-8}}(127 \times 10^{-6} - 1.56 \times 10^{-6}) = 12.5\,\text{k}\Omega \qquad (9.40)$$

The closest 10% value to this would be 15 kΩ, but in order to see the effects of varying the damping factor on the circuit, we will select a variety of resistors for R_2 and calculate the damping factor for each one. The results are listed in Table 9.1.

So you can see that for values of R_2 much smaller than R_1, the damping factor is roughly proportional to R_2. To complete the design process, we will select $R_2 = 12\,\text{k}\Omega$ to use for checking our results. Going above a damping factor of 0.7 does not improve transient behavior significantly and increases the loop bandwidth, so we will be satisfied with a damping factor of 0.59.

Now that we've decided on C and R_2, we should recheck what our natural frequency ω_n has changed to:

$$\omega_n = \sqrt{\frac{K_C K_P}{(R_1 + R_2)C}} = \sqrt{\frac{640 \times 10^3}{(692\,\text{k})(10\,\text{nF})}} = 9617\,\text{s}^{-1} \qquad (9.41)$$

or in Hz, 9617/2π = 1.53 kHz. That isn't much larger than the 1.5 kHz we started with. Because R_2 is so much smaller than R_1, varying R_2 from 1 to 12 kΩ changes the natural frequency hardly at all. from about 1.53 to 1.55 kHz.

So the important conclusion from this exercise is that when you fix C arbitrarily, choosing the value of R_1 mainly influences the natural frequency ω_n, and varying R_2 mainly influences the damping factor ζ, as long as R_2 is much smaller than R_1.

To check the validity of this design, it was built and tested in the lab. We allowed it to lock to a 50-kHz reference signal that we frequency-modulated at a 100-Hz rate with a square wave. The peak deviation of the frequency modulation was 1 kHz, just inside the specified variation range for the PLL. This meant that 100 times a second, the input frequency jumped from 50 − 1 = 49 to 50 + 1 = 51 kHz and back. This is not the same as a step-function change in phase, but the PLL will respond in a similar fashion either to changes in phase or in frequency. The results as measured from the PLL control voltage are shown in Figure 9.13, not only for the chosen value of $R_2 = 12\,\text{k}\Omega$, but for 1 and 6.2 kΩ too (a 1.6-kHz 1-pole lowpass filter was inserted between the VCO voltage at pin 9 and the scope to make the waveforms easier to see).

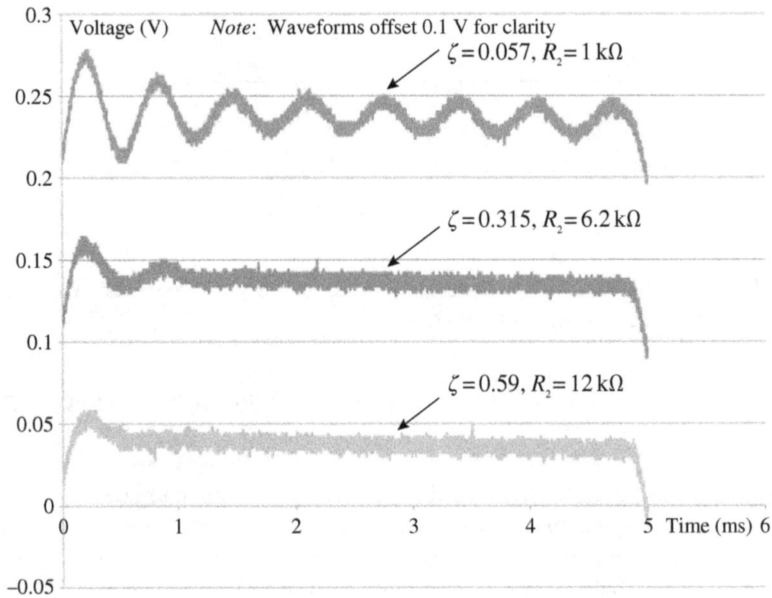

FIGURE 9.13 Actual measured PLL VCO voltage responses to 1-kHz-deviation square-wave modulation on 50-kHz carrier.

As you can see, the transient waveforms on the VCO control voltage bear a strong resemblance to the theoretical waveforms shown in Figure 9.6. While it was somewhat difficult to measure the loop bandwidth using sine-wave frequency modulation, indications were that the bandwidth was on the order of 1.6 kHz. Given the limitations of linear analysis for use in such a basically nonlinear circuit as a PLL, this level of agreement is acceptable.

The complete circuit of the PLL including the divide-by-2 J–K flip-flop IC (a Schottky TTL-type 74LS73) is shown in Figure 9.14. Because the CD4046B is a CMOS device running with V_{DD}=+15 V and the 74LS73 must operate at a supply voltage of +5 V, we need a couple of **level-shifting** circuits to interface between the two ICs. The reason is basically that the CMOS IC will not recognize an input voltage level as a logical HI (1) until the voltage exceeds about $V_{DD}/2$ or +7.5 V for a supply voltage to the CMOS device of +15 V. So if we just sent the +5-V logical HI from the 74LS73 device to the CMOS device, it would always see a logical LO and the circuit wouldn't work.

To drive the CMOS device's phase-detector input at pin 3 from the TTL output, we have inserted a simple discrete-device inverter, consisting of two resistors and an npn transistor (2N3904). Because the transistor's 3.3-kΩ collector resistor is connected to +15 V, its output waveform goes between +15 V and ground, more than high enough to drive the CD4046B input. As long as the system is not operating at a frequency much faster than 1 MHz, this type of circuit will work adequately. But it has a speed limitation due to collector-base capacitance that slows down its

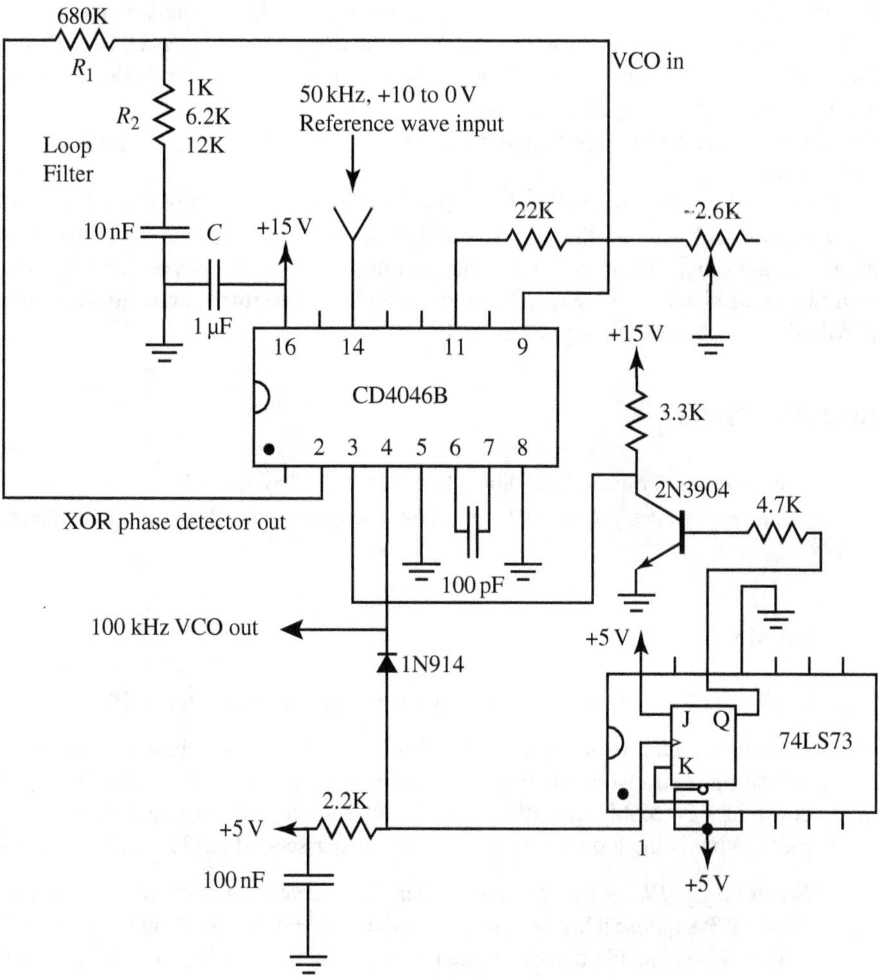

FIGURE 9.14 Complete schematic diagram of frequency doubler PLL.

response time for fast signals. There are specialized level-shifting ICs available that translate between different logic levels and can handle much faster clock speeds than 1 MHz, and these should be used for more complex and higher-speed applications.

Because the CMOS IC puts out +15 V for a logical HI, that voltage is high enough to lead to possible damage of the TTL circuit. The level-shifting circuit that takes the CMOS-level output from the VCO at pin 4 of the CD4046B and sends it to the clock input (pin 1) of the 74LS73 is very simple: a 1N914 diode and a 2.2-kΩ resistor. TTL inputs need to have a low resistance to ground in their LO state but can have a higher resistance facing them when connected to the HI level (+5 V). When the CMOS output is LO, the 1N914 diode becomes forward biased and pulls pin 1 of the 74LS73 IC down to about 0.7 V, which the TTL device recognizes as a LO. The 2.2-kΩ resistor allows only about 2 mA to flow into the CMOS device (plus whatever the TTL pin 1 supplies), which it can handle. When the CMOS output goes HI (to +15 V), the diode's anode cannot rise to a voltage higher than +5 V, so the diode becomes reverse biased and disappears from the circuit. The TTL input (pin 1) then just sees the resistor connected to +5 V, which it recognizes as a HI, and everything works fine from then on.

You will also notice a couple of power-supply bypass capacitors (100 nF on the +5-V supply and 1 μF on the +15-V supply), which were needed to stabilize the circuit's operation. Otherwise, transients coupled through the power-supply leads caused various kinds of problems. This is an example of **electromagnetic interference** (**EMI**) that will be discussed in Chapter 12.

BIBLIOGRAPHY

Egan, W. F. *Phase-Lock Basics*, 2nd Edition. Hoboken, NJ: Wiley-Interscience, 2008.

Stensby, J. L. *Phase-Locked Loops: Theory and Applications.* Boca Raton, FL: CRC Press, 1997.

PROBLEMS

Note: Problems of above-average difficulty are marked with an asterisk (*).

9.1. *Loop bandwidth for first-order PLL.* Suppose a first-order phase-locked loop, whose phase transfer function $H(s)$ is given by Equation 9.12, has the VCO constant $K_C = 52{,}000$ rad s^{-1} V^{-1} and its phase-detector constant is $K_P = 1.2$ V rad^{-1}. What is the loop bandwidth in radians per second (ω)? In Hz (f)?

9.2. *Second-order PLL loop filter design.* For the passive loop filter circuit shown in Figure 9.3, suppose it has been determined that $\omega_1 = 5700$ rad s^{-1} and $\omega_2 = 46{,}000$ rad s^{-1}. Using standard 10%-tolerance values only (see Chapter 6 for a list of what these are), select a value for the capacitor C and resistors R_1 and R_2 so that the geometric mean $[R_1R_2]^{1/2}$ of the two resistors is as close to 10 kΩ as possible. This choice will avoid both extremely high and extremely low resistor values.

9.3. *Solving for natural frequency ω_n and damping factor ζ given passive filter component values.* Suppose a PLL circuit using the CD4046B IC with a +15-V power supply is used to produce a clean 1-MHz signal from a noisy 1-MHz source. Assume the phase-detector constant $K_P = 4.8$ V rad^{-1}. Using a value for $C_{VCO} = 47$ pF, find

(a) R_{VCO} so that $\omega_{VCO} = 2\pi$ (1 MHz) when $V_9 = 11$ V,

(b) values for the damping factor ζ and natural frequency ω_n (in both rad s^{-1} and Hz) if the passive loop filter of Figure 9.3 has values $C = 1$ nF, $R_1 = 750$ kΩ, and $R_2 = 22$ kΩ.

***9.4.** *Solving for passive loop filter component values given ω_n and ζ.* Suppose a PLL circuit using a CD4046B IC is used in a frequency synthesizer that covers a frequency range from 1 to 10 kHz in 1-kHz steps. The reference frequency used is $f_{REF} = 1$ MHz. The frequency synthesizer circuit shown in Figure 9.12 is used, and so the frequency division factor N varies from $N = 1$ for a 1-kHz output to $N = 10$ for a 10-kHz output.

(a) Find the value for the reference frequency division factor M (which is constant).

(b) Because N varies from 1 to 10, the effective VCO constant K_V also varies by a factor of 10, which will affect the values for ω_n and ζ. Assuming C in the passive loop filter of Figure 9.3 is 100 nF and the product $K_P K_C =$ 33,500 rad s^{-1} before division by N, find values for R_1 and R_2 (nearest standard 10%-tolerance values) that will keep the natural frequency ω_n in the range $2\pi(25\text{ Hz}) < \omega_n < 2\pi(100\text{ Hz})$ and the damping factor ζ in the range $0.5 < \zeta < 2.0$ for the entire range of possible Ns from 1 to 10.

9.5. *Exclusive-OR phase detector with asymmetrical input waveform.* Suppose an exclusive-OR type of phase-detector circuit is used with two inputs, numbered 1 and 2. Input no. 1 is a square wave with a 50% duty cycle, but input no. 2 is a rectangular wave with a 10% duty cycle (see Fig. 9.15).

Assume that the phase detector's output voltage V_{OUT} is 1.0 V when it is at a logical HI and 0 V when it is at logical LO. On a graph similar to the one shown in Figure 9.15 for the average value $\langle V_{OUT} \rangle$ versus phase ϕ, sketch the phase detector's transfer function $\langle V_{OUT} \rangle$ versus ϕ for the input waveforms shown, assuming ϕ is positive when waveform 2 leads waveform 1, measuring from the positive-going leading edge of each pulse. Is the transfer function linear or nonlinear? What problems might such a transfer function cause if you

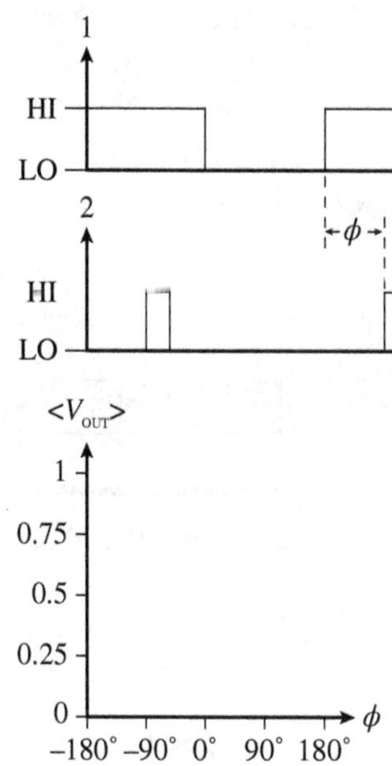

FIGURE 9.15 Exclusive-OR phase-detector input waveforms for Problem 9.5.

attempted to use it in a PLL that was designed assuming the phase-detector output was linear?

9.6. *Bode plot of passive loop filter.* If $R_1 = 99R_2$ in the passive loop filter shown in Figure 9.3, sketch a general Bode plot of the magnitude (in dB) of $F(s)$ in Equation 9.16 versus log frequency, indicating break points at frequencies ω_1 and ω_2. (See Chapter 5 for information about how to draw Bode plots).

***9.7.** *Deriving signal for synchronizing PLL for PSK clock recovery.* Two popular types of digital modulation are **binary phase-shift keying** (abbreviated **BPSK**) and **quadrature phase-shift keying** (abbreviated **QPSK**). In BPSK, a constant-frequency carrier is transmitted with a phase shift of either 0° or 180°, while in QPSK, there are four phase-shift choices: −90°, 0°, +90°, and 180°. In order to demodulate the digital signal encoded in these forms of modulation, it is necessary to **recover** a steady, constant-frequency and constant-phase clock signal, which should have the same frequency and phase as the original unmodulated carrier before the phase shifts were made. One approach to **clock recovery** uses electronic analog multiplier circuits, a PLL, and a digital frequency divider as shown in Figure 9.16.

Each box labeled V^2 squares the instantaneous voltage presented to its input, to within an arbitrary constant. The coupling capacitors C remove any DC component from the squared signal, leaving only an AC part to be either squared again (in the case of the QPSK circuit) or fed to the reference input of a PLL circuit.

Using either algebra or carefully drawn sketches, show that the phase and frequency of the PLL reference inputs V_{B2} and V_{Q4} remain the same for every possible input phase to the BPSK and QPSK circuits. To do this, find the shapes of the intermediate voltages V_{B1}, V_{Q1}, V_{Q2}, and V_{Q3}, and show that the desired constant phase reference voltage results in both cases.

***9.8.** *Demodulator for FSK signal.* A popular method of transmitting analog data from industrial sensors is known as a "4–20-mA current loop." This loop is a

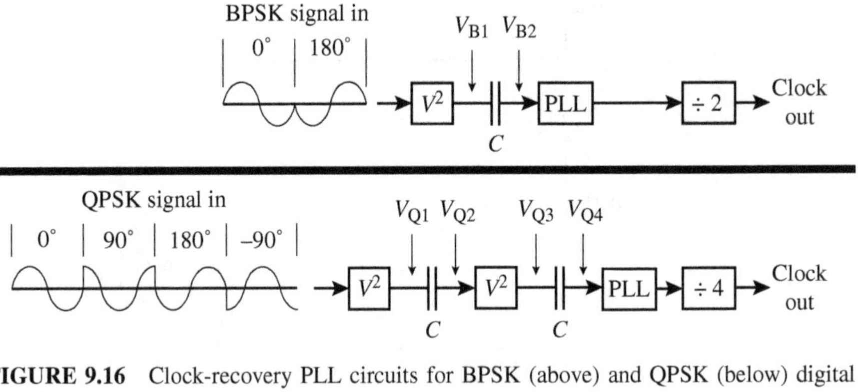

FIGURE 9.16 Clock-recovery PLL circuits for BPSK (above) and QPSK (below) digital modulation.

series circuit containing a power supply, a sensor that acts as a current source whose current varies with the quantity being sensed, and a receiver that measures the current. A current of 4 mA indicates a measured quantity of 0 and 20 mA indicates full scale. Because the system is based on current amplitudes, it is relatively insensitive to long cable runs and is installed in thousands of industrial locations around the world. A digital modulation scheme called the **HART communications protocol** allows users to double the data capacity of existing systems without the expense of adding additional wires. (HART is short for **Highway Addressable Remote Transducer**.) The HART protocol overlays on the DC 4–20-mA signal an audio-frequency carrier modulated with *FSK*, which stands for **frequency-shift keying**. In HART FSK, a frequency of 1200 Hz indicates a binary 1 and 2200 Hz indicates a binary 0. The number of bits per second (which is roughly equal to the number of times per second that the frequency can shift back and forth between 1200 and 2200 Hz) is also 1200, which means that sometimes only one cycle of the 1200-Hz frequency will be transmitted.

Suppose you are tasked with the design of a second-order PLL using a CD4046B IC to demodulate a HART FSK signal, providing an output at pin 10 of the IC that will be at least 1 V (peak to peak) when the circuit receives a signal consisting of alternating 1s and 0s (101010…) at a rate of 1200 bits per second. The supply voltage to the IC is +5 V and ground. Choose an appropriate natural frequency ω_n and damping factor ζ so that the PLL will be likely to stay in lock as the carrier frequency shifts rapidly between 1200 and 2200 Hz. Choose values for the VCO components R_{VCO} and C_{VCO} so that the VCO constant K_{VCO} is large enough to provide the required 1-V (peak-to-peak) output at pin 10. Also, choose values of the components C, R_1, and R_2 that will provide the required natural frequency and damping factor. (*Note*: As in many open-ended design problems, there is not a unique solution to this problem, and some judgment will be necessary. If facilities are available, build and test your design to see if it meets the specifications.)

For further resources for this chapter visit the companion website at
http://wiley.com/go/analogmixedsignalelectronics

10

POWER ELECTRONICS

10.1 INTRODUCTION

When the output of an electronic system or the load controlled by the system exceeds about 1 W, the system falls into the category of **power electronics**. All electronic systems that draw electrical power must dissipate some of that power as heat, but in low-power (milliwatt-level) systems, the heat produced in resistors and active devices is often small enough to be neglected. But in power electronics, the question of **efficiency** arises. The most basic definition of efficiency, which is customarily denoted by the Greek letter η (pronounced "*ay*-ta" with the same *ay* sound as in "gate"), is the ratio of useful output power P_{OUT} produced by a system divided by the total input power P_{IN}:

$$\eta = \frac{P_{OUT}}{P_{IN}} \tag{10.1}$$

Unless a system has its own internal source of electrical energy, the law of conservation of energy implies that the average efficiency of any system will be a fraction less than 1. Efficiency is often cited in terms of a percentage, so that η (%) = 100η (fraction). Either way, as the total amount of delivered power P_{OUT} from a system increases, the efficiency of the system becomes more and more important.

Analog and Mixed-Signal Electronics, First Edition. Karl D. Stephan.

Companion Website: http://wiley.com/go/analogmixedsignalelectronics

There are several reasons why the efficiency of a power-electronics system is a critical performance criterion. Here are a few:

1. *Maximum temperature of electronic systems*. Inefficient systems dissipate the difference between total input power and net output power as heat, and that heat must be conducted away somehow to the ambient environment. Lower efficiency for a given output means more heat dissipated in the system, and more heat means a higher working temperature, other things being equal. Every electronic device has a maximum temperature above which its performance is not guaranteed. For example, most silicon devices are rated to operate at a **junction temperature** of less than 150°C. Above that temperature, enough free electrons and holes are produced by thermal excitation that leakage current increases and device performance suffers. Excessive leakage can cause a phenomenon called **thermal runaway** in which leakage causes higher power dissipation, which raises the temperature, which causes more leakage in a vicious circle that can rapidly destroy the device. Passive devices such as capacitors, resistors, and inductors also incur damage from excessive temperature, although they tend to be more robust than active devices.

   Even below a system's maximum rated temperature, the fact that the rate of most chemical processes doubles for every 10°C rise in temperature means that aging, oxidation, and other potentially harmful processes occur more rapidly at higher temperatures. The net result is that the **reliability** of electronic systems declines rapidly at higher temperatures, making premature failure much more likely. Other things being equal, a power-electronics system with low efficiency will run hotter than a more efficient system, and its **mean time to failure** (abbreviated as **MTTF**) will be shorter.
2. *Energy demand from power source*. Today's **green designs** take energy consumption into consideration from the start. The amount of energy consumed from power supplies contributes to the total lifetime cost of any electronic system and is increasingly being taken into consideration by energy-conscious consumers and manufacturers. Power electronics with a higher efficiency rating will consume less total power for a given output power, and so efficiency is an important factor in green designs. Low energy consumption and high efficiency are especially important for battery-powered systems, which form a growing portion of the total volume of electronic systems sold annually. Improvements in performance for a fixed battery capacity often depend vitally on improving the power efficiency of a system before any additional features can be added.
3. *Thermal design and device ratings*. An inefficient power-electronics design can be made to work, but in order for the devices used to operate within their temperature ratings, the devices must have a large power-dissipation ability. Such devices are usually more expensive and bulky than the same type of device with a lower maximum power dissipation. Also, the **thermal design** of a system that must dissipate a large amount of power as heat is more difficult. While ambient air or **forced-air** (fan) cooling can deal with medium power

levels and power densities, the removal of large amounts of heat from a small physical area may require expensive heat-transfer systems such as **liquid cooling**, using a coolant circulating through a radiator by means of a pump. A design that has high intrinsic efficiency to begin with will dissipate less heat and require a less elaborate thermal design than a less efficient circuit will.

While efficiency is not the only important criterion of power-electronics design, it should always be borne in mind and estimated during the design process.

10.2 APPLICATIONS OF POWER ELECTRONICS

The applications of power electronics can be divided into two broad categories: (1) power supplies and (2) amplifiers, if the category of amplifiers is considered to include systems such as motor controllers and drives. The distinction between the two categories is discerned by deciding whether a system provides a constant (usually DC) voltage or whether the output voltage or current varies in response to a low-level input signal or command.

Power supplies are the unsung workhorses of the electronics world. Every electronic system needs a power supply, and power-supply failures generally guarantee that an entire system will fail. But too often, designers deal with the issue of power-supply design as an afterthought and do not consider it as an integral part of the overall system. The good analog system designer will consider issues relating to the power supply as an integral part of the design process. Many problems can be avoided if the power source is appropriately sized and matched to its loads. This rule applies to battery-operated systems too. All but the simplest battery-powered systems now include power-management electronics that keep track of battery conditions and performance, so the fact that a system is battery powered does not eliminate the need for good power-supply design.

Power amplifiers are used in a great variety of applications ranging from sound systems to digital displays, industrial processes, and communications systems involving high-power radio-frequency (RF) transmitters. While conventional design techniques are adequate for output frequencies up to 1 MHz or so, special **RF** design techniques are needed at frequencies much higher than that. (These techniques are the subject of Chapter 11.) Even below 1 MHz, special circuits and approaches have been devised to permit high-efficiency high-power amplifiers to be developed for a variety of applications, and we will describe several of these in this chapter.

10.3 POWER SUPPLIES

10.3.1 Power-Supply Characteristics and Definitions

Generally speaking, a power supply is any system that delivers electric power to a load. In analog electronics, the term is usually restricted to mean the system or systems that provide reasonably constant-voltage (or, rarely, constant-current) DC power to be used by the remainder of the system for a useful purpose. The actual source of electrical energy may be included in the system (as in battery power supplies) or may be external to the system (as in power supplies operated from an AC power line). Regardless of the source of energy, all power supplies can be characterized by certain performance characteristics, which we will now describe.

As an example, we will use a hypothetical power supply that uses a power-line input voltage $\vec{V}_{AC}$ and converts it to an output voltage V_{DC}, as shown in Figure 10.1.

We assume that the input power is a single-frequency AC sine wave and that the AC voltage and current are phasor RMS values. For example, in the United States, the nominal utility voltage is about 120 V (RMS) at a frequency of 60 Hz, but other frequencies (e.g., 50 Hz) and voltages (e.g., 100 or 240 V) are standards in other parts of the world. The designer should also bear in mind that AC utility voltages can vary by as much as ±10%, so a **universal** power supply that can be used with any standard line voltage in the world must be able to work with a line voltage ranging from 90 V (100 V – 10%) to 264 V (240 V + 10%).

Whatever the nominal voltage, the *real* input power P_{IN} to the power supply is

$$P_{IN} = \mathrm{Re}\left(\frac{\vec{V}_{AC}\vec{I}^{*}_{AC}}{2}\right) = \left|\vec{V}_{AC}\right|\left|\vec{I}_{AC}\right|\cos\theta \tag{10.2}$$

where Re means "take the real part of," * means complex conjugation, and θ is the phase angle between the AC voltage and the AC current. The term $\cos\theta$ is called the **power factor** of the power supply. Ideally, it should be 1, but it is often less than 1, and a large load with a power factor considerably below 1 can cause problems in utility networks. This analysis also assumes that the current drawn is sinusoidal, which is not true in general. But most AC power-line sources of reasonable quality are basically well modeled by a pure sine wave.

The output power P_{OUT} delivered by the supply to its load is simply the product

$$P_{OUT} = V_{DC}I_{DC} \tag{10.3}$$

and so the efficiency η immediately follows from Equations 10.2 and 10.3 as

$$\eta = \frac{V_{DC}I_{DC}}{\left|\vec{V}_{AC}\right|\left|\vec{I}_{AC}\right|\cos\theta} \tag{10.4}$$

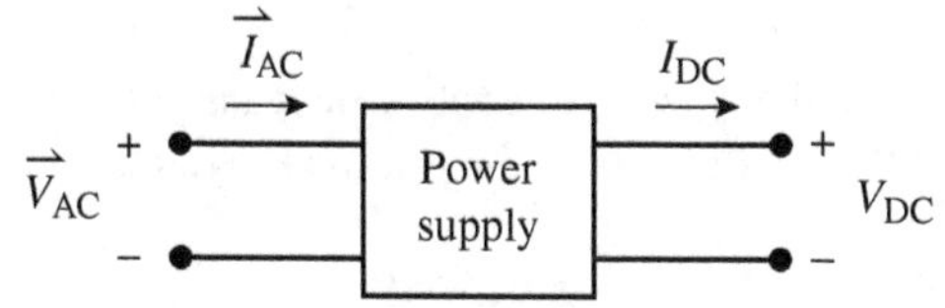

FIGURE 10.1 Generic power supply showing input and output voltages and currents.

Note that if I_{DC} is zero, the efficiency is also zero unless the power supply ceases to draw any power from the source. Usually, a power supply consumes a small amount of power when idle for internal uses, so efficiency varies from zero with no load to a maximum value when the load current is the optimum value that the supply is designed for. This variation in efficiency needs to be considered when a supply is provided for loads that vary a considerable amount, as in laboratory or bench power supplies.

In addition to efficiency, power supplies can be characterized with respect to how well they maintain a constant output voltage as either the primary power source voltage V_{AC} varies or as the load current I_{DC} varies. If a power supply's output voltage V_{DC} remains constant as the primary input voltage varies, it is said to have perfect **line regulation**. If its output voltage remains constant as the load current varies, it has perfect **load regulation**. Various authorities define these concepts quantitatively in different ways, but the following definitions are in fairly common use. Referring to Figure 10.1, assuming I_{DC} is held constant, the fractional line regulation D_{LINE} for a change $\Delta|V_{AC}|$ in the power-line voltage magnitude that produces a change ΔV_{DC} in the DC output voltage is

$$D_{LINE} = \frac{\Delta V_{DC} / V_{DC}(\text{nom})}{\Delta |V_{AC}| / |V_{AC}(\text{nom})|}, \tag{10.5}$$

where (nom) means the nominal or average quantity. For example, if the DC output voltage averages 5 V and changes by 50 mV for a nominal AC input voltage of 120 V that changes by 1.2 V, the line regulation D_{LINE} is 1 or 100%. Obviously, a power supply with perfect line regulation has $D_{LINE} = 0$. Both line and load regulation can be improved by the use of a **voltage regulator** circuit, which we will describe in Section 10.3.

One of the most common definitions of load regulation D_{LOAD} is the difference between the no-load and the full-load voltage, divided by the nominal or average output voltage, for a constant primary input voltage. Every power supply has a maximum or full output load current I_{MAX} it is designed to produce under normal operating conditions. The output voltage will tend to fall to a level V_{MIN} when the output current is I_{MAX} and will rise to a no-load voltage V_{MAX} when $I_{DC} = 0$. Taking the nominal output voltage to be the average of V_{MAX} and V_{MIN}, the equation for load regulation D_{LOAD} as a fraction is

$$D_{LOAD} = 2\frac{V_{MAX} - V_{MIN}}{V_{MAX} + V_{MIN}} \tag{10.6}$$

So, for example, a nominal 5-V power supply whose output falls from $5.02\ \text{V} = V_{MAX}$ at no load to $4.98\ \text{V} = V_{MIN}$ under full load has a load regulation of

$$D_{LOAD} = 2\frac{5.02 - 4.98}{5.02 + 4.98} = 0.008 \tag{10.7}$$

or 0.8%. An alternative way to specify load regulation is to state the power supply's **internal resistance** R_{INT}, which can be defined as the series resistance in the Thévenin equivalent-circuit model of the power-supply output circuit.

Besides line and load regulation, another important specification for power supplies operated from AC primary power is the AC content of the DC output voltage, generally referred to as **ripple**. The maximum ripple output voltage V_{RIPPLE} is usually defined as the peak-to-peak voltage of the AC waveform that is superimposed on the DC output and may be expressed either as a percentage of the nominal DC output voltage or in absolute terms (e.g., 2 mV maximum peak-to-peak ripple). A ripple voltage can cause problems such as noise or hum in high-gain amplifiers and erratic performance in digital circuits as well, so for a given power-supply design, the ripple specification should be considered carefully. On the other hand, loads such as lamps, heaters, motors, and other power equipment are relatively insensitive to ripple. Because it takes additional resources to reduce ripple to very low values, the ripple specification should be no lower than necessary for a given power-supply design.

10.3.2 Primary Power Sources

The design of a power supply depends on the nature of the primary power source from which it draws energy to be converted into the desired electrical form. We will assume that the primary power source provides an electrical output that varies over specified limits. While renewable sources such as wind and solar energy are increasingly important, they are outside the scope of this chapter, and we will consider only the two main types of primary power sources in common use: batteries and AC utility-line power.

10.3.2.1 Batteries A **battery** is a device that stores energy in chemical form and converts it to electrical energy through an electrochemical process. (Strictly speaking, the term **battery** should be reserved for devices composed of two or more electrochemical **cells**, but single-cell units are often called batteries anyway.) A battery is characterized by its basic chemical makeup (e.g., lead acid or lithium ion), its nominal output voltage V_{NOM}, and its **capacity** Q, which is a number measured in **ampere-hours** (**AH**) and has the dimensions of charge. When discharged at the rate of current specified in the AH rating, the battery will deliver a total amount of power approximately equal to QV_{NOM}, although under actual conditions of use, the delivered power will typically be somewhat less than the theoretical value. The circuit model of a battery consists of the simple Thévenin equivalent circuit shown in Figure 10.2: an ideal voltage source V_B in series with an internal resistance R_{INT}. Both V_B and R_{INT} can vary with time over the battery's discharge cycle, but it is clear from the equivalent-circuit model that every battery has a maximum short-circuit current that it can provide, limited by the battery's internal resistance. Shorting a battery is always a bad idea, especially in the larger sizes and capacities, because it can cause a heat buildup that can rupture the battery case and cause further damage or even personal injury. For this reason, some batteries have internal fuses that open when the battery is shorted, permanently disabling the unit but preventing a rupture.

10.3.2.2 AC Power Sources The utility outlets provided in residences and business establishments typically provide three contacts for the lowest-voltage service (100V in Japan, 120V in the United States and certain other countries, and 240V in the rest of the world). While single-phase service can be provided by only two wires (and was provided that way, for many years), the third wire—termed the **ground wire**—was added to electrical code requirements for safety reasons. A voltage as low as 48V can be fatal under the wrong conditions, such as if you are standing in bare feet on a damp concrete floor and touch a live terminal with a sweaty hand. For this reason, the utility wiring of most locations can be approximated by the equivalent circuit shown in Figure 10.3.

The series impedance of most AC utility outlets is so low (milliohms to about 1 Ω) that it can be neglected under most circumstances, so the AC voltage source in Figure 10.3 is shown as ideal. For safety reasons (viz., to prevent the high distribution voltage from entering a house in the event of a distribution-transformer insulation failure), one side of the equivalent voltage source is always connected to a physical earth ground at the service-entry point. This ground connection is carried throughout the utility wiring in the form of a third wire, or sometimes as a conductive conduit or enclosure. Either way, the ground connection (labeled **ground** in Fig. 10.3) appears at the power outlet along with the ungrounded side of the AC source (labeled **hot** in Fig. 10.3) and the grounded side (labeled **neutral**). Note that for various reasons having to do with distribution of loads and the nature of the power wiring, the neutral wire is not necessarily at ground potential, because it is separated from ground by

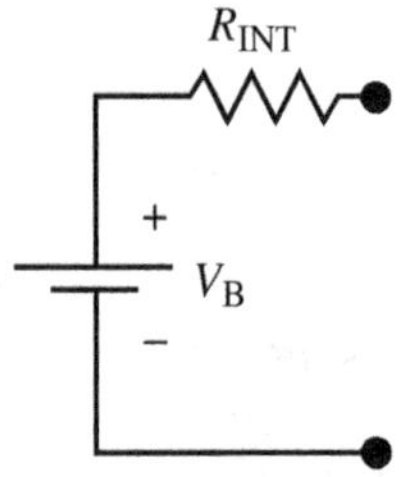

FIGURE 10.2 Equivalent circuit of battery.

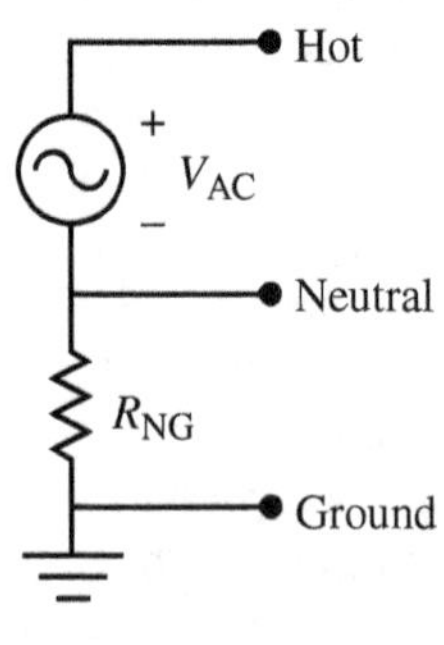

FIGURE 10.3 Equivalent circuit of AC utility power source.

a low but nonzero resistance R_{NG}. In US wiring, the standard wire-insulation color codes for these three connections are that hot is black, neutral is white, and ground is green. It is good practice for internal equipment wiring to adhere to this color code inside any system that is connected to an AC power line.

At first glance, the ground wire seems superfluous, because the power circuit is completed when a load is connected between the hot and the neutral wires, and it would appear that the ground wire is not needed. The usefulness of the ground wire in preventing accidents is shown in Figure 10.4. Suppose a load (anything from a lamp to a computer) is housed in a conductive enclosure and a fault with resistance R_{LEAK} develops in the insulation between the hot lead of the primary power source and the enclosure. Without the third ground wire, this leakage resistance would raise the voltage of the enclosure to the power-line voltage of 100 V or more. If a person happened to touch both the enclosure and a solid grounded object such as a workbench, he or she could receive a fatal shock. But if the conductive enclosure is connected to the ground wire as shown, the leakage current is drained harmlessly to ground and the enclosure remains at a safe ground potential. It is not good practice to eliminate the ground wire and, for example, connect the enclosure to the neutral terminal, because neutral is not necessarily grounded, and it is easy to reverse neutral and hot wires by mistake, leading to the very dangerous condition that the enclosure would be connected to the hot lead.

As long as the ground wire is connected to all parts of the system that should be grounded, including its enclosure, any insulation faults from the primary power line will not raise the enclosure to dangerous potentials. If the leakage current becomes large enough, it will trip the primary power circuit breaker or other high-current protection device. And if the utility outlet is further protected by a device called a **ground-fault circuit interrupter** (**GFCI**), also known as a **residual-current device** (**RCD**), the device will detect the leakage current, which is the difference between

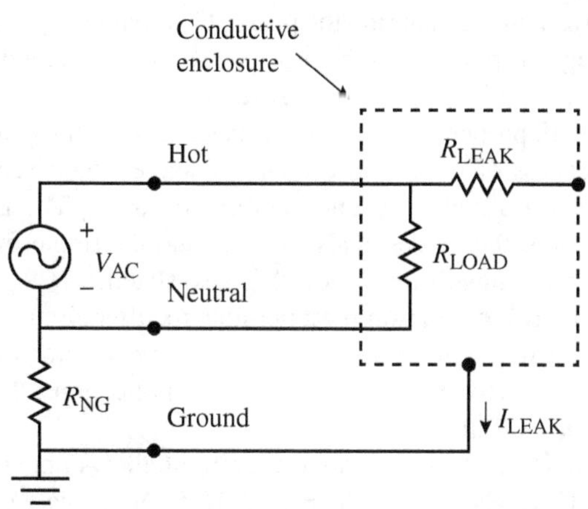

FIGURE 10.4 AC power-line source connected to load with leakage resistance R_{LEAK}, showing how presence of ground lead conducts leakage current I_{LEAK} harmlessly to ground.

the current flowing through the hot and neutral wires, and open the circuit in response. Such safety devices can operate with a leakage current as small as a few mA, which can nevertheless be enough to cause injury. Because of the safety issues involved, any design that connects to an AC utility source should be carefully checked from a safety standpoint, both in terms of electrocution hazards and in terms of overcurrent protection, lack of which can lead to extreme overheating and a fire in the case of a high-current fault in the circuit.

For stand-alone loads exceeding about 2–5 kW total, single-phase AC power at the standard residential voltage (100–220 V) becomes impractical because of the excessively large conductors required, and typically, **three-phase** power is used, often at multiples of the standard voltages (e.g., 240 or 480 V instead of 120 V). The three-phase lines are designated A, B, and C and have identical voltage and current capabilities except that each sine wave is offset in phase by 120° from the other two. Certain improvements in efficiency are obtained through the use of multiphase AC power sources, and these efficiencies can also be obtained in locally generated multiphase power supplies as well as utility-provided ones.

10.3.3 AC-to-DC Conversion in Power Supplies

Any DC power supply that operates from AC must first convert the AC to DC. This is true not only for line-operated supplies but for supplies that perform an AC-to-DC conversion and then reconvert to AC for further processing. In either type of system, typically, diode rectifiers are used for conversion, although active devices can also be employed in a technique called **synchronous rectification**. Unless the load can use the so-called raw DC, which is the pulsating DC that results from straight rectification with no filtering, most DC power supplies also use some type of **filter circuit** that reduces the AC ripple voltage to an acceptable level. Further reductions in ripple can be achieved by means of **voltage regulator** circuits to be described later in this chapter.

Four popular types of rectifier circuits are shown in Figure 10.5. In addition to the circuit's name and schematic diagram, the peak DC output voltage is shown (neglecting any diode voltage drop, which is about 0.7 V per diode for silicon devices). Every diode in a rectifier circuit is subjected to a **reverse voltage** V_R that it must withstand if the circuit is to work properly. Diodes for rectifier service must have a peak reverse voltage rating in excess of the actual reverse voltage encountered in the circuit to allow for transient peaks and other unusual circumstances. The maximum reverse voltage V_R presented to the diodes is also shown, and the ripple frequency f_{RIPPLE} in terms of the AC input voltage's frequency f_{IN} is also stated.

The **half-wave rectifier** is the simplest possible rectifier circuit. As we will see in the following text, it requires a relatively large filter capacitor in order to deliver low-ripple DC, but in low-current and high-voltage applications, the half-wave rectifier is sometimes the best choice.

Adding three more diodes to the half-wave circuit allows one to construct a **full-wave bridge rectifier**, also shown in Figure 10.5. Note that with the full-wave bridge, the AC voltage source and the DC load resistance *cannot* share a common terminal (e.g., both cannot be grounded at one end), because such a connection shorts

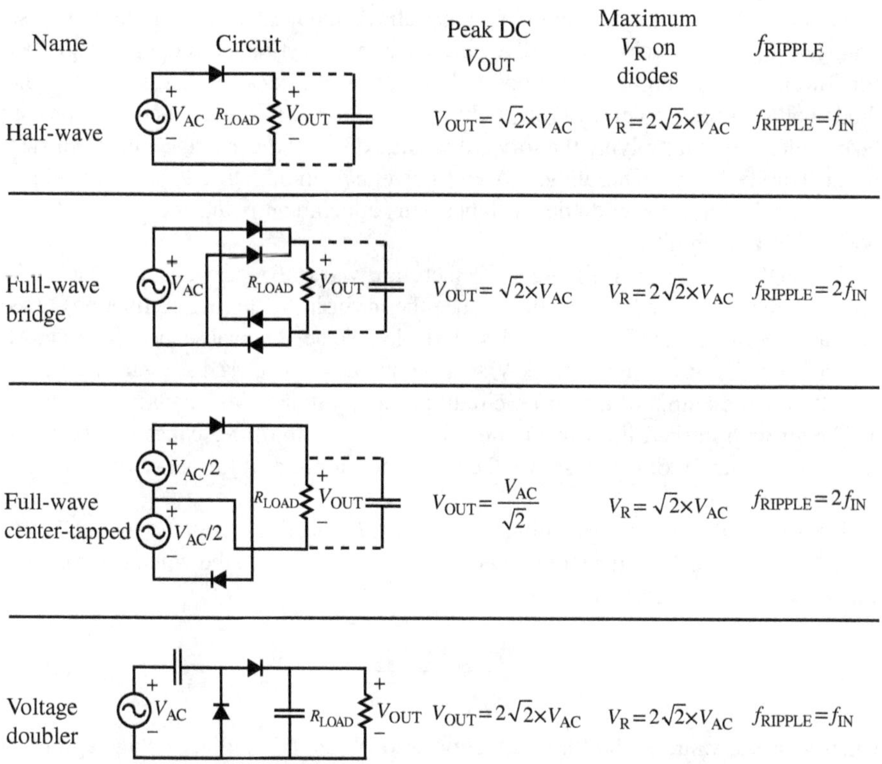

FIGURE 10.5 Four popular diode rectifier circuits, showing each circuit's name, diagram, peak DC output voltage, peak reverse voltage presented to rectifiers, and ripple frequency. A filter capacitor can be connected as shown in dashed lines. V_{AC} is RMS value.

out one of the diodes, and the circuit will not work. A **transformer-operated** power supply can operate with an ungrounded transformer secondary, and so full-wave bridge circuits are often used with transformer-operated supplies.

The **full-wave center-tapped** rectifier circuit uses a **center-tapped** voltage source such as a transformer with a connection to the center of the winding. Both full wave circuits have a ripple frequency that is ***twice*** the AC input frequency, because both the positive and the negative halves of the input sine wave appear with the same polarity at the output. This frequency doubling makes it easier to filter the output for a given maximum ripple voltage requirement.

The **voltage-doubler rectifier** produces a voltage that is *twice* the peak AC input voltage. The basic principle of the voltage doubler can be carried further to multiply the AC input voltage by many times in so-called **voltage multiplier** circuits. These circuits are sometimes used in low-current high-voltage power supplies, for which their relatively poor load regulation is not a problem.

It should be noted that all the output voltages given in Figure 10.5 should be reduced by one diode forward voltage drop for the half-wave and full-wave center-tapped circuits and by two diode voltage drops for the full-wave bridge and voltage-doubler circuits. For output voltages above about 12 V, the omission of these voltage drops makes less than a 10% difference in the estimated output voltage, but for lower-voltage outputs, the diode voltage drop becomes more significant. The diode voltage drop also plays a role in the power dissipated by each diode, which can be calculated by multiplying the forward voltage drop by the average current carried by each diode. For currents above 1 A, this power can amount to several watts, which is why special high-power diodes with heat sink attachment points are used for high-power rectifier circuits.

Although not shown in Figure 10.5, there are ways of rectifying multiphase AC power (three phases and higher) that reduce the intrinsic ripple (before filtering) to as little as 12% of the peak DC value or less, reducing the need for subsequent filtering. At higher power levels above about 3 kW, such techniques can be used advantageously.

Filtering the output of a rectifier circuit is usually done with a capacitor attached as shown with dashed lines in Figure 10.5. The way the filter capacitor works is shown in Figure 10.6, which shows the idealized output of a full-wave rectifier circuit in dashed lines.

We assume that a constant average DC current I_{DC} is drawn from the supply. From the peak of the AC waveform until a point near the next peak, the voltage across the capacitor will fall linearly at a rate

$$\frac{dV(t)}{dt} = -\frac{I_{\text{DC}}}{C}, \tag{10.8}$$

where C is the value of the filter capacitor used. When this falling voltage approximately equals the rising edge of the full-wave rectifier's output voltage, the capacitor recharges and its voltage follows the rectifier's output voltage to the next peak. Assuming this charging interval is negligibly small compared to the ripple period $1/f_{\text{RIPPLE}}$, the peak-to-peak AC ripple voltage superimposed on the DC output voltage is approximately

$$V_{\text{RIPPLE}}(\text{p}-\text{p}) \approx \frac{I_{\text{DC}}}{Cf_{\text{RIPPLE}}} \tag{10.9}$$

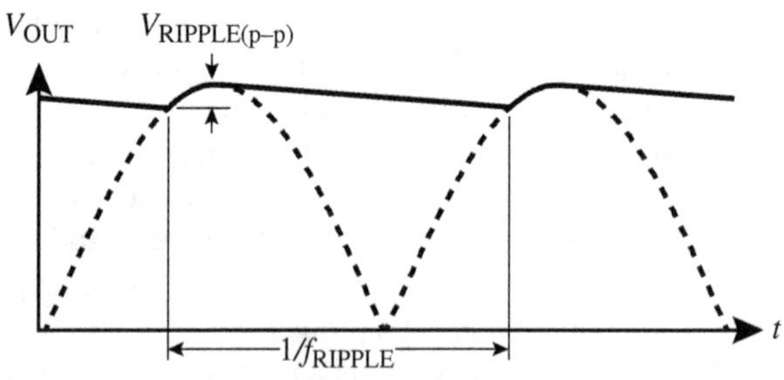

FIGURE 10.6 Estimation of peak-to-peak ripple voltage $V_{\text{RIPPLE(p–p)}}$ of full-wave rectifier circuit with single-capacitor filter.

Low values of ripple voltage (below a few mV) can be obtained solely with a practical size of filter capacitor operating on power-line frequencies only for small values of DC current, less than 100 mA or so. For example, to reduce the ripple of a 1-A DC supply operating with $f_{RIPPLE} = 120$ Hz to below 10 mV requires a filter capacitor

$$C \approx \frac{I_{DC}}{V_{RIPPLE} f_{RIPPLE}} = \frac{1\,\mathrm{A}}{(10\,\mathrm{mV})(120\,\mathrm{Hz})} = 833\,\mathrm{mF} \tag{10.10}$$

or nearly 1 F! While such capacitors are available, and in the form of **supercapacitors** are reasonably compact, they are costly and not available in higher voltage ratings. Such **brute-force** filtering is not necessary if the raw filtered DC output of the rectifier circuit is processed by a voltage regulator circuit, an example of which is discussed in the following section. Voltage regulators are also needed with other types of power sources, including renewable ones such as solar and wind power.

10.3.3.1 Other Primary Power Sources Besides batteries and the AC utility supply, recent years have seen greatly increased use of alternatives such as **renewable energy** (solar, wind, etc.) and an activity termed **energy harvesting**, which uses such sources for small autonomous devices. One important requirement that most of these energy sources share is the need for **power conditioning**, an operation that takes the raw form of power as supplied by the source and transforms it into a steady usable form. For example, solar-cell panels produce their greatest output near the middle of a cloudless day and zero output at night. Wind generators produce power only when the wind is blowing. For use either directly in DC form by electronic systems or in AC form for introduction to the general AC power grid, the power from such sources must be conditioned by means of storage (usually batteries) and sometimes DC-to-AC conversion so that it can be used for its intended purpose. Power conditioning for these types of energy sources is beyond the scope of this text, but many of the design elements needed for such systems are treated in the following sections.

10.3.4 Linear Voltage Regulators for Power Supplies

Once AC is converted to DC, it may not be at the voltage level required by the load or loads. The conversion from one DC voltage to a lower voltage can be performed by either a linear circuit called a **linear voltage regulator** or by a nonlinear circuit called a **switching power supply**. We will defer discussion of switching power supplies to the next section. In this section, we will describe linear voltage regulators, which are not as efficient as switching supplies but are still useful for many applications.

The DC output voltage from a rectifier and filter circuit operating from unregulated AC power will change in proportion to any changes in the AC voltage supplied. Although electric utility companies make a reasonable attempt to maintain the voltage available at power outlets within a few volts of the nominal value, unavoidable fluctuations do occur because of local changes in load, weather conditions, and other factors. These changes can amount to ±10% or more and are passed on without reduction except for any voltage step-down that the power supply provides. In other words, the **line regulation** of an unregulated supply is usually about 1, or 100%. In addition, the resistance of transformers and rectifier diodes used in the rectifier circuit, as well as increased ripple at higher drain currents, will cause a drop in the average DC output voltage as current drain increases, leading to poor **load regulation**. Both of these characteristics can be greatly improved through the use of a voltage regulator circuit.

10.3.4.1 Voltage References Most voltage regulators depend for their operation on an internal **voltage reference**: a circuit that will provide a nearly constant reference voltage despite variations in temperature, primary power voltage, and other factors. One of the most common types of voltage-reference circuits used in discrete circuits employs a **Zener diode**, a type of diode described in Chapter 2 that shows a nearly constant breakdown voltage in reverse bias over a wide range of currents. Zener diodes are available in voltages ranging from below 4 to 100 V or more, although their **temperature coefficient** (variation of Zener voltage with temperature) tends to be smallest around 5.6 V. Another type of voltage reference popular in integrated-circuit (IC) designs is the **Brokaw bandgap reference**, which bases its operation on the forward voltage drop across a silicon diode and compensates for the temperature variation of that voltage drop. The bandgap reference circuit provides a voltage of about 1.22 V, which is the **bandgap voltage** of silicon.

10.3.4.2 Series-Pass Voltage Regulator Whichever type of voltage-reference circuit is used, a voltage regulator is basically a feedback circuit that compares a scaled version of its actual output voltage with the internal reference voltage. The difference between these two values is an error voltage, which is used to adjust the output voltage in a direction so as to maintain a nearly constant output voltage. A block diagram of a typical voltage regulator circuit is shown in Figure 10.7.

The source of raw DC must be sufficiently filtered so that its minimum voltage under maximum load conditions is still sufficient to operate the power supply's internal circuits, including the voltage reference and the feedback amplifier. However, it can still show a substantial amount of ripple. The voltage reference produces a constant voltage V_{REF} that is compared to V_{FB}, a scaled-down version of the output voltage V_O. The scale factor is set by the voltage-divider resistors R_1 and R_2 so that

$$V_{FB} = V_O \frac{R_2}{R_1 + R_2} \tag{10.11}$$

The differential feedback amplifier adjusts its output voltage to control the **series-pass** device or devices so that the output voltage is maintained at a nearly constant value, despite changes in load current. "Series-pass" simply means that the device regulating the main current to the load is in series between the raw power source and the load. (While this is the most common way to regulate low-voltage supplies, certain types of high-voltage, low-current power supplies are more easily regulated with a variable current sink in shunt with the output, which is called a **shunt regulator**.) Standard feedback-loop analysis can be applied to this type of circuit, and considerations of frequency response and stability should be taken into account. However, the loop bandwidth of most power-supply regulators extends only into the kHz region, because frequencies higher than that can be effectively dealt with by means of small bypass capacitors across the output terminals between the power supply and the load. The overall goal is to approach the ideal of presenting a perfect constant-voltage source to the load, and such a source has zero impedance at all frequencies. A low output impedance is guaranteed by the feedback circuit at frequencies within the circuit's loop bandwidth, but at higher frequencies, passive capacitive bypassing can take over this task.

To illustrate how even a very simple discrete voltage regulator can incorporate all the elements of Figure 10.7, we will analyze the basic discrete-component regulator circuit shown in Figure 10.8. This circuit is not presented because it represents the

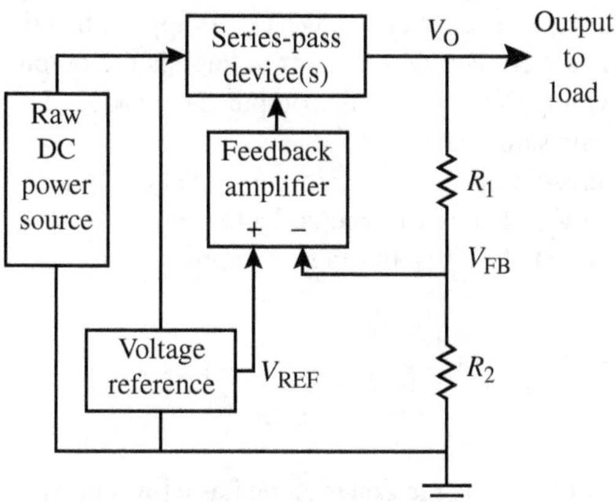

FIGURE 10.7 Block diagram of typical series-pass voltage regulator circuit.

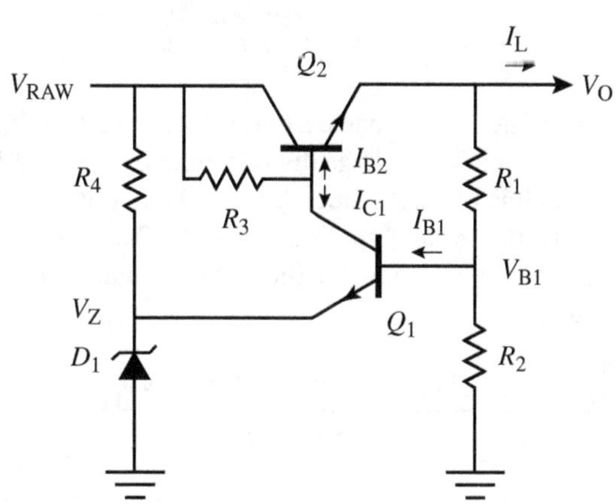

FIGURE 10.8 Discrete-component series-pass voltage regulator circuit.

current state of the art (IC voltage regulators are the preferred choice for small fixed-output regulators), but because it demonstrates all the essential functions of a voltage regulator in a circuit that is easily analyzed.

The raw DC input is used both to bias the Zener-diode voltage reference D_1 through R_4 and to supply current to the series-pass transistor Q_2, which performs the regulating function. (Zener diodes provide the Zener breakdown voltage when reverse biased as shown.) The two inputs of the differential amplifier are transistor Q_1's emitter and base leads. In operation, suppose the output voltage V_O rises because the load current decreases. A scaled amount of this positive change will appear at Q_1's base. Because Q_1's emitter voltage is held approximately constant by the voltage-reference Zener diode, this positive change will cause Q_1's base current to increase. If β_1 is the ratio of Q_1's DC collector current to its base current and β_1 is on the order of 100, a much larger change in its collector current will cause a larger voltage drop across resistor R_3. That causes the base voltage of regulator transistor Q_2 to fall and given that its base–emitter voltage V_{BE2} is approximately constant, its emitter voltage will fall as well. But Q_2's emitter voltage is the output voltage V_O, so we see that an initial voltage increase at the output has produced a negative-feedback voltage decrease at the same output.

The analysis of this circuit can use a simple model of the transistors in which the base–emitter voltage V_{BE} is constant, and collector current is simply β times the base current. We begin the analysis by finding an expression for the base current I_{B1} of transistor Q_1:

$$I_{B1} = \frac{V_O(R_2/(R_1+R_2)) - (V_Z + V_{BE1})}{R_1 \| R_2} \tag{10.12}$$

In this analysis, we will assume the Zener diode has a low enough dynamic resistance that it behaves like an ideal voltage source with voltage V_Z. We next write an expression for output voltage V_O in terms of the raw DC input voltage V_{RAW} and the voltage drops across R_3 and the base–emitter junction of Q_2:

$$V_O = V_{RAW} - R_3(I_{C1} + I_{B2}) - V_{BE2} \tag{10.13}$$

The base current I_{B2} is related to the load current I_L by β_2, the β of Q_2 (we will neglect the current drawn by the sampling voltage divider formed by R_1 and R_2 in comparison to the load current, which is typically much larger). And I_{C1} is related to I_{B1} by β_1, so we can make these substitutions and use Equations 10.12 and 10.13 to derive the following "hand-calculation" expression for the output voltage in terms of the other variables and constants:

$$V_O = \frac{V_{RAW} - V_{BE2} + \beta_1(R_3(R_1+R_2)/R_1R_2)(V_Z + V_{BE1}) - (R_3 I_L/\beta_2)}{1 + \beta_1(R_3/R_1)} \tag{10.14}$$

Equation 10.14 has several interesting features. First, the form of the denominator is 1 plus a dimensionless constant proportional to β_1, the current gain of feedback transistor Q_1. In the limit of infinite current gain (β_1 goes to infinity), the expression for output voltage would simplify to

$$V_O\big|_{\beta_1 \to \infty} \approx \frac{R_1 + R_2}{R_1}(V_Z + V_{BE1}) \tag{10.15}$$

This much simpler expression shows that, ideally, the output voltage would depend only on the sum of the reference voltage V_Z and Q_1's base–emitter voltage, which is equal to the output voltage V_O scaled down by the R_1–R_2 voltage divider. This is consistent with the negative-feedback principle that a differential amplifier with sufficiently high gain in a negative-feedback circuit causes the difference between its input voltages to approach zero.

In actuality, β_1 is only 100 or so, and the need to pass sufficient base current to operate the series-pass transistor Q_2 at the maximum load current means that R_3 is typically a fairly small resistance compared to R_1. Nevertheless, the product ($\beta_1 R_3/R_1$) can still be made much larger than 1, and this allows good but not perfect line and load regulation, as we will now show.

Once we have Equation 10.14, we can easily determine the theoretical line and load regulation by taking the derivative of V_O with respect to V_{RAW} and I_L, respectively. These operations give the following expressions:

$$\frac{dV_O}{dV_{RAW}} = \frac{1}{1 + \beta_1 (R_3 / R_1)} \tag{10.16}$$

$$\frac{dV_O}{dI_L} = \frac{-(R_3 / \beta_2)}{1 + \beta_1 (R_3 / R_1)} \tag{10.17}$$

As you can see, both of these numbers will be improved (made smaller) as the feedback transistor's current gain β_1 increases.

To complete this exposition of the circuit, we designed an example including component values and specific devices and modeled it on Multisim™ to compare the theoretical hand-calculated performance with the software's circuit model that uses much more detailed device models. The design goals were to provide a nominal output voltage of $V_O = 5.0$ V as the load current I_L varies from 0 to 3 A and the power source voltage V_{RAW} varies from 7 up to 12 V. The modeled circuit, complete with component values and specific device type numbers, is shown in Figure 10.9.

The 1N4728A Zener diode has a nominal voltage of 3.3 V. The 2N4401 feedback transistor is a medium-current device with a plastic *TO-92* package and can safely dissipate up to 625 mW at an ambient temperature of 25°C without an external heat sink. The 2N3055 series-pass device is encased in an all-metal *TO-3* enclosure, which can dissipate up to 6.5 W at 25°C without a heat sink. With a maximum output current of 3 A, the 2N3055 is operating safely within its maximum collector-current rating of 7 A. At the end of the design exercise, we will check the power dissipation of critical components to see whether external heat sinks are needed.

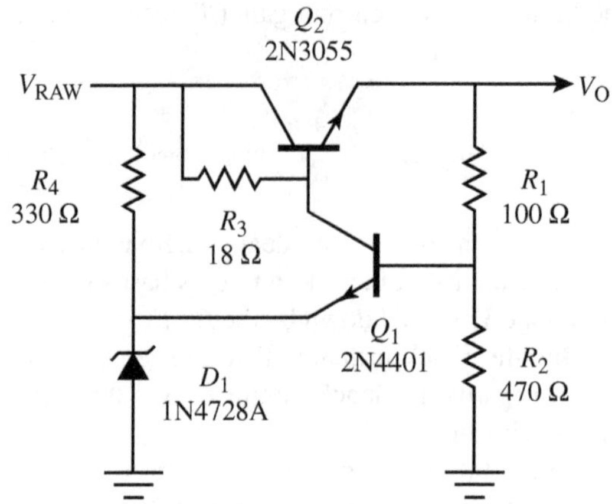

FIGURE 10.9 Voltage regulator circuit of Figure 10.8 with devices and component values designed for $V_{OUT} = 5$ V at up to 3 A.

Under the conditions stated, the DC β of the Multisim model for transistor Q_1 is approximately 100, and that of Q_2 is about 88. We will assume all transistors have a base–emitter voltage $V_{BE}=0.7$ V. If we use these values and the component values of Figure 10.8 in Equation 10.14, the calculated output voltage for $V_{RAW}=8$ V and $I_L=1$ A is $V_O=4.96$ V. The Multisim model of Figure 10.9 under these conditions produces $V_O=5.002$ V, which is in quite good agreement with the hand calculation of Equation 10.14.

To test the circuit's line regulation, we held the output current constant at $I_L=1$ A and varied the DC input voltage V_{RAW} from 5 to 12 V. (It is usually a good idea to try a circuit under conditions beyond its specifications to learn what its failure modes are.) The plot of output voltage V_O as V_{RAW} varies over this range for the hand calculation of Equation 10.14 is linear, as shown with the dashed line in Figure 10.10. The more sophisticated Multisim model agrees fairly well with the hand calculation for V_{RAW} above about 6 V. When the input voltage falls below this level, however, the series-pass transistor begins to **saturate**, and its collector–emitter voltage falls to about 0.2 V as the circuit attempts to pass all available voltage to the output. This demonstrates a general principle: a linear series-pass voltage regulator requires a minimum difference voltage between its input and the desired regulated output in order to function properly. For this circuit, the minimum difference is about 1.5 V, and so the raw DC input cannot fall much below $(5+1.5)=6.5$ V for the circuit to work properly.

For the load regulation test, V_{RAW} was held at 8 V and the load current I_O was varied from 0 to 5 A, beyond the specified maximum of 3 A. The results of both the hand calculation and the Multisim model results are shown in Figure 10.11. Up to a load current of 3 A, the output voltage remains within about 2% of the nominal value of 5 V. But for currents higher than that, the output voltage falls rapidly. Once the output voltage falls below about 4.8 V, the feedback transistor Q_1 has insufficient base voltage to conduct and cuts off, leaving the series-pass

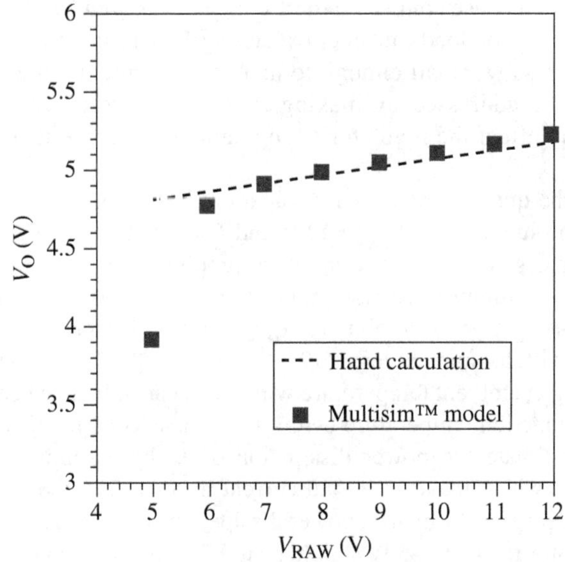

FIGURE 10.10 Line regulation of example design in Figure 10.9 with constant I_L = 1 A. Hand calculation shown as dashed lines, and Multisim data shown as square data points.

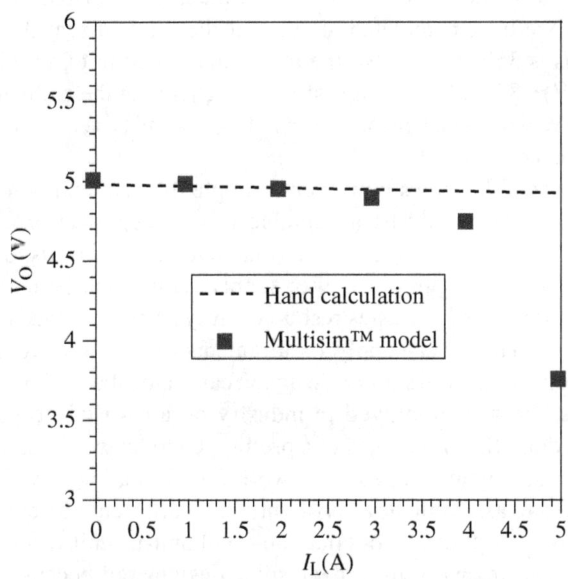

FIGURE 10.11 Load regulation of example design shown in Figure 10.9 with constant V_{RAW} = 8 V. Hand calculation shown as dashed lines, and Multisim data shown as square data points.

transistor Q_2 with its base connected to its collector through 18-Ω resistor R_3. The current through R_3 for load currents exceeding 3 A is insufficient to keep the voltage drop across Q_2 small enough to maintain an output voltage of 5 V. This problem could be addressed by making R_3 smaller, but as we have seen, that would adversely affect the regulator's loop gain and degrade its performance in other ways.

To address the question of power dissipation, we choose a "maximum-stress" condition for the supply with $V_{RAW} = 12$ V and $I_L = 3$ A. In general, any series-pass regulator dissipates the most power when supplying the maximum rated load current for the maximum rated raw input voltage. The most critical components from a power-dissipation viewpoint are Q_1, Q_2, and R_3. Under the maximum-stress conditions, Q_1 will dissipate at least (12 – 5 V)(3 A) = 21 W, well above its 6.5-W maximum rating at ambient temperature without a heat sink. So a heat sink capable of keeping the junction temperature below 150°C is needed for Q_2 in this application. We can estimate the power dissipation of R_3 by assuming that Q_2's base voltage is 5.7 V, which makes the total current through R_3 to be (12 – 5.7 V)/(18 Ω) = 350 mA. Using $P = VI$ gives a power dissipation for R_3 of 2.2 W. For a margin of safety, a resistor rated at 4–5 W should be used for this component. While power resistors at lower power levels (below about 10 W) do not necessarily need heat sinks, they can cause heating in adjacent components through radiation and convection, and sufficient air space should be available around them to prevent undesirable heat buildups in operation. Finally, transistor Q_1 will experience the greatest power dissipation under the condition of maximum V_{RAW} (12 V) and minimum I_L, because then it will have to draw the largest current to keep Q_2's base at about 5.7 V. We have calculated above that the total current through R_3 under these conditions is 350 mA, and so the power dissipated in Q_1 in this case is (350 mA)(5.7 – 3.3 V) = 840 mW, which is slightly higher than the ambient-temperature rating of 650 mW without a heat sink. A small "clip-on" type of heat sink would be adequate for this power level.

This design example shows how even a simple discrete-component circuit can deliver regulation that would be acceptable in some applications. The designer should have a clear idea of how good the voltage regulation needs to be for a given application, because designing a power supply with much better specifications than the application calls for wastes resources. Any system design should be tested to see what its actual power-supply requirements are in order to set reasonable specifications on the power-supply design. Years ago, the author neglected this precaution when he was employed in industry on a rush job to get a consumer product out the door. Several copies of a prototype model were tested with a single prototype power supply and appeared to work fine. But a few days later, when the factory attempted to assemble dozens of units with different power supplies, many of them failed to work. It turned out that the small unit-to-unit variations in power-supply voltage, less than 100 mV, made some designs fail because they were too dependent on having an exact value of power-supply voltage. A hasty redesign fixed the immediate problem but at the cost of a tedious hand assembly of added components.

10.3.4.3 IC Voltage Regulators The basic series-pass voltage regulator circuit is available in a number of IC forms. The LM7805 is typical of these. It is a three-terminal regulator that provides 5 V at up to 1 A from a source voltage ranging from 8 to 18 V, with a basic output-voltage accuracy as good as 2%. It is housed in a **TO-220** package, which consists of a metallic mounting plate designed to contact the surface of a heat sink. The plate is covered by a plastic housing, leaving only one metal side exposed to the heat sink, and the three leads (power in, power out, and ground) emerge from one end of the package. The TO-220 package is a popular housing for many types of medium- and high-power semiconductor devices and ICs. Using a prepackaged voltage regulator circuit such as the LM7805 is a simple and cost-effective way to obtain regulated voltages from a power-supply system.

Many of the three-terminal regulator ICs are designed for specific voltages, but if a particular fixed or adjustable voltage is needed for which an IC is not available, the circuit in Figure 10.12 can be used to obtain a selectable voltage from a fixed-voltage IC regulator. The IC regulator is designed to maintain its design output voltage V_{REG} between its output terminal (terminal 3 in Fig. 10.12) and ground (terminal 2 in Fig. 10.12). If a voltage divider R_1–R_2 is interposed between terminal 2 and ground and arranged so that the desired voltage V_O appears between terminal 3 and ground when V_{REG} appears between terminals 2 and 3, the overall circuit produces a regulated voltage of V_O when the regulator is maintaining the smaller voltage V_{REG} across its terminals. No exact design guidelines for the voltage divider can be given, although it should not consume more than 5% or so of the total load current supplied by the power supply. Some current flows through terminal 2 to operate the IC, and this must be taken into account in the design. Although this circuit will work with almost any three-terminal regulator IC, some ICs are available that are designed to be used this way. The bypass capacitor C keeps terminal 2 at AC ground and preserves the ripple-rejection ability of the regulator.

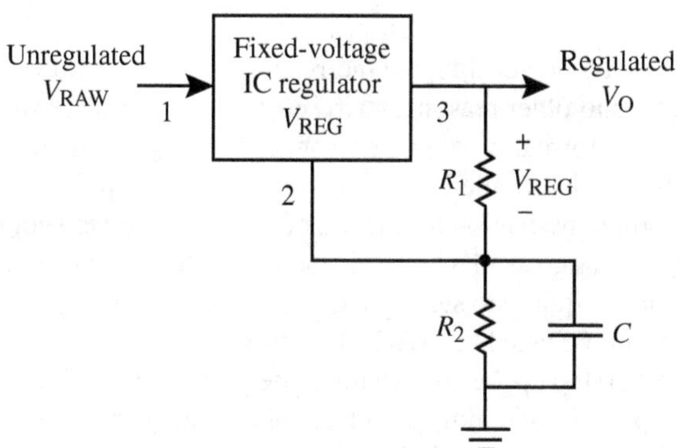

FIGURE 10.12 User-selectable output voltage obtained from fixed-voltage IC voltage regulator.

10.3.4.4 Current Limits and Other Voltage Regulator Features One important feature of many IC voltage regulators and power-supply systems is a **current-limit function**. The output current from a given regulator can be sensed by a small series resistance and used to impose a limit on the maximum current that the power supply will provide to the load. This is an important safety feature that can protect both a load that draws excessive current and the power-supply circuit itself, because otherwise excessive load current can cause overheating and may damage or destroy components, perhaps even causing a fire. While the so-called overcurrent protection is a good safety feature to have, it should be supplemented by additional protective devices such as fuses and circuit breakers, because sometimes even the limited current allowed by overcurrent protection circuits can cause problems.

Another feature provided by some power-supply systems is **remote voltage sensing**. If the power supply is separated from the load by a long wiring run, this wiring will have resistance that will cause a voltage drop at the load when the load draws higher currents. Maintaining the voltage constant at the power supply will still allow voltage to drop at the load, possibly to an unacceptable degree. A power supply that provides for remote voltage sensing can sense the voltage at the load through additional sense wiring, at a point after the voltage drop incurred by the long wiring leads. When connected this way, the power supply will automatically increase its output voltage at its terminals in order to compensate for the voltage drop along the main leads from the supply to the load.

These features can also be provided by **switching power supplies**, which will be our next topic in the following section.

10.3.5 Switching Power Supplies and Regulators

While linear voltage regulators can provide very low levels of ripple, we have seen that their efficiency is intrinsically limited by the fact that the voltage drop between the raw input voltage and the regulated output voltage is achieved by dissipating power in the series-pass device or devices. For low and medium power levels, this heat loss can be acceptable, but for high-power applications or those where efficiency is critically important to the design, the inefficiency of a linear power supply is a disadvantage. For this and other reasons, **switching power supplies** were developed to provide efficient transformation of power from one voltage to another and regulation of the output voltage without the losses incurred in linear supplies.

A second important motivation for the use of switching power supplies arises when the primary supply voltage available is DC and lower than the desired output voltage. A nonlinear circuit is required to step up DC voltage, and typically, a switching circuit is used to convert low-voltage DC to AC. Once this conversion is made, either a transformer can be used to step up the voltage for subsequent rectification to higher-voltage DC, or certain types of switching power supplies can increase DC input voltages without a transformer, although an inductor or other energy-storage component is typically used. Regardless of the circuit details, the higher efficiency of the switching supply compared to the linear supply results from the fact that the active devices in it are either turned on or off completely and spend only a short amount of time in regions of voltage and current combinations that result in large power dissipation. To see in detail how this is achieved, we will describe the parameters surrounding a typical switching operation.

10.3.5.1 Active Devices as Switches Many types of devices—bipolar junction transistors (BJTs), power field-effect transistors (FETs), insulated-gate bipolar transistors (IGBTs), and even vacuum tubes—can be used as switches. Whatever the nature of the device, for purposes of analysis in a switching circuit, we will make certain assumptions about its operation as a switch. The waveforms and variables we will use in the following analysis are shown in Figure 10.13. These waveforms do not show the worst-case situation of the current waveform's position with respect to the voltage waveform, which happens when the voltage stays at V_{MAX} until the current reaches I_{MAX} and vice versa. But the analysis shown here represents typical waveforms and is a good starting place on which to base more detailed analyses.

When the device is turned off, we will assume that the current flow through it is zero and the voltage across it is a value V_{MAX}, which the device is designed to withstand without breakdown or other problems. When the device turns on, we assume that the current through it rises linearly from zero to a maximum value I_{MAX} in a time $t_{S(ON)}$, which we will term the **turn-on time** (also sometimes referred to as the **rise time**). Simultaneously, the voltage across it falls linearly from V_{MAX} to a low voltage termed V_{MIN}. Once the device is on, the current I_{MAX} continues to flow, and the voltage across the device remains at the small but nonzero voltage V_{MIN}. The length of time that the device is fully on is termed t_{ON}. When the device is commanded to turn off, the current falls linearly from I_{MAX} to zero, and the voltage rises linearly from V_{MIN} to V_{MAX}. Both of these transitions take place during a **turn-off time** $t_{S(OFF)}$ (also referred to as the **fall time**). The time that the device is fully off is t_{OFF}.

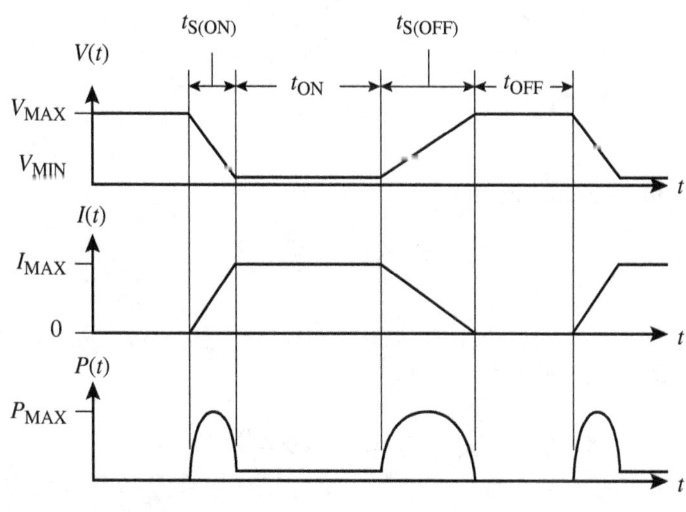

FIGURE 10.13 Voltage, current, and power waveforms of a switching device over one period. Rise and fall times are conventionally defined between 10 and 90% of the full waveform height. This detail has been omitted in this figure for clarity.

The switch-waveform period T_S is obviously the sum

$$T_S = t_{S(ON)} + t_{ON} + t_{S(OFF)} + t_{OFF}, \tag{10.18}$$

the switching frequency f_S is

$$f_S = \frac{1}{T_S}, \tag{10.19}$$

and the fractional switching duty cycle D_S will be defined as

$$D_S \equiv \frac{t_{S(ON)} + t_{ON}}{T_S} \tag{10.20}$$

If we determine the total amount of energy U_{TOT} dissipated in the device during one switching cycle, then the average dissipated power P_{DISS} in the device (neglecting any power associated with the control circuit that drives the device) will be

$$P_{DISS} = f_S U_{TOT} \tag{10.21}$$

The energy U_{TOT} consists of three parts: (1) energy dissipated during the off-to-on transition, (2) energy dissipated while the device is fully on, and (3) energy dissipated during the on-to-off transition. The two transitions can be treated the same, and so we will assume simply that the transition lasts a time t_S, where t_S can be either $t_{S(ON)}$ or $t_{S(OFF)}$.

Taking the off-to-on transition first, we can write expressions for the voltage and current as a function of the time t, which we will take as the time elapsed from the beginning of the off-to-on transition:

$$V(t) = V_{MAX}\left(1 - \frac{t}{t_S}\right) \tag{10.22}$$

$$I(t) = I_{MAX}\frac{t}{t_S} \tag{10.23}$$

If we make the substitution

$$x = \frac{t}{t_S}, \tag{10.24}$$

we can write an expression for the energy $U_S(t_S, V_{MAX}, I_{MAX})$ (in joules) dissipated during one transition lasting a time t_S:

$$U_S = t_S V_{MAX} I_{MAX}\int_0^1 x(1-x)dx = \frac{t_S V_{MAX} I_{MAX}}{6} \tag{10.25}$$

This energy will be dissipated both during the off-to-on transition lasting $t_{S(ON)}$ and the on-to-off transition, lasting $t_{S(OFF)}$. In addition to the energy dissipated during the transitions, there will be energy dissipated during the on time t_{ON} due to the nonzero voltage V_{MIN} that is present when the device is conducting. The on-time energy U_{ON} is

$$U_{ON} = t_{ON} V_{MIN} I_{MAX} \tag{10.26}$$

(This analysis assumes current falls to zero between the transitions and the on period, which is not quite true, but the error caused by this assumption is usually very small.) The total energy dissipated per cycle U_{TOT} is thus

$$U_{TOT} = U_{S(ON)} + U_{ON} + U_{S(OFF)} \tag{10.27}$$

and so Equations 10.21–10.27 can be used to derive the total power dissipation involved in a device with the switching waveform of Figure 10.13 as

$$P_{DISS} = f_S I_{MAX} \left[t_{ON} V_{MIN} + \frac{(t_{S(ON)} + t_{S(OFF)}) V_{MAX}}{6} \right] \tag{10.28}$$

The quantity in brackets has two terms: the dissipation due to the small voltage V_{MIN} across the device during the time it is on and the dissipation due to the two transition periods. One or the other of these terms tends to dominate the expression in most cases. In low-current, high-voltage systems, the effect of V_{MAX} makes the transition power dissipation larger, so fast switching speeds (i.e., small values of $t_{S(ON)}$ and $t_{S(OFF)}$) are needed for good efficiency. On the other hand, in high-current, low-voltage applications, the minimum *on* voltage V_{MIN} may be the determining factor in efficiency, and it should be held as low as possible.

The efficiency advantage of using power devices in switching circuits as opposed to linear circuits can be made apparent by the following example. Suppose the 2N3055 power BJT used in the previous section's example of a linear power supply is now used in a switching supply that achieves voltage reduction by varying the duty cycle of a rectangular pulse. To be specific, suppose the input voltage V_{RAW} is 10 V and the desired output voltage is 5 V at a current of 3 A. In the linear circuit, the series-pass device must dissipate a constant power of $P_{LINEAR} = (10 - 5\text{ V})(3\text{ A}) = 15$ W. Now, suppose the device is used instead in a switching power supply to switch the 10-V input voltage on and off at a duty cycle of $D_S = 50\%$ so that the average voltage is $(10 + 0\text{ V})/2 = 5$ V. Because of energy conservation, the current flow during the on period must be twice the average output current, which means $I_{MAX} = 6$ A. The typical on voltage for a turned-on BJT is $V_{MIN} = 0.7$ V, assuming it is nearly in saturation. According to the 2N3055 spec sheet, the turn-on time is $t_{S(ON)} = 6$ μs and the turn-off time is $t_{S(OFF)} = 12$ μs. Supposing that a relatively low switching frequency of $f_S = 20$ kHz is used, the current waveform that results is shown in Figure 10.14. The turn-off time of 12 μs is a large fraction of the off time, which is generally not good practice, but even for this relatively slow device, the advantage of using it as a switch rather than a linear element is significant. Applying Equation 10.28 to the switch-mode case gives a total device power dissipation of

$$P_{DISS} = (20\text{ kHz})(6\text{ A})[(13.3 + 15) \times 10^{-6}\text{ V-s}] = 3.4\text{ W} \tag{10.29}$$

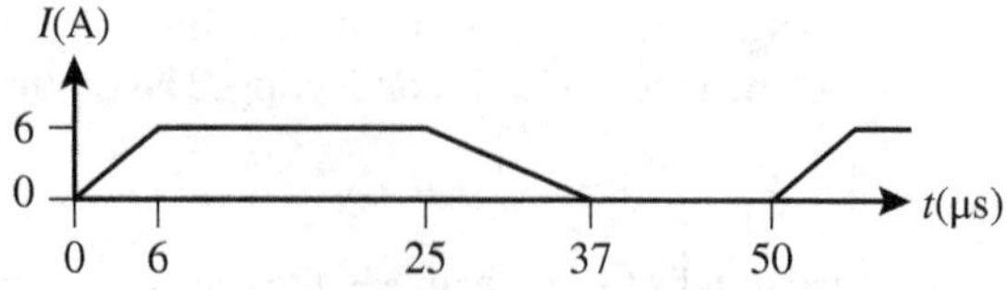

FIGURE 10.14 Switching waveform used in hypothetical switch-mode power supply to deliver 5 V at 3 A from a 10-V input voltage.

The 13.3×10^{-6} V-s terms originates in the on-time dissipation, while the 15×10^{-6} V-s term comes from the switch-transition times. In a better design, one of these terms would dominate the other. But even in this nonoptimum design, we find that the dissipated power of 3.4 W when the device is used in a switch-mode power supply is much less than the 15 W the device must dissipate in the linear power-supply circuit. While it is true that some additional filtering is needed for the switching circuit, the task of filtering is made much easier when the fundamental frequency to be filtered is 20 kHz rather than 50 or 60 Hz, and a relatively small capacitor could be used to achieve a small value of ripple in the output. The next section describes a variety of switch-mode power-supply circuits that can be used both to lower and raise the input DC voltage from its initial value. In addition, some switching circuits use only a single inductor rather than a transformer.

10.3.5.2 Switch-Mode Power Supplies When one or more power devices are used as a switch, a number of possible circuit configurations can be used to convert DC power at one voltage to an output voltage that is either higher or lower than the original input. By varying the duty cycle of the switch waveform, the output voltage can be varied in a way that can compensate for line-voltage or load-current changes, performing the function of a linear voltage regulator but without the large power losses associated with linear regulator circuits. While no switch-mode circuit is 100% efficient, a well-designed switch-mode power supply's efficiency can easily exceed 90%. In addition, a switch-mode supply packs a large power output capability into a small and economical package that shows high **power density** measured in watts per unit volume or weight, with modest heat-sinking requirements.

While there are many circuit topologies used in switch-mode power supplies, they all share some common features. They use DC input power whose voltage can vary between certain limits to produce an alternating waveform that is applied to a reactive element or elements such as an inductor, a transformer, or (rarely) a capacitor. The energy stored in the reactive element is used to alter the original input voltage to a different value, and in the process, the current is also transformed in accordance with the conservation of energy. Following these operations, additional filtering may be required to reduce the output voltage ripple to a desired level. The switching waveform is produced by an independent circuit that varies the waveform's duty cycle either to maintain a constant output voltage or to vary the output voltage in response to an external control signal.

The four types of switch-mode power-supply circuits we will examine in detail are the **buck converter**, the **boost converter**, the **buck–boost converter**, and the

push–pull converter. All but the last of these use a single inductor as the main energy-storage element, while the push–pull converter uses a transformer. The buck converter always delivers a lower output voltage than its input voltage, the boost converter raises the input voltage to a higher output value, and the buck–boost converter can either raise or lower the input voltage. The push–pull converter can isolate the load from the source by means of its transformer and can either step up or step down the input voltage.

While there are many other switch-mode power-supply configurations, most of them simply add more switches to achieve various special functions for certain applications.

In the explanations to follow, the switching devices will be shown as simple on–off zero-resistance switches. This idealization will simplify the analysis, but in an actual design, practical matters such as device on-resistance, losses in the drive circuit for the switching devices, and other factors must be considered. We will also assume that the output filter capacitor is large enough to maintain the DC output voltage essentially constant, which is how we will show it in waveform plots. When we calculate the output ripple voltage, we will assume it is too small to show up on the output voltage waveform, although not too small to estimate with an approximate calculation.

Buck Converter. The **buck converter** "bucks" or reduces the raw input voltage to a lower level by means of a variable-duty-cycle switch waveform that draws current from the power source only a fraction of the time. An inductor in series with the switch stores energy during the switch's on time and releases it during the off time. Ideally, the voltage is decreased by the duty-cycle ratio

$$D_{\text{BUCK}} = \frac{t_{\text{ON}}}{T} \tag{10.30}$$

and the current is increased by the ratio $1/D_{\text{BUCK}}$, as the following analysis will show. The basic buck converter circuit is shown in Figure 10.15. A series switch S periodically sends the input voltage V_{RAW} to the inductor L whenever S is closed. Because an inductor tends to oppose any change in current flow, when the switch opens, current I_L continues to flow through a path that now includes the (essential!) diode D, whose forward voltage drop we will neglect. With the presence of ripple filter capacitor C, the voltage across the load is maintained practically constant despite variations in the inductor current I_L.

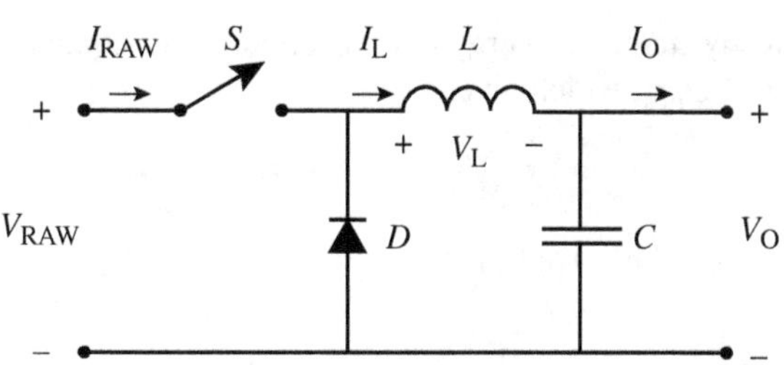

FIGURE 10.15 Basic buck converter circuit.

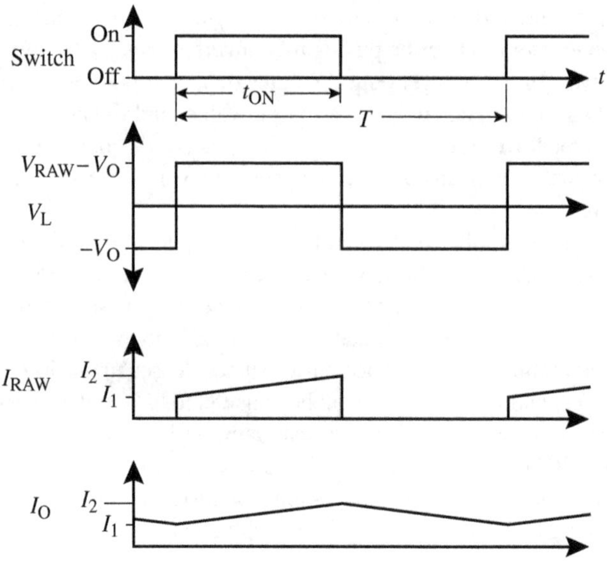

FIGURE 10.16 Voltage and current waveforms in buck converter shown in Figure 10.15.

The important voltage and current waveforms in the buck converter are shown in Figure 10.16, where it is assumed that the circuit has reached a steady state with constant input and output voltages. Neglecting coil, switch, and diode losses, we can apply the principle of energy conservation to find the relations among the input voltage V_{RAW}, input current I_{RAW}, output voltage V_O, and output current I_O. For a single period of length T, the energy U_{IN} absorbed by the circuit from the power source is

$$U_{IN} = t_{ON}V_{RAW}\left(\frac{I_1 + I_2}{2}\right) = t_{ON}V_{RAW}I_O, \tag{10.31}$$

if we realize that the output current I_O is the average of the minimum current I_1 and the maximum current I_2 through the inductor. During the same period T, the circuit delivers energy to the load equal to

$$U_{OUT} = TV_OI_O \tag{10.32}$$

Because in the steady state, the energy into the circuit must equal the energy coming out, we can set $U_{IN} = U_{OUT}$, which gives

$$t_{ON}V_{RAW}I_O = TV_OI_O \tag{10.33}$$

and Equation 10.33 can be solved for V_O to give

$$V_O = V_{RAW}\frac{t_{ON}}{T} = V_{RAW}D_{BUCK} \tag{10.34}$$

So we see that the ratio of output voltage to input voltage is indeed the duty cycle D_{BUCK} of the switch, which is always less than 1. This result is independent of the inductor value, but the value of L has a significant effect on the ripple current, as we will now show.

In the steady state, the increase in inductor current I_L during the switch-on part of the switching cycle equals the decrease in I_L during the time the switch is off, so either one can be used to calculate the peak-to-peak current ripple $I_2 - I_1 = \Delta I_{BUCK}$. Using the definition of inductance as the ratio of voltage to the rate of current change with time gives

$$\Delta I_{BUCK} = t_{ON}\left(\frac{V_{RAW} - V_O}{L}\right) \tag{10.35}$$

But V_O and t_{ON} can be expressed in terms of V_{RAW}, switching frequency f, and the switching duty cycle D_{BUCK}, so Equation 10.35 becomes

$$\Delta I_{BUCK} = \frac{V_{RAW}}{fL} D_{BUCK}(1 - D_{BUCK}) \tag{10.36}$$

Equation 10.36 shows that the current ripple varies with the duty cycle and is inversely proportional to the value of inductance L for a given switching frequency f. One can show that the peak-to-peak ripple voltage $\Delta V_{BUCK(p-p)}$ for a constant-current load (the worst case for ripple voltage) is

$$\Delta V_{BUCK(P-P)} = \frac{V_{RAW} D_{BUCK}(1 - D_{BUCK})}{8Cf^2 L} \tag{10.37}$$

Because the current waveform is a linear ramp, the ripple voltage waveform, which is the integral of current, has a parabolic shape that closely approximates a sine wave.

The advantage in size and weight for switch-mode power-supply components is clear if we sketch a design for a switch-mode buck converter to accomplish the same purpose as the linear voltage regulator discussed earlier, which required a nominal input voltage of $V_{RAW} = 8$ V to provide $V_O = 5$ V at up to 3 A. Suppose the peak-to-peak ripple voltage $\Delta V_{BUCK(P-P)}$ is required to be less than 50 mV. If we choose a switching frequency $f = 50$ kHz (on the low side of the typical range of values for switch-mode supplies) and require that the filter capacitor C be limited to 0.1 μF, we can find the various operating parameters required such as the duty cycle D_{BUCK} and the inductor value L as follows.

The duty cycle follows immediately from the values of the input voltage (8 V) and output voltage (5 V). Using Equation 10.34, we find

$$D_{BUCK} = \frac{V_O}{V_{RAW}} = \frac{5\,\text{V}}{8\,\text{V}} = 0.625 \tag{10.38}$$

Once D_{BUCK} is known, we can solve for the inductance L using Equation 10.37:

$$L = \frac{V_{RAW} D_{BUCK}(1 - D_{BUCK})}{8Cf^2 \Delta V_{BUCK(P-P)}} = 4.68\,\text{mH} \tag{10.39}$$

or about 5 mH. A 5-mH inductor with a current-carrying capacity of 3 A occupies only a few cubic centimeters. To use the buck converter as a regulator, it is necessary to vary the duty cycle with a mixed-signal circuit that compares the actual output voltage with a scaled reference and adjusts the duty cycle to maintain an approximately constant output voltage despite changes in load. We will deal with such techniques later in this chapter, when switch-mode power amplifiers are discussed.

The efficiency of the switch-mode regulator depends on the switching time and on-resistance of the device used, but can easily exceed 90%, depending on the device and the actual duty cycle used. Using the buck converter to achieve a voltage step-down ratio of more than 10:1 (e.g., 10 V in and 1 V out) is best achieved in two stages, because the switching times of most devices become an appreciable part of the on time for duty cycles of less than 10%. If the current requirement is very large (more than 30–40 A), a **multiphase buck converter** can be used. This circuit consists of several (usually three or more) buck converters in parallel and operated so that the phases of their switch waveforms are spaced evenly around a 360° circle (e.g., 120° apart for a three-phase converter). This configuration distributes the total load current among more components and decreases the ripple level compared to what could be achieved with a single converter.

From the current waveform for I_O in Figure 10.16, it is clear that if the load current falls sufficiently, the current through the inductor may go to zero for part of the cycle. Although the converter can operate in this **discontinuous mode**, as it is called, the analysis becomes nonlinear and the required duty cycle for a given output approaches zero, making it difficult to operate in this region. So for most designs, operation in the **continuous mode** is preferred, although if the load current falls below a certain minimum value, operation in the discontinuous mode may not be avoidable.

Boost Converter. The **boost converter** does the opposite of the buck converter, and raises the input voltage level to a higher output value while lowering the average current drawn from the power source. Again, an inductor is used as the primary storage element, but the output capacitor performs a more important function in the boost converter than the buck converter, as we will see.

The circuit of a basic boost converter is shown in Figure 10.17. The inductor L is connected between the low-voltage input V_{RAW} and a switch that periodically connects the opposite end of the inductor to ground. As the waveforms in Figure 10.18 show, when the switch is closed, the inductor current charges up at a

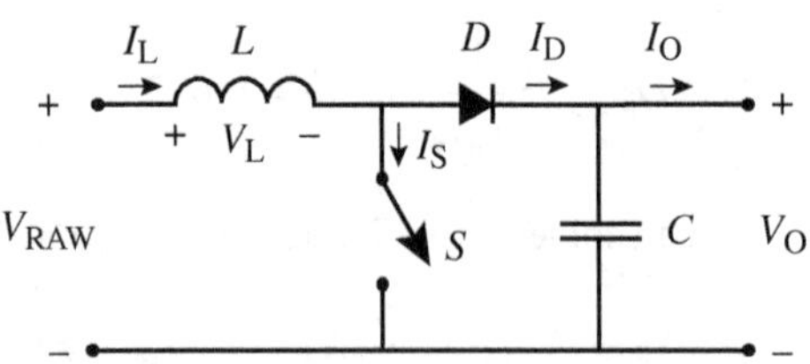

FIGURE 10.17 Basic boost converter circuit.

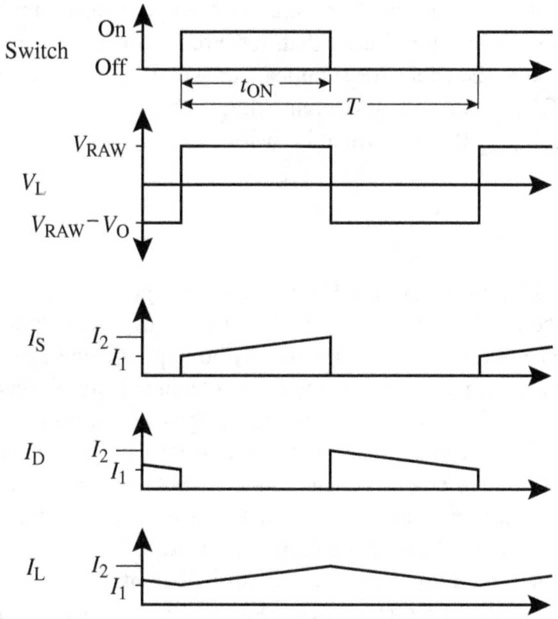

FIGURE 10.18 Voltage and current waveforms in the boost converter of Figure 10.17.

rate determined by the input voltage V_{RAW} and the inductance value L. During this time, diode D is reverse biased and prevents any current from flowing between the switch and the circuit's output. When the switch opens, the current through L cannot change instantaneously and instead raises the right-hand end of L to a potential that is high enough to forward-bias diode D and send a current pulse into the output circuit at the nominal output voltage V_O. As in the buck converter, the inductor's current varies in a triangle-wave fashion between a minimum value I_1 and a maximum value I_2. However, unlike the buck converter, the boost converter's output is connected to the inductor only when the switch is open. Therefore, filter capacitor C must supply all the output energy for the load during the switch-on time t_{ON}, because diode D is reverse biased in that interval and the inductor cannot supply current.

Using an energy-conservation argument similar to the one employed in the buck converter analysis, it is easy to show that the output voltage for the boost converter is

$$V_O = \frac{T}{t_{ON}} V_{RAW} = \frac{V_{RAW}}{1 - D_{BOOST}} \tag{10.40}$$

where D_{BOOST} is the duty cycle of the boost converter switching waveform. As with the buck converter, this simplified analysis applies best for duty-cycle values of 0.9 or less, because switch delays and other inefficiencies begin to degrade the circuit's performance if too great a step-up ratio is attempted in one circuit.

Because current is delivered to the output in interrupted pulses, the filter capacitor C must be larger for a boost converter than for a buck converter with similar output current. If we assume the capacitor supplies a constant output current I_O to the load during the time TD_{BOOST}, its voltage decrease during this time is also the peak-to-peak ripple voltage $V_{RIPPLE(p-p)}$ (boost), which is easily calculated as

$$V_{RIPPLE(p-p)}\,(\text{boost}) = \frac{I_O D_{BOOST}}{fC} \tag{10.41}$$

where f is the switching frequency in Hz. In a good design, the filter capacitor will be chosen to limit the ripple voltage to be 10% or less of the DC output voltage V_O. As with the buck converter, voltage regulation may be implemented by varying the duty cycle in accordance with an error voltage from a feedback circuit that compares the output voltage to a reference voltage. Some designs use *current feedback*, which monitors both the output voltage and the inductor current to achieve a more rapid response to input voltage changes than simple output voltage monitoring can provide. However, the design of current-feedback converter regulators is complex and involves subtleties that are beyond the scope of this text.

The boost converter may also operate in a discontinuous mode if the load current falls below a certain minimum, and a more complex analysis is called for in that case.

Buck–Boost Converter. Yet another arrangement of the same elements in the buck and boost converters—a switch, an inductor, a diode, and a capacitor—produces the **buck–boost converter**. The only differences between the buck–boost converter and the boost converter are that the output voltage is taken across the inductor instead of the switch and the ground reference is different. But the waveforms in both converters for the same switch duty cycle are substantially the same, and so the waveforms shown in Figure 10.18 apply equally to the boost converter of Figure 10.17 or to the buck–boost converter shown in Figure 10.19.

One advantage of the buck–boost converter is that its output polarity is opposite to the input voltage's polarity. That is, if the input voltage is positive with respect to ground, the buck–boost converter's output will be negative with respect to ground. This can be convenient when a dual-polarity power supply is needed (e.g., for powering op amps) but a power source of only one polarity is available.

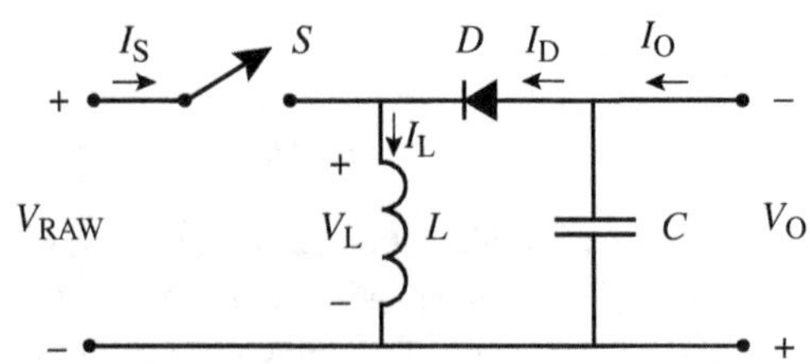

FIGURE 10.19 Basic buck–boost converter.

Applying energy conservation to the buck–boost converter yields the following expression for its output voltage:

$$V_O = V_{RAW} \frac{D_{BB}}{1 - D_{BB}}, \tag{10.42}$$

where D_{BB} is the duty cycle of the switching waveform used. Equation 10.42 shows that for a duty cycle of 50%, the input and output voltage magnitudes are equal (minus any diode voltage drop and losses, of course). Because most DC–DC converters involve a fairly large voltage ratio between input and output, the buck–boost converter may not be used as often as the more specialized circuits, but its advantage in reversing the source's polarity may be helpful in certain applications. Its ripple voltage is given by Equation 10.41, the same expression applicable to the boost converter, except for the substitution of D_{BB} for D_{BOOST}.

10.3.5.3 Dielectric Isolation and Isolation Transformers The rectifiers and DC–DC converters discussed so far do not provide dielectric isolation between the power source and the load. This is especially important when the power source is the AC utility line, one side of which is grounded to the earth. The essential problem that arises if such isolation is not provided is illustrated in Figure 10.20.

In general, any DC supply that operates from the AC utility will have some impedance Z_H from the "hot" side of the line to one side of the DC power-supply output and an impedance Z_N from the "neutral" side of the line to the other DC output terminal. (This does not include all possible impedances, but is adequate for this illustration.) For example, if a simple buck converter shown in Figure 10.15 is used to lower 120 to 12 VAC, one side of the power line will be in common with one side of the converter's output. Therefore, either Z_H or Z_N will be about zero. The hot and neutral wires can mistakenly be exchanged, and so the worst case occurs if Z_H is zero and a person comes in contact with the positive terminal of the power-supply output. The entire line voltage appears across the person, and fault current I_1 will flow through the circuit, possibly leading to injury or death. Hazardous voltages can exist even when Z_H is high and Z_N is zero, because the neutral wire is not necessarily at

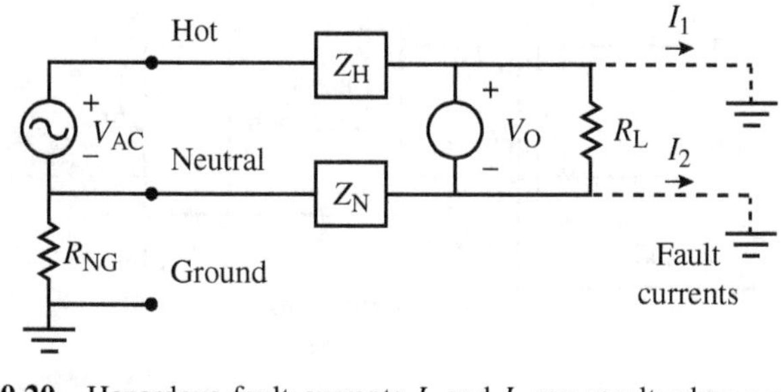

FIGURE 10.20 Hazardous fault currents I_1 and I_2 can result when a power supply is not dielectrically isolated from AC utility line.

ground potential, and a low-resistance connection between the negative power-supply terminal and ground can also allow substantial fault current I_2 to flow.

The basic lesson learned from this example is that the fault currents I_1 and I_2 are limited by the impedances Z_H and Z_N between the hot and neutral line conductors, respectively, and the power-supply circuit. In both rectifier-only AC–DC conversion circuits and DC–DC converters that use a single inductor, at least one of these impedances must be low: there must be either a direct connection, a capacitor that carries substantial line-frequency current, an inductor, or a rectifier diode present. For this reason, it is a hazardous practice to operate such power supplies directly from the power line unless they are insulated so that a person cannot possibly make contact between any part of the circuit and ground. This means surrounding the entire system with insulation that can withstand up to 600 V, which is not always easy.

The usual alternative to insulating the entire circuit is to provide an **isolation transformer** between the AC utility line and the power supply, or at least the output portion of the power supply. An isolation transformer allows the power supply to be separated from the AC utility supply by high impedances that are typically the capacitances of the transformer's secondary winding to an isolating ground shield. If the secondary is a center-tapped one, the center tap can be connected to the power-supply ground and the internal ground of the system, as shown in Figure 10.21.

The input current from the AC utility line flows through the primary winding of transformer T, which can be used for stepping up or stepping down the AC utility voltage as well as for isolation. In this transformer, the primary and secondary windings are physically separated by a conducting **interwinding shield**, which is typically connected to the third-prong wire (green) on the power cord that goes to a true earth ground. The shield reduces the mutual capacitance between the primary and secondary windings almost to zero, although there is some capacitance C_{W1} and C_{W2} between each winding and the shield.

The secondary winding's capacitance C_{W2} is evenly distributed between the two halves of the secondary in a properly designed transformer. This means that any voltage developed across the transformer is symmetrical with respect to the center tap, placing the voltage of the center tap at zero with respect to the shield. This is the critical feature of the isolation transformer: it enables one side of the power-supply

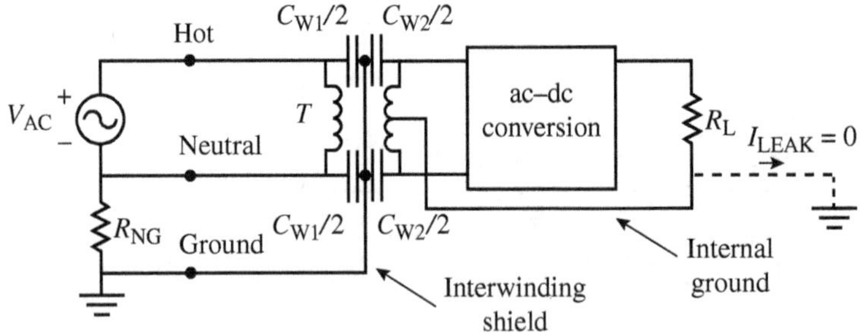

FIGURE 10.21 An isolation transformer T used to isolate a DC power supply and internal ground from the AC utility lines.

circuit driven by the transformer secondary winding to be connected to a point that is at zero volts with respect to true earth ground. If the center tap is made to be the load circuit's ground or common lead, this connection eliminates the fault current that would otherwise flow when the system's ground is connected to an external earth ground, and allows such connections to be made safely. The secondary interwinding capacitance C_{W2} varies but is usually less than 1 μF, meaning that even if the center tap connection opens up, the fault current is limited only to a mA or so at power-line frequencies, which is usually harmless except in critical applications such as certain medical devices.

An isolation transformer can also be incorporated into a DC–DC converter that operates from high-voltage DC derived directly from the AC power line, typically by means of a bridge or voltage-doubler rectifier circuit. As long as the portion of the power supply connected directly to the line is dielectrically isolated from the rest of the circuit, these systems can be made safe for consumer and household use. The isolation transformer can then operate at a high conversion frequency (50 kHz or more), and such high-frequency operation requires much smaller cores and windings than a 50- or 60-Hz transformer of the same power capacity does. The last DC–DC converter circuit we will describe uses this type of isolation transformer and is useful for applications that require such isolation.

10.3.5.4 Push–Pull Converter The **push–pull converter** is one of several types of converters that use a transformer, which provides for dielectric isolation between circuits that are connected to separate **windings** of the transformer. The type of transformer in the push–pull converter uses **center-tapped** primary and secondary windings. A center tap is simply a connection to the center of a given winding and allows the use of symmetrical circuits that have certain advantages in various applications. One advantage of the push–pull converter is that it draws current continuously from the raw DC power source, rather than in interrupted pulses. If this power source is derived from the AC power line (e.g., through a full-wave rectifier), the fact that the input current flows continuously means that there are fewer **harmonics** produced by the circuit. Harmonic currents drawn from the AC power line can cause problems with other systems powered by the same line and sometimes can even violate regulations pertaining to **radio-frequency interference** (**RFI**), which is one type of **electromagnetic interference** (**EMI**). On the other hand, the push–pull converter is not well adapted to making major adjustments of its output voltage by means of varying the switch waveforms, so it is more suitable for applications in which close regulation of output voltage by the converter is not needed. This regulation can be provided by a separate buck or boost converter operated from the DC output of the push–pull converter, if required.

A basic push–pull converter circuit is shown in Figure 10.22. Two switches S_1 and S_2 alternately connect to the minus supply lead either end of the primary of transformer T, whose center tap is connected to the positive terminal of the V_{RAW} DC power source. The dots on each winding of the transformer in Figure 10.22 indicate terminals that have the same voltage polarity. For example, when the dotted end of one winding is positive with respect to its nondotted end, the dotted ends of all the

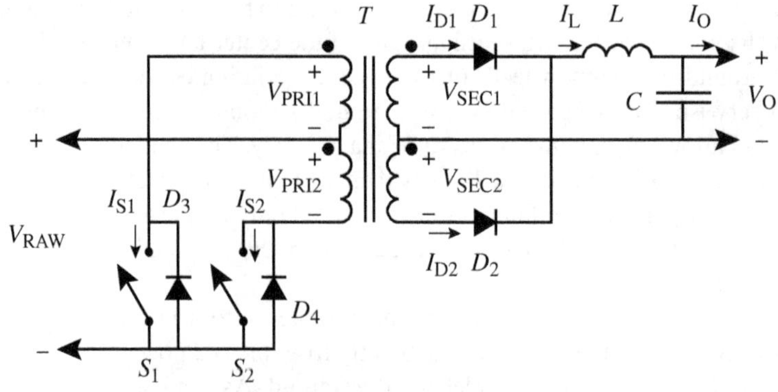

FIGURE 10.22 Basic push–pull DC-to-DC converter circuit.

other windings are positive with respect to their nondotted ends. A simple way of thinking about how the circuit works is that the switches present an approximate AC square wave to the primary, which is full-wave-rectified by the secondary.

As you probably know, the voltages across the primary and secondary of a transformer are proportional to the number of turns in each winding. The center-tapped transformer can step the primary voltage either up or down, depending on the ratio of secondary turns N_S to primary turns N_P:

$$\frac{V_{SEC1}}{V_{PRI1}} = \frac{V_{SEC2}}{V_{PRI2}} = \frac{N_S}{N_P} \tag{10.43}$$

By choosing a suitable turns ratio, almost any nominal voltage can be obtained from the secondary, although for very high or very low voltages, other types of circuits are more suitable than the push–pull circuit.

The waveforms in the push–pull circuit are shown in Figure 10.23. Because the on time t_{ON} is shorter than the off time t_{OFF}, there is a gap or *dead time* between the two on times of length t_{DT} during which *neither* switch is on. This dead time is a safety precaution to ensure that there is never a time when both switches are on simultaneously. If that happens, a condition called **shoot-through** occurs, because the transformer is then effectively shorted out and the full source voltage V_{RAW} appears across both closed switches. This undesirable condition will quickly lead to switch failure and is a common difficulty that arises in push–pull and other symmetrical power switching circuits. Shoot-through is usually averted by careful design of switching times to ensure that it never occurs for any combination of duty cycles and delay times. In the push–pull converter, the dead zone is also useful for other reasons, but its main purpose is to prevent shoot-through.

In operation, suppose switch S_1 closes and S_2 is open. The voltage across S_1 falls to zero, and in a seesaw fashion, the voltage across S_2 rises to $2V_{RAW}$, twice the raw input voltage. This voltage doubling must be taken into account when choosing the maximum voltage rating for the switching devices used. In the meantime, the current through the primary winding connected to S_1 rises as shown. Because the voltages

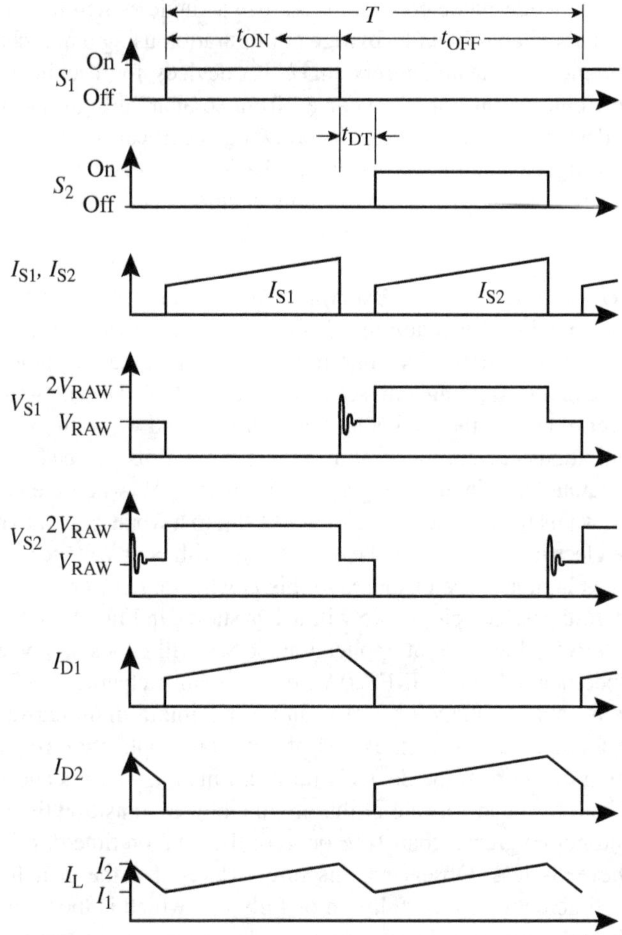

FIGURE 10.23 Switch, voltage, and current waveforms in the push–pull converter of Figure 10.22.

and currents in the secondary windings are scaled versions of the corresponding voltages and currents in the primary windings, the output current through diode D_1 also rises during this time.

When S_1 opens, the dead time t_{DT} begins. The secondary current I_{D1} that was formerly increasing when the switch was on now begins to fall and decreases until switch S_2 closes. When S_2 closes, the voltages across the primary and secondary change abruptly, and the result is that D_1 becomes reverse biased and D_2 becomes forward biased. The same process now occurs except that the opposite-phase switch and diode are now involved, and the cycle repeats. The filter capacitor needed for reducing ripple to a desired level can be calculated by means of an analysis similar to the one leading to Equation 10.41 for the buck converter's ripple.

Other types of switch-mode converters can use a single switch in a resonant circuit or two to four switches in an **H-bridge** configuration using a non-center-tapped transformer. Numerous manufacturers make IC devices for use in switch-mode power supplies, eliminating the need for a "from-scratch" design and making the power-supply designer's task a matter of following specification-sheet instructions. However, the designer should always have a basic understanding of how a given power-supply circuit works in order to apply it intelligently to a specific design problem.

10.3.5.5 Buck Converter Design Example We will conclude this section with a design example showing how a nominal 12-V automotive power source can be converted down to 3.6 V, suitable for charging a low-voltage battery-powered device such as a mobile phone drawing a maximum power of 2 W. We will assume that the minimum power drawn by the load at any time is 1 W, and we want the converter to stay in the continuous-current mode at all times. We will use a buck converter that will need to operate from input voltages ranging from 9 V (when the car battery is dying, one still wants the mobile phone to work!) up to a worst-case maximum of 30 V. Automotive electrical systems can be quite noisy, with poor voltage regulation and voltage spikes as high as 30 V or more, so this is why we have chosen such a wide range of input voltages. The circuit we will use is shown in Figure 10.24.

This is a relatively low-current application, so we will choose a power FET that has modest specifications. The IRF820A device is an n-channel FET rated for a maximum drain–source voltage of 200 V and a maximum drain current of 2.5 A. Both of these numbers are well in excess of what our application requires. Under typical operating conditions, the device's turn-on time $t_{S(ON)} = 20\,\text{ns}$ and the turn-off time $t_{S(OFF)} = 30\,\text{ns}$. As a general rule of thumb, the longest transition time (turn on or turn off) should not be greater than 10% or so of the total on time or off time of the switch, whichever is less. Observing this rule ensures that the switch spends the majority of its time either turned fully on or fully off, which is the main reason for using a switch-mode converter in the first place. For a given transition time, this rule will set an upper limit on the switching frequency. The IRF820A also has a built-in

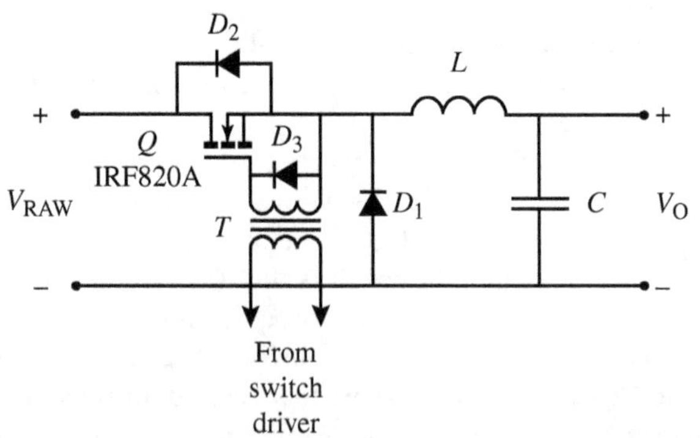

FIGURE 10.24 Circuit of buck converter design example to convert 9–30 VDC to 3.6 VDC.

protection diode D_2 from drain to source. This diode prevents any reverse voltages from appearing across the device and is actually an intrinsic feature of the way many n-channel power FETs are made.

This particular FET requires a V_{GS} of at least 9 V to completely turn on. Because the transistor's source terminal is "floating" in this circuit rather than tied to ground, we have chosen to couple the switch-drive pulse to the gate via a **pulse transformer** *T*. Typical pulse transformers have a turns ratio of 1:1 and are often used for isolating the primary and secondary circuits when the circuits cannot share the same ground connection. The transformer's secondary behaves as a floating two-terminal voltage source, so we can safely connect one terminal to the transistor's source terminal and know that the gate–source voltage will not be affected by fluctuations in the source terminal's voltage with respect to ground. Because the minimum value of V_{RAW} is 9 V, this voltage is available to form a 9-V rectangular pulse to be applied to *T*'s primary from the switch driver circuitry (not shown). In order to clamp the negative-going portion of the transformer's output waveform at −0.7 V or so, diode D_3 is connected across the pulse transformer. Details of gate driver circuits can vary, but the goal is to provide a clean, fast, positive-going rectangular pulse with a height of at least 9 V to turn the FET fully on during the on portion of the switch cycle.

The maximum load current I_{MAX} required by the load is easily calculated as

$$I_{MAX} = \frac{P_{MAX}}{V_O} = \frac{2\,\text{W}}{3.6\,\text{V}} = 555\,\text{mA} \tag{10.44}$$

The minimum current at a power of 1 W output would be half of that value, namely, $I_{MIN}=278$ mA. The range of duty cycles D_{BUCK} required for this load can be calculated from Equation 10.34, which gives

$$D_{BUCK} = \frac{V_O}{V_{RAW}} \tag{10.45}$$

The worst-case input voltage extremes are V_{RAW} (minimum)=9 V and V_{RAW} (maximum)=30 V, which give, respectively, D_{BUCK} (maximum)=0.4 and D_{BUCK} (minimum)=0.12. The minimum value of 0.12 is close to the recommended single-stage minimum of 0.1, but because such a high source voltage is expected to be only a momentary or transient condition, the circuit will not typically operate with such a low duty cycle for any length of time.

The case when $V_{RAW}=30$ V will produce the shortest on time, namely, 12% of the switch period *T*. If we set *T* so that the turn-off time of 30 ns is 10% of (12% of *T*), we can solve for *T*; thus,

$$t_{S(OFF)} = 0.1(0.12T) = 30\,\text{ns} \tag{10.46}$$

which gives

$$T = \frac{30\,\text{ns}}{(0.1)(0.12)} = 2.5\,\mu\text{s}, \tag{10.47}$$

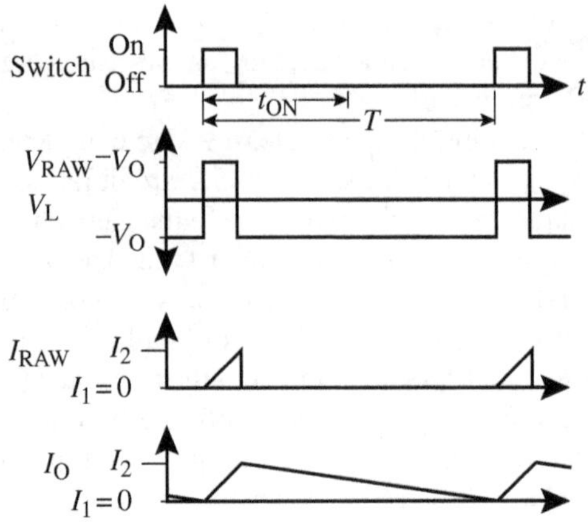

FIGURE 10.25 Waveforms of design example buck converter circuit of Figure 10.24 when $V_{RAW} = 30$ V and $D_{BUCK} = 0.12$, at limit of continuous-current mode.

and of course, a period of 2.5 μs amounts to a frequency $f = 1/T = 400$ kHz. So we will choose 400 kHz as the switching frequency. Obviously, a device with longer transition times would have required a lower frequency, so we are using the relatively fast transition times of this device to push the switching frequency into a range where the associated inductor can be quite small.

No ripple voltage specification was given, but the condition that the minimum output current is 278 mA sets a lower limit on the inductance value if the circuit is to operate in the continuous-current mode at all times. Recall that the expression for the peak-to-peak current ripple in the buck converter is Equation 10.35, which is

$$\Delta I_{BUCK} = t_{ON}\left(\frac{V_{RAW} - V_O}{L}\right) \tag{10.48}$$

If the minimum load current I_{MIN} is drawn and the ripple current ΔI_{BUCK} is at a maximum, we wish to choose inductance L so that the circuit is still in the continuous-current mode. This will happen if we choose L so that the negative-going peak of the current ripple waveform is just at zero for the worst-case ripple condition, which is when $V_{RAW} = 30$ V and $D_{BUCK} = 0.12$. This situation is shown in Figure 10.25, which shows that the 12% duty cycle produces a narrow interrupted-sawtooth current wave for input current I_{RAW} and an output ripple waveform that barely reaches 0 at its minimum current point.

We can use Equation 10.48 to solve for the inductance L required by first realizing that no matter what the duty cycle is, the average output current for the waveform in Figure 10.25 is $I_O = I_2/2$. Because I_2 is also the peak-to-peak ripple current ΔI_{BUCK}, we have

$$I_2 = 2I_O = \frac{t_{ON}(V_{RAW} - V_O)}{L} \tag{10.49}$$

If we substitute $t_{ON}=0.12T$ for the high-input-voltage worst case and solve Equation 10.49 for L, we find

$$L=\frac{0.12T(V_{RAW}-V_O)}{2I_O} \tag{10.50}$$

Based on transition-time considerations, we have already chosen $T=2.5$ μs, so for the maximum value of $V_{RAW}=30$ V and minimum output current $I_O=278$ mA, we find that the inductance L is

$$L=\frac{0.12(2.5\times10^{-6})\text{s}(30-3.6)\text{V}}{2(278\,\text{mA})}=14.2\,\mu\text{H}, \tag{10.51}$$

a very convenient low value for an inductor, especially one that carries substantial current. Inductors that carry significant amounts of DC current must have magnetic cores that are designed to keep from saturating even with DC flowing through the winding. These cores are typically larger than cores used in AC-only inductors, so lower inductance values will compensate somewhat for the size of the core needed. Higher DC current also means that the wire used in the winding cannot be below a certain minimum size if heat losses in the wire resistance are not to be excessive, so that is another reason for using as low an inductance as possible.

For purposes of battery charging, a relatively large ripple such as shown in Figure 10.25 is not usually a problem, so strictly speaking, no output filter capacitor C is necessary. But if the device is in use while charging is in progress, the large AC current into the battery might cause interference problems. Also, switching a large amount of current at a 400-kHz rate could cause problems with RF harmonics (multiples of the switch frequency), some of which lie within the AM broadcast-band range of about 550–1600 kHz. There are waveshaping and filtering techniques to avoid these problems, although details will have to wait for discussions in Chapter 12.

10.4 POWER AMPLIFIERS

Besides power supplies that provide various levels of DC power, the other major use of power electronics is in power amplifiers, broadly defined. A power amplifier is any system that delivers a changing level of electrical power above a watt or so, to a load in accordance with a lower-power input signal. As we will see, the input signal may be in either analog or digital form.

If the input is in analog form and the desired output is simply a magnified but otherwise faithful reproduction of the input signal, the system is called a **linear power amplifier**. The term "linear" here applies to the system's input-to-output or **transfer function**, not to the details of how the circuit or its devices operate. Not all linear power amplifiers use linear circuits. As we found in the case of power supplies, a circuit that uses power devices in their linear range of operating characteristics tends to be very inefficient, because the device must spend a significant amount of time in regions where it absorbs considerable power and converts it to waste heat.

So even most linear power amplifiers use power devices in various nonlinear ways: switching them off part of the time or even using them solely as on–off switches, as is the case for switch-mode power supplies. The most efficient linear power amplifiers use switch-mode power devices and ensure that the devices spend most of their time either fully on or fully off, yet manage to deliver a reasonably linear input-to-output voltage or current relationship under realistic conditions of use.

Nonlinear power amplifiers are driven by waveforms that do not have to be reproduced exactly. For example, motor controllers are driven by digital switching pulses that are often simply signals to turn on and off various motor windings, and the voltage waveforms across the windings that result may not resemble the switch waveforms except with regard to timing. Certain types of RF power amplifiers reproduce an input signal's frequency but compress or strip away any amplitude modulation present while preserving the signal's frequency and phase characteristics. Nevertheless, the required output power in the desired form is achieved by the proper utilization of power devices, and all these systems fall into the category of power amplifiers.

We will start this section with a discussion of the earliest type of power amplifier circuit to be developed, the **class A** amplifier. We continue with discussions of **class AB**, **class B**, and **class C** amplifiers, each of which can show higher efficiency than the previous class for a single-frequency sine-wave output waveform. Then we discuss design approaches for various types of **switch-mode power amplifiers**, which use techniques similar to those we have studied in switch-mode power supplies to deliver analog output waveforms with very high efficiency.

10.4.1 Class A Power Amplifier

A power amplifier can be categorized by its **class**, which is a letter ranging from A to at least D, with less well-known circuits making claims to classes F, G, and even H. We will be concerned with only the first four classes (A to D) in this text, because they are the most well-known and commonly used types. An amplifier's class describes the power device's mode of operation in the circuit, which includes the fraction of time the devices are operating in their linear range, as opposed to being turned fully off or fully on. The assumed input waveform to be reproduced is a sine wave in all cases.

10.4.1.1 Resistance-Coupled Class A Amplifier A class A amplifier's power device conducts current during the entire cycle (or 360°) of the input waveform. A low-power version of a class A amplifier is shown in Figure 10.26, which is a circuit that should be familiar to anyone who has taken an undergraduate electronics class. (Although a BJT is used for the active device in this circuit, a FET can be used as well.) The basic operation of this amplifier is very simple. In the absence of an input signal, resistors R_1 and R_2 form a **bias circuit** that fixes the transistor's DC emitter voltage V_E at approximately 1/3 of the DC power-supply voltage V_{CC}. Emitter bypass capacitor C_2 is large enough to maintain the emitter at essentially V_E even when an AC signal is applied. Collector load resistor R_3 is chosen so that

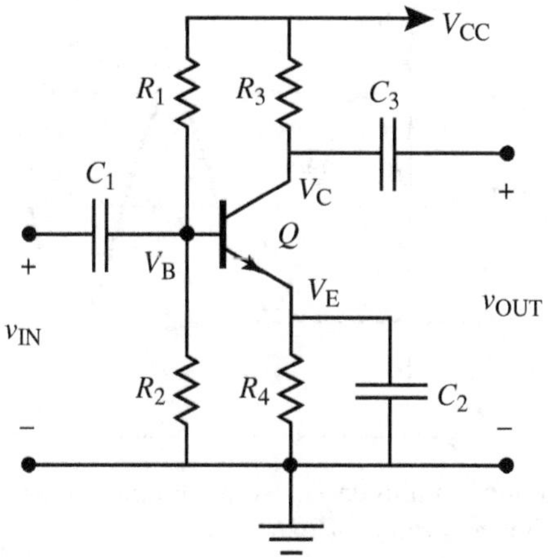

FIGURE 10.26 Low-power class A amplifier using BJT.

the DC collector voltage V_C is about 2/3 of V_{CC}. When an AC input voltage v_{IN} appears, it passes through input coupling capacitor C_1 and produces a variation in transistor Q's base current that depends on the transistor's AC base–emitter resistance. This variation in base current is multiplied by the device's AC β (=h_{fe}), which is typically on the order of 100, and appears as a much larger collector-current variation. This collector-current variation causes the instantaneous collector voltage to vary, and this AC output voltage is conveyed to the external load via the output coupling capacitor C_3.

Using the equivalent circuit for a BJT given in Figure 2.9b, we can easily calculate the small-signal voltage gain of this circuit (with no load attached) as

$$\frac{v_{OUT}}{v_{IN}} = A_V = h_{fe}\frac{R_3}{r_{be}} \tag{10.52}$$

Recalling that the AC base resistance $r_{be} = V_T/I_C$, for a collector current of a few mA, the base resistance with $V_T = 25$ mV is on the order of 5–50 Ω. Allowing a reasonable collector voltage in the range of 5–15 V means that the collector resistor's value is in the low kilohm range, and so theoretical voltage gains well in excess of 100 (>40 dB) can be obtained from this circuit.

However, if one attempts to draw any significant power from it, problems occur. We have assumed that the only load on the collector so far is the collector resistor R_3. Suppose we wish to draw 1 W of power from this circuit with a DC power-supply voltage V_{CC} of 10 V. The first question that arises is: what is the maximum undistorted voltage swing the circuit can deliver at the output?

With no load other than R_3, that question is easy to answer. Suppose that the circuit is biased so that $V_E = 3.3$ V and $V_C = 6.6$ V. The reason these "rule-of-thumb" voltages were chosen is that the resulting DC **operating point** (also called the **quiescent point** or **Q-point**) for the collector voltage is close to halfway between the DC emitter voltage V_E and the power-supply voltage V_{CC}. For this bias

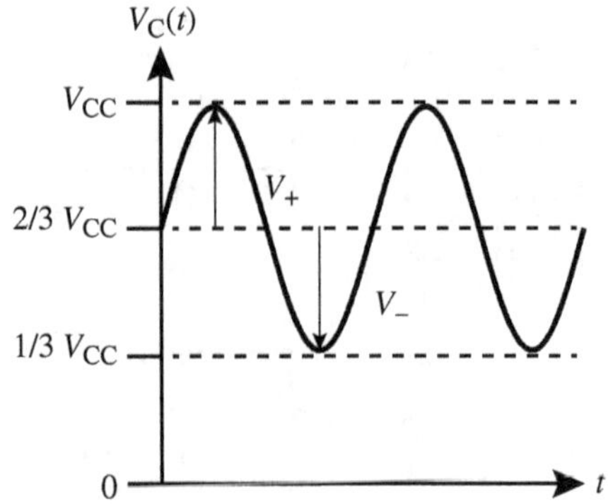

FIGURE 10.27 Maximum-amplitude sine-wave output from properly biased class A small-signal amplifier shown in Figure 10.26.

condition, the collector voltage can "swing" (as the colloquial phrase goes) both positively and negatively from its no-signal point with approximately equal amplitudes, as Figure 10.27 shows.

The limiting factor in the positive-peak amplitude V_+ happens when the transistor goes from its active region into **cutoff**, which means that the total base current (and therefore the total collector current) falls to zero. In the absence of an external load, cutoff makes the voltage drop across collector resistor R_3 zero, which raises the output voltage to V_{CC}. The factor that limits the negative-peak amplitude V_- is the **saturation** of the BJT. In saturation, the base current becomes so large that the collector–base junction goes from reverse bias (the normal condition in the active region) to forward bias, and the collector–emitter voltage falls to a low value of about 0.2 V or less. In either cutoff or saturation, the transistor fails to behave as a linear amplifier, and so these regions define the limits of a class A amplifier's output range if anything approaching linear operation is to be expected.

It stands to reason that unless the DC operating point of the collector is situated halfway between the positive- and negative-peak extremes, one peak of the sine wave will encounter its limit (become **clipped**) before the other peak does. If the goal is to produce the maximum-amplitude undistorted sine wave for the DC emitter voltage V_E chosen and V_{CC} value available, centering the DC collector voltage V_C between the two extreme limits is the best choice.

Even if that is done, the maximum power capability of the small-signal class A amplifier is very limited. The equivalent output circuit of a BJT is essentially a current source, so the output resistance of the amplifier is simply the value of the collector resistor R_3. Recalling our goal of obtaining 1 W of output power from this amplifier and assuming the highest undistorted peak sine-wave voltage we could ever get from it is 3.3 V, we can easily calculate the output current needed to deliver 1 W (rms). Converting 3.3 V peak into RMS volts, we find

$$V_{RMS} = \frac{V_{PK}}{\sqrt{2}} = \frac{3.3\,V_{PK}}{\sqrt{2}} = 2.33\,V_{RMS} \tag{10.53}$$

The RMS current required to deliver 1 W at 2.33 V is $I_{RMS} = P/V_{RMS} = 1/2.33 = 428.5$ mA. However, in order to ensure undistorted operation at the peaks of the waveform, we should calculate the *peak* output current required, which is

$$I_{PK} = \sqrt{2}(I_{RMS}) = \sqrt{2}(428.5\,\text{mA}) = 606\,\text{mA} \tag{10.54}$$

Clearly, a collector resistor value of a few kilohms will not allow anything close to this much output current to flow. To retain the class A nature of the amplifier while raising the linear output power requires reconfiguring the output circuit.

One source of inefficiency in the class A circuit of Figure 10.26 is the DC power dissipated in the collector resistor. The average DC power thus dissipated is constant and entirely wasted, because it contributes nothing to the useful AC output power. If we could reduce or eliminate this wasted DC power, the circuit's efficiency would be somewhat improved. One way to do this is to replace the output capacitor C_3 and collector resistor R_3 with an **output transformer** in a **transformer-coupled** output circuit.

10.4.1.2 Transformer-Coupled Class A Amplifier Transformer coupling of the class A amplifier's output is shown in Figure 10.28.

Instead of the collector load resistor, a transformer T is connected with its primary between the collector of transistor Q and the power-supply bus at V_{CC}. The transformer's turns ratio N_1:N_2 determines the AC impedance seen by the collector. We will designate this impedance appearing at the primary (with N_1 turns) R_{CL}, where *CL* stands for "collector load." Because of the way transformers work, the impedance R_{CL} is related to the secondary's load resistance R_L by Equation 10.55:

$$R_{CL} = \left(\frac{N_1}{N_2}\right)^2 R_L \tag{10.55}$$

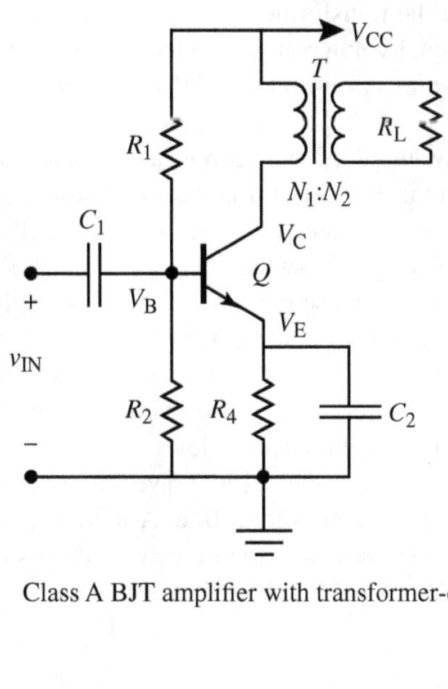

FIGURE 10.28 Class A BJT amplifier with transformer-coupled output circuit.

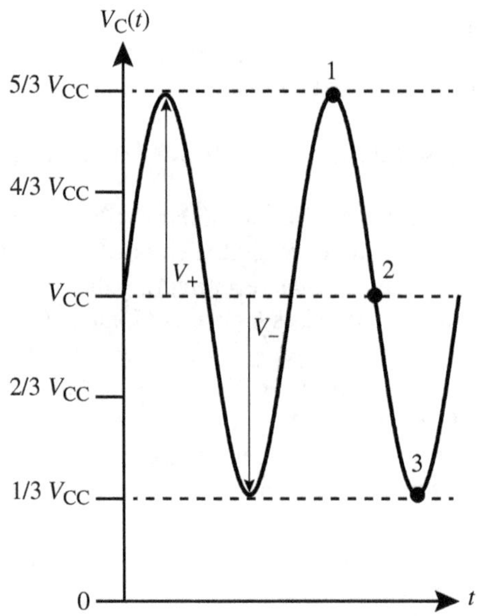

FIGURE 10.29 Maximum-amplitude sinusoidal waveform $V_C(t)$ appearing at collector of Q in transformer-coupled class A amplifier shown in Figure 10.28. Points 1–2–3 correspond to points in load-line graph in Figure 10.30.

If the load is a speaker voice coil, for example, R_L is typically a few ohms. The question of what value should be chosen for R_{CL} (and thus the turns ratio N_1:N_2) depends on the **large-signal** conditions at the collector, which are shown in Figure 10.29.

Still assuming that we have biased the emitter at a constant value $V_E = 1/3V_{CC}$, we find that if we assume the transformer is ideal (no AC or DC losses or internal resistance), then the no-signal voltage at Q's collector is simply V_{CC}, even when the transistor is drawing DC bias current at its collector. With a transformer-coupled output, the allowable one-way voltage swing at the collector is now $2/3V_{CC}$, because if the transistor draws enough additional current to become saturated, its collector voltage falls to $1/3V_{CC}$. And if the collector current decreases toward cutoff, the end of the transformer's primary connected to the collector will rise *above* V_{CC} to a voltage equal to the negative-going swing, namely, $V_{CC} + 2/3V_{CC} = 5/3V_{CC}$. It may seem paradoxical that a circuit whose highest DC power-supply voltage is V_{CC} can produce an AC voltage exceeding V_{CC}, but with an inductor or transformer in a circuit, this can happen. So one advantage of the transformer-coupled output circuit is already clear: it allows the peak-to-peak voltage swing at the collector to be twice the value that a resistor–capacitor output circuit allows.

To choose R_{CL} so as to maximize the output power for a given bias condition, we may resort to a bit of analysis called a **load line**. A load line is simply a graphical solution of Kirchhoff's voltage law over the range of currents and voltages present across a load impedance. Kirchhoff's voltage law says that the sum of the collector

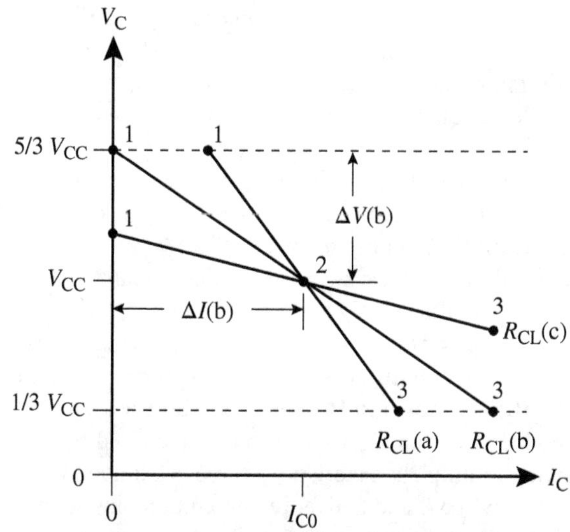

FIGURE 10.30 Load lines for class A amplifier circuit of Figure 10.28 showing points 1–2–3 corresponding to similarly numbered points on waveform of Figure 10.29, with various values of collector load resistance R_{CL}.

voltage and the voltage across the collector load equals the power-supply voltage. In the case of the transformer-coupled class A amplifier, the relevant load impedance is R_{CL}, the load seen by the collector looking into the transformer primary. In a load-line plot, typically, the device voltage is the vertical axis, and the device current is the horizontal axis. If the load is a linear resistance, the load line takes the form of a downward-sloping line whose slope is the inverse of the load resistance seen by the device.

Figure 10.30 shows how we can use the load-line concept to choose the value of collector load resistance R_{CL} that will maximize the output power delivered to the load for a given value I_{C0} of no-signal DC collector current. As we will show in the following text, it turns out that the condition under which a class A device dissipates the most power is when there is *no signal*. Consequently, the no-signal bias condition in which $V_C = V_{CC}$ and $I_C = I_{C0}$ is also the condition of maximum continuous power dissipation. The device Q will have to dissipate a power $P_{DISS}(\max) = V_{CC}I_{C0}$ safely in order to be used in this class A power amplifier.

Assuming that this can be done, the next problem is to choose R_{CL} so as to obtain the greatest AC power from the circuit. If no power is lost in the transformer, then the amount of AC power delivered will be proportional to the AC collector-voltage swing ΔV_C times the AC collector-current swing ΔI_C. These swings are limited by the requirement that they have equal amplitudes above and below the DC operating point, and the design-rule limit that $V_E = 1/3\,V_{CC}$, which sets the lower limit of the voltage swing.

Within these constraints, we can choose a variety of collector load resistances, but Figure 10.30 shows that there is one unique choice that will maximize the linear output power for the circuit. Suppose resistance R_{CL}(a) is chosen. This value will allow the full collector-voltage swing $\Delta V_C = 2/3V_{CC}$, but when the collector current is at a minimum and voltage at a maximum (point 1 on the waveform of Fig. 10.29), the collector current is above zero, and so the collector-current swing could be larger.

If resistance R_{CL}(c) is selected, the full collector-current swing $\Delta I_C = I_{C0}$ is available, but when the collector current goes to zero, the maximum collector-voltage swing is less than the maximum possible value of $2/3V_{CC}$.

Clearly, for both collector-voltage swing and collector-current swing values to reach their maximum values simultaneously, the load line must intersect two points: (1) the point where $I_C=0$ when $V_C=5/3V_{CC}$ and (2) the DC no-signal operating point where $V_C=V_{CC}$ and $I_C=I_{C0}$. This condition is met by the intermediate value of collector load resistance R_{CL}(b), as shown in Figure 10.30. So for a given value of V_{CC} and permissible DC collector current I_{C0} (which in turn are fixed by the device's power-dissipation and other limiting characteristics), there is a unique value R_{CL}(b) that will maximize the product $\Delta V_C \Delta I_C$ and deliver the maximum output power to the load. In practice, this value is selected for a given fixed load resistance R_L by a proper choice of the turns ratio $N_2{:}N_1$.

As we will now show, the class A amplifier is a relatively inefficient circuit from the viewpoint of maximum AC sine-wave power delivered for a given DC power input. If the emitter bias voltage is eliminated (which can be done with special bias circuits) and a simplified analysis is performed on an idealized class A resistance-coupled amplifier as shown in Figure 10.31, the maximum efficiency will occur when the active device barely reaches $V_{AD}=0$ across it on the negative voltage peaks and barely cuts off ($I_{AD}=0$) on the positive peaks.

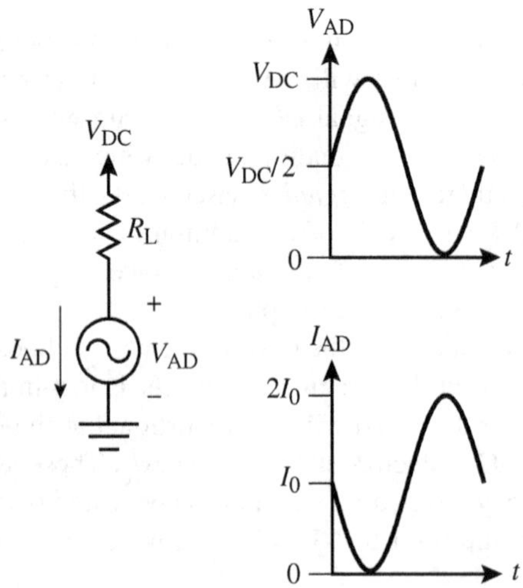

FIGURE 10.31 Circuit and waveforms for idealized resistance-coupled class A amplifier efficiency analysis.

In that case, the average DC input power is

$$P_{IN}(R\text{-coupled class-}A) = V_{DC} I_0, \tag{10.56}$$

where I_0 is the no-signal quiescent DC bias current through the active device. The useful AC output power that appears across the load R_L is

$$P_{OUT}(R\text{-coupled class-}A) = i_{RMS} v_{RMS}, \tag{10.57}$$

where i_{RMS} and v_{RMS} are the RMS values of AC current and voltage, respectively. From the waveforms shown in Figure 10.31, it is clear that

$$i_{RMS} = \frac{I_0}{\sqrt{2}} \tag{10.58}$$

and

$$v_{RMS} = \frac{1}{\sqrt{2}}\left(\frac{V_{DC}}{2}\right) \tag{10.59}$$

Equations 10.56–10.59 lead immediately to the maximum theoretical efficiency of a resistance-coupled class A amplifier:

$$\eta(R\text{-coupled class-}A) = \frac{\left(\frac{I_0}{\sqrt{2}}\right)\left(\frac{V_{DC}}{2\sqrt{2}}\right)}{I_0 V_{DC}} = 0.25 \tag{10.60}$$

or 25%. Recall that this is the *maximum* efficiency the circuit can show, which happens when the output is at its maximum amplitude just before clipping occurs. Many types of waveforms, including audio signals, do not spend much time at maximum amplitude, so the actual average efficiency in use will be much lower.

The transformer-coupled class A amplifier has somewhat better efficiency because no DC power is continually wasted in the resistance of the load, as it is in the resistance-coupled circuit. The idealized circuit and waveforms for an analysis of the transformer coupled circuit's efficiency are shown in Figure 10.32.

As we found earlier, the voltage at the junction between the AD and the transformer can swing twice as much for a given power-supply voltage V_{DC} as the resistance-coupled circuit's voltage can. This leads to an increased AC power output because the peak voltage is doubled and the result is that the maximum efficiency is doubled as well, resulting in a transformer-coupled class A efficiency of

$$\eta(T\text{-coupled class-}A) = \frac{\left(\frac{I_0}{\sqrt{2}}\right)\left(\frac{V_{DC}}{\sqrt{2}}\right)}{I_0 V_{DC}} = 0.5, \tag{10.61}$$

or 50%. Still, a theoretical maximum efficiency of only 50% is very costly in terms of waste-heat management, the cost of devices with power-dissipation ratings high enough to operate safely while dissipating large amounts of heat, and primary power

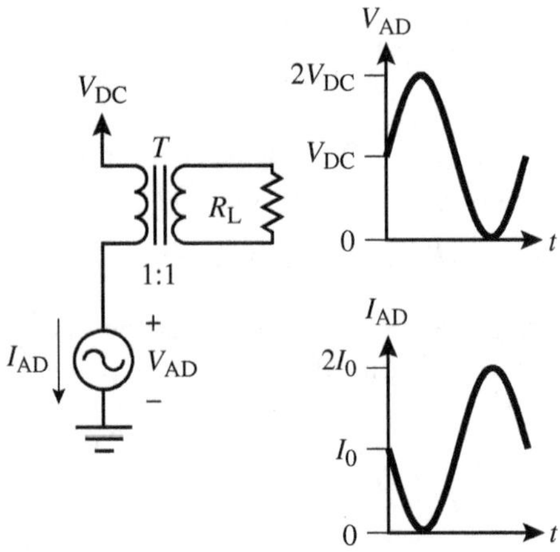

FIGURE 10.32 Circuit and waveforms for idealized transformer-coupled class A amplifier efficiency analysis.

usage. Only when matters of taste or other considerations outweigh these disadvantages should one consider a high-power class A circuit as a practical design. Even though the transformer-coupled version theoretically has better efficiency, the transformer used must be specially designed to deal with the large unbalanced DC current that flows through it, leading to early saturation unless special precautions are taken. For most applications, amplifiers with efficiency better than the class A configuration should be considered for medium- and high-power applications.

10.4.2 Class B Power Amplifier

A class of linear amplifier that is much more efficient than the class A circuits is called the **class B** amplifier. Class B amplifiers require two active devices arranged in a way so that one device amplifies the positive portion of the AC input waveform and the other device amplifies the negative portion. This is most easily achieved in solid-state circuits by using **complementary** devices, which are designed to have similar characteristics except for polarity. Two complementary BJTs, for example, have similar values for β, I_S, and so on, but one device is an npn type and the other is pnp. N-channel and p-channel FETs are also available as complementary pairs. Class B amplifiers can also use transformers, but this is not necessary.

A simple class B power amplifier is shown in Figure 10.33. This circuit is an **emitter-follower** amplifier that provides current gain but no voltage gain. (Voltage gain is easily obtainable in low-power amplifiers preceding the power output stage.)

Both the AC input signal and DC bias are provided by the signal source v_i, which must produce an output of at least ±0.7 V to turn on the output devices. Positive input voltages above 0.7 V will forward-bias npn transistor Q_N and produce an output current i_N that will follow the input voltage at a level of $V_{BE} = 0.7$ V below the input voltage, as shown in the waveforms of Figure 10.34. Similarly, the negative-going

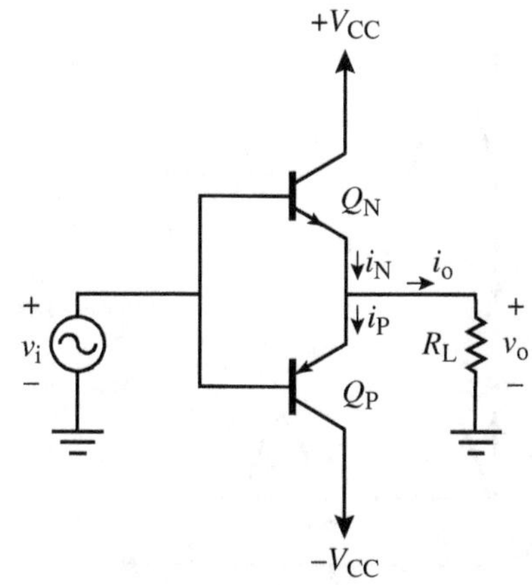

FIGURE 10.33 Simple class B emitter-follower power amplifier.

part of the input waveform below −0.7 V will produce a current i_P in the pnp transistor Q_P and will follow the input waveform at a level of 0.7 V higher than the input.

One advantage that the class B amplifier has over the class A amplifier is the fact that its DC power consumption in the absence of an input signal is *zero*. No current flows unless a signal is applied. This is in contrast to the class A amplifier, in which maximum power consumption and minimum efficiency occur with no signal input. As we will show in the following text, the maximum-signal efficiency of a class B amplifier is also better than the highest possible efficiency of a class A circuit.

The resulting output waveform v_O in the class B circuit of Figure 10.33 is a reasonably good reproduction of the input except near the point where the waveform changes sign. However, because of the 1.4-V-wide "dead zone" in which neither device is turned on, the circuit of Figure 10.33 will show a severe form of **crossover distortion**, as indicated in Figure 10.34. Because the devices turn on and off abruptly, crossover distortion will contain many harmonics of the input waveform. For some applications involving motor drives or other constant-amplitude waveforms, crossover distortion may be acceptable. But in applications such as audio amplifiers, in which low distortion at all signal levels up to the maximum output is desirable, crossover distortion of this magnitude is usually intolerable.

10.4.3 Class AB Power Amplifier

To remedy this problem, a bias network can be added to the class B circuit so that both devices conduct a small DC **quiescent current** I_Q in the absence of an input signal. If the output devices conduct for more than half the period of an AC sine wave, the circuit is no longer strictly class B, which is why the following circuit is known as a **class AB** amplifier. One simple form of bias network that will do this is illustrated in Figure 10.35. The R_1–D_1–D_2–R_2 bias circuit uses two diodes whose **scale current** I_S (defined later in the text) is similar to that of the complementary BJT pair Q_N and Q_P.

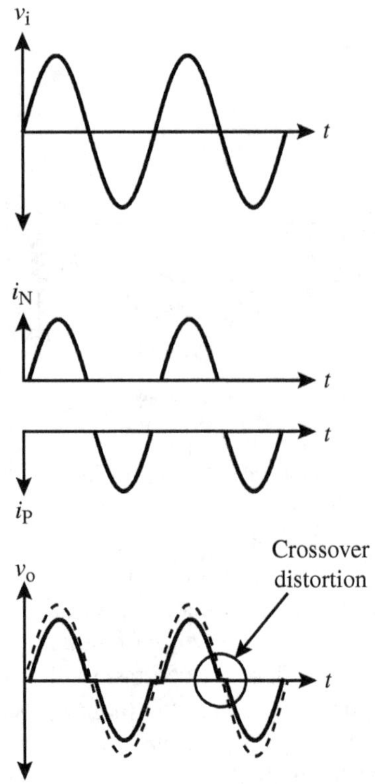

FIGURE 10.34 Voltage and current waveforms in class B amplifier of Figure 10.33, showing crossover distortion in output voltage waveform.

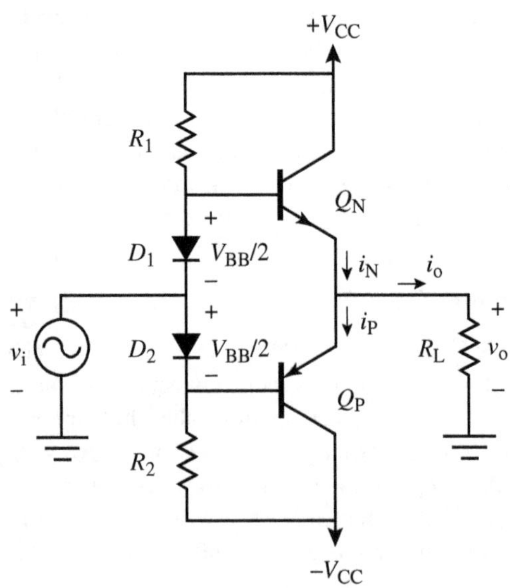

FIGURE 10.35 Class AB amplifier with bias circuit to reduce crossover distortion.

Assuming R_1 and R_2 are equal, the entire circuit is symmetrical with respect to ground, and so in the absence of an input signal ($v_i = 0$), a DC voltage $V_{BB}/2$ is present across each diode. By symmetry, $i_N = i_P$ and the output current $i_o = 0$ as well. However, a quiescent current I_Q flows in accordance with this equation that relates the base–emitter voltage of a BJT to its emitter current:

$$i_N = i_P = I_S \exp\left(\frac{V_{BB}}{2V_T}\right), \tag{10.62}$$

where $V_T = 25$ mV at room temperature (300 K). I_S is a fixed device parameter called the **scale current** and is a function of the structure and physical area of the base–emitter junctions.

When an input signal appears, the addition of input voltage v_i and output voltage v_o leads to the following expressions for the emitter currents:

$$i_N = I_S \exp\left(\frac{v_i + \frac{V_{BB}}{2} - v_o}{V_T}\right) \tag{10.63}$$

$$i_P = I_S \exp\left(\frac{-v_i + \frac{V_{BB}}{2} + v_o}{V_T}\right) \tag{10.64}$$

The output current i_o is given by the difference between the two emitter currents:

$$i_o = i_N - i_P = \frac{v_o}{R_L} \tag{10.65}$$

When Equations 10.63 and 10.64 are substituted into Equation 10.65, we obtain

$$\frac{v_o}{R_L} = I_S \exp\left(\frac{v_i + \frac{V_{BB}}{2} - v_o}{V_T}\right) - I_S \exp\left(\frac{-v_i + \frac{V_{BB}}{2} + v_o}{V_T}\right) \tag{10.66}$$

or

$$\frac{v_o}{R_L} = I_S e^{\frac{V_{BB}}{2V_T}} \left[\exp\left(\frac{v_i - v_o}{V_T}\right) - \exp\left(-\frac{v_i - v_o}{V_T}\right)\right] = 2I_Q \sinh\left(\frac{v_i - v_o}{V_T}\right) \tag{10.67}$$

where we have used Equation 10.62 to substitute in I_Q.

We would like to find an explicit relation between v_i and v_o so as to get an idea of how crossover distortion depends on the value of DC quiescent current I_Q. If I_Q is set too low, the class AB circuit approaches the bias condition of the pure class B circuit and will show severe crossover distortion. But if I_Q is set too high, the no-signal power dissipation of the circuit will rise and reduce its average efficiency.

Fortunately, we can solve Equation 10.67 for v_i explicitly in terms of output voltage v_o:

$$v_i = v_o + V_T \sinh^{-1}\left(\frac{v_o}{2R_L I_Q}\right) \tag{10.68}$$

The inverse hyperbolic sine function $\sinh^{-1}(x)$ begins at small values of x as a linear function and grows very slowly, approaching the rate of the function $\log_e(x)$ for large values of x. So as long as we make sure that the ratio of V_T to $2R_L I_Q$ is much less than 1, the inverse hyperbolic sine term will be negligible compared to v_o itself, and crossover distortion will be minimized.

This effect can be illustrated if we *normalize* all voltages to V_T and plot the dimensionless normalized output voltage

$$y = \frac{v_o}{V_T} \tag{10.69}$$

as a function of normalized input voltage

$$x = \frac{v_i}{V_T} \tag{10.70}$$

for various values of the dimensionless parameter

$$k = \frac{V_T}{2R_L I_Q} \tag{10.71}$$

The value of k depends on the designer's choice of quiescent current I_Q: a higher quiescent current means a lower k. The effect of quiescent current on the normalized input–output relationship (the amplifier's **transfer function**) is shown in Figure 10.36, which shows how, for small values of quiescent current ($k=10$), the output fails to follow the input until the input is several times larger than V_T, on the order of the turn-on voltage for a BJT. For $k=1$, the crossover distortion is reduced but is still visible, while for $k=0.1$, the transfer function appears to be almost perfectly linear.

To be a bit more quantitative, we can calculate the absolute magnitude of error e encountered due to crossover distortion as a function of the normalized input voltage x. If we define the fractional error

$$e \equiv \left|\frac{y-x}{x}\right| \tag{10.72}$$

and plot e versus x for $k=10$, 1, and 0.1 as shown in Figure 10.37, we find that the deviation from linearity never exceeds 10% for $k=0.1$.

Depending on the requirements for linearity, this amount of error may or may not be acceptable, but using $k=0.1$ appears to eliminate the majority of crossover distortion for all but the lowest-level signals. The effect of biasing a class AB amplifier with quiescent current I_Q is shown in Figure 10.38, which indicates how the crossover transitions are smoothed out as both devices operate near the crossover region.

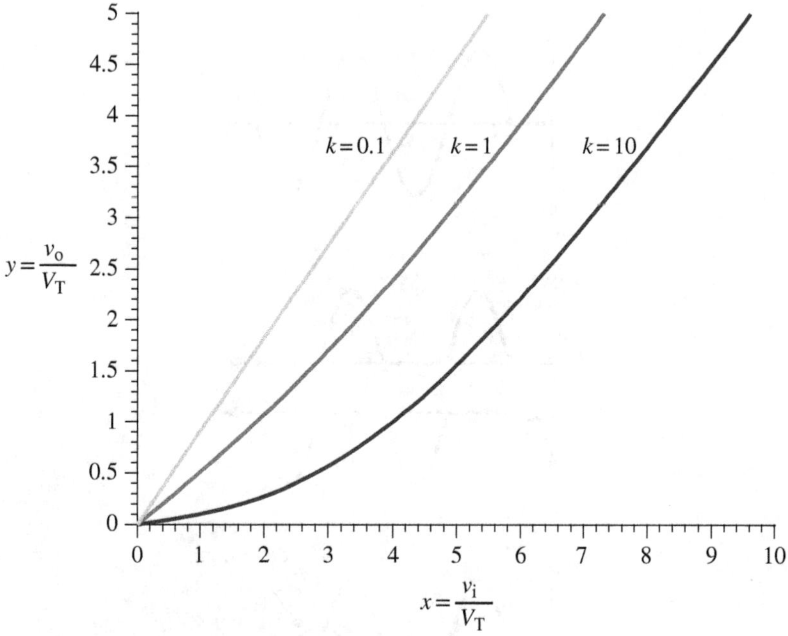

FIGURE 10.36 Normalized transfer function of class AB amplifier as value of k varies from 10 to 0.1.

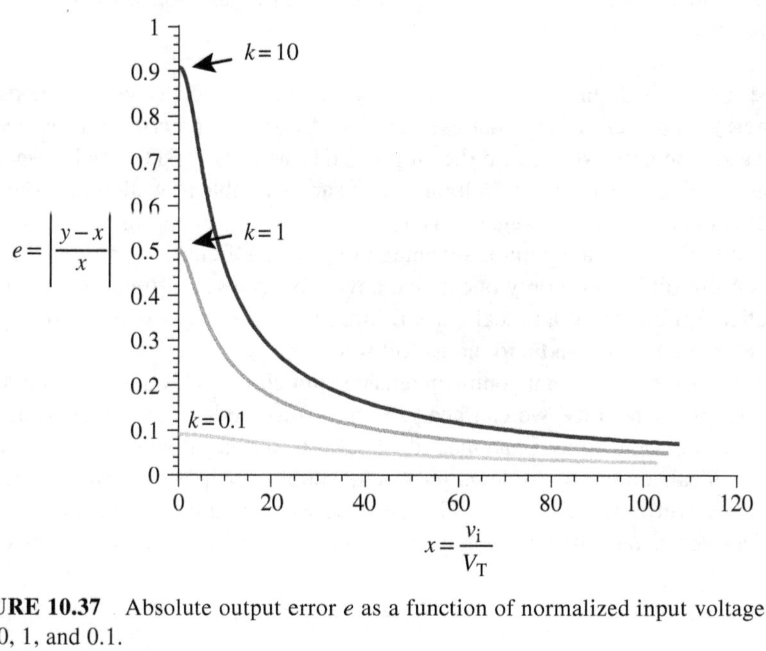

FIGURE 10.37 Absolute output error e as a function of normalized input voltage x for k = 10, 1, and 0.1.

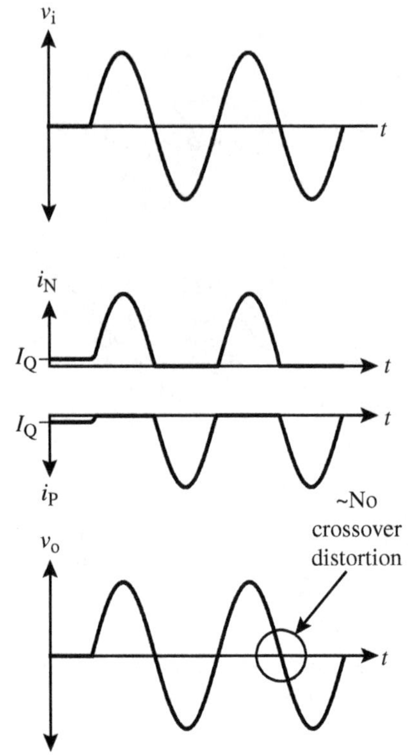

FIGURE 10.38 Waveforms of class AB amplifier in Figure 10.35, showing reduced crossover distortion with quiescent-current bias.

The circuit in Figure 10.35 is only one example of many **complementary-symmetry** amplifier circuits that use devices of opposite polarity to provide both positive and negative voltages at the output. FETs as well as BJTs can be used, and common-collector and common-drain circuits are available as well as the common-emitter connection shown. Regardless of the specific circuit configuration, all class B-type circuits share a common advantage that their efficiency can exceed that of a class A circuit because only one active device operates at a time. The maximum theoretical efficiency of an ideal class B circuit (with no crossover distortion and no quiescent current) can be found as follows.

Because each device in a complementary-symmetry class B amplifier works the same except for polarity, we can analyze the positive-side output circuit only and know that the same analysis applies identically to the negative side. Figure 10.39 shows an idealized positive-side output circuit supplied by power-supply voltage V_S.

As the waveforms in Figure 10.40 show, we assume that the output is the positive half of a sine wave whose peak voltage is v_{pk} and whose peak current is i_{pk}. Of course,

$$i_{pk} = \frac{v_{pk}}{R_L} \tag{10.73}$$

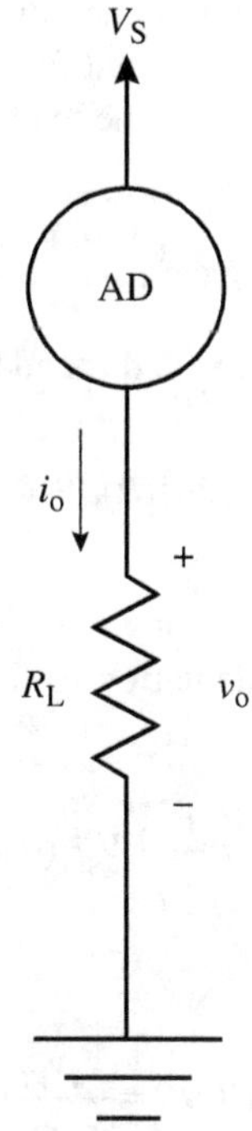

FIGURE 10.39 Idealized equivalent circuit of positive side of class B amplifier output circuit for efficiency calculation.

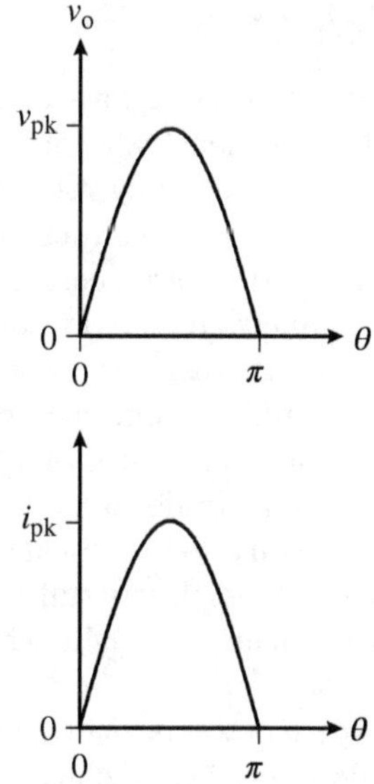

FIGURE 10.40 Waveforms for class B amplifier efficiency analysis.

Because time is arbitrary in this situation, the horizontal axis is in radians instead of seconds, which will simplify the dimensional form of the result.

The average output power delivered to the load over the sine-wave's half period is

$$\langle P_{\text{out}} \rangle = \frac{1}{\pi} \int_{\theta=0}^{\theta=\pi} v_{\text{o}}(\theta) i_{\text{o}}(\theta) d\theta \tag{10.74}$$

Substituting in the sine functions for v_{o} and i_{o} and using Equation 10.73 gives

$$\langle P_{\text{out}} \rangle = \frac{1}{\pi} \int_{\theta=0}^{\theta=\pi} \frac{v_{\text{pk}}^2}{R_{\text{L}}} \sin^2(\theta) d\theta = \frac{v_{\text{pk}}^2}{\pi R_{\text{L}}} \frac{\pi}{2} = \frac{v_{\text{pk}}^2}{2R_{\text{L}}} \tag{10.75}$$

While the current consumed by the circuit varies sinusoidally, the voltage V_{S} supplied does not, so we need a different calculation to find the average input power to the circuit:

$$\langle P_{\text{in}} \rangle = \frac{1}{\pi} \int_{\theta=0}^{\theta=\pi} V_{\text{S}} i_{\text{o}}(\theta) d\theta, \tag{10.76}$$

which leads to

$$\langle P_{\text{in}} \rangle = \frac{1}{\pi} \int_{\theta=0}^{\theta=\pi} V_{\text{S}} \frac{v_{\text{pk}}}{R_{\text{L}}} \sin(\theta) d\theta = \frac{V_{\text{S}} v_{\text{pk}}}{\pi R_{\text{L}}} [-\cos(\theta)]_0^{\pi} = \frac{2V_{\text{S}} v_{\text{pk}}}{\pi R_{\text{L}}} \tag{10.77}$$

Consequently, the efficiency η is a function of the peak output voltage v_{pk}:

$$\eta = \frac{\langle P_{\text{out}} \rangle}{\langle P_{\text{in}} \rangle} = \frac{(v_{\text{pk}}^2)/(2R_{\text{L}})}{(2V_{\text{S}} v_{\text{pk}})/(\pi R_{\text{L}})} = \frac{v_{\text{pk}}}{V_{\text{S}}} \frac{\pi}{4} \tag{10.78}$$

The maximum theoretical efficiency for a sine wave occurs when $v_{\text{pk}} = V_{\text{S}}$, which produces a maximum efficiency of π/4 or about 78.5%. The efficiency decreases linearly to zero as the peak voltage falls below the supply voltage, but the power consumption decreases as well, in contrast to the class A amplifier.

Where linearity is important and low-noise operation without spurious signals is desired, class AB or class B amplifiers are very useful for medium-power applications. The residual nonlinearity due to crossover distortion after a quiescent current is added, as discussed in Section 10.4, can be further reduced by applying feedback to the amplifier. While feedback can reduce distortion far below the level a circuit would show in **open-loop** operation (no feedback), if less distortion is present before feedback is applied, less distortion will also be present after applying a given amount of feedback. So class B and class AB amplifiers still find a wide variety of applications in consumer and industrial systems, especially where the ultimate in efficiency is not needed.

However, advances in high-speed power devices capable of switching at very fast rates have made it possible to develop types of power amplifier circuits that use the power devices exclusively as switches. When a device is used as a switch, the time that it spends in its linear region is minimized. Since the linear region is where both substantial voltage appear across it and substantial current flow through it, the less time spent in the linear region the better, as far as efficiency is concerned. So any device used as a switch will show higher average efficiency than when it is used in a purely linear circuit. In the next section, we will describe several types of switching power amplifiers that use power devices in this way.

10.4.4 Class D Power Amplifier

The basic idea of a class D power amplifier[1] is that the analog output signal level is controlled by varying the **duty cycle** of a rectangular wave whose **switching frequency** is much higher than the highest-frequency analog output signal to be reproduced. A passive filter circuit then averages the output voltage to eliminate the switching frequency and its harmonics, leaving only the desired analog output voltage.

10.4.4.1 External-Clock Class D Amplifier A simplified block diagram of one type of class D power amplifier using an external clock is shown in Figure 10.41, and the important waveforms in the circuit are illustrated in Figure 10.42.

This mixed-signal system is driven by a constant-frequency clock generator that produces a switching frequency f_S. The switching frequency is chosen so that it is at least twice as high as the highest-frequency component of the analog input signal v_i (in accordance with the Nyquist sampling criterion), but not so high that the power output devices cannot switch rapidly enough to follow it. The clock waveform is transformed into a symmetrical AC **triangle wave** v_t, which is applied to the inverting input of a comparator. The analog input signal v_i to be amplified appears at the comparator's noninverting input.

As the waveforms in Figure 10.42 show, the comparator's digital output v_{ci} is HI whenever the analog input signal v_i exceeds the triangle-wave input v_t. Otherwise, the

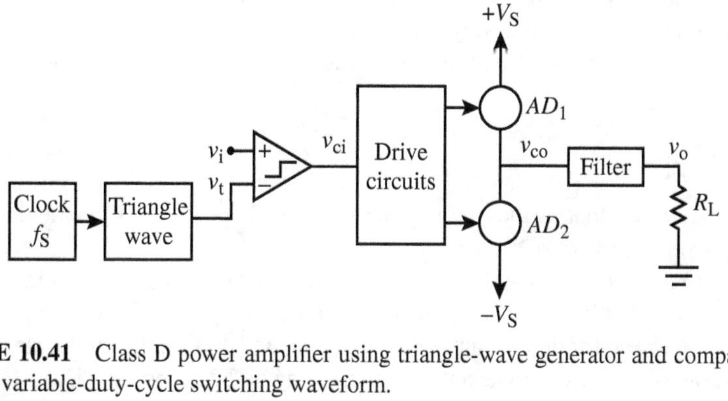

FIGURE 10.41 Class D power amplifier using triangle-wave generator and comparator to produce variable-duty-cycle switching waveform.

[1]As we went from class B to class D, you may be wondering what happened to class C. Class C power amplifiers are used primarily in radio-frequency (RF) circuits and will be covered in chapter 11.

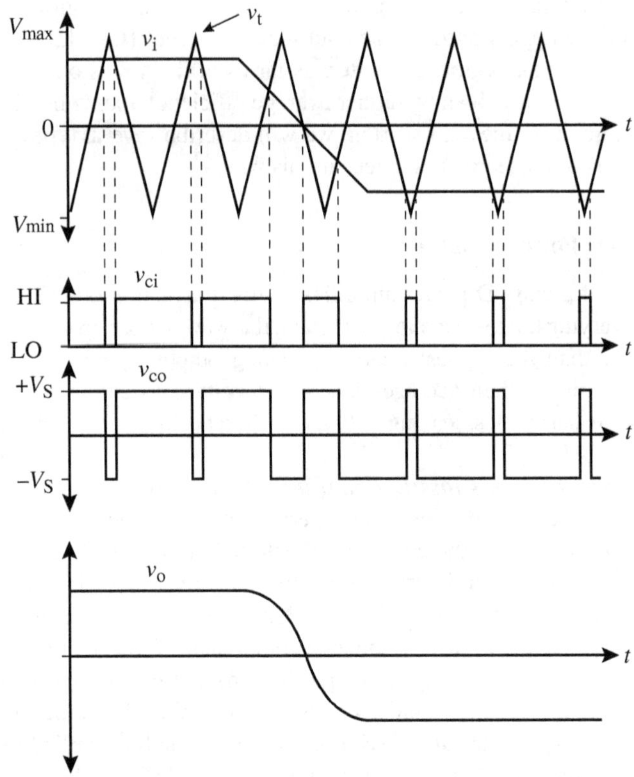

FIGURE 10.42 Waveforms showing operation of class D amplifier illustrated in Figure 10.41.

output is LO. The triangle wave is designed to maintain a constant upward slope from a minimum negative voltage V_{min} to a maximum positive voltage V_{max} during the first half of the clock cycle. Then it reverses direction and falls at a rate equal to its rise rate. Considering the input voltage v_i to be approximately constant during a single cycle of the triangle waveform, the **duty cycle** *d* of the comparator output v_{ci} is

$$d(v_i) = \frac{v_i - V_{min}}{V_{max} - V_{min}} \tag{10.79}$$

You can easily see that as the input signal v_i varies from V_{min} to V_{max}, the duty cycle will vary from 0 to 1 (0 to 100%).

The output of the comparator is a *digital* waveform, that is, its level is either a logical HI or LO, with fast transitions between the two states. This is the ideal type of waveform to use for driving active devices AD_1 and AD_2 in the output stage of the circuit as on–off switches. (The active devices may be BJTs, power FETs, IGBTs, or other power devices.)

The drive circuits amplify the low-level digital waveform v_{ci} coming from the comparator output to a level sufficient to drive the output devices either fully on or fully off. In addition, the drive circuits shift the switch-on and switch-off times of the output devices slightly in order to avoid the **shoot-through** problem, which was discussed earlier in the section on switching power supplies. Shoot-through occurs whenever both output devices AD_1 and AD_2 are on at the same time and allows a low-resistance short circuit from the positive supply $+V_S$ to the negative supply $-V_S$, which causes excessive power dissipation, lower efficiency, and ultimately device failure. So an important aspect of the driver circuit design is to arrange the timing of the two independent drive signals to the output devices so as to avoid shoot-through while minimizing output-waveform distortion.

The drive signals cause the output devices to turn on and off alternately so that most of the time, the unfiltered output voltage v_{co} is either at $+V_S$ or $-V_S$. A Fourier analysis of the resulting waveform will show that the original input signal v_i is present in amplified form. Along with the desired output signal, the switching frequency f_S, harmonics of f_S, and intermodulation products of f_S and the frequencies in the input waveform are also present. To eliminate (or substantially reduce) all frequency components except those originally present in v_i, a filter (usually a passive circuit) is connected between the raw output v_{co} of the amplifier and the load resistance R_L. This filter reduces the level of f_S and its harmonics to an acceptable degree and allows the output signal v_o to be delivered to the load, as the waveform v_o shows in Figure 10.42.

10.4.4.2 Self-Oscillating Class D Amplifier When a class D amplifier's output circuit switches states, large currents flow, and if the currents change in a very short time (as they must for high efficiency), a large number of harmonics of the switching frequency are generated. These harmonics can cause **EMI** to other systems or even to parts of the amplifier itself. This type of interference is often more noticeable if the originating frequency is **stable** (i.e., it does not change much over time), because if a particular harmonic happens to be present at the same frequency as a much-used radio channel, for example, it can render that channel useless.

Another type of class D amplifier called a **self-oscillating** amplifier minimizes the problems that can arise from a constant clock frequency by allowing the switching frequency to vary with the input signal. The self-oscillating amplifier also has a simpler block diagram, because no separate external clock is required. One simple type of self-oscillating class D amplifier is shown in Figure 10.43.

The drive circuits, output devices, and output filter are identical to those used in the external-clock amplifier of Figure 10.41. However, the comparator used in the self-oscillating circuit is a type with **hysteresis**, indicated by the squared-off "S" shape with a double vertical line. A circuit showing hysteresis has memory, in that its present state depends not only on its inputs at the present time, but what its inputs were in the past as well.

The specific type of hysteresis shown by the comparator in Figure 10.43 is illustrated in the graph in Figure 10.44. A normal comparator with no hysteresis changes its output state when the voltage difference $v_+ - v_-$ between its two inputs changes sign, and it doesn't matter which direction the voltage difference is going (from plus to minus or from minus to plus). This is expressed by saying the **threshold voltage** is zero. But the comparator with hysteresis in Figure 10.43

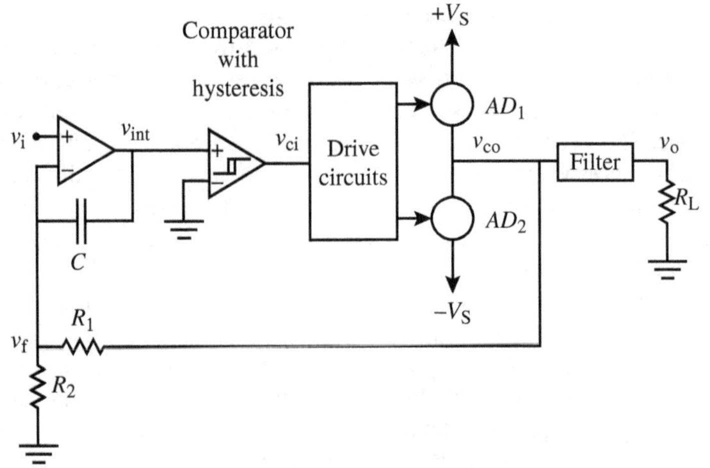

FIGURE 10.43 Simplified self-oscillating class D amplifier schematic.

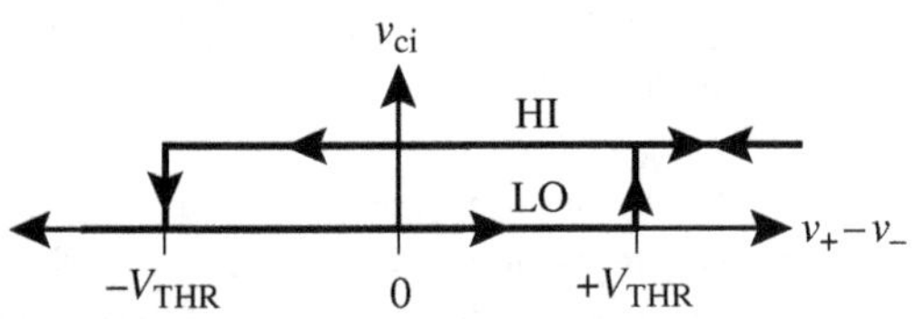

FIGURE 10.44 Input–output characteristic of comparator with hysteresis having positive-going threshold at $+V_{THR}$ and negative-going threshold at $-V_{THR}$.

has *two* threshold voltages, and which one applies depends on whether the comparator's output is initially HI or LO.

If the comparator's output is initially LO, the threshold $+V_{THR}$ applies. This means that in order for the comparator's output to switch from LO to HI, the input voltage difference $v_+ - v_-$ must exceed $+V_{THR}$. When this happens, the comparator output goes HI and stays there until the input voltage difference goes negative to the second threshold value of $-V_{THR}$. Then it switches back from HI to LO. Another name for a circuit with this general type of hysteretic behavior is a **Schmitt trigger**, which was discussed in Chapter 5.

Suppose initially that the input signal voltage v_i to the self-oscillating class D amplifier of Figure 10.43 is zero. A scaled-down version v_f of the raw output waveform from the output stage is produced by the voltage divider consisting of resistors R_1 and R_2, whose parallel equivalent resistance is $R_p = R_1 \| R_2$. This voltage is integrated by the input-stage op amp in conjunction with capacitor C. The noninverting input of the op amp is connected to the input signal v_i, so the rate of change of the integrator's output voltage v_{int} is

$$\frac{dv_{int}}{dt} = \frac{v_i - v_f}{R_p C} \tag{10.80}$$

The slope is therefore dependent on the difference between the input voltage and the feedback voltage v_f, whose magnitude is always

$$|v_f| = V_S \frac{R_2}{R_1 + R_2} \tag{10.81}$$

but whose sign is either positive or negative, depending on the state of the comparator output.

The circuit will oscillate at a frequency determined by how long it takes the integrator output v_{int} to go from the negative threshold $-V_{THR}$ to the positive threshold $+V_{THR}$, and vice versa. With a proper choice of R_p and C for a given power-supply voltage V_S, the circuit can be designed so that its self-oscillating frequency in the absence of an input signal is high enough to be filtered out by the output filter and is well above the highest signal frequency to be amplified. This situation with no input signal is shown by segment *a* of the circuit's waveforms shown in Figure 10.45. If the circuit is designed with the proper symmetry, the comparator's output signal v_{ci} has a duty cycle of exactly 50%, and the net output voltage from the circuit is zero, as it should be.

When the input voltage v_i rises above zero, the rising slope from the integrator increases and the falling slope decreases, as shown in region *b* of Figure 10.45. The result is that the duty cycle of the comparator output increases above 50%, resulting in an average positive voltage at the output. When the input voltage falls below zero, the opposite happens: the integrator's rising slope is lower than its falling slope, the comparator's duty cycle falls below 50%, and the average output voltage is less than 0. As long as the circuit values are proportioned properly and the input voltage is not so great that the integrator's slope goes to zero during part of a cycle, the output voltage will be a reasonably good reproduction of the input voltage.

A more careful analysis of these class D circuits will reveal certain shortcomings that produce various types of distortion in the output waveform. A commonly used criterion for the distortion performance of amplifiers, especially in the audio range, is the **total harmonic distortion** (**THD**) for a given set of operating conditions, a number usually expressed as a percentage. This is usually measured with a pure sine-wave input signal at a level to produce the rated output power from the amplifier. Under those conditions, assume V_1 is the RMS voltage of the fundamental (input) frequency at the output. If $V_2, V_3, \ldots, V_n$ are the harmonics of V_1 at a measurable level (above the noise floor of the instrumentation), then the THD in percent is defined as

$$\text{THD}(\%) = \frac{\sqrt{V_2^2 + V_3^2 + \cdots + V_n^2}}{V_1} \times 100\% \tag{10.82}$$

Achieving THD of less than 1% in a class D amplifier requires modifications to the basic circuits shown earlier which are beyond the scope of this text, including feedback circuitry and other linearizing features. But the essential idea of using the output devices as switches remains the same regardless of the means used to lower distortion.

Besides the class D circuit, there are various other classes with higher letters (E, F, G, etc.) that have been developed for special purposes. But the basic class D

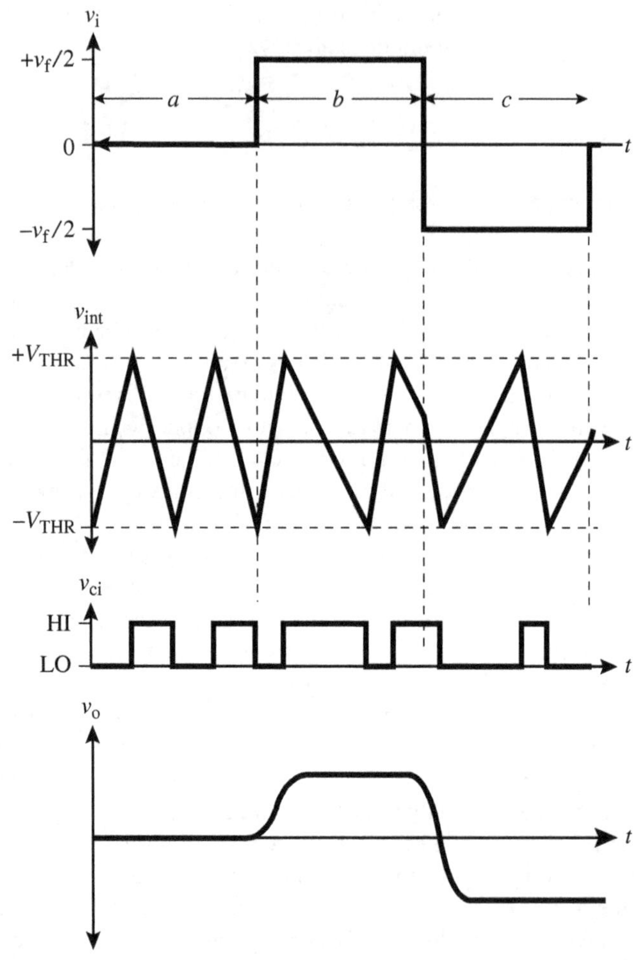

FIGURE 10.45 Assumed input signal voltage waveform v_i, resulting integrator waveform v_{int}, comparator output waveform v_{ci}, and output signal waveform v_o for self-oscillating class D amplifier shown in Figure 10.43.

approach is quite often used with a variety of digital waveforms that can be made to have an average value that is proportional to the desired analog output signal, which can include **delta-sigma modulation** waveforms, the output of **one-bit digital-to-analog converters**, and others.

10.5 DEVICES FOR POWER ELECTRONICS: SPEED AND SWITCHING EFFICIENCY

While the power devices we will describe in this section were treated briefly in Chapter 2, in this section, we will discuss their characteristics with particular regard to their use in power electronics.

10.5.1 BJTs

The **BJT** was the earliest type of solid-state power amplifier device to become available, and it still appears in a large share of applications that require inexpensive medium-power circuits in which efficiency is moderately important. Because BJTs require a base current drive that is a small percentage of the collector current, the drive power requirements for BJTs can be substantial, especially for very large output currents. Also, charging and discharging the base–emitter junction rapidly require high peak currents, with the result that a BJT with a given power output capability (maximum current times maximum open-circuit voltage) cannot switch as rapidly as other solid-state devices can. Specially designed BJTs can withstand collector voltages of up to 1.5 kV, but this voltage maximum is also exceeded by other types of devices described in the following text.

10.5.2 Power FETs

A **power FET** is a field-effect transistor specially designed to carry large currents and to withstand a large drain–source voltage in the off state. Most power FETs are **metal–oxide–silicon (MOS)** devices, meaning that the gate is electrically insulated from the rest of the device. This makes the gate primarily a capacitive load to the driver circuit, which simplifies the power and current requirements for the driver. Nevertheless, high-power FETs can have gate capacitances of 1000 pF or more. For the power device to turn on and off at a speed high enough to preserve good efficiency in a switch-mode power supply or amplifier, the driver circuit must provide peak currents to the gate that charge and discharge it rapidly.

Power FETs are available with maximum currents per device in the tens of amperes and maximum drain–source voltages in the multi-kV range. Because of their relatively simple mode of operation, they tend to be the fastest type of solid-state switch for a given power range, but they also often cost more. In recent years, power FETs using materials other than silicon have been introduced. Silicon carbide (SiC) devices show lower $\boldsymbol{R_{DS}}$**(on) resistance** (the drain–source resistance when the gate turns the device fully on) than silicon power FETs of comparable power capability and also have improved temperature characteristics relative to silicon devices.

10.5.3 IGBTs

The **IGBT** combines some of the characteristics of the power FET with some characteristics of the BJT. Structurally, it resembles a FET-input circuit that drives an internal BJT-like structure. Like the power FET, its gate is insulated and therefore draws no DC current. However, its gate or control electrode does have capacitance, which must be taken into consideration when designing the driver circuit. IGBTs have the advantage over power FETs that they behave in the on state more like a forward-biased BJT, showing a more or less constant forward voltage drop of a fraction of a volt that is almost independent of current output. By contrast, the equivalent-circuit model of a turned-on FET is the on-resistance R_{DS}(on), and so the voltage drop across a conducting power FET is proportional to its current output. On the other hand, IGBTs have additional junctions that must be charged and discharged for each turn-on cycle, so they tend to switch more slowly than power FETs. Of all the common solid-state power devices available, IGBTs can handle the highest power levels. The so-called bricks of individual IGBT dice bonded together in a common package can switch currents of up to 1200 A at 3.3 kV. Designers using high-power IGBTs should be aware that these devices are designed exclusively for switch-mode operation and cannot be operated in a linear mode without destroying the device.

10.5.4 Thyristors

The active devices used in the switch-mode power systems discussed earlier were completely controllable by means of a third terminal (the base or gate terminal), in that the device could be turned either on or off by means of the input terminal. A class of solid-state devices called **thyristors** behave differently. One form of thyristor is the **silicon-controlled rectifier (SCR)**, shown in Figure 2.11. Its symbol resembles that of an ordinary two-terminal rectifier diode except for the addition of a third terminal called the **gate electrode**. This terminal does not behave like the gate of a FET, however. With no current to the gate, the SCR does not conduct in either direction. However, if the anode is made positive with respect to the cathode and a small positive current is supplied to the gate, the SCR turns on and remains in a conducting state as long as there is forward current flowing through it, regardless of gate current. The only way to turn off an SCR is to reduce the current between its anode and cathode to zero.

This trigger-like behavior makes it difficult to use SCRs in systems with DC power supplies, but in systems with AC supplies, the SCR automatically turns off when the power-supply voltage reverses. Consequently, SCRs and a related device called a **triac** are used in certain types of AC power supplies and power-conditioning systems, in which low-power gate driver circuits can control high-power loads connected to the thyristors. Thyristors are relatively slow devices used mainly in power-line-frequency applications, but they are available in power ratings up to the multikilowatt range.

10.5.5 Vacuum Tubes

At this point in the history of electronics, vacuum tubes are employed in only a few special circumstances for which solid-state devices are still not available. All but the highest-power amplifiers for RF and microwave frequencies can use semiconductor power devices, so vacuum tubes are found only in multikilowatt transmitters, industrial heating applications, and situations where cultural factors are more important than strictly technical ones. For reasons having as much to do with taste and esthetics as physics, audio engineers, musicians, and some others for whom electronic sound reproduction is important continue to design, make, and buy certain types of audio gear using vacuum tubes. The higher operating voltages of vacuum tubes compared to typical small-signal semiconductors mean that vacuum tubes do have a slight advantage with regard to linearity when low-level signals are involved, although this advantage comes at the price of the inefficiency, fragility, and bulkiness of vacuum tubes compared to solid-state devices. Audio applications of vacuum tubes are primarily in high-end studio equipment such as microphones and preamplifiers, as well as power amplifiers for guitar amps and similar musical applications. This market is an example of "technology becoming culture," in that a technology may long outlast its useful life as a rational alternative if it is adopted as a cultural symbol. Another good example of this phenomenon is the use of candles for ceremonial purposes such as birthdays and religious occasions.

BIBLIOGRAPHY

Mohan, N. *Power Electronics: A First Course*. Hoboken, NJ: John Wiley & Sons, Inc., 2012.

Ramshaw, R. S. *Power Electronics: Semiconductor Switches*, 2nd Edition. London: Chapman & Hall, 1993.

Self, D. *Audio Power Amplifier Design Handbook*. New York, NY: Focal Press, 2013.

PROBLEMS

Note: Problems of above-average difficulty are marked with an asterisk (*).

10.1. *Power-supply efficiency.* A certain power supply operates from an AC line and delivers 48 VDC (essentially constant) at a current ranging up to 25 A. As the 48-VDC load current increased from 0 to 25 A, data was taken on the power factor and current drawn from the constant-voltage 120 VAC (rms) source. This data is presented in Table 10.1. For each line of the table, calculate

(a) net $P_{IN} = V_{IN}(\text{rms})I_{IN}(\text{rms})(\text{power factor})$,

(b) $P_{OUT} = V_{OUT}I_{OUT}$,

(c) percent efficiency $\eta = (P_{OUT}/P_{IN}) \times 100\%$, and

(d) power dissipated in the supply $P_{DISS} = P_{IN} - P_{OUT}$. For what value of output current is the efficiency highest? For what value of output current is the dissipated power highest?

TABLE 10.1 Power-supply Efficiency Test Results

AC current (rms)	Power factor (%)	Load current (A)
0.1	75	0
0.5	80	0.6
5	85	8.5
10	90	19.1
13.9	90	25

TABLE 10.2 Power-Supply Regulation Test Results

AC voltage in (rms)	DC load current (A)	DC output voltage (V)
120	0	6
120	5	5.7
130	0	6.05
110	0	5.95

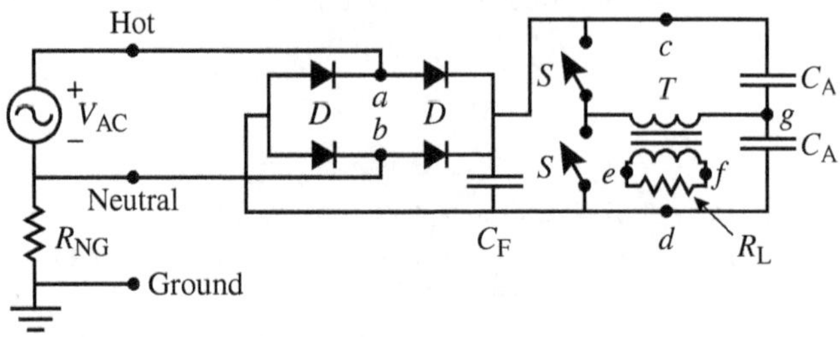

FIGURE 10.46 AC-line-operated high-frequency switching power supply, with nodes labeled for safety evaluation.

10.2. *Line and load regulation of power supply.* Suppose a certain power supply operates from an AC power line with a nominal voltage of 120 V (rms) and produces a nominal DC output voltage of 6 V when no load current is being drawn. The supply is placed in a test circuit in which both the AC power-line voltage and the load current can be varied. The test conditions and results are shown in Table 10.2. Based on the data in Table 10.2, calculate

(a) the line regulation D_{LINE} as a percentage,
(b) the load regulation D_{LOAD}, expressed as a percentage, and
(c) the load regulation in terms of the supply's internal resistance R_{INT}.

10.3. *AC-line safety.* The circuit in Figure 10.46, abstracted from a commercial design, shows an AC-line-operated power supply designed to produce high-frequency AC at the secondary of transformer *T*. DC power is obtained directly from the AC line by means of a full-wave rectifier using diodes *D* and filter capacitor C_F. Electronic switching devices *S* produce an AC square wave applied across the primary of transformer *T*, whose "cold" end is connected to a pair of capacitors C_A in a half-H-bridge configuration. For each node labeled *a* through *g*, estimate whether a person standing on a good connection to an earth ground (such as a wet surface in bare feet) could safely contact that node without experiencing a hazardous AC or DC voltage. If the answer is "no" (meaning it is *not* safe to touch), state why that node would be hazardous to touch. (*Hint*: Only two nodes are probably safe to contact.)

10.4. *Full-wave rectifier.* Design a full-wave bridge rectifier (see Fig. 10.5) to operate from an AC power source of 240 V(rms) ±10% with a frequency $f = 60$ Hz to

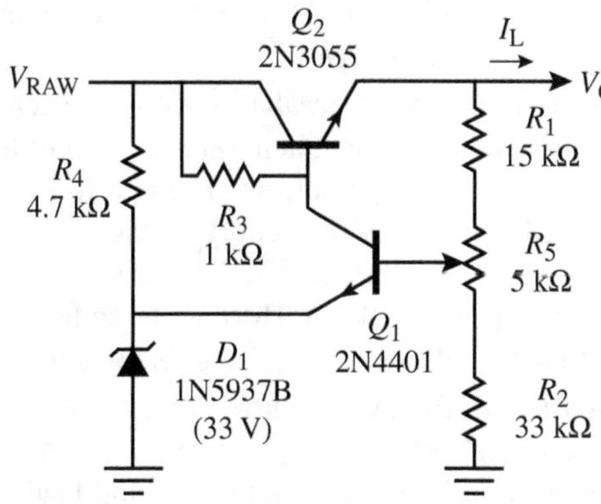

FIGURE 10.47 Discrete-component linear voltage regulator redesigned for $V_O = 48$ V.

supply a maximum DC current $I_{MAX} = 2.0$ A with a maximum peak-to-peak ripple voltage of 1 V.

(a) the minimum peak reverse voltage (PRV) that each rectifier diode must withstand (stated as a multiple of 100 V),

(b) the minimum filter capacitance required to meet the ripple specification (in multiples of 10,000 μF), and

(c) the minimum voltage rating of the capacitor under worst-case (no DC load, maximum AC voltage) conditions (as a multiple of 100 V).

10.5. *Voltage doubler.* Design a voltage-doubler power supply (see Fig. 10.5) to operate from an AC power source of 120 V(rms) ±10%. Calculate

(a) the minimum peak reverse voltage (PRV) required for each diode (in multiples of 100 V),

(b) the maximum voltage (DC plus peak AC) that appears across the capacitor connected directly to the AC source, and

(c) the maximum voltage (DC plus peak AC) that appears across the capacitor connected in parallel with the load. Can a polarized (electrolytic) capacitor be used in both locations? Why or why not?

10.6. *Discrete voltage regulator.* In Figure 10.47, the discrete-component linear voltage regulator circuit in Figure 10.9 has been redesigned to regulate a primary supply voltage of V_{RAW} ranging from 51 to 59 V down to a constant output of $V_O = 48$ V, at an output current of up to 100 mA. To allow for precise setting of the output voltage, a 5-kΩ potentiometer has been added to the voltage-sensing voltage divider. Assume the divider ratio is set so as to produce the nominal output voltage of 48 V with a raw input voltage of 55 V.

(a) Use hand calculations to estimate the line regulation dV_O/dV_{RAW} and load regulation (actually, the negative of the AC output resistance) dV_O/dI_L for this circuit. Use Equations 10.16 and 10.17, assuming $\beta_1 = 100$ for the 2N4401 and $\beta_2 = 50$ for the 2N3055.

(b) If you have access to circuit simulation software such as Multisim, simulate this circuit and adjust the potentiometer R_4 until the output voltage is close to 48.0 V with no-load current. Then perform two tests: (1) with a load current of 50 mA, vary the supply voltage V_{RAW} from 51 to 59 V by increments of 1 V; (2) with $V_{RAW} = 55$ V, vary the load current from 0 to 200 mA by increments of 20 mA. How do the actual slopes of these curves compare with the hand calculations you obtained in part (a)?

10.7. *IC voltage regulator.* In the circuit shown in Figure 10.12, a fixed-voltage IC voltage regulator is used to supply a regulated output voltage greater than its designed output by connecting its ground lead to a voltage divider consisting of R_1 and R_2. Suppose you need a way to provide regulated 12 VDC at 1 A from a supply voltage V_{RAW} that varies from 16 to 20 VDC. The only IC voltage regulator you have is designed to provide 5 VDC at a current of 1 A with a maximum voltage drop across the regulator of 10 V. Under such a condition, the current in the ground lead (terminal 3) is 2.0 mA. Choose values for resistors R_1 and R_2 (nearest 5%-tolerance values) that will provide a regulated DC output voltage near 12 V. Select the resistors so that the current through the divider is at least 10 times the current through terminal 3 of the IC. Then select a capacitor (also nearest 5%-tolerance value) so that at the ripple frequency of 60 Hz, its reactance is no more than 10% of the voltage divider's net resistance to AC ground (which is $R_1 \| R_2$).

10.8. *Current limit and power dissipation.* In designing a power supply for a load with limited power-dissipation capability, you need to provide the load with a voltage of 24 VDC. The power supply itself regulates its raw input voltage $V_{RAW} = 48$ V (maximum) by a series-pass transistor that can dissipate a maximum of 30 W. Your design includes a feature that automatically limits the maximum output current from the supply to a value I_{MAX}.

(a) If the power-supply output is shorted ($V_O = 0$ V), what is the highest current-limit setting I_{MAX} that will still keep the series-pass transistor below its maximum power-dissipation rating?

(b) Given the I_{MAX} you found in part (a), what is the maximum output power that the supply can provide at its regulated output voltage of 24 V?

10.9. *Buck converter design.* Design a buck converter to reduce $V_{RAW} = 48$ VDC to $V_O = 24$ VDC. Use a switch frequency $f = 200$ kHz and a filter capacitor C no larger than 10 nF. With a constant-current load, the maximum ripple voltage $\Delta V_{BUCK(P-P)} = 100$ mV. Use the circuit shown in Figure 10.15.

(a) Assuming the switch S is ideal, specify values for D_{BUCK}, L, and the peak inverse voltage required for diode D.

(b) Now, instead of an ideal switch S, assume that the switch is an IGBT with the following characteristics: t_{ON} = 50 ns, t_{OFF} = 170 ns, and V_{CE}(on) = 2.7 V. If the power supply is providing a current of 1.0 A, calculate the total power P_Q dissipated in the device for V_{RAW} = 48 VDC and V_O = 24 − 2.7 = 21.3 VDC. Include power dissipated during the turn-on and turn-off transitions as well as power dissipated while the device is on.

10.10. *Buck–boost converter design.* A variable-voltage-output supply is to be designed to use a regulated input voltage V_{RAW} = +12 VDC to provide a negative output voltage V_O in the range of −6 to −18 VDC. The maximum current drawn from the supply output is 0.5 A, and the largest tolerable ripple voltage V_{RIPPLE} is 50 mV (peak to peak).

(a) Find the values for duty cycle D_{BB} that will provide the needed output voltage variation, and indicate which duty-cycle value corresponds to which output voltage, assuming an ideal switch is used.

(b) If a switch frequency f = 100 kHz is used, find the value of C required to keep the peak-to-peak output ripple voltage less than 50 mV.

10.11. *Class AB amplifier design.* Design the output stage of a class AB amplifier using complementary pnp and npn BJTs that can dissipate up to 50 W each. Use the diagram shown in Figure 10.35. The load resistance R_L is a 4-Ω speaker, and the power-supply voltages $\pm V_{CC}$ are ±40 V.

(a) For low distortion, choose $k = 0.1$ in Equation 10.71. Using $V_T = 25$ mV, find I_Q and then values for $R_1 = R_2$, given that diodes D_1 and D_2 have the same current–voltage relationship as the transistors, namely, $I(V) = I_S e^{V/2V_T}$, where I_S = 21.7 fA (1 fA = 10^{-15} A) and V is the total voltage across *both* diodes (in series). Calculate the power P_R dissipated in each resistor with no signal present. This power is approximately equal to the quiescent power P_Q dissipated in each transistor under no-signal conditions.

(b) Using Equation 10.78 to calculate maximum efficiency when the peak output voltage equals the power-supply voltage of 40 V, calculate the maximum average power P_Q (max) dissipated in each device (ignoring the quiescent power dissipation with no signal present). Are the devices operating within their power rating?

***10.12.** *Class D amplifier switch timing.* In addition to the switching turn-on time $t_{S(ON)}$ and turn-off time $t_{S(OFF)}$, many power FETs have appreciable **delay times**, denoted as $t_{D(ON)}$ and $t_{D(OFF)}$. These delay times indicate how long it takes after the control voltage V_{IN} transitions at the control terminal input (base or gate, as the case may be) for the input to take effect at the output, measured by the delay time between the 50% point of the input voltage waveform and the 50% point of the output voltage waveform. A more precise definition of rise and fall times is the time it takes for a rectangular

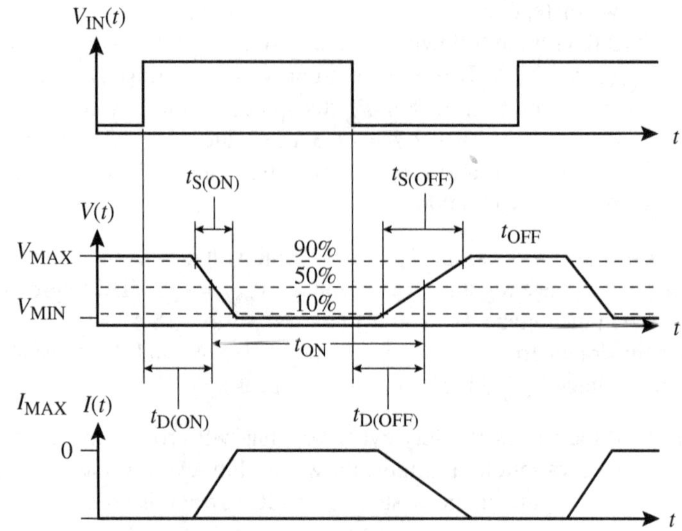

FIGURE 10.48 Definitions of switching times $t_{S(ON)}$ and $t_{S(OFF)}$, delay times $t_{D(ON)}$ and $t_{D(OFF)}$, and on time t_{ON}.

TABLE 10.3 Power FET Specifications

Device	$t_{D(ON)}$(ns)	$t_{S(ON)}$(ns)	$t_{D(OFF)}$(ns)	$t_{S(OFF)}$(ns)	$R_{DS(ON)}$(mΩ)
FQP50N06 (n-channel)	40	220	130	140	22
FQP47P06 (p-channel)	110	910	210	400	26

two-level waveform to go from 10% of its total transition voltage change to 90% of the change. These definitions are incorporated in Figure 10.48, which shows their meanings graphically.

The complementary 60-V power FET pairs FQP50N06 and FQP47P06 have the following worst-case specifications, as shown in Table 10.3.

Suppose these devices are used as the switches AD_1 and AD_2 in the class D power amplifier shown in Figure 10.41. Suppose the power-supply voltages are ±30 V and the load resistance is $R_L = 1.5\ \Omega$. If the switching frequency $f_S = 150$ kHz, find

(a) the dead time T_{+-} (time that both devices are off) needed to avoid shoot-through when the n-channel device is turned off and the p-channel device is turned on

(b) the dead time T_{-+} (time that both devices are off) needed to avoid shoot-through when the p-channel device is turned off and the n-channel device is turned on. In calculating the dead times, allow enough time for one device to turn off completely (delay time plus switch time) before the second device receives a signal to turn on. This will produce a dead time longer than strictly necessary, but will definitely prevent shoot-through.

(c) What fraction of the total period $T = 1/f_S$ do the two dead times occupy (if added together)? If this fraction is too large, the devices may be operating at an f_S that is too high.

(d) What is the approximate efficiency of the amplifier? Estimate the efficiency by assuming the amplifier delivers a 29.5-V (peak) sine wave into a 1.5-Ω load and that the input power is the power delivered to the load plus the power dissipated in the devices. Calculate power dissipated in the devices using Equation 10.28, assuming

$$t_{ON} = \frac{0.5}{f_S} - (T_{+-} + T_{-+})$$

For further resources for this chapter visit the companion website at **http://wiley.com/go/analogmixedsignalelectronics**

11

HIGH-FREQUENCY (RF) ELECTRONICS

11.1 CIRCUITS AT RADIO FREQUENCIES

Most electronic systems deal with voltages and currents that change with time. The significant frequency components of these signals can lie anywhere from the near-DC range (<1 Hz) up through the audio range (16 Hz to 20 kHz) and beyond. Most of the systems we have described up to this point were assumed to operate in a range of frequencies below a few megahertz. In this range, op amps and active devices have enough gain to allow the design of systems with feedback determined by complex networks of passive components, and most passive components can be treated like simple **lumped elements**. A lumped element is completely described by the current–voltage relationship between its terminals, which is often a simple function such as resistance, capacitance, or inductance.

If you try to design a circuit to operate at a frequency higher than, say, 5 MHz or so, you will find that some of the assumptions made for low-frequency designs gradually cease to be valid as the frequency involved increases. Capacitors can begin to act like inductors, inductors like capacitors, and resistors can change their values and behave more like inductors or capacitors than resistors. Most op amps have a gain–bandwidth product of only a few MHz, so they become useless at frequencies higher than their gain–bandwidth product. And any attempt to build a high-frequency circuit that is physically large (e.g., for reasons of power dissipation) will run into the fact that no signal can travel faster than the speed of light,

Analog and Mixed-Signal Electronics, First Edition. Karl D. Stephan.

Companion Website: http://wiley.com/go/analogmixedsignalelectronics

which is about $3 \times 10^8\,\text{m}\,\text{s}^{-1}$. In one nanosecond ($10^{-9}$ s), therefore, a signal can travel only a distance of

$$d = u \cdot t = (3 \times 10^8 \text{ m s}^{-1})(10^{-9} \text{ s}) = 0.3 \text{ m}, \tag{11.1}$$

or 30 cm (about a foot), where u is the velocity and t is the time. So, for example, if a circuit has a clock frequency f of 1 GHz ($=10^9$ Hz), the period of one clock cycle is $t = 1/f = 1$ ns. If the clock waveform were visible like a water wave, and we had some way of slowing down time so that we could see the wave as it travels, you could see the waves of the clock signal moving along their conductors, and the distance from one rising edge to the next would be less than 30 cm. This is because (as we will see) the speed of signals along conductors and **transmission lines** is always less than the speed of light in a vacuum, which is a universal speed limit for signals. This is one reason that as we go to higher frequencies, it is no longer safe to assume, as we have implicitly done in all discussions up to now, that signals travel *instantaneously* to every part of a circuit.

A sinusoidal voltage or current source at a single frequency f can produce an **electromagnetic wave** whose **wavelength** λ (e.g., measureable as the distance from one positive peak to the next positive peak) depends on the velocity u in the medium in which the wave is traveling. As you probably know from physics, the relationship between the frequency, the wavelength, and the propagation velocity in a **nondispersive** medium (one in which the wave velocity is independent of frequency) is given by

$$\lambda = \frac{u}{f} \tag{11.2}$$

In a vacuum, electromagnetic waves travel at the speed of light, which is customarily denoted by the letter c. So if we are dealing with the special case of electromagnetic waves traveling in a vacuum, $u = c$, and the wavelength equation becomes $\lambda = c/f$. But generally speaking, to solve for the wavelength, you should first determine what the signal velocity u is in the medium in question and then use Equation 11.2, which is true for any nondispersive medium.

One generally accepted term for electronic design techniques that take into account the finite speed of signals at HF is **radio-frequency** or **RF** design, which is the term we will adopt from this point onward. Just because a circuit deals with frequencies above 5 MHz or so does not necessarily mean that you must use RF design techniques. Many digital systems with clock speeds of 100 MHz or more can be designed without too much concern for RF design, although some knowledge of high-frequency effects can be helpful. But analog circuits are much more affected by the often deleterious phenomena that arise at high frequencies, so any attempt to deal with analog signals that have significant frequency components above about 5 MHz should at least consider what may be happening at these frequencies because of RF effects.

11.2 RF RANGES AND USES

When Heinrich Hertz (1857–1894) demonstrated that high-frequency electricity could be used to generate waves that behaved like light waves except for their longer wavelengths, it took a while for people to realize the potential that these **radio waves** held for communications. When radio communications systems began to be developed shortly after 1900, inventors such as Guglielmo Marconi (1874–1937) used waves that were many miles long, simply because they worked well over long distances and equipment was available to generate large amounts of power at frequencies below about 300 kHz. But when amateur radio operators discovered in the 1920s that **shortwaves** with wavelengths below 200 m or so (corresponding to a frequency of 1.5 MHz) could propagate around the world, the technology was developed to produce and receive signals at frequencies up to 30 MHz or so.

During World War II (1939–1945), an intense effort was mounted in many combatant nations to develop **RAdio Detection And Ranging** (now called **radar**). In order to make radar antennas small enough to fit on airplanes, engineers had to use much shorter wavelengths than were generally available at the time, and methods were found to produce high-power radar transmitters in the **microwave** range, which consists of frequencies between about 1 and 30 GHz. (The wavelengths corresponding to these frequencies are 30 cm down to 1 cm.) Wavelengths shorter than 1 cm (which correspond to frequencies higher than 30 GHz) are termed **millimeter waves** and are not as useful for communications as lower-frequency waves because the atmosphere begins to absorb such short wavelengths excessively. But because the need for more wireless communications is so great, one can expect that more use will be made of millimeter-wave radiation for communications in the future.

Because two or more different radio communications systems that use the same frequency range can interfere with each other, the entire range of useful wavelengths called the **radio spectrum** is a limited natural resource, rather like real estate. There is only so much to go around, and so there are national and international agreements that **allocate** different parts of the radio spectrum for various private and public uses. An exhaustive radio spectrum chart for a given region is a very complicated thing, but Figure 11.1 shows a simplified version of the radio spectrum in the United States, where the rules are promulgated by the **Federal Communications Commission** (**FCC**). Corresponding charts for other countries are basically similar, though details vary from place to place. In order not to clutter the chart with frequency data, Table 11.1 gives the numerical values of the various spectrum limits shown.

Because the radio spectrum covers such a wide range of frequencies and wavelengths, the horizontal axes in Figure 11.1 are **logarithmic**, which distorts an important feature called **bandwidth**. The bandwidth of a given frequency range is the difference between its high-frequency limit and its low-frequency limit in Hz. For example, the bandwidth of the amplitude modulation (AM) broadcast band is (1705–535) kHz = 1170 kHz, or 1.17 MHz. Just as a larger plot of land costs more than a smaller plot because you can build more on it, the value of a frequency allocation range for communications varies with the bandwidth,

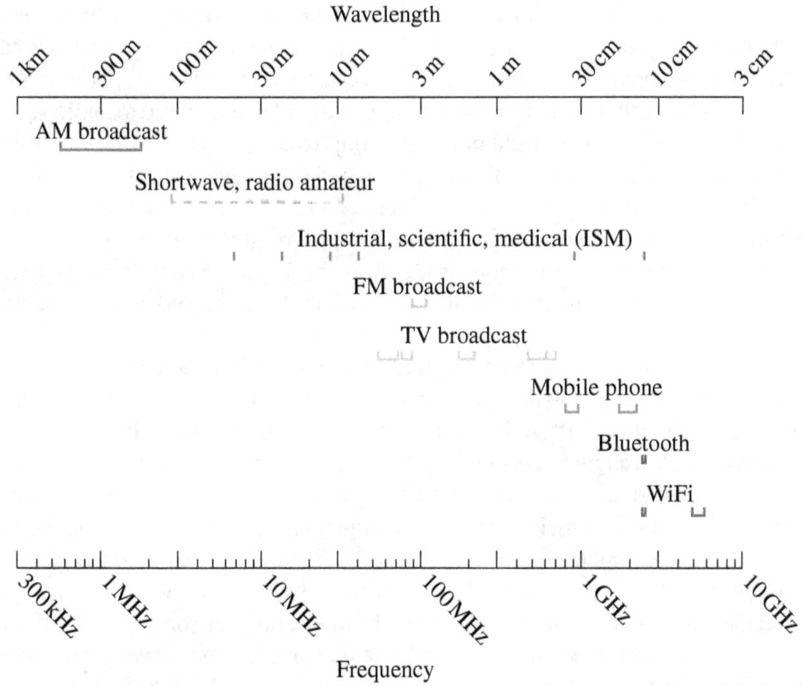

FIGURE 11.1 Simplified version of frequency allocation chart in the US radio spectrum, showing selected allocations.

TABLE 11.1 Frequencies of Allocations Shown in Figure 11.1

Purpose	Frequency range
AM broadcast band	535–1705 kHz
Industrial, scientific, medical (ISM) bands	6.78 MHz, 13.56 MHz, 27.12 MHz, 40.68 MHz, 915 MHz, 2.45 GHz
FM broadcast	88–108 MHz
TV broadcast	54–72 MHz (Channels 2–4), 76–88 MHz (Channels 5–6), 174–216 (Channels 7–13), 470–608 MHz (Channels 14–36), 614–698 MHz (Channels 38–51)
Bluetooth wireless technology standard	2.4–2.485 GHz
WiFi wireless technology standard	2.4–2.485 GHz, 4.9–5.8 GHz
Mobile-phone bands (approximate)	806–960 MHz, 1.71–2.2 GHz
Shortwave broadcast, radio amateurs	Numerous small bands in 3–30 MHz range

because the rate of information transfer that a given frequency range can handle is proportional to bandwidth, other things being equal.

The need for more bandwidth is an important factor that has driven engineers to use higher and higher frequency bands over the years. Although the logarithmic scales in Figure 11.1 somewhat disguise the fact, the higher WiFi band at 4.9–5.8 GHz has 900MHz of bandwidth, hundreds of times more than the entire AM broadcast band. Wider bandwidth can be used either to improve the quality of transmission, as the FM broadcast band was designed to do compared to the AM band, or to transmit formats that inherently use more bandwidth, such as video compared to audio signals. As digital communications technology has improved, it has become possible to squeeze more information transmission per second into a given spectrum allocation. This is one reason that a large chunk of frequencies formerly allocated to **analog television (TV) broadcasting** in the **ultrahigh-frequency (UHF)**[1] range was reallocated to other uses beginning in 2008 in the United States, because **digital TV broadcasting** does more with less bandwidth than the old analog-standards technology could.

The **propagation** characteristics of radio waves influence the uses to which they are put. Propagation refers to how waves travel: in straight lines or along paths altered by reflection, refraction, diffraction, and absorption, similar to the different ways that light waves travel. Longer waves in the AM broadcast band can propagate dozens of miles in the daytime and thousands of miles at night, due to the varying influence of the earth's **ionosphere**, which is a layer of charged particles in the upper atmosphere several hundred kilometers above the ground. The same ionosphere effectively reflects waves in the shortwave bands back to earth, so that worldwide propagation is easily achieved with shortwave transmitters of modest power (on the order of a few kilowatts or less). But above about 30MHz, radio waves are not usually reflected by the ionosphere and increasingly tend to travel only in straight lines between reflections by earth-bound objects. Either direct **line-of-sight** paths are necessary at higher frequencies or else numerous reflections (e.g., in buildings and cities) are required to establish communication between a transmitter and receiver in the very high frequency (VHF) range above 30MHz. The only long-distance communication possible at these wavelengths is with **communications satellites**, including the **geosynchronous** satellites that appear to float stationary in the sky, allowing **earth stations** to maintain continuous contact with them from remote locations. Other satellites in lower orbits are used for the **Global Positioning System (GPS)**, an increasingly popular way to determine a user's exact location anywhere in the world to within a few meters. All these satellite systems use frequencies that are typically in the lower microwave region (1–10GHz).

This brief review of RF uses has not mentioned the noncommunications applications of radio waves. These include industrial heating, microwave ovens, power transmission, and other nonsignaling purposes. Most of these applications use frequencies in the **industrial, scientific, and medical (ISM)** bands, which are very narrow slices of the spectrum scattered in the HF to the microwave ranges. For example, consumer-type microwave ovens use a frequency of 2.45GHz to cook your pizza. Astronomers construct elaborate sensitive receivers for **radio astronomy** in order to detect radiation from interstellar space. However, these noncommunications uses are outnumbered by the vast majority of RF electronic designs for communications purposes.

[1]You will sometimes see references to broadly defined frequency ranges as follows: **high frequency (HF)** is 3–30MHz, **very high frequency (VHF)** is from 30 to 300MHz, and **ultrahigh frequency (UHF)** is from 300MHz up to about 1GHz.

11.3 SPECIAL CHARACTERISTICS OF RF CIRCUITS

RF circuit designers must take the special characteristics of RF circuits and devices into consideration. As we have already seen, the fact that signals do not travel instantaneously to all parts of a circuit means that certain familiar assumptions of circuit theory, such as Kirchhoff's voltage law and Kirchhoff's current law, are no longer universally valid. For example, the voltage law involves adding up voltages measured around a closed path along interconnected circuit elements. If the path involves voltages whose frequencies are high enough, the distance represented by the path may be comparable to the distance that the highest frequency travels in one period, and Kirchhoff's voltage law breaks down. For this and other reasons, voltages and currents become increasingly difficult to measure unambiguously above around 100 MHz. Instead, more accurate measurements can be made if the quantity measured is a **wave**, especially if the wave is traveling along a well-defined **transmission line**, which we will discuss in more detail later in Section 11.4.

Another effect at HF that makes the straightforward analysis of circuits difficult is the fact that lumped-element components such as resistors, capacitors, and inductors no longer behave the way they do at lower frequencies. In Chapter 3, we described how inductors have **parasitic capacitance** that initially increases the device's effective or equivalent inductance as frequency increases. But above a critical frequency called the **self-resonant frequency**, the inductor begins to behave like a capacitor! A similar phenomenon occurs with capacitors at RF. The terminal impedance of a capacitor falls below its nominal value at higher frequencies until its capacitance resonates with its **parasitic inductance**, and above the capacitor's self-resonant frequency, it behaves like an inductor. Resistors can also show variable impedance at higher frequencies as well and can behave more like lossy inductors or capacitors than like resistors. These effects can be modeled and accounted for, and ignoring them completely can cause problems.

A third issue that arises when dealing with RF circuits is the possibility of **electromagnetic radiation** from the circuit itself. This is not a type of **ionizing radiation** associated with radioactivity and X-rays, because the energy associated with RF waves is much too small to dislodge an electron from a normal atom or molecule. Electromagnetic radiation (a radio wave) results when a circuit carrying an RF signal behaves like a transmitting antenna. If any current-carrying part of a circuit has a physical dimension more than a small fraction of a wavelength (say, 5% or more), you should consider the possibility that it will radiate some of the energy it is carrying, in the form of radio waves. These waves can cause interference with other circuits and also appear in the circuit as a loss, similar to an unexpected resistance. There are ways of preventing a circuit from radiating, and they include **shielding** (a conducting cover or enclosure) and **absorption** (the use of lossy material to absorb radiated energy). But the best way to avoid radiation is to make sure that your circuit is smaller than 5% or so of the shortest wavelength of the signals involved.

Finally, active devices all have frequency limitations beyond which their specifications are not guaranteed. We have already mentioned that typical op amps have a gain–bandwidth product of only a few megahertz. This means that no matter what you do with the external circuit, the product of the gain and the bandwidth of the op amp will never exceed a certain value. If the gain–bandwidth product of a certain op amp is 2 MHz, for example, it is impossible to derive any gain from it above a frequency of 2 MHz.

For this reason, there are special-purpose active devices such as diodes, rectifiers, and transistors designed especially for good performance in RF circuits. RF transistors are available with gain–bandwidth products exceeding 1 GHz, and integrated circuits (ICs) that operate in that range are available as well. As you might expect, the layout and design of an RF circuit's **printed circuit board** (**PCB**) have a great influence on how well it operates. Long conductors that can be treated as simple connections at low frequencies can become significant circuit elements at RF, causing inefficiencies and poor performance. The presence of **ground planes** (large areas of grounded conductive material) is important for the proper operation of many RF circuits, but no amount of good passive-circuit design effort will overcome an intrinsic bandwidth limitation in the active device used. In other words, no circuit can perform better than the active devices it uses. High-power RF devices are usually mounted in special packages that show low parasitic inductance and capacitance as well as low thermal resistance and are often fairly costly.

With these considerations in mind, you can appreciate some of the reasons that RF circuits are designed the way they are. Even if you never actually design one yourself, you may have to deal with an RF transmitter or receiver unit as part of a larger system, and it will be helpful to understand some of the basics of RF design in order to deal with these systems intelligently.

11.4 RF TRANSMISSION LINES, FILTERS, AND IMPEDANCE-MATCHING CIRCUITS

11.4.1 RF Transmission Lines

A **transmission line** is a structure designed to carry an electromagnetic wave along a specified path while minimizing losses due to dissipation, radiation, or reflections. You are probably familiar with certain types of transmission lines already. If your residence is served by cable TV, you have seen the **coaxial cable** that attaches to the set-top box or receiver. A close-up photo of the inner structure of a typical coaxial cable is shown in the photograph in Figure 11.2.

To show how a signal at a given frequency travels down this type of structure in the form of a wave, we will assume a simplified structure for the cable shown in Figure 11.3. In this figure, the outer braided conductor is modeled by a hollow perfect conductor whose inner radius is b meters. The inner conductor is modeled by another cylindrical perfect conductor with a radius of a meters, and the two conductors are separated by a lossless dielectric insulator.

Assume we have a **semi-infinite** length of cable, meaning that we can make a connection to one end of the cable that resembles Figure 11.3, but the other end is indefinitely far away. What happens if we connect a voltage source to drive the near end of the cable? Specifically, what is the voltage and current at a point along the cable a distance z away from the driven end?

A coaxial cable cannot be modeled with the usual lumped-element components that we have used in circuit models up to now. The reason is that the cable has both

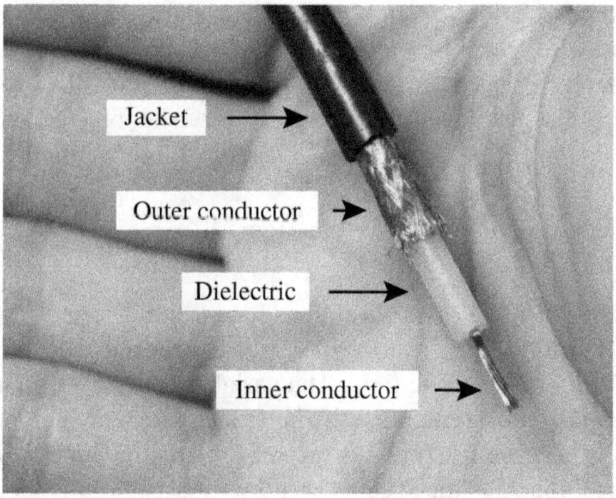

FIGURE 11.2 Photo of typical coaxial cable showing insulating jacket, braided outer conductor, dielectric, and stranded inner conductor.

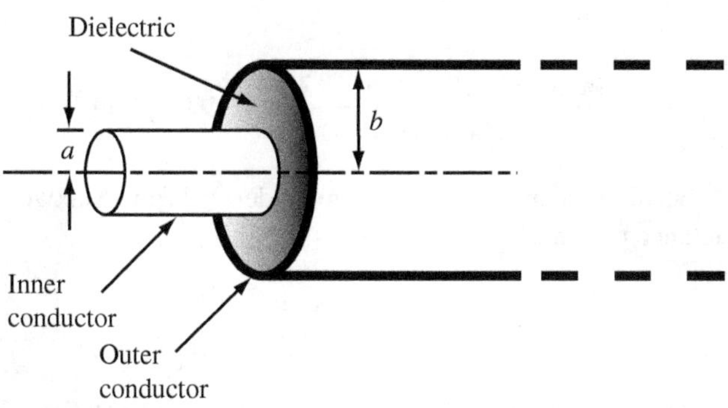

FIGURE 11.3 Idealized geometry of coaxial transmission line.

inductance and capacitance at every point along its length. If we model a piece of the cable simply as a capacitor, we ignore the inductance and vice versa. The inductance and capacitance are said to be **distributed** along the entire transmission line, instead of **lumped** in one place as in a single inductor or capacitor.

The best we can do is to examine a small section of the cable that is much shorter than a wavelength at the frequency of the signal in question. (This assumes that we know what the wavelength is, and we really don't know that yet, but we can always check later to see if our assumption was correct.) Such a short section of length Δz will act pretty much like a pair of small lumped-element components, because the time delay from one end to the other will be fairly small (but not zero). This small section of coaxial line Δz long will have both a capacitance C and an inductance L, in general.

Next, we are going to pull a rabbit out of a hat, so to speak. The following equations for capacitance per unit length C' and inductance per unit length L' cannot be derived from anything in this book. They are based on the electromagnetic equations discovered by James Clerk Maxwell (1831–1879), whose discovery inspired Hertz to perform his successful search for the waves we now know as radio waves. Explaining Maxwell's equations is outside the scope of this text, so you will simply have to take my word for it that the distributed capacitance C' per meter of a section of coaxial line is given by

$$C' = (8.854 \times 10^{-12}\ \text{F m}^{-1}) \frac{2\pi\varepsilon_r}{\ln(b/a)} \tag{11.3}$$

The constant 8.854 pF m^{-1} is often designated as ε_0 and is termed the **permittivity of free space**. Permittivity is that property of a material that allows it to store energy in the form of an electric field. The variable ε_r is the **relative permittivity** of the dielectric (insulator) separating the inner and outer conductors. It is a dimensionless number, because it expresses the ratio of the dielectric's permittivity to that of free space, much as the specific gravity of a material is the dimensionless number that compares the material's density to the density of water. So, for example, if a coaxial cable has an inner radius of $a = 1$ mm and an outer radius of $b = 3.44$ mm and the dielectric is a polymer with a relative permittivity of $\varepsilon_r = 2.2$, the cable has a capacitance of

$$C' = \frac{2\pi(8.854 \times 10^{-12}\ \text{F m}^{-1})(2.2)}{\ln(3.44\ \text{mm}/1\ \text{mm})} = 99.06\ \text{pF m}^{-1} \tag{11.4}$$

As for the distributed inductance, another formula derived from Maxwell's equations gives the inductance per unit length L' as

$$L' = (4\pi \times 10^{-7}\ \text{H m}^{-1}) \frac{\mu_r}{2\pi} \ln\left(\frac{b}{a}\right) \tag{11.5}$$

The leading constant in Equation 11.5 is termed μ_0, the **permeability of free space**, and measures the tendency of a material to store energy in the form of a magnetic field. The **relative permeability** μ_r is the permeability of a substance relative to that of free space. Roughly speaking, relative permeability measures the ease with which a material may be magnetized by an external current. Magnetic materials such as iron and certain ceramics called **ferrites** have very large relative permeabilities, ranging into the thousands, which is why coils and inductors often have **cores** made of iron or ferrite material that allow large amounts of magnetic energy to be stored in a small space. But most plastics and other nonmagnetic substances (including nonmagnetic metals) have a relative permeability of about 1. Applying Equation 11.5 to the same coaxial cable that we found the capacitance for in Equation 11.3, we find that the cable has an inductance per meter of

$$L' = (12.566 \times 10^{-7}\ \text{H m}) \frac{1}{2\pi} \ln\left(\frac{3.44\ \text{mm}}{1\ \text{mm}}\right) = 247.1\ \text{nH m}^{-1}, \tag{11.6}$$

or about a quarter of a microhenry in a 1-m length.

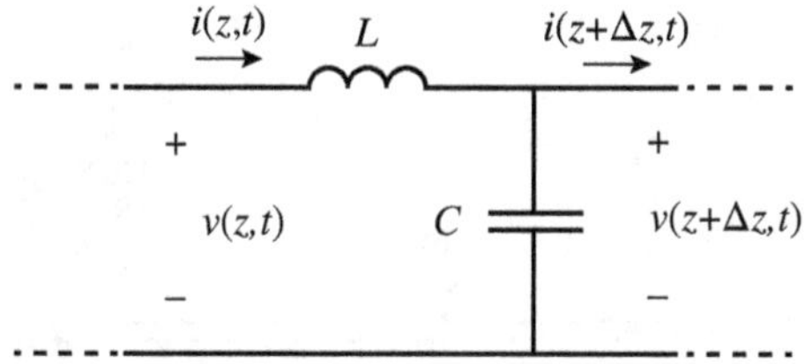

FIGURE 11.4 Voltage and current changes in small section of transmission line of length Δz, with distributed capacitance C and inductance L.

Now, back to the question of what the voltage and current look like along the line a distance z from the voltage source at the end. We can write a pair of differential equations to describe the situation, if we consider a small section of transmission line Δz and estimate what the change in voltage and current is from one end of this small section to the other, as illustrated in Figure 11.4.

Assuming a voltage $v(z,t)$ at the left end of the section and a current $i(z,t)$, this current will cause a voltage drop across the inductor L that amounts to the difference between the voltage at the left and the voltage at the right:

$$v(z+\Delta z,t) = v(z,t) - L\frac{di(t,z)}{dt}, \tag{11.7}$$

where we have used the definition of inductance to obtain the voltage across the inductance as a function of the time derivative of the current. This says that the voltage on the right between the center conductor and ground is diminished by whatever voltage drop occurs across the line's inductance. Similarly, not all the current that enters the line on the left leaves it; some is shunted to ground through the line's capacitance. This fact leads to a second differential equation that relates the currents along the line at the left and right ends of the section:

$$i(z+\Delta z,t) = i(z,t) - C\frac{dv(z,t)}{dt} \tag{11.8}$$

We can now use Equations 11.7 and 11.8 to obtain a pair of **simultaneous partial differential equations** in both time and distance by letting the length Δz approach zero:

$$\frac{\partial v(z,t)}{\partial z} = -L'\frac{\partial i(z,t)}{\partial t} \tag{11.9}$$

$$\frac{\partial i(z,t)}{\partial z} = -C'\frac{\partial v(z,t)}{\partial t} \tag{11.10}$$

In Equations 11.9 and 11.10, we have substituted the inductance and capacitance per unit length (L' and C', respectively) for the original values L and C, by using the facts that $L = L'\Delta z$ and $C = C'\Delta z$.

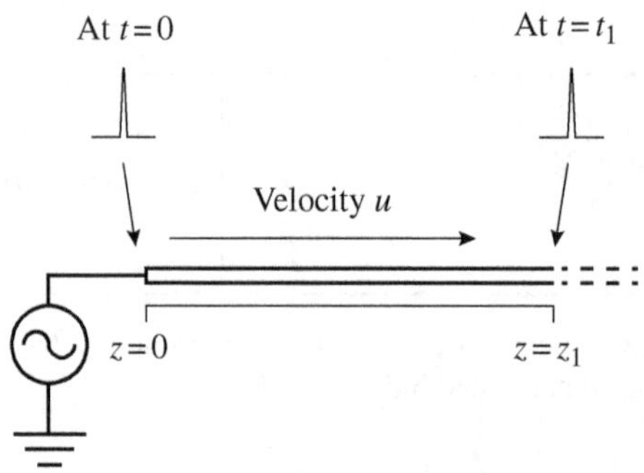

FIGURE 11.5 Showing how a voltage impulse traveling along a transmission line moves a distance z_1 in a time $t_1=z_1/u$, where u is the wave velocity.

Now, suppose we assume that the signal voltage is a function of time $V(t)$. We will now show how a voltage function of time (only) becomes a **traveling wave** along the transmission line.

A traveling wave must be a function of both time and distance. If it travels at a velocity u meters per second, the function that expresses this looks like $V(t-(z/u))$. For example, suppose the applied voltage is just a sharp pulse that occurs at $t=0$. At a location at a distance z_1 along the line, if the wave travels at velocity u, the expression $t_1-(z_1/u)$ equals zero at a time t_1, which is exactly the time it takes for the pulse to travel from $z=0$ to $z=z_1$ (see Fig. 11.5). So we can use the same voltage function mathematically, and all we have to do is to make its **argument** (the expression it is a function of) depend on both distance z and time t. That is what the function $V(t-(z/u))$ does.

This is all very well if it works, but is that really what happens? In particular, what is the relationship between the **wave velocity** u (m s^{-1}) and the distributed inductance L' and capacitance C'? We can find that out by using the PDEs (Eqs. 11.9 and 11.10) and our assumed voltage function of time and distance.

If the assumed voltage wave really exists, we will have the following relationship between its (partial) time derivative and its (partial) space derivative:

$$\frac{\partial V(t-z/u)}{\partial t} = -u\frac{\partial V(t-z/u)}{\partial z} \tag{11.11}$$

All this says is that if the wave has a certain shape and derivative in space (along the z-axis), the wave's changes with time (as it passes a given point) will be faster the faster it moves (the larger velocity u is). The same will be true of the current I as a function of time and space as well.

Equation 11.11 allows us to replace the time partial derivatives in Equations 11.9 and 11.10 with space partial derivatives, leading to

$$\frac{\partial v}{\partial z} = L'u\frac{\partial i}{\partial z} \tag{11.12}$$

and

$$\frac{\partial i}{\partial z} = C'u\frac{\partial v}{\partial z}, \tag{11.13}$$

where we have dropped the explicit functions of time and space for simplicity. If we substitute in the expression for the space derivative of current given by Equation 11.13 into Equation 11.12, we get the following result:

$$\frac{\partial v}{\partial z} = L'u(C'u)\frac{\partial v}{\partial z}, \tag{11.14}$$

which can be true for a function that changes in space $(\partial v/\partial z \neq 0)$ only if

$$L'C'u^2 = 1 \tag{11.15}$$

And *voila!* Equation 11.15 lets us find the relationship between the distributed reactances and wave velocity to be

$$u = \pm\frac{1}{\sqrt{L'C'}} \tag{11.16}$$

Notice that the wave velocity u can be either positive or negative. This means that we can have waves traveling in either the $+z$ or the $-z$ direction. For the particular case of a semi-infinite line, there is no source at the other end because the other end is infinitely far away, and so there can be only forward-traveling waves. But in general, real transmission lines can carry waves in both directions. This property allows **standing waves** to appear as the interference pattern between forward- and reverse-traveling waves, as we will see.

Notice also that we have placed no restrictions on the voltage function, other than it must be differentiable. This means that, in principle at least, almost any sort of wave shape and frequency will be transmitted without change along a lossless and dispersionless transmission line. While this analysis eventually breaks down at extremely high frequencies, most coaxial lines do exhibit a very wide bandwidth, extending from DC into the low GHz range or higher.

Along with the voltage wave, there must be a current wave if the transmission line actually transmits power. An analysis similar to the one we used to find the velocity u shows that for a forward-traveling wave, the ratio of instantaneous voltage at a point z to the current at that same point is

$$\frac{v(z,t)}{i(z,t)} = Z = \sqrt{\frac{L'}{C'}} \tag{11.17}$$

This impedance is characteristic of any transmission line having a particular set of dimensions and materials and is therefore termed the **characteristic** (or **surge**) **impedance**. For the coaxial line shown in Figure 11.3, the characteristic impedance is

$$Z_0 = \sqrt{\frac{L'}{C'}} = \sqrt{\frac{247.1 \text{ nH m}^{-1}}{99.06 \text{ pF m}^{-1}}} = 49.94\ \Omega, \tag{11.18}$$

or very close to 50 Ω. Many types of transmission lines are designed to have a characteristic impedance of 50 Ω, because this standard impedance allows interconnections among transmission lines, sources, and loads in a way that matches impedance. This maximizes efficient power transfer and minimizes reflections that can lead to inefficiency. As long as we are calculating numerical values, we can use Equation 11.16 to find what the wave's velocity is along this transmission line:

$$u = \frac{1}{\sqrt{L'C'}} = \frac{1}{\sqrt{(247.1 \text{ nH m}^{-1})(99.06 \text{ pF m}^{-1})}} = 2.02 \times 10^8 \text{ m s}^{-1} \tag{11.19}$$

or about two-thirds of c, the speed of light in a vacuum ($c = 3 \times 10^8 \text{ m s}^{-1}$). In fact, you will find if you put the vacuum values μ_0 for L' and ε_0 for C' into Equation 11.19, the result is exactly the speed of light in a vacuum, as it better be!

From a circuit point of view, what is the equivalent circuit of the near end of the semi-infinite transmission line? The characteristic impedance equation applies to each point along the line, including $z = 0$ at the near end, so we must conclude that the ratio of voltage to current at the end is 50 Ω. Because there is no reverse-traveling wave, the result is that the semi-infinite transmission line appears to external circuitry to be a *resistor* whose value is $Z_0 = 50\ \Omega$.

Normally, power going into a resistor is **dissipated** as heat. But in the case of the semi-infinite transmission line, the power going into the near end simply travels along the line forever. Although we can't build semi-infinite transmission lines in the lab, we can make one that behaves like a semi-infinite line, by **terminating** it in a resistor whose value equals the line's characteristic impedance. This situation is shown in Figure 11.6, where we have replaced the semi-infinite line in Figure 11.5 with a line of finite length z_1 and terminated it with a resistor R_0 whose value equals the transmission-line characteristic impedance Z_0.

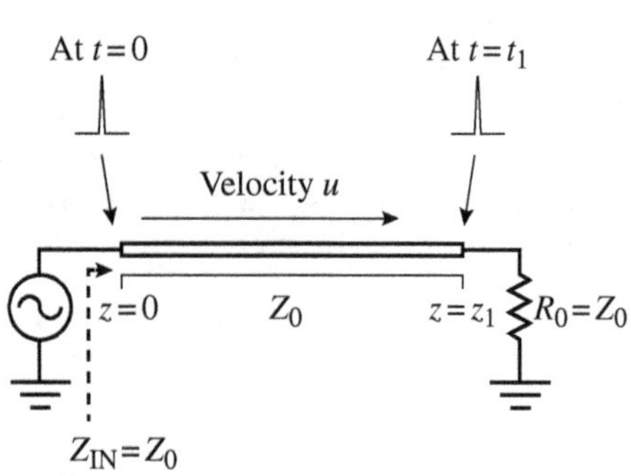

FIGURE 11.6 Pulse transmitted from voltage source along transmission line of characteristic impedance Z_0, which is terminated in matched load $R_0 = Z_0$.

Observe that the voltage wave (or current wave—both have the same shape) impressed on the left-hand end appears unchanged at the right-hand end but after a *delay* of t_1 seconds. This is the basic ideal function of a transmission line: to transfer a signal without changing its amplitude or frequency characteristics, except for imposing an unavoidable delay caused by the physical separation between two different locations in a circuit or system.

Many digital and mixed-signal systems use clock and signal waveforms that have significant frequency components well into the hundreds of megahertz. You can gain a rough estimate of the highest-frequency component of a digital signal by dividing 1 by the shortest pulse that can appear in it. For example, if a bit stream has pulses as short as 20 ns, the Fourier transform of the signal has significant energy up to a frequency of at least 1/(20 ns) = 50 MHz. And generally speaking, the highest fundamental frequency in a digital system is the clock signal, so the system's clock frequency can be taken as a guide, bearing in mind that a square-wave signal has significant harmonics as well.

If a system has significant high-frequency components in its signals, both the time delays and the way the signals are conveyed around the circuit can affect how well the circuit performs or even if it will function at all. Simply connecting point A to point B with a circuit-board trace does not ensure that the waveform at point A will be faithfully reproduced at point B, unless care is taken to create conditions similar to the ideal transmission-line situation shown in Figure 11.6. This includes consideration of the distributed inductance and capacitance of the line, a calculation of its characteristic impedance, and termination of the line (preferably at both ends) with its characteristic impedance.

If a transmission line is *not* terminated in its characteristic impedance, some of the incident wave's energy is not absorbed by the load, but is instead **reflected** and becomes a **reverse-traveling wave**. In moving backward toward the source of the signal, the reverse wave interferes with the forward wave, leading to **standing waves**. The net effect of this is to make the *total* voltage at any point along the line *vary* with position. In extreme cases, the voltage at one point on the line can be as much as four or five times lower than the voltage at another point. This situation is clearly unacceptable for digital circuitry and can cause problems with analog systems as well. This is why RF designs that involve signal runs of any significant length must take into account the transmission-line character of such runs and deal with them appropriately to minimize reflections.

In digital systems, specialized ICs called **line drivers** and **line receivers** are used to transmit pulses along transmission lines. Many of these lines are designed to be **balanced lines**. A balanced transmission line consists of two wires, each of which has the same impedance relationship to a third ground wire or ground reference. Figure 11.7 illustrates an ideal balanced transmission line with transmitter and receiver circuits designed to minimize reflections along the line. In contrast to an **unbalanced line** such as a coaxial transmission line, a balanced transmission line consists of *three* conductors: a pair of signal conductors and a ground conductor. In use, a balanced transmission line is driven symmetrically by two voltages of equal

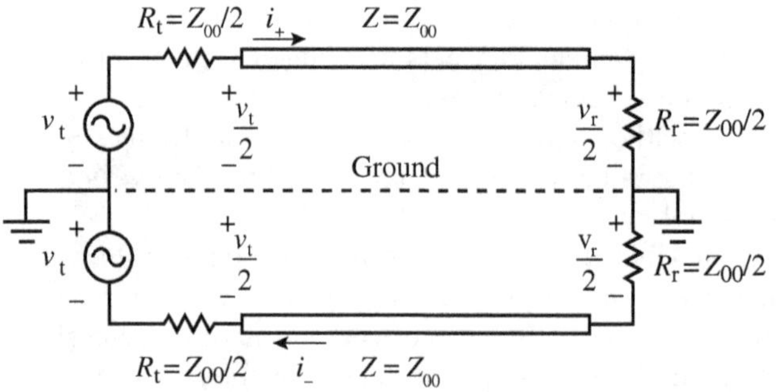

FIGURE 11.7 Balanced transmission line showing transmitter and receiver circuits matched to transmission-line impedance Z_{00}.

magnitude and opposite phase, as the v_t sources are doing in Figure 11.7. When driven this way, a balanced transmission line shows a characteristic impedance called the **odd-mode impedance** Z_{00}, defined as

$$Z_{00} \equiv \left. \frac{v_t}{i_+} \right|_{\text{odd mode}}, \tag{11.20}$$

where the **odd mode** subscript means that equal and opposite voltages are applied to the two ungrounded conductors. The voltage v_t is the total voltage measured from one conductor to the other conductor, and the current i_+ is the current entering the positive conductor.

When a balanced line is excited in this way and terminated in a total resistance equal to Z_{00}, the incident voltage wave at the transmitter will propagate without reflections through the balanced line and disappear into the receiver load resistances. What is just as important is the fact that in such a balanced line, the currents along the two ungrounded conductors have *equal* and *opposite values* everywhere along the line. If you are far enough away from the balanced line to treat it as a *single* conductor, the *net* current it seems to carry is *zero*.

This balancing of currents makes balanced transmission lines less prone to **couple** to other circuits magnetically, because the magnetic field set up by one current-carrying conductor tends to cancel the field set up by the other one. One common form of balanced transmission line is called a **twisted pair** and is often used in telephone and data cables. A twisted-pair cable consists of many pairs of conductors, each pair of which is twisted together in a spiral pattern of a few twists per meter. The twisting tends to average out magnetic coupling from each of the pair of wires. Many twisted pairs may share a common ground shield in such a cable, and although there is some coupling among the pairs, it is less than would occur if the pairs were not twisted. Twisted pairs and other types of balanced transmission lines are used extensively in digital signal cables and also high-quality audio cables used for microphones and other sensitive gear, because of their higher rejection of magnetic coupling to other systems such as power lines.

To summarize what we've said about transmission lines, we have found that if a conductor has a constant value of distributed capacitance and distributed inductance per meter, it can present a resistance to a wideband signal voltage source, and the resistance is called the transmission-line characteristic impedance Z_0. Any signal entering a transmission line travels along the line and appears at the far end undistorted except for a delay as long as the line is terminated in its characteristic impedance. If any of these conditions are not met—if the distributed inductance or capacitance is not uniform, or if the line is not terminated in Z_0—reflections can occur that lower transmission efficiency and cause the frequency response of the line to be nonuniform. Although the coaxial line is the only structure we analyzed in detail, there are many other types of conductors that can be used as transmission lines. For example, a standard trace on double-sided fiberglass circuit-board material with a wide ground plane on the opposite side can be made to act as a transmission line whose impedance is a function of the trace's width. The wider the line, the lower the characteristic impedance. For example, if a circuit board about 1.5 mm thick has a solid conducting ground plane on one side and has a relative permittivity of 4.5, a conducting strip about 2.8 mm wide on the top side will show a characteristic impedance of about 50 Ω. Because the dielectric of such a transmission line is not **homogeneous** (uniform) in this structure's cross section (some is fiberglass and some is air), the impedance and the wave velocity will vary somewhat with frequency, making it slightly **dispersive**, but these effects are small for all but the most critical applications.

11.4.2 Filters for Radio-Frequency Interference Prevention

A **filter** for frequencies above 50 MHz or so is difficult to construct with active devices. As we explained previously, the performance of most IC op amps degrades so much at high frequencies that they are no longer useful, and most discrete devices are not much better. For this reason, many filters for use at RF use only passive devices—inductors, capacitors, and resistors—although for special applications, active devices can be used in filters with frequencies as high as the microwave range.

Because RF signals can radiate and travel long distances, an entire engineering specialty has developed that is concerned with **radio-frequency interference** (**RFI**). Chapter 12 covers this subject in detail, but in this section, we will describe some of the simpler ways of dealing with RFI, namely, passive filters.

The goal of an RFI filter is to block the passage of *unbalanced* RF currents while allowing the passage of desired power or signals. We specify unbalanced currents, because balanced currents in properly designed balanced transmission lines do not radiate nearly as much as unbalanced currents do. This fact is the basis for one of the simplest types of RFI filter: the ferrite-choke or ferrite-bead filter.

You have probably seen cylindrical objects a few centimeters long (see Fig. 11.8) attached to cables that carry digital signals, such as USB cables between computers and printers. This type of filter consists of a hollow cylinder of **ferrite** material, which has a very high relative permeability at RF. If an *unbalanced* current happens to be flowing along any part of a cable that passes through such a filter, the current

FIGURE 11.8 Ferrite-choke RFI filter on USB cable carrying digital signals.

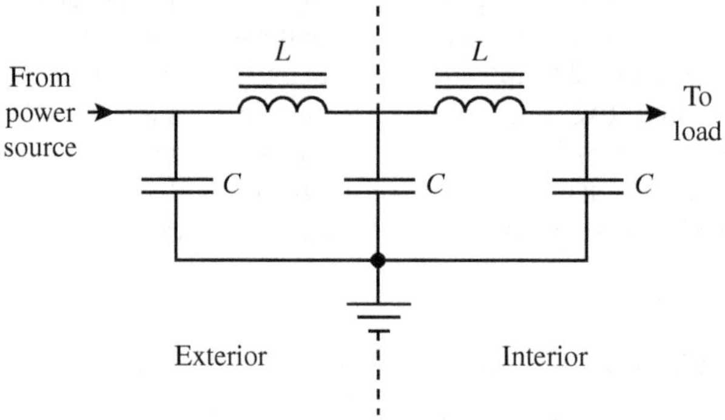

FIGURE 11.9 Power-line RFI filter circuit with five poles.

"sees" a large inductive impedance. Other things being equal, the inductive impedance will reduce the amount of unbalanced current flowing along the cable, which also lessens the chance that the current will radiate or otherwise couple to other systems. Because the desired digital signals are all transmitted along *balanced* transmission lines in such cables, the desired signal currents produce virtually no magnetic field outside the cable. So the ferrite-choke filter is invisible to them and they pass through it unaffected.

Another type of common RFI filter is the **power-line RFI filter**. The main purpose of this type of filter is to ensure that RF energy produced in a power-line-operated piece of equipment does not travel back along the AC cord set into the building wiring, where it could follow the same path into other systems and cause problems. Figure 11.9 shows the schematic diagram of a better-quality filter of this type.

It consists of two inductors L and three capacitors C. The inductors usually have cores made of ferrite or other high-permeability material, indicated by the double lines above the inductance symbols. The capacitors must be large enough to present a low impedance at the lowest interference frequency for which the filter is designed, but small enough not to conduct significant power-line current. This is a five-pole lowpass filter, because each reactive component in such a filter contributes a pole to the circuit's transfer function. Above the filter's **cutoff frequency**, usually in the low kilohertz range, each pole can contribute about 20 dB per decade of attenuation, so the five-pole circuit in Figure 11.9 could provide as much as 100 dB of attenuation at a decade higher than the cutoff frequency, although parasitic reactances usually limit the maximum to 60–70 dB or so. The capacitors must be designed to withstand the highest expected power-line voltage and the inductors must carry the largest expected power-line current without **saturating**. Such filters are most effective when they are installed so they penetrate a **metallic shield** that (ideally) completely encloses the system that generates the RF energy. This shield is symbolized by the dashed line in Figure 11.9 that separates the interior of the system from the outside. In this way, any stray RF currents that are traveling along the inside of the shield stay there, rather than passing to the exterior through an opening.

Besides digital circuits, many other types of electronic systems can produce RF currents that interfere with other systems. Switching power supplies (discussed in Chapter 10) produce high-current transients that have many RF harmonics, and these currents can cause interference unless means such as power-line filters are used to prevent it.

11.4.3 Transmitter and Receiver Filters

Besides their use in preventing unwanted RF currents from escaping to cause interference, passive RF filters find many uses in RF communications systems, specifically transmitters and receivers. Because many RF power devices operate in a nonlinear mode, they produce **harmonics** (multiples) of the desired RF signal. These harmonics, if radiated, can cause interference and represent loss of useful power as well. For these reasons, passive RF filters are used in most transmitters ranging from low-power devices such as used with the Bluetooth standard up to high-power systems used for FM and TV broadcasting. At the lower and middle ranges of the RF spectrum, these filters are usually simple lowpass types with a circuit resembling the power-line filter of Figure 11.9, except that the component values are chosen to be much smaller to pass the desired transmitted frequency with little or no attenuation. Because the second harmonic is twice the frequency of the fundamental, such filters need only discriminate against signals that are at least an octave above (twice the frequency of) the fundamental. Using the approximate rule that each pole of a filter provides attenuation at the rate of 6 dB per octave above the cutoff frequency, a five-pole filter such as that shown in Figure 11.9 could provide about $(5 \times 6) = 30$ dB of attenuation to the second harmonic of a signal and more for higher harmonics.

At the receiver end, one important task is usually to **select** a relatively narrow band of frequencies to be processed by the receiver electronics while rejecting all signals outside that band. This is a task for a **bandpass filter**, designed to pass a range of frequencies around a center frequency f_0. Simple bandpass filters were discussed in Chapter 6, and one of the simplest types consists of an *L–C–R* circuit shown in Figure 6.14 and reproduced here as Figure 11.10.

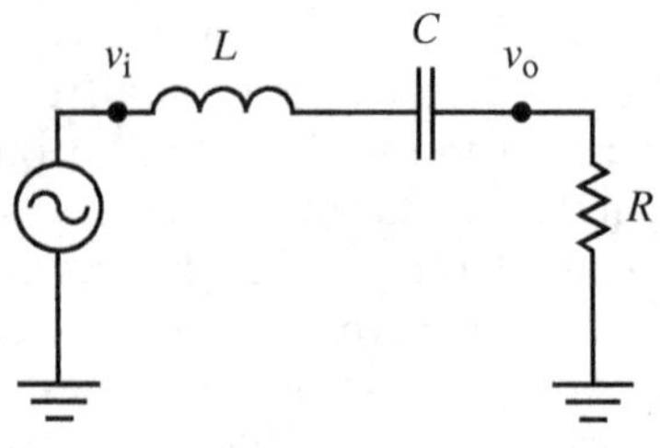

FIGURE 11.10 Simple series *L–C–R* bandpass filter.

The basic idea of a bandpass filter is simple. A single frequency (or range of frequencies) is to be passed with minimum attenuation, while all frequencies outside the desired range are to be blocked. At the resonant frequency f_0 of the L and C in Figure 11.10, which is

$$f_0 = \frac{1}{2\pi\sqrt{LC}}, \tag{11.21}$$

their reactances are equal in magnitude but opposite in sign, so they cancel, allowing any signal at f_0 to pass through with no attenuation. Frequencies removed from f_0 encounter a net series impedance that is larger for signals farther removed from f_0.

The ability of a bandpass filter to transmit the desired frequency range while rejecting signals outside that range is called the **selectivity**. A simple **two-pole** filter such as the one in Figure 11.10 provides only about 20 dB of rejection for every decade that the interfering signal is removed from the passband center frequency f_0. While this amount of rejection may be adequate for some purposes, many radio receivers and other systems require much better selectivity than this. Accordingly, filters with more than two poles can be designed to have nearly flat response for a passband of desired frequencies, with extremely steep "skirts" that fall by as much as 80–100 dB in a few kilohertz. An example of a bandpass filter response of this type is shown in Figure 11.11. This filter is designed to pass a signal with a bandwidth of about 3 kHz centered at a frequency $f_0 = 500$ kHz. In this range, the signal is transmitted with a relative loss (compared to the minimum loss, which is scaled to 0 dB) of less than 6 dB. However, for frequency offsets larger than ±4 kHz away from the center frequency, the filter provides attenuation of almost 80 dB. Such filters are useful in the **intermediate-frequency** (**IF**) amplifiers of radio receivers, because they can select one of many channels of signals from the radio spectrum while effectively rejecting all the others.

Bandpass filters can be designed to use passive circuits made with inductors and capacitors, but modern designs use less bulky and more efficient approaches. Quartz-crystal or ceramic resonators can be used in bandpass filter designs, and **microelectromechanical system** (**MEMS**) devices have recently been developed for similar purposes. And of course, most filtering functions, at least at lower frequencies, can be carried out much more efficiently and flexibly than by analog circuits with **digital signal processing** (**DSP**) circuitry after transforming the RF

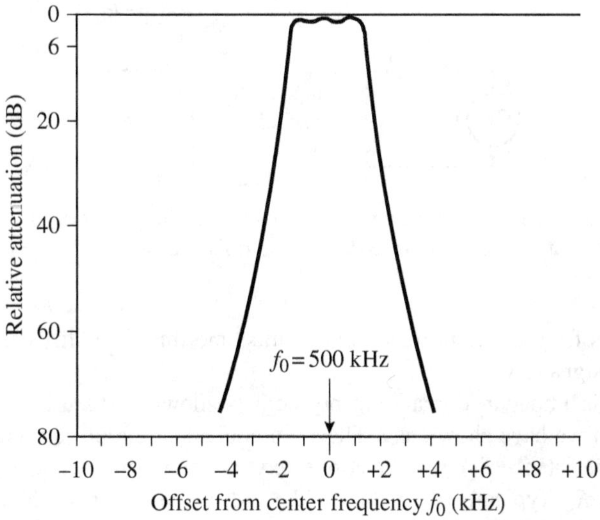

FIGURE 11.11 Response of f_0 = 500-kHz narrowband bandpass filter used for communications receivers.

signal into digital form. However, in situations where the signal is either too weak or noisy to be converted into digital form or too powerful (as in filters for transmitter outputs), passive analog filters may be the best option.

11.4.4 Impedance-Matching Circuits

For several reasons, proper impedance matching is especially important in RF circuit design. In transmitters, generating RF energy is costly in terms of circuit complexity and expense for special-purpose RF devices. A circuit that generates 100 W of expensive RF energy only to lose 80 W in an inefficient mismatched output circuit is far from realizing its ultimate potential. In receivers, an important performance criterion is the **signal to noise ratio (SNR)**, which determines the noise level in analog systems and affects the **bit error rate (BER)** in digital systems. When part of the energy of an incoming signal has been lost due to impedance mismatches or other factors, it can never be recovered, and no amount of subsequent signal processing or amplification will get it back, other things being equal. Finally, poor impedance matching to transmission lines can cause reflections that result in **frequency distortion**—an uneven or irregular frequency response over the desired band of signal frequencies. For these and other reasons, good impedance matching is an essential part of RF design.

The topics of impedance matching and filters overlap, because every impedance-matching circuit has a certain frequency response, and the response of every passive filter circuit can be affected by its source and load impedances. For the purposes of this section, we will assume that the required impedance transformation is required to work at only a single frequency f_0, and any filtering action is secondary to the impedance-matching task. As we will see, one can obtain lowpass or highpass filter

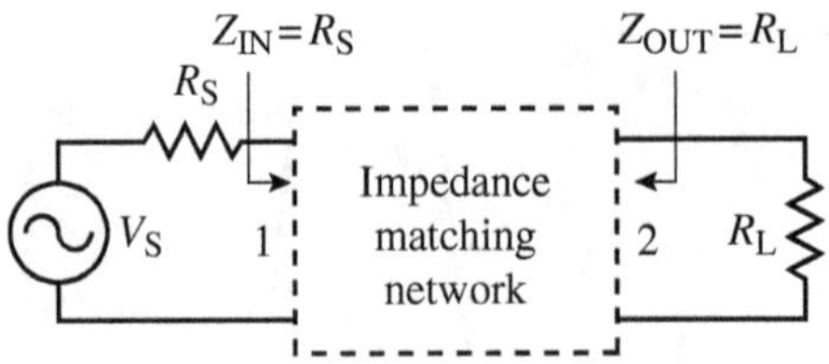

FIGURE 11.12 Operation of general impedance-matching network at a single design frequency f_0, showing source resistance R_S and load resistance R_L.

characteristics from different matching circuits, and this flexibility can be useful in RF circuit designs as well.

The generic impedance-matching problem is shown in Figure 11.12. A signal source, which we have shown as a Thévenin equivalent circuit (a voltage source in series with a resistance R_S), is to be connected to a load resistance R_L whose value is different than R_S. Typically, the values of R_S and R_L are not under the control of the designer. For example, R_S might be the output resistance of a solid-state RF power device, which can be as low as a few ohms, and R_L might be the input impedance of a 50-Ω transmission line. (In reality, the source and load circuits may present **complex** impedances at f_0, but this can be dealt with by **absorbing** the reactive part in the network, as we will describe in the following.)

While there is an infinite variety of passive circuits that can perform impedance matching, we will concentrate only on those with two or three lumped-element components (inductors and capacitors). We will assume the matching elements are ideal (no losses), although losses can be accounted for if present. The type of impedance-matching circuit using two elements is called an **L network**, while the two types that use three elements are called the **pi network** and the **T network**. All these names derive from the resemblance of the respective circuit to the letter it is named for, although the *L* of the *L* network is upside-down!

11.4.4.1 L-Network Matching Circuits There are two types of *L*-network matching circuits, a **lowpass** and a **highpass** type, named for the type of filtering action they perform at frequencies away from the design frequency f_0. At a given design frequency with known source and load resistances, a two-component *L* network has no free parameters. That is, once the type of network is chosen (highpass or lowpass), the component values for a given transformation and frequency are completely determined. For either the lowpass or the highpass *L* network, there is a series element connecting input to output and a shunt element to ground. The shunt element always goes between ground and the terminal that connects to the *higher* of the two impedances to be matched. So the upside-down *L* of the *L* network faces one way or the other, depending on whether the source impedance is lower or higher than the load impedance.

All these matters are taken into account in the formulas presented in Figure 11.13. To show how to use the formulas, we will work through an example.

Suppose a certain RF filter has a nominal output resistance R_S of 1 kΩ at a center frequency of $f_0 = 455$ kHz. (This is a frequency that is often used in the IF amplifiers of

L network formulas

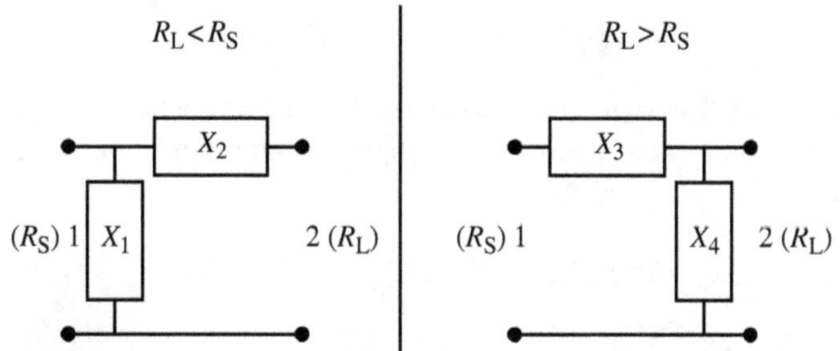

$$Q_D = \sqrt{\frac{R_S}{R_L} - 1} \qquad\qquad Q_U = \sqrt{\frac{R_L}{R_S} - 1}$$

$$X_1 = \frac{R_S}{Q_D} \quad X_2 = Q_D R_L \qquad\qquad X_3 = Q_U R_S \quad X_4 = \frac{R_L}{Q_U}$$

$$\text{LP: } C_1 = \frac{1}{\omega_0 X_1} \quad L_2 = \frac{X_2}{\omega_0} \qquad\qquad \text{LP: } L_3 = \frac{X_3}{\omega_0} \quad C_4 = \frac{1}{\omega_0 X_4}$$

$$\text{HP: } L_1 = \frac{X_1}{\omega_0} \quad C_2 = \frac{1}{\omega_0 X_2} \qquad\qquad \text{HP: } C_3 = \frac{1}{\omega_0 X_3} \quad L_4 = \frac{X_4}{\omega_0}$$

FIGURE 11.13 Formulas for design of lowpass (LP) and highpass (HP) L-network matching circuits between source resistance R_S and load resistance R_L.

radio receivers.) Let's say we need to transform the 1000-Ω resistance to a lower load resistance R_L of 50 Ω, in order to send the output through a 50-Ω transmission line. And because it is generally good practice to attenuate any harmonics that may be present unless there is a reason not to, we will choose a lowpass filter configuration for the task.

Because the load resistance is lower than the source resistance ($R_L < R_S$), we will use the equations on the left-hand side of Figure 11.13. The first quantity to calculate is Q_D (the D stands for "down" because the circuit is stepping *down* the source resistance). This quantity is easily found to be

$$Q_D = \sqrt{\frac{R_S}{R_L} - 1} = \sqrt{\frac{1000\ \Omega}{50\ \Omega} - 1} = \sqrt{19} = 4.359 \tag{11.22}$$

The next step is to find the reactances X_1 and X_2. The values of these reactances (in Ω) are the same regardless of whether a highpass or lowpass configuration is chosen. Using the equations given in Figure 11.13, we find that

$$X_1 = \frac{R_S}{Q_D} = \frac{1000\ \Omega}{4.359} = 229.42\ \Omega \tag{11.23}$$

and

$$X_2 = Q_D R_L = 4.359 \cdot 50\,\Omega = 217.94\,\Omega \tag{11.24}$$

At this point, we choose to use the lowpass (*LP*) equations, which will give a shunt (grounded) capacitor value C_1 and a series (input-to-output) inductor value L_2. Using the fact that

$$\omega_0 = 2\pi f_0 = 2\pi(455\ \text{kHz}) = 2.8588 \times 10^6\ \text{s}^{-1}, \tag{11.25}$$

we find that

$$C_1 = \frac{1}{\omega_0 X_1} = \frac{1}{(2.8588 \times 10^6\ \text{s}^{-1})(229.42\,\Omega)} = 1.5247\ \text{nF} \tag{11.26}$$

and

$$L_2 = \frac{X_2}{\omega_0} = \frac{217.94\,\Omega}{2.8588 \times 10^6\ \text{s}^{-1}} = 76.233\ \mu\text{H} \tag{11.27}$$

If these values are used in the lowpass step-down *L* network shown in Figure 11.14, the input impedance looking into port 1 will be exactly 1000 Ω at the design frequency of 455 kHz, assuming the load resistance is truly 50 Ω and vice versa.

A good way to check your calculations for an *L* network is to imagine cutting the circuit open at the junction of the two reactive components and calculating the impedance seen looking both ways. The results should be complex conjugates of each other, indicating that maximum power transfer occurs across the junction. To show how to do this, we will carry out this check with the example earlier. Figure 11.15

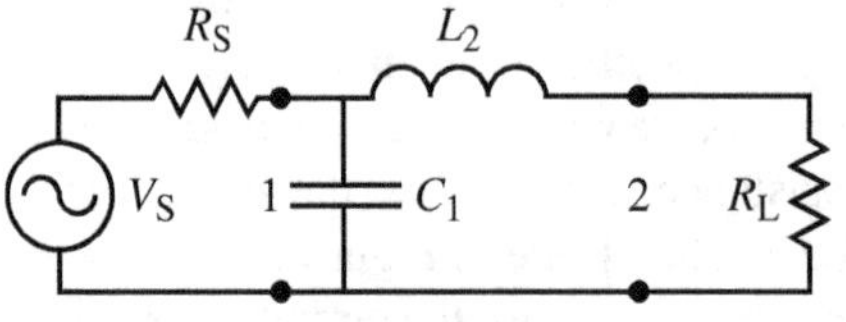

FIGURE 11.14 Lowpass step-down *L* network used in example where $R_S = 1\ \text{k}\Omega$, $R_L = 50\,\Omega$, $f_0 = 455\ \text{kHz}$, $C_1 = 1.5247\ \text{nF}$, and $L_2 = 76.233\ \mu\text{H}$.

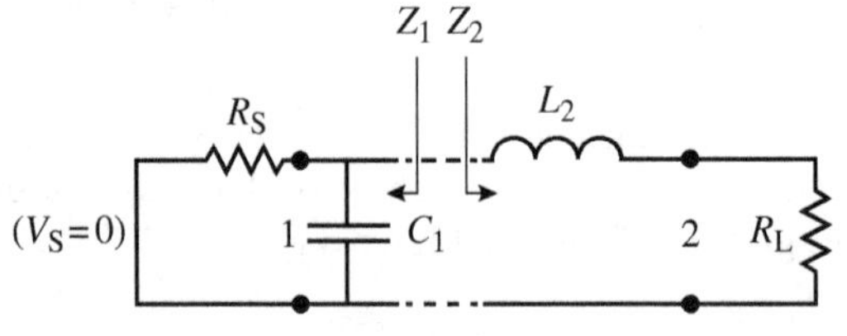

FIGURE 11.15 Checking example of Figure 11.14 to ascertain whether impedances Z_1 and Z_2 are complex conjugates.

shows the matching network split in the center so that we can calculate the impedance Z_1 that the parallel combination of source resistance R_S and capacitor C_1 presents to the series combination of inductor L_2 and resistance R_L:

$$Z_1 = \frac{1}{1/R_S + j\omega_0 C_1} = \frac{1}{(1 + j4.3588)\ \text{mS}} = (50.002 - j217.95)\ \Omega \qquad (11.28)$$

(The voltage source V_S is set to zero for this calculation, leaving a short circuit in its place.) The impedance Z_2 that Z_1 faces is easily calculated as

$$Z_2 = R_L + j\omega_0 L_2 = 50 + j217.95 \qquad (11.29)$$

These check calculations confirm that the complex-conjugate condition $Z_1 = Z_2^*$ is met almost precisely.

There are some practical considerations you should be aware of when using *L* networks for impedance matching. Strictly speaking, the match is exactly correct at only the design frequency $f_0 = \omega_0/2\pi$ and gradually falls away from the maximum-power-transfer condition for frequencies increasingly removed from f_0. However, the rate at which this occurs is proportional to the circuit Q (Q_U or Q_D, as the case may be). And for practical impedance-matching ratios, which rarely exceed about 20–1, the Q is less than 5, meaning that for a 10–20% bandwidth near f_0, the loss due to mismatch is less than 3 dB. All the foregoing discussions assume that the Q of each matching component (inductors and capacitors) is much higher than the circuit Q. This is usually true of capacitors, but inductors for use at RF can have fairly low Q values of 20 or less (the Q of an inductor expresses all losses that occur in it at a given frequency). The effect of a finite component Q on a matching network is to increase the loss through the network, which shows up both as mismatch loss and dissipative loss. But as long as the Q of each component is at least 10 times the circuit Q, these losses can usually be neglected.

11.4.4.2 Pi and T Matching Networks While the values for an *L* network can always be found to allow a match between two different resistances, there are no free parameters that the designer can adjust, and so for a particular pair of resistances to be matched, there is exactly one set of unique component values for each type of *L* network (lowpass or highpass). But if the source or load is an **impedance**, having reactance as well as resistance, an *L* network may not be able to **absorb** the source or load reactance, because of its inflexibility. In this context, "absorbing" means to incorporate the reactance of the source or load in the matching circuit itself by reducing the value of a capacitor in parallel (or an inductor in series) so as to achieve the desired circuit values while including the reactance. An example will illustrate this process.

Suppose in the problem of matching a 1000-Ω source to a 50-Ω load, the source is not a pure resistance. Instead, its equivalent circuit is a 1000-Ω resistance in parallel with a 200-pF capacitance. In this case, the 200-pF capacitance can be absorbed in the 1525-pF matching-network capacitor by

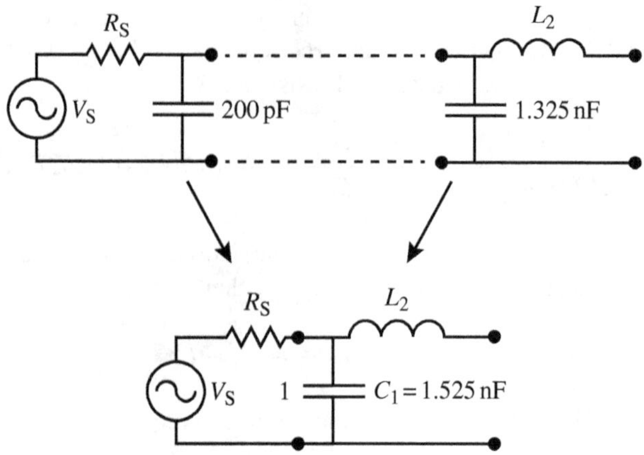

FIGURE 11.16 Showing absorption of 200 pF of source capacitance in L network's C_1.

reducing the matching-network capacitor by 200 pF to 1.325 nF (see Fig. 11.16). When the reduced capacitor is connected to the source, the source's 200 pF creates a total capacitance of 1.525 nF at that junction, which is what the design calls for. But if the extra capacitance had been across the 50-Ω load, the lowpass L network would not have been able to absorb it, because the L network for that matching problem does not allow for any capacitance at its output.

In this situation and many others, therefore, a three-element matching network is called for. The two possible configurations of three-element networks are called the *pi* network and the T network. As its name implies, the *pi* network consists of two shunt elements to ground on either side of a single series element. For the lowpass *pi* network, the shunt elements are capacitors and the series element is an inductor, while the reverse is true for the highpass *pi* network.

The design of a *pi* network can be approached by considering it to be two L networks connected back-to-back in cascade. Instead of directly matching the source resistance R_S to the load resistance R_L, each L network matches to a (hypothetical) intermediate resistance R_{IP}, which is *lower* than either R_S or R_L. The value of R_{IP} is a free choice of the designer. Once R_{IP} has been chosen, the two L networks are designed independently, and then their series elements X_2 and X_3 are combined to form $X_P = X_2 + X_3$, as shown in the top half of Figure 11.17. The resulting *pi* network has a total Q_P of

$$Q_P = \sqrt{\frac{R_S}{R_{IP}} - 1} + \sqrt{\frac{R_L}{R_{IP}} - 1} \tag{11.30}$$

This is simply the sum of the Q's of the two L networks making up the *pi* network, so Q_P is higher than that of either L network by itself. It is easy to see from Equation 11.30 that as the intermediate resistance R_{IP} becomes lower, the value of Q_P rises. Generally, one should not use a Q_P higher than about 10–15 or so, because otherwise

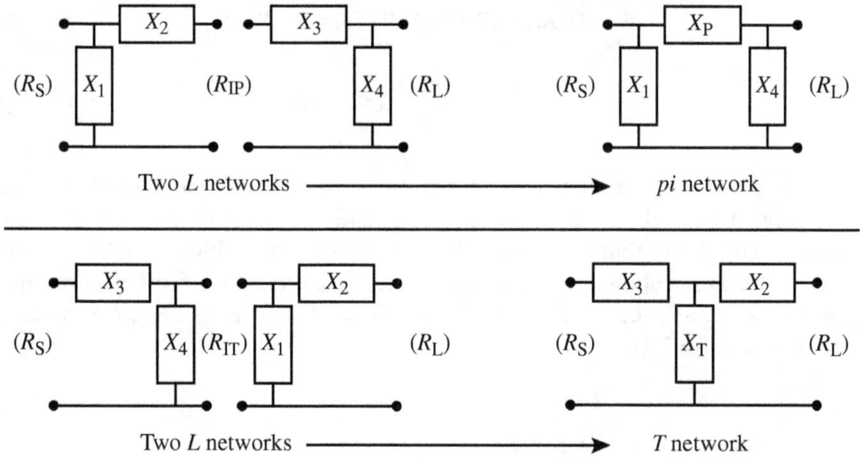

FIGURE 11.17 Showing how two cascaded *L*-type matching networks can be combined to form either a *pi* network (top) or a *T* network (bottom).

the losses in the matching-network components may lower the circuit's efficiency unacceptably.

One advantage of the *pi* matching network is that it has capacitors at both the source and load terminals, which makes it possible to absorb source and load capacitances and still achieve a good match.

One way to begin the design of a *pi* matching network is to start with the value of an unavoidable shunt reactance and determine its *L*-network matching ratio to find R_{IP}. For example, suppose we want to match a resistive 200-Ω source to a 50-Ω load at 13.56 MHz, and the 50-Ω load has a capacitive reactance of 200 pF in parallel with it. The Q of the load is

$$Q = \frac{R}{X} = \omega_0 CR = 2\pi(13.56\text{ MHz})(200\text{ pF})(50\ \Omega) = 0.852 \qquad (11.31)$$

Setting this Q equal to Q_U in the equations of Figure 11.13, solving for $R_S = R_{IP}$ in terms of Q_U and R_L, and letting $R_L = 50\ \Omega$ give us the value of R_{IP}:

$$R_{IP} = \frac{R_L}{Q_U^2 + 1} = 28.97\ \Omega \qquad (11.32)$$

With this choice for R_{IP}, we can find X_3, X_2, and X_1 from the formulas in Figure 11.13 (X_4 is already known from the 200-pF capacitor in parallel with the load resistance):

$$X_3 = Q_U R_{IP} = (0.852)(28.97\ \Omega) = 24.68\ \Omega \qquad (11.33)$$

$$Q_D = \sqrt{\frac{R_S}{R_{IP}} - 1} = \sqrt{\frac{200\ \Omega}{28.97\ \Omega} - 1} = 2.43 \qquad (11.34)$$

$$X_2 = Q_D R_{IP} = (2.43)(28.97\ \Omega) = 70.39\ \Omega \tag{11.35}$$

$$X_1 = \frac{R_S}{Q_D} = \frac{200\ \Omega}{2.43} = 82.31\ \Omega \tag{11.36}$$

If the lowpass configuration for both L networks is chosen, the shunt reactance X_1 is capacitive and the two reactances X_2 and X_3 in series become a single inductor. The steps required to get from the initial matching problem to the finished circuit are shown in Figure 11.18. Because the step-down L network has $Q_D = 2.43$ and the step-up L network has $Q_U = 0.852$, the entire *pi* network's $Q = 2.43 + 0.852 = 3.28$.

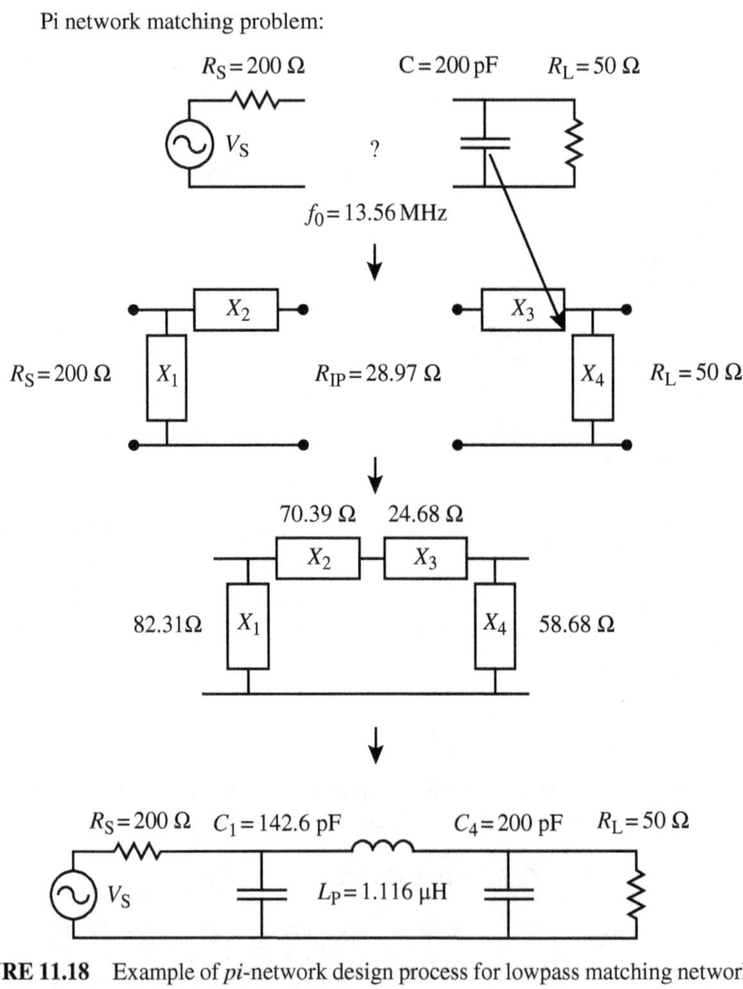

FIGURE 11.18 Example of *pi*-network design process for lowpass matching network incorporating capacitance across 50-Ω load.

Similar steps are performed for the design of a *T* network, except that the intermediate hypothetical resistance R_{IT} for a *T* network is *greater* than either the source resistance R_S or the load resistance R_L. This places the center terminal of the *T* network at a high impedance, which might cause high-voltage stress problems in high-power circuits. But for low-power transmitting and receiving circuits, this is generally not a problem.

11.4.4.3 Transformers and Baluns While the various two- and three-component matching networks perform well for single-frequency and narrow-bandwidth systems, many communications applications involve a relatively wide band of frequencies. If a band of frequencies of interest extend from a lower limit f_L to an upper limit f_H, the **percentage bandwidth** *P* can be expressed by the ratio

$$P = \frac{f_H - f_L}{(f_H + f_L)/2} \times 100\% \qquad (11.37)$$

For percentage bandwidths greater than about 20%, more complex discrete-component matching circuits than the simple two- and three-component circuits shown earlier must be used to maintain a fairly flat response over the entire range of frequencies. While there are ways to design such circuits, they become increasingly complicated as the bandwidth increases. For wideband amplifiers covering up to an **octave** (2:1 ratio) or more of bandwidth, transformers are often used in addition to inductors and capacitors for matching among active devices, sources, and loads.

The design of RF transformers is beyond the scope of this text, involving selection of core materials (iron powder, ferrites of many types, or others), calculation of maximum flux density to avoid saturation, choice of wire diameter and number of turns, and other factors. But from the user's point of view, every RF transformer has a frequency range or bandwidth whose limits are determined by the elements of a transformer's equivalent circuit. An equivalent circuit suitable for a general-purpose RF transformer with two separate windings, called the **primary** and the **secondary**, is shown in Figure 11.19.

At the heart of the equivalent circuit is an **ideal transformer**, a mathematical fiction that is nevertheless a good model for many transformers in the region of

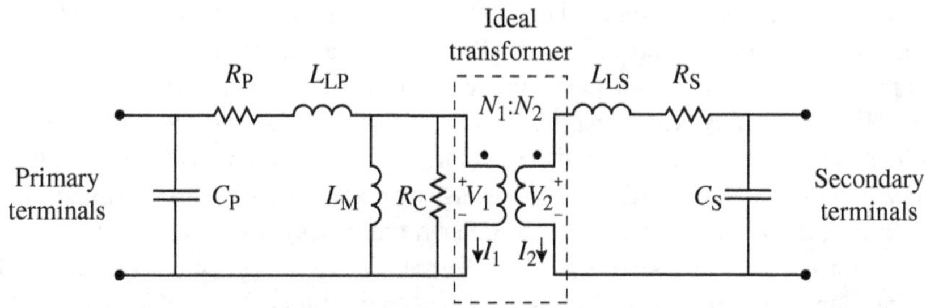

FIGURE 11.19 Equivalent circuit of RF transformer with primary and secondary windings.

operation for which they were designed (including power transformers at low and medium frequencies). If V_1 and V_2 are AC voltages and I_1 and I_2 are the corresponding AC currents with the polarities and directions shown in Figure 11.19, the following relations apply for a transformer having N_1 turns in its primary winding and N_2 turns in its secondary:

$$\frac{V_1}{V_2} = \frac{I_2}{I_1} = \frac{N_1}{N_2} \tag{11.38}$$

Equation 11.38 says that the voltage ratio is in direct proportion to the turns ratio, while the current ratio is in *inverse* proportion. The polarity of the secondary voltage with respect to the primary voltage is indicated on schematics by the dots shown in Figure 11.19, with the dotted terminals having the same polarity.

A consequence of Equation 11.38 is that inserting an ideal transformer between a source and a load transforms the load impedance by a factor that is the *square* of the turns ratio. If the secondary of an ideal transformer has N_2 turns and is connected to a load Z_2, the impedance Z_1 looking into the transformer primary with N_1 turns is

$$Z_1 = \left(\frac{N_1}{N_2}\right)^2 Z_2 \tag{11.39}$$

It is this property that makes transformers so useful for broadband impedance matching in RF circuits.

While transformers can be designed to operate in a nearly ideal fashion over two or more decades of bandwidth, all transformers have low- and high-limitations, as the equivalent circuit in Figure 11.19 indicates.

The low-frequency limit of a transformer is determined mainly by the value of L_M, sometimes called the **magnetizing inductance**. Roughly speaking, this is the inductance that the transformer would have if only the primary winding were present and the component was considered as an inductor. For efficient operation, the impedance of the magnetizing inductance must be much larger than the impedance of the circuit that the transformer is used in. At a sufficiently low frequency, however, the inductive reactance of L_M falls and starts to bypass current away from the ideal-transformer portion of the equivalent circuit. Ultimately, of course, at DC, no transformer can operate, but in practice, the low-frequency limit of a transformer is determined by the value of magnetizing inductance. This is one reason why low-frequency transformers require larger windings and cores than RF designs: a transformer for use at 60 Hz must have a magnetizing inductance on the order of henries, while an RF transformer for 30 MHz can get by with a magnetizing inductance in the microhenry range.

Most transformers use a magnetic **core** that increases the amount of stored magnetic energy in a given volume. All core materials show some loss, and one way to model this loss (which typically increases at higher frequencies) is with an equivalent core loss resistor R_C. Loss also occurs in a transformer due to the nonzero resistance of the windings, and these resistances are represented by resistors R_P in the primary and R_S in the secondary. You can measure the DC resistance of a transformer winding to check whether it is open or not, but at higher frequencies, the AC resistance of wires can actually be higher than at DC due to a phenomenon called the **skin effect**. The values of R_P and R_S will never be lower than the DC winding resistance, however.

The high-frequency limit of a transformer is established by the values of the leakage inductances L_{LP} and L_{LS} together with the winding capacitances C_P and C_S. These components form lowpass filters whose cutoff frequencies determine the upper frequency limit of the transformer. Leakage inductance is caused when the primary and secondary windings do not share all their magnetic flux lines in common and is minimized by good design and techniques such as **bifilar windings**, which are achieved by winding a transformer with a double wire so that both primary and secondary wires follow the same path. Winding capacitance is unavoidable, but can be reduced by good design approaches and minimizing the number of coil turns. Many RF transformers incorporate bifilar windings that act like **transmission lines**, with the result that the winding capacitance is distributed uniformly along the winding. The distribution of this capacitance in this way can greatly extend the upper frequency limit of the transformer with proper design. Such transformers can easily cover a frequency range of 1000–1 or more. Figure 11.20 is a pictorial drawing of a 4:1 step-up transformer showing how the paired transmission-line wires are wound around a ferrite ring-shaped core and connected to a source at B–C and a load at A–C, as shown in the more conventional circuit diagram in Figure 11.21. Strictly speaking, the device in Figure 11.21 is an **autotransformer**, meaning that the input and output circuits share one or more windings in common. But additional windings can be added to provide DC isolation between primary and secondary and balanced-to-unbalanced conversion, as mentioned briefly in Chapter 5. A transformer that produces symmetrical (equal and opposite) output voltages with respect to ground when provided with an unbalanced input voltage with respect to ground at a single input

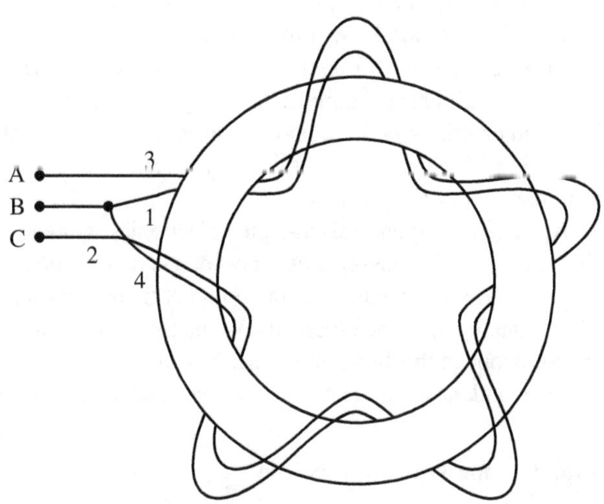

FIGURE 11.20 Pictorial drawing of RF transformer using transmission-line bifilar winding to perform a 4:1 impedance transformation. Paired wires are wound as shown on ferrite ring. Adapted from Ruthroff (1959).

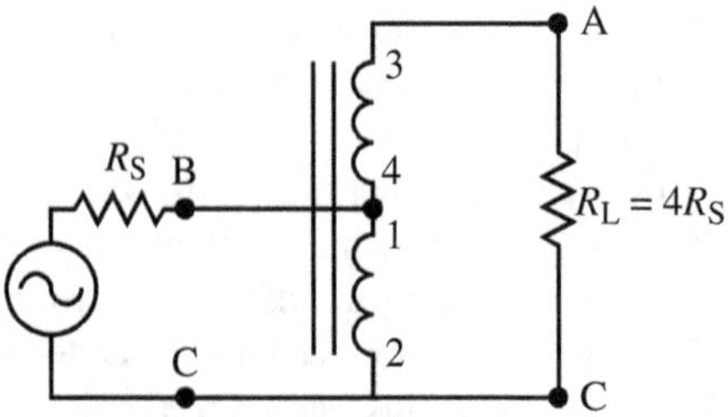

FIGURE 11.21 Conventional schematic diagram of 4:1 impedance-matching transformer shown in Figure 11.20.

terminal is called a **balun** and is useful in driving both balanced transmission lines (as used in many digital circuits) and balanced amplifier stages.

11.5 RF AMPLIFIERS

Amplifying RF signals is a more challenging task than at lower frequencies for the reasons already stated. Especially for high-power systems, the physical dimensions of the circuits and devices used may become comparable to a wavelength at the highest frequency of operation, and the associated phase-delay effects must be taken into account in the design. All conventional electronic devices operate less efficiently at higher frequencies, so special devices are designed expressly for RF use. And because these devices can be costly, RF designers must pay added attention to efficiency. RF energy is expensive, so a design that uses a costly component to produce 100 W of raw RF power and then proceeds to burn up 70 W in circuit losses is a bad design, no matter what else it does. And for low-power circuits such as radio receivers, poor circuit efficiency can dissipate weak incoming signals and degrade the all-important **signal-to-noise ratio** of a system. In analog systems, a lower signal-to-noise ratio raises the noise level for a constant amplitude of signal, and in digital systems, it increases the BER, other things being equal. So efficient designs that transmit the highest possible fraction of energy while dissipating or reflecting a minimum amount are increasingly important for HF designs.

As with other types of analog and mixed-signal electronics, **systems on a chip** are available for RF designs as well. But even when most of a system is contained within one or more ICs, the system designer needs to know certain basic principles of RF design in order to apply RF ICs intelligently and efficiently. While the finer details of RF amplifier design are beyond the scope of this book, in this section, you will learn some basic principles that will aid both understanding of existing designs and development of new ones.

11.5.1 RF Amplifiers for Transmitters

In the context of RF design, a **transmitter** is a system that produces an RF signal intended to be transmitted through space from a **transmitting antenna** to one or more **receiving antennas**. Connected to each receiver antenna is a **receiver.** Nearly

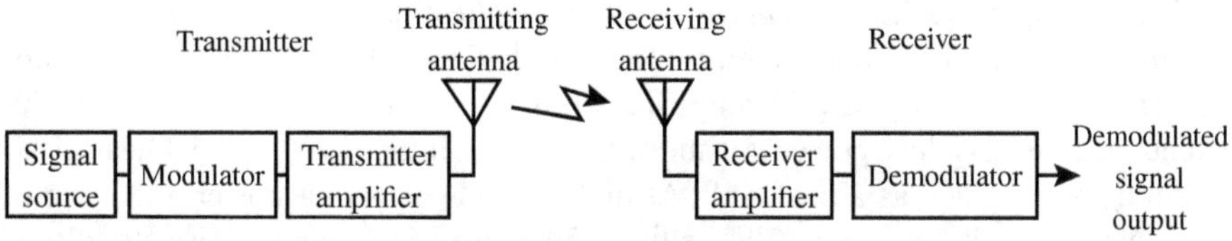

FIGURE 11.22 Block diagram of generic RF radio link system to transmit data over a wireless link.

all "wireless" systems that use RF signals (as opposed to infrared or visible-light-wavelength signals) use some variation of the basic block diagram shown in Figure 11.22. The goal is usually to send information from one location to another without interconnecting wires. The signal source (which can be either analog or digital) sends its output to a **modulator**. The modulator impresses the signal's information on one or more **carriers**. The carriers are frequencies in the RF band chosen (and authorized by the appropriate regulating authority) for transmission of the information through space. For example, Table 11.1 and Figure 11.1 show that Bluetooth devices all produce carriers in the range of 2.4–2.485 GHz. Because transmission of RF energy through space is an inefficient process, a great deal more power must be transmitted from the transmitting antenna than is received at any receiver. The reason for this is simple: most low- and medium-cost RF systems cannot use highly directional antennas, so the energy is radiated more or less evenly in all directions, and only a tiny fraction of it ever reaches a receiving antenna. This **space loss** is rarely less than 60 dB and can easily reach 120 dB or more. The space loss is represented by the lightning-like symbol connecting the transmitting and receiving antennas in Figure 11.22.

At the receiver, the RF signal is usually very feeble, with an amplitude of microvolts to millivolts. Such small signals are not easily handled by conventional digital circuitry, so they are often amplified with an RF receiver amplifier specially designed not to add noise to the already weakened signal. The amplified RF signal is sent next to a **demodulator**, which performs operations inverse to that of the modulator and produces a more or less faithful replica of the original signal at the transmitter. The demodulated signal is then sent to the rest of the system for use.

You can see how this same basic description covers most wireless systems: mobile phones, **wireless local area networks (WLANs)**, all radio and TV broadcasting, and more exotic applications such as satellite links and signals from interstellar space probes. While the details of each block in the block diagram can vary a great deal, the same basic ingredients are nearly always present. As digital technology advances, larger portions of the transmitter and receiver are implemented with digital circuitry, but the operations themselves remain the same. For example, modulators and demodulators are now almost always digital circuits, although the circuits directly connected to the antennas are usually analog in nature.

With this background in mind, we will proceed with the description of RF amplifiers for transmitters.

11.5.1.1 Amplifiers for Transmitters Many applications for local transmission of signals over distances of a few meters require only a few milliwatts of power to provide efficient data links. Any RF amplifier whose total output power is less than 1 W can be regarded as a low-power transmitter, requiring little, if any, special heat-dissipation measures such as a heat sink. At higher levels of output power, the issue of efficiency arises because heat dissipation and power consumption rise rapidly as efficiency falls. There are various definitions of efficiency for power amplifiers. One that is commonly used is termed **power-added efficiency** η_{PA} and is defined as

$$\eta_{PA} = \frac{P_{OUT(RF)} - P_{IN(RF)}}{P_{DC}} \times 100\% \tag{11.40}$$

This definition of efficiency takes into account the fact that an RF amplifier with relatively low gain may require a substantial amount of RF power to drive it. The power-added efficiency considers only the *net* amount of RF power added by the amplifier, not the total output power, which may not be all that much greater than the input power in the case of low-gain amplifiers.

RF amplifiers for transmitters require different design approaches depending on the bandwidth of the amplified signal. Narrowband amplifiers can be designed for a single center frequency and will usually perform well enough to be used in a 10% or so bandwidth around that frequency. An example of a discrete-component medium-power RF amplifier suitable for a narrowband application is shown in Figure 11.23.

A notable feature of the amplifier in Figure 11.23 is that uses its BJT device in a **common-base** configuration. Bipolar transistors can amplify with any one of their three terminals grounded (used as the common terminal between input and output): the emitter, the base, or the collector. The most frequently used configuration for amplifiers is the common-emitter one, but at RF, the **collector–base capacitance** conducts enough current to interfere with gain. This capacitance forms an undesirable feedback path that either reduces gain or causes undesirable oscillations unless measures are taken to **neutralize** it. (This problem also arises with most FETs, in which the drain-gate capacitance causes problems.)

The common-base configuration is desirable for RF amplifiers because the base terminal can be grounded (as is done in Figure 11.23 through the base bypass capacitor C_2). With the base grounded, any current flowing through the collector–base capacitance due to the output voltage simply flows harmlessly to ground, without causing undesirable feedback into the input circuit (the emitter).

One reason the common-base configuration is not often used at lower frequencies is that the input impedance is that of a forward-biased diode, namely, the emitter-base junction. This is often very low, 100 Ω or less, and such a low impedance is not often used at audio frequencies. But in RF designs, a 50-Ω impedance level is useful, and proper choice of device Q_1 and bias conditions can make the input impedance of the common-base amplifier quite close to 50 Ω. It should be noted that this low-level amplifier is biased for **class A** operation, meaning that collector current flows during the entire 360° of a sine-wave input signal (see Chapter 10 for more information about amplifier classes).

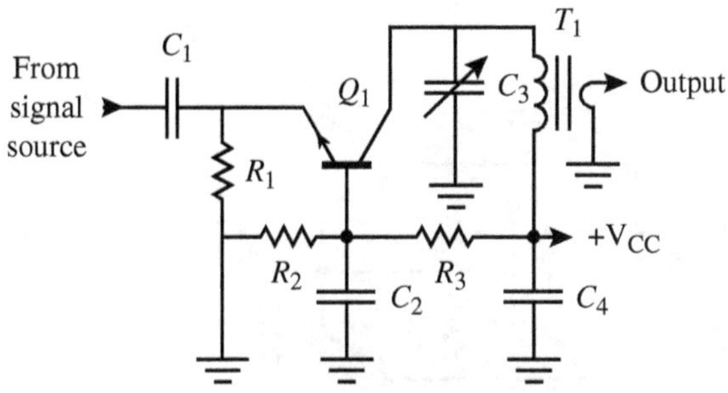

FIGURE 11.23 Narrowband RF amplifier using a discrete BJT in a common-base configuration.

Another feature of the circuit in Figure 11.23 that is commonly found in many RF amplifiers is the presence of a **tuned circuit** (sometimes called a **tank circuit**). A tuned circuit is one in which circuit capacitance (including device capacitance) is tuned out or canceled at a given design frequency by inductance, yielding a pure resistance at the design frequency. In Figure 11.23, the collector terminal is the high-impedance point of the tuned output circuit. The capacitance in the resonant circuit is a combination of the device's collector-base capacitance and the **variable capacitor** C_3. The inductance is provided by the **leakage inductance** of the RF transformer T_1, which is purposely designed to have enough leakage inductance to **resonate** with the circuit capacitance. Because device and layout capacitance can vary significantly from one circuit to the next, it is often necessary to *tune* such circuits manually so that the tuned circuit resonates (presents a real impedance) at the designed center frequency. In this case, tuning is accomplished by adjustment of the variable capacitance C_3 for maximum output through the transformer secondary winding. Bypass capacitor C_4 ensures that the power-supply end of transformer T_1 is at RF ground.

In general, bypassing and circuit layout are especially important features of RF circuit design. Because a wire even a few centimeters long can show significant inductance of a few nH at RF, it is important to minimize lead lengths for all conductors that carry RF current. Also, small amounts of stray capacitance that would not cause problems for an audio-frequency design may conduct enough current at RF to severely compromise circuit operation or prevent it from working altogether. That is why the standard multihole spring-loaded **protoboard** type of breadboard often cannot be used for RF circuit prototyping, because the stray and distributed capacitance of the interconnections causes problems in RF circuits. Instead, most successful RF circuit designs use carefully designed double-sided PCBs that incorporate extensive **ground planes** for shielding and limiting stray coupling capacitance. Layout practices for RF circuits are beyond the scope of this work, but you should be aware that the physical layout of an RF circuit is an important part of the overall design. Two circuits with identical schematics (interconnections) but with different physical layouts may operate very differently at RF for these reasons.

Broadband transmitter amplifiers require a different design approach than narrowband amplifiers. Instead of circuits tuned to a single frequency, such designs require

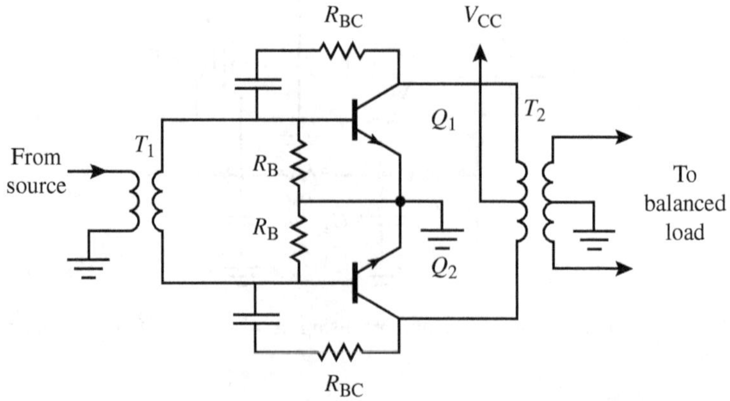

FIGURE 11.24 Balanced push–pull type of medium-power RF amplifier operating in class C mode.

components and configurations that amplify a broad range of frequencies. In addition, high-power designs often use devices in a **push–pull** configuration such as the one shown in Figure 11.24.

A push–pull circuit uses a pair of active devices whose input waveforms are 180° out of phase. Accordingly, so are the output waveforms, but if the two outputs are applied to opposite ends of a center-tapped transformer that is wound correctly (i.e., one with symmetrical primary windings properly phased), the magnetic fluxes add and the total power output from both devices is available from the transformer secondary winding. One advantage of a balanced push–pull circuit is that odd harmonics (third, fifth, seventh, etc.) of the fundamental input frequency are suppressed, though not completely canceled. This feature is helpful when the devices are operated in **class C**, meaning that they conduct for less than 180° of a sine-wave input waveform.

Class C operation is very commonly used for RF amplifiers, because it allows higher efficiency than class A or B operation, in terms of useful RF output power for a given amount of DC power consumed. While class C operation inherently generates many harmonics of the input waveform, RF circuits commonly employ tuned circuits, filters, and other frequency-selective features that can reduce harmonics to an acceptable level. You will note in the push–pull amplifier of Figure 11.24, while the collectors are biased at the power-supply voltage V_{CC}, there is no base or emitter bias provided. With no input, the bases are at DC ground and the circuit draws no current. Only when a sufficiently large input signal is supplied does the circuit draw current, and then it operates in the efficient class C mode. While such operation can lead to **envelope distortion** (a nonlinear relation between the input power level and the output power level), proper circuit design (including feedback techniques) can minimize problems resulting from this effect.

Broadband operation is achieved in the circuit of Figure 11.24 by the use of broadband transformers T_1 and T_2, as well as the use of **swamping resistors** R_B and R_{BC}.

The addition of losses to any tuned circuit tends to widen its frequency response, at the expense of efficiency. Swamping resistors, typically with values of 100 Ω or less, are often used for this purpose in wideband amplifiers, along with other frequency-response control approaches. The use of swamping resistors is a compromise between achieving the desired wideband response and maintaining circuit efficiency.

Another design issue that is of critical importance in RF amplifiers is the avoidance of undesirable oscillations. The high intrinsic gain of many RF devices means that they are capable of sustaining oscillation at frequencies that may be far removed from the design frequency, either higher or lower. Such devices are called **conditionally stable**, meaning that under some conditions of source and load impedances, they can become unstable in the sense of producing oscillation. Designing an RF amplifier with conditionally stable devices is a complex and exacting task, using techniques that are too involved to discuss here. But anyone using RF amplifiers should be aware that the stability of such circuits may depend on critical choice and placement of components that, if disturbed, will cause the circuit to oscillate. And such oscillations can produce strange and hard-to-understand effects: noise superimposed on desired signals, overheating, and even **body capacity** effects such as changes in performance if you bring your hand near the circuit. The only way to be sure that a circuit is oscillating is to observe its output with an oscilloscope or spectrum analyzer whose maximum frequency is greater than the oscillation frequency. If you note such effects in an RF circuit, you will save time by looking for oscillation with such equipment, although getting rid of it once you find it can be a difficult problem.

Like any other type of analog amplifier, an RF amplifier has a useful **dynamic range**. For class C RF power amplifiers, there is a certain minimum RF input level below which the devices will not turn on and no output will result. A **transfer curve** (different from a **transfer function**) of output power versus input power for a hypothetical class C RF amplifier is shown in Figure 11.25. For this amplifier, a minimum input power of about +6 dBm[2] is needed to turn on the devices and reach the linear portion of the input–output power characteristic. The amplifier appears to be linear up to an input power of about +47 dBm (which is about 50 W), at which level the output has fallen 1 dB below the **linear asymptote** (the straight line the output would follow if it continued to be perfectly linear). At this input level, the amplifier is said to be in **1-dB compression**, the term "compression" referring to the falling off of the output level below the ideal linear asymptote. Typically, an amplifier operating in compression begins to generate not only excessive amounts of undesirable harmonics but other nonlinear products as well, such as **cross-modulation** products between components of a narrowband signal. If the input signal's **envelope** (average waveform's peak amplitude) varies with time, compression is undesirable because it can distort the spectrum of the input signal. For truly linear operation, the input level would need to stay within the +6 to +47 dBm range shown as "input dynamic range" in Figure 11.25.

[2]The decibel-milliwatt is an absolute unit of power that expresses the ratio in decibel of the power in question to 1 mW: $\text{Power(dBm)} = 10\log_{10}\left(\frac{\text{Power(W)}}{1\,\text{mW}}\right)$.

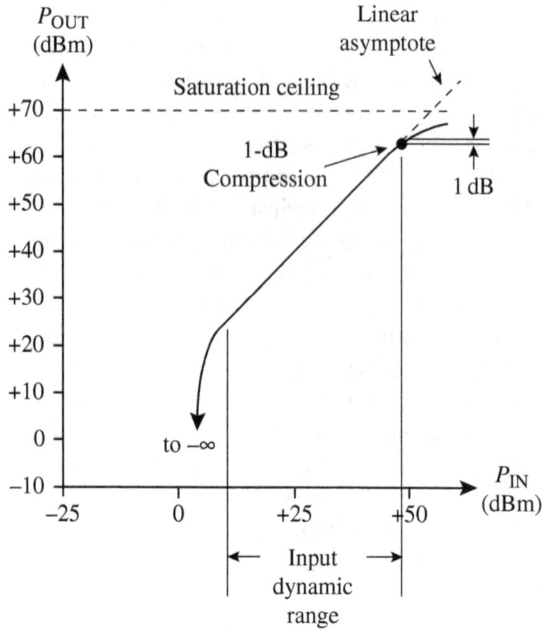

FIGURE 11.25 Power-level transfer curve for hypothetical class C RF amplifier showing falloff below minimum input level of +6 dBm and 1-dB compression at input level of +47 dBm.

However, certain types of digital signals and **frequency-modulated** (**FM**) analog RF signals are intrinsically **constant-envelope** signals. This means that although the frequency and phase can vary, the amplitude is essentially constant. Such signals can be amplified with little or no distortion by an amplifier that is operated far in the compression region. In fact, this is commonly done, because a class C amplifier's efficiency is often best in its compression region. A given amplifier design should be optimized to the particular type of waveform being amplified, which is why RF amplifier design needs to be integrated into the design of the overall system.

11.5.2 RF Amplifiers for Receivers

To understand the importance of receiver amplifiers in RF systems, the concept of **signal-to-noise ratio** is vital. This concept (often abbreviated as **SNR**) was introduced in Chapter 3, where various physical sources of noise were discussed and ways were given to calculate how much noise a circuit will produce.

The reason that the signal-to-noise ratio is important in communications systems involving RF (wireless) links is that the system will fail to operate within specifications if the received signal is so weak that the SNR falls below a certain minimum level. The same basic principle applies to a whispered conversation: if the person whispering to you is too far away, your ears and brain simply cannot extract intelligence from what you hear, and no communication takes place.

As a received signal becomes weaker, analog and digital systems tend to react differently.

Many analog systems are roughly linear with respect to noise, in the sense that the SNR at the system's output is proportional to the SNR at the receiver input. AM radios and (now obsolete) analog TV receivers showed this effect. If a received signal gradually faded away, you could hear (or see) the desired sound or image become weaker as the random noise (hiss in the case of sound, and random speckles of **snow** in the case of video) took over and eventually dominated the signal, making it inaudible or invisible.

Most digital communications systems do not behave this way. Because of error-correction coding and other factors, a digital communications link provides essentially perfect transmission of the original data for a range of input signal levels that goes down to a minimum level often called the **threshold**. If the received signal strength falls below the threshold, sometimes by as little as 1 or 2 dB, the quality of reception crashes abruptly and nothing gets through. So designers do everything they reasonably can to ensure that the received signal level is high enough to be well above threshold for all normal circumstances. And the quantity that determines whether the system is operating above threshold or not is the SNR at the input of the receiver's **demodulator** section, which operates on the raw signal to turn it into the desired output data format.

Two things obviously affect the SNR: the level of the incoming signal and the level of noise. Anything that makes the signal power higher or lowers the noise power will improve SNR. If a signal has been transmitted as an electromagnetic wave through space, it is received by an **antenna**. At the output of the antenna, there is a certain level of signal power P_s (in watts) and also a certain noise level P_n (in watts) in a bandwidth B (in hertz). The reason we mention bandwidth (which is the bandwidth utilized by the receiver) is that there is a simple relation between the noise power P_n an antenna receives and the bandwidth B. It is given in Equation 3.42, namely, $P_n = k_B TB$.

In this expression, T is the temperature (in degrees K) of the antenna's environment (not necessarily the physical temperature of the antenna itself). For most antennas used for receiving terrestrial signals (e.g., as opposed to radio astronomy antennas), the appropriate T to use is room temperature, 290 K. The quantity k_B is *Boltzmann's constant*, a physical constant equal to about 1.38×10^{-23} J K^{-1}. Because nothing much can be done at the receiver about the temperature of the environment or the power received from the signal source, the SNR of a received signal as it comes from the antenna terminals is fixed. The best a receiver designer can do is to make sure that the SNR does not become much worse as it is amplified. And that is what low-noise RF amplifiers do.

Amplification of an RF signal received from an antenna is almost always needed. The digital systems in modern receiver demodulators require digital-level signals, which are in the range of hundreds of millivolts to volts. In terms of power, levels in the range of 0 to +15 dBm are required—1 to 50 mW or so. However, it is unusual for a received RF signal from an antenna to be stronger than a few **microwatts**, and often the received power is much less than this. Therefore, power amplification is required between a receiving antenna and the demodulator or **back-end** electronics, as it is sometimes called. (By contrast, the part of a receiver that the incoming signal encounters first is called the **front end**.)

An ideal RF amplifier for receiver use would amplify the signal without adding any noise whatsoever. But for fundamental physical reasons, it is impossible to do this, and so every real amplifier adds some amount of noise. Because the noise added by an amplifier cannot be easily removed later, it is important to quantify a receiver amplifier's noise performance, and there are several methods of measuring the noise added by an amplifier.

Perhaps the easiest noise-characterization quantity to understand is the **effective input noise temperature** T_e of an amplifier. This concept is illustrated in Figure 11.26.

First, we imagine we could perform the following experiment. We take a source impedance Z_S that has the same value as the output impedance of the antenna or other signal source normally used with the amplifier in question. If the impedance can deliver real power (and it must be able to), it will have a real part; that is, if $Z_S = R_S + jX_S$, R_S is greater than zero. We imagine lowering the temperature of the source impedance to absolute zero (0 K) so that it produces no thermal noise of its own, and we connect it to the amplifier under test. Under these conditions, the only noise emerging from the amplifier output is that due to its own internal noise sources: resistors, active

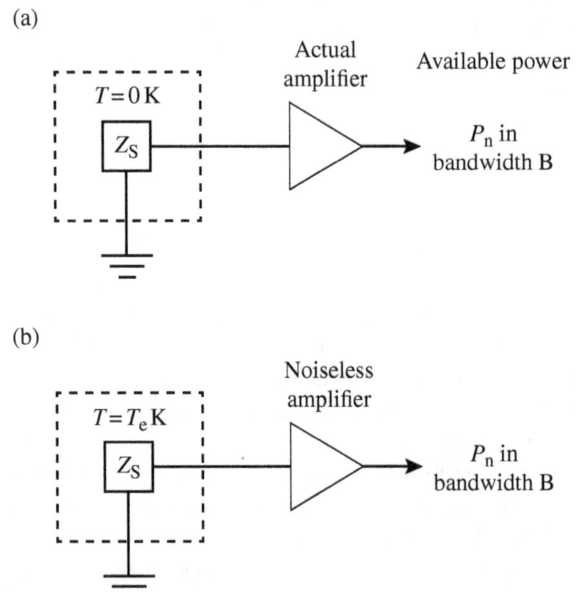

FIGURE 11.26 (a) Actual RF amplifier connected to (hypothetical) source impedance Z_S at 0 K, producing available output noise power P_n. (b) Hypothetical noiseless amplifier with same gain characteristics producing same P_n with source impedance Z_S at effective input noise temperature T_e.

devices, and other components. (Obviously, the amplifier must be powered on during this test!) The output power in a bandwidth B under this no-input-noise condition is called P_n. This test is shown in Figure 11.26a.

Next, we imagine that we have a (hypothetical) amplifier that is identical to the real one in every respect (gain, bandwidth, etc.) except for one thing: it generates *no* internal noise of its own. We connect this ideal noiseless amplifier to the same source impedance as the real one uses, but now instead of cooling the source impedance to 0 K, we warm it up until the noise power coming out of the (noiseless) amplifier P_n is *equivalent* to the P_n in the case of the real amplifier and the 0-K source. At that point, the temperature of the source driving the noiseless amplifier is termed T_e, the effective input noise temperature of the real amplifier.

It should be clear that the higher the internal noise level generated in the real amplifier, the hotter we will have to make the source in Figure 11.26b in order to equal the noise produced by the real amplifier. In other words, the noisier the amplifier, the higher T_e will be.

When this concept of effective input noise temperature was first developed in the 1940s, the typical receiver had a noise temperature of several thousand degrees K. That does not mean that any physical component in the receiver was that hot; it simply means that early receivers were very noisy. Today, fairly low-cost amplifiers are available with values of T_e well under 100 K (room temperature is about 300 K). Again, that does not mean that anything in the receiver is physically that cold (although physically cooling an amplifier is a good, but expensive, way to lower its noise temperature). It simply means that good engineering has reduced or eliminated most sources of noise in the amplifier, so it more closely approaches the ideal of amplifying a signal without adding any noise.

The effective input noise temperature is not the only important feature of an amplifier. It must also provide some gain, but if you connect the output of a **low-noise amplifier** (sometimes abbreviated **LNA**) to the input of a second amplifier that adds a lot of noise, you may lose whatever advantage you hoped to gain with the LNA. This is why it is important to be able to calculate the **system noise temperature** T_{SYS} of a cascade of several amplifiers, each having its own effective input noise temperature.

The system noise temperature T_{SYS} is defined in exactly the same way as effective input noise temperature in Figure 11.26, except that a linear system is substituted for the amplifier. The linear system may have lossy components such as cables or filters as well as amplifiers, and we will show how to deal with lossy components in calculating system noise temperature in the following.

Figure 11.27 shows a cascade of three amplifiers, each with its associated gain G_i and effective noise temperature T_{ei}. The system's overall effective noise temperature T_{SYS} is found by performing the thought experiment shown in Figure 11.26. First, we calculate the total noise power P_{SYS} by adding the contributions of each amplifier. The third amplifier's contribution will be $k_B B G_3 T_{e3}$, because the power from the equivalent temperature T_{e3} is increased by the gain of the amplifier. The contributions of the second and first amplifiers are calculated similarly, except that their

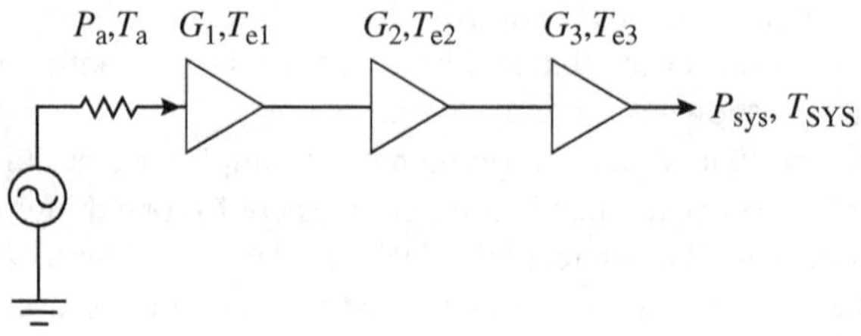

FIGURE 11.27 Three cascaded amplifiers illustrating system effective noise temperature T_e(sys).

contributions are amplified more, and the source's temperature is also amplified by all three amplifiers. If the source temperature and power are zero (at 0 K), the system's power output will be

$$P_{SYS} = k_B B[G_1 G_2 G_3 T_{e1} + G_2 G_3 T_{e2} + G_3 T_{e3}] \tag{11.41}$$

Next, we ask what temperature would the source T_a have to be raised to, in order to produce this same power P_{SYS} from a noiseless system with the same gain. This temperature will be T_{SYS}. Setting $P_{SYS} = k_B B G_1 G_2 G_3 T_{SYS}$, we can now solve for T_{SYS} in terms of the gains and noise temperatures of the amplifiers. Treating all the amplifiers in cascade as a single block with a gain $G_c = G_1 G_2 G_3$, we find that the system noise temperature is

$$T_{SYS} = T_{e1} + \frac{T_{e2}}{G_1} + \frac{T_{e3}}{G_1 G_2} \tag{11.42}$$

Note that the noise temperatures of the second and third amplifiers in the cascade are divided by the gains G_1 and $G_1 G_2$, respectively. In a well-designed low-noise receiver system, these gains are large enough in comparison to the later stages' noise contributions so that T_{e1} is the major contributor to the overall system noise power, and the contributions of later stages are insignificant. In this way, special low-noise design techniques can be used for the first amplifier (sometimes called the **preamplifier** or **preamp**), and if its gain is large enough, the noise contributions of the second and subsequent amplifiers are not important, allowing more conventional higher-noise designs to be used for them without as much attention being paid to their noise performance.

Besides the concept of system noise temperature, there are other measures of a system's noise performance. Recall that the goal of low-noise amplification is to degrade the incoming signal-to-noise ratio (SNR) by as little as possible. A measure that directly specifies the ratio of input SNR to output SNR under a standard test condition is the **noise figure** F_{dB} of an amplifier. This quantity is defined as follows.

Suppose the effective noise temperature of a source connected to an amplifier is set to the standard value of $T_0 = 290$ K (16.85°C or 62.33°F). The noise power N_{in} available from the source in a bandwidth B will therefore be

$$N_{in} = k_B T_0 B \tag{11.43}$$

Suppose the signal power available from the source is S_{in} (all powers are measured in watts). At the amplifier output, the signal power is S_{out} and the noise power is N_{out}. Under these specific conditions, the noise figure F_{dB} is defined to be

$$F_{dB} = 10 \log_{10}\left[\frac{S_{in}/N_{in}}{S_{out}/N_{out}}\right]\Bigg|_{T_0=290\text{ K}} \tag{11.44}$$

in which the condition that the source's noise temperature be 290 K is noted.[3]

An ideal noiseless amplifier will have the same signal-to-noise ratio at its input and its output, so the best (and lowest) noise figure possible is 0 dB (assuming that the amplifier does not filter out any noise). Noise figure F_{dB} and effective input noise temperature T_e convey exactly the same information and are related by the following equation:

$$F_{dB}(T_e) = 10 \log_{10}\left(\frac{T_e}{290\text{ K}} + 1\right) \tag{11.45}$$

An ideal amplifier that adds no noise has an effective noise temperature of 0 K, which is a noise figure of 0 dB, as it should be.

Passive structures and devices can also be considered to have a noise figure, in the sense that they can degrade the signal-to-noise ratio of a low-level signal passing through them. It can be shown from thermodynamic principles that the effective noise temperature of a passive circuit at a physical (ambient) temperature of T_A is

$$T_e(\text{atten}) = T_A\left(1 - \frac{1}{L}\right), \tag{11.46}$$

where L is the *loss ratio* of the passive circuit:

$$L = \frac{P_{in}}{P_{out}} = \frac{1}{G} \tag{11.47}$$

The loss ratio is the inverse of gain, so that an attenuator with a loss ratio of 2 can be considered as an "amplifier" with a gain of 0.5. Equation 11.46 shows that as loss increases, the effective noise temperature rises from zero (for a lossless component in which $L = 1$) and approaches the physical temperature T_A of the passive circuit as the loss approaches infinity.

[3] In the original paper in which H. T. Friis defined noise figure (H. T. Friis, "Noise Figures of Radio Receivers," *Proceedings of the IRE*, vol. 32, pp. 419–422), he defined noise figure without using decibels as $F \equiv \left(\frac{S_{in}}{N_{in}}\right) / \left(\frac{S_{out}}{N_{out}}\right)$. This quantity F (without dB) has come to be referred to as **noise factor**, while **noise figure** F_{dB} is now defined as $10 \log_{10}(F)$. We distinguish noise figure from noise factor by the use of the decibel subscript.

11.5.2.1 Example Calculation of Link Loss and Threshold The usefulness of these quantities can be shown by the following example calculation of the **uplink** part of a typical mobile-phone system. As you probably know, mobile phones work by sending a signal from the portable device to a **base station** nearby. The base station receives the signal, demodulates it, and sends it in digital form to the **landline** telephone network, where it is routed to the appropriate phone number. If that number is another mobile phone, another base station sends a radio signal to the appropriate mobile phone through another radio connection called the **downlink**.

The important parts of the uplink are shown in Figure 11.28. The mobile phone contains a transmitter that emits a signal in one of the authorized mobile-phone bands. In the following example, the 2.4-GHz band will be used. The radiated signal travels a distance of anywhere from a few meters to a kilometer or more to the base station antenna, typically on a tower. Although a straight line is shown for the uplink path in Figure 11.28, it is unusual to have a direct **line-of-sight** path between the phone and the base station. Far more commonly, the path involves multiple reflections from intervening objects such as buildings, trees, and hills, and a great deal of energy is scattered and absorbed along the way. At the base station tower, a receiving antenna picks up the signal and sends it in this example to a LNA situated close to the antenna. After amplification, the signal travels down the downlead coaxial cable to the receiver, where it is demodulated and sent into the telephone network.

In Figure 11.29, we have reduced the pictorial description of the uplink in Figure 11.28 to its essential electronic elements. The calling phone is represented by a transmitter that produces a power P_t watts and radiates it from an antenna, which is part of the phone housing. A typical value for the power produced by a mobile-phone transmitter is $P_t = 250\,\text{mW}$.

If an antenna can direct the energy it radiates in a preferred direction, the signal intensity in that direction is increased over what would result if the power was just spread out equally in all directions (a type of radiation called **isotropic**, meaning "equally in all directions"). Isotropic antennas are also sometimes called **omnidirectional**, because they transmit and receive more or less equally well in every direction.

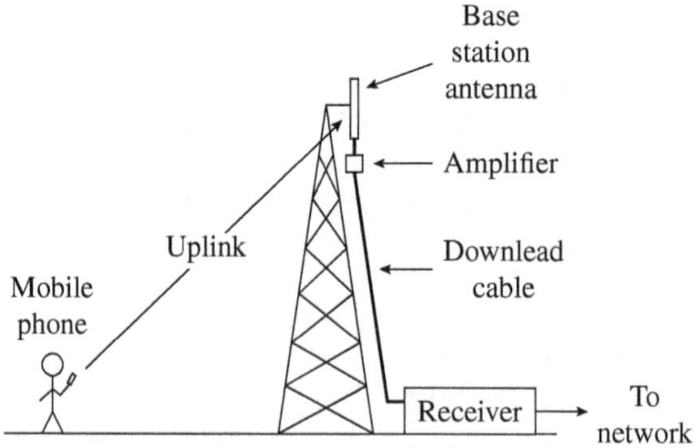

FIGURE 11.28 A typical mobile-phone uplink showing how the signal is transmitted from the phone through space to a base station antenna, amplifier, cable, and receiver.

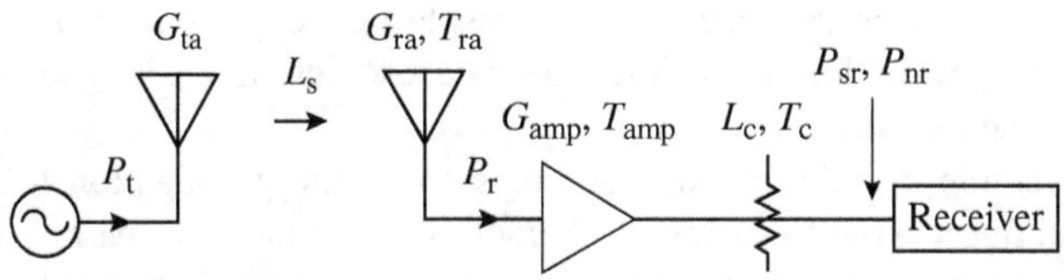

FIGURE 11.29 Schematic diagram of essential elements of uplink portrayed in Figure 11.28.

The amount by which an antenna increases radiation in a preferred direction over the isotropic case is called the **antenna gain**. There is not much room on a typical mobile phone to construct an antenna with appreciable gain, so the gain G_t of the mobile-phone antenna is about 1. (In this example, we will use *numeric ratios* for all gains and losses, although antenna gains are more commonly cited in dB.)

Once the signal energy is radiated into space in a given direction, it spreads out like a patch of color on a balloon being inflated. The way a particular radiated wave travels in a complex environment such as an urban street or neighborhood is the subject of a science called **radio propagation**. The only type of radio-wave propagation loss that is relatively easy to calculate is called **free-space propagation**. The free-space loss L_{FS} encountered by a wave can be calculated from a knowledge of the distance R between transmitter and receiver and the wavelength λ of the transmitted wave.[4] The free-space loss (a number >1 because it is a loss) is then

$$L_{FS} = \left(\frac{4\pi R}{\lambda}\right)^2 \tag{11.48}$$

Free-space propagation has very special conditions: both the transmitter and the receiver must be literally floating in free space, away from all other objects. The only situation for which this condition is strictly met is communication between objects in outer space.

So, for example, if a mobile phone got loose from a space station and drifted 1 km away and it used the standard mobile-phone frequency of 2.4 GHz, the corresponding wavelength is $\lambda = c/f = 0.125$ m. The theoretical free-space loss would be

$$L_{FS} = \left(\frac{4\pi(1\text{ km})}{0.125\text{ m}}\right)^2 = 10.1\times10^9, \tag{11.49}$$

which means the signal strength through space is *reduced* by a factor of about 10^{10}, which is 100 dB. This loss results from the fact that most of the radiation goes in directions other than directly from the transmitter to the receiver. Viewed from the transmitter, a given receiver "looks" smaller and smaller as a fraction of the sphere surrounding the transmitter, so less and less energy is received as the distance

[4]Equation 11.48 is derived from fundamental thermodynamics, which is beyond the scope of this book. See the references at the end of this chapter for more details.

between the transmitter and receiver increases. For exactly the same reason, a point source of light appears dimmer as it recedes because light intensity goes inversely as the square of the distance.

In this example calculation, we are going to let the propagation loss L_s be the unknown and figure it out from known characteristics of the other parts of the system. In this way, we will be able to state the maximum propagation loss that the uplink can tolerate. The whole operating principle of the cell-phone system is based on the fact that the maximum usable distance between a phone and a base station is limited. In this way, the same band of frequencies can be reused in other cells some distance away without encountering interference between phones in different cells. And the maximum usable distance is related to the maximum propagation loss for a given distance, so the maximum propagation loss is an important quantity to calculate.

In contrast to the transmitting antenna, which is basically isotropic, the receiving antenna of a mobile-phone base station is designed to be more sensitive to radiation at directions near and slightly below the horizon. It does this at the expense of sensitivity to waves coming in at high angles, but because virtually all mobile phones are used at ground level, this **directional** antenna pattern increases the received level of desired signals. In the example we have chosen, the receiving antenna gain is $G_{ra}=63$.

The receiving antenna also has an equivalent temperature T_{ra}, which is basically the temperature of its physical surroundings. Although this will vary some with weather conditions, we will assume that $T_{ra}=290\,\text{K}$.

After reception, the signal travels to a low-noise **masthead amplifier** whose noise figure is assumed to be $F_{dB}=2\,\text{dB}$ and whose gain is 15 dB. In numeric terms, a power gain of 15 dB is a numeric ratio of $10^{15/10}=31.62=G_{amp}$. Transforming these values into numeric forms, we solve Equation 11.45 for effective noise temperature T_e in terms of F_{dB} and obtain

$$T_e(F_{dB}) = 290\ \text{K}(10^{F_{dB}/10} - 1) \tag{11.50}$$

Using Equation 11.50, we find that a noise figure of 2 dB amounts to an effective noise temperature for the amplifier of $T_{amp}=169.6\,\text{K}$.

The coaxial downlead cable contributes a loss of 3 dB. As we noted earlier, a lossy element at room temperature contributes noise as well as reducing signal strength, so first we calculate the loss L_c of the cable:

$$L_c = 10^{3/10} = 1.995, \tag{11.51}$$

which is essentially 2. (Note that if a component is characterized by a positive dB number indicating loss, the resulting ratio is >1, as it should be for a loss term.) The effective noise temperature T_c of the cable is given by Equation 11.46 for $T_A=290\,\text{K}$:

$$T_c = 290\ \text{K}\left(1 - \frac{1}{1.995}\right) = 144.6\ \text{K} \tag{11.52}$$

For a given bandwidth and modulation type, a receiver will require a certain minimum SNR to operate without excessive errors or noise. Some of the digital modulation types used for mobile phones can operate successfully when the total signal power in the bandwidth used is less than the total noise power. In this example, the minimum SNR required by the receiver is −7 dB, which is a numeric ratio of $10^{-0.7}=0.1995$ (about 0.2), in a bandwidth $B=360$ kHz. So as long as the signal power is greater than about 1/5 of the noise power in a 360-kHz bandwidth, the receiver will be able to demodulate the uplink signal with acceptable quality.

Now that we have values for all the essential variables needed to calculate the maximum propagation loss, we can find the SNR at the receiver by finding the signal power level P_{sr} and the noise power level P_{nr} at the receiver input and setting their ratio equal to the required minimum $SNR_{min}=0.2$.

First, the signal power is obtained by a simple multiplication of the transmitter power P_t by all the gains the signal encounters and division by all the losses:

$$P_{sr}=P_t\frac{G_{ta}G_{ra}G_{amp}}{L_sL_c} \tag{11.53}$$

To find the noise power P_{nr} at the receiver input, we will calculate the noise contribution of each element of the system:

$$P_{nr}=k_BB\left[(T_a+T_{amp})\frac{G_{amp}}{L_c}+T_a\left(1-\frac{1}{L_c}\right)\right] \tag{11.54}$$

Now that we have expressions for both the signal power and the noise power at the receiver input, we can take their ratio and solve for the unknown, which is the maximum permissible propagation loss L_s:

$$\text{SNR}=\frac{P_{sr}}{P_{nr}}=\frac{(P_tG_{ta}G_{ra}G_{amp})/(L_sL_c)}{k_BB[(T_a+T_{amp})(G_{amp}/L_c)+T_a(1-(1/L_c))]} \tag{11.55}$$

If we set the SNR found in Equation 11.55 equal to the minimum acceptable SNR(min) = 0.2, we can solve explicitly for the maximum acceptable propagation loss L_s(max):

$$L_s(\text{max})=\frac{P_tG_{ta}G_{ra}G_{amp}}{\text{SNR(min)}L_ck_BB[(T_a+T_{amp})(G_{amp}/L_c)+T_a(1-(1/L_c))]} \tag{11.56}$$

Equation 11.56 shows the relative importance and effects of the different variables. Higher antenna gains anywhere along the line—in transmitter or receiver—will increase the allowable propagation loss L_s, because they are all in the numerator. On the other hand, a higher minimum SNR, a larger cable loss L_c, a wider bandwidth B, or higher noise temperatures will all decrease the allowable propagation loss. Note, however, that if the amplifier gain G_{amp} is large enough and the amplifier noise temperature T_{amp} is low enough, cable loss L_C can have a minimal effect on the allowable propagation loss.

To show how effective the LNA is at reducing the effects of the downlead cable loss, we will calculate the system noise temperature T_{SYS} of the system consisting of the LNA and the downlead cable:

$$T_{SYS} = T_{amp} + T_a \frac{L_c - 1}{G_{amp}} = 169.6\ \text{K} + (290\ \text{K})\frac{1.995 - 1}{31.62} = 178.7\ \text{K} \qquad (11.57)$$

Note that the second term, which is the cable-loss contribution, is divided by the gain of the amplifier, so that its contribution to the system temperature is negligible.

Replacing the remaining variables by their respective values in the example, we find the maximum acceptable propagation loss from Equation 11.56 is

$$L_s(\text{max}) = \frac{498\ \text{W}}{14.72 \times 10^{-15}\ \text{W}} = 33.82 \times 10^{15}, \qquad (11.58)$$

which in dB terms is $10\log_{10}(33.82 \times 10^{15}) = 165.3$ dB. Interference and other factors may reduce this figure slightly, but it is a fairly typical value for an average mobile-phone uplink. It is impressive to note that this large uplink propagation loss is more than 60 dB greater than the free-space loss that occurs over a distance of 1 km. The large margin of safety is needed because mobile phones are expected to operate inside buildings and even metal-enclosed elevators, which greatly attenuate the amount of energy radiated through them. And the low-noise masthead amplifier is crucial in achieving this large allowable propagation loss. In the problem set at the end of this chapter, you will have an opportunity to calculate the same quantity if the amplifier is located *after* the downlead cable, and the degradation in allowable loss is significant.

11.6 OTHER RF CIRCUITS AND SYSTEMS

While filters, impedance-matching networks, and amplifiers comprise a large portion of RF circuits and systems, there are other important operations on HF signals that require more specialized circuits and systems. For transmission through space, an RF signal must have a frequency range that fits within the legally allocated spectrum band, which is often as high as the microwave region (above 1 GHz). However, signal processing (either analog or digital) becomes more difficult and costly at higher frequencies, so it is advantageous to carry out signal processing at lower frequencies when possible. In order to shift a band of signal frequencies higher or lower, circuits called **mixers** are used. (The term *mixer* in electronics is ambivalent: it can mean either a nonlinear RF circuit used to convert frequencies of a signal up or down, or a linear audio-frequency circuit used to combine audio signals. We are discussing the former meaning in this section.) In order to impose a **baseband** signal (e.g., the audio signal from a microphone or the data stream from a sensor) onto an RF signal, circuits called **phase shifters** or **modulators** are employed. Some systems need low-loss **RF switches** to allow common use of an antenna by a transmitter and receiver, for example. To produce RF sine waves in the first place, **oscillators** and **frequency multipliers** are commonly used. RF signals can be transformed from electrical form into light signals and back again via **photonic transducers** such as lasers and photodiodes. And finally, an electrical signal on a transmission line must pass through an **antenna** in order to become an electromagnetic wave traveling through free space. Although there is not space here to give more than a brief description of each of these circuits or systems, we will now describe their basic operating principles in turn.

11.6.1 Mixers

A **mixer** (in the RF sense, not the audio sense) is used to add or subtract a constant frequency (termed the **local oscillator (LO)** frequency) to or from every frequency component in an input signal. This action is useful when, for example, a 2.4-GHz signal received from a mobile phone must be processed by digital circuitry that can handle no frequency higher than about 500 MHz.

Suppose that a band of signals that extends from $f_L = 2.40$ GHz to $f_H = 2.52$ GHz is to be converted with a mixer to a lower frequency range, namely, around 400 MHz. One way to do this is to generate a LO frequency at a frequency $f_{LO1} = 2.0$ GHz and send the band of signals to the mixer's **RF input**. The LO sine wave goes into a second input, appropriately termed the **LO input**. The mixer's output is at (usually lower) **IF** and appears at the terminals of the **IF output**. Every frequency component in the original RF signal appears at the IF output unchanged except for the fact that its frequency has been reduced by an amount f_{LO1}. For example, if a 2.4-GHz carrier frequency was present in the original RF signal, a reduced-frequency carrier at (2.4–2.0) = 0.4 GHz or 400 MHz will appear at the mixer's IF output and similarly for any other frequency components. If the mixer is operating properly, all the amplitudes and phase relationships in the original RF signal will be preserved in the IF signal band, and the only difference will be that each frequency component is 2 GHz lower than it was originally. The RF and IF spectra for this hypothetical example are shown in schematic form in Figure 11.30a.

Another set of frequency relationships that will accomplish a similar **downconversion** is shown in Figure 11.30b. The RF signal range is the same, but instead of choosing a **low-side** LO frequency of 2.0 GHz (so called because it is lower than the RF input band), we choose to place the LO frequency 400 MHz *higher* than the low edge of the RF band, at $f_{LO2} = 2.8$ GHz, using **high-side** downconversion. For an RF carrier at exactly 2.4 GHz, the effect of high-side downconversion is the same: an output at (2.8–2.4) = 0.4 GHz. But with high-side downconversion, the RF input spectrum as a whole is *inverted*: that is, the frequencies that were originally highest in the RF range are now lowest in the IF range and vice versa. Depending on the type of modulation employed, the spectrum inversion may or may not have to be taken into account in the demodulation process. Usually, this is not a significant problem, and either low-side or high-side downconversion can be chosen. The schematic symbol for a mixer is a circle with an *X* superimposed on it, and the

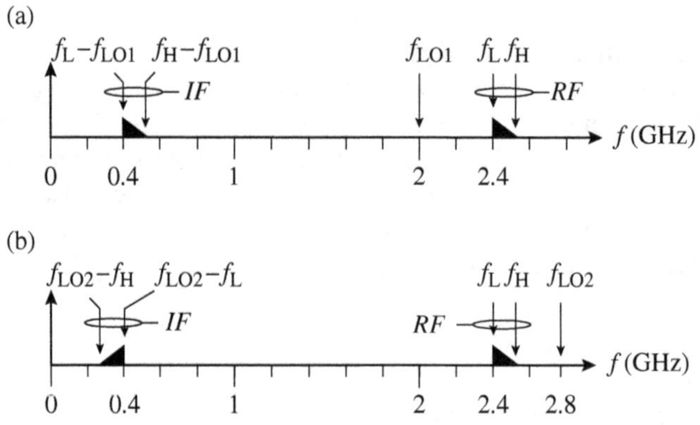

FIGURE 11.30 Downconversion of 2.4-GHz RF signal to IF signal in 400-MHz range using (a) low-side LO of 2.0 GHz or (b) high-side LO of 2.8 GHz.

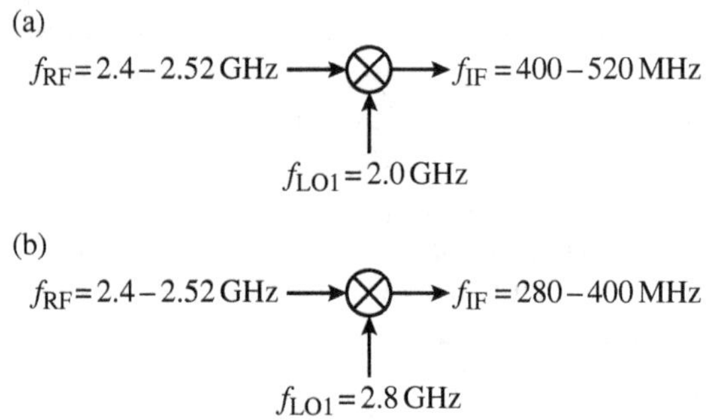

FIGURE 11.31 Schematic representation of RF mixer showing (a) low-side downconversion and (b) high-side downconversion.

operations shown in spectrum diagrams in Figure 11.30 are shown in schematic form in Figure 11.31.

All RF mixers perform their operations by means of nonlinear transfer functions. In principle, almost any type of nonlinearity can be used to convert frequencies, and in early mixer circuits, the existing nonlinearity of active components such as rectifier diodes and transistors was used. However, more efficient mixing with lower distortion is possible when components and circuits are explicitly designed for the purpose.

The IF output of an ideal RF mixer circuit is proportional to the algebraic *product* of the RF and LO inputs. Special ICs called **four-quadrant analog multipliers** are available, which perform an analog multiplying function with good accuracy, and can be used as mixers up to their frequency limit of 100 MHz or so. Beyond that frequency, passive circuits using diodes and transformers are commonly used, although passive mixers always introduce some power loss from the RF input to the IF output.

Here is how a multiplier-type RF mixer changes a high RF to a lower IF by subtracting the LO frequency from it. Suppose the RF is represented by $v_{RF}(t)$, which is a sine wave at a (radian) frequency ω_{RF}:

$$v_{RF}(t) = V_{RF}\sin(\omega_{RF}t) \tag{11.59}$$

The LO tone or sine wave is represented by $v_{LO}(t)$ at a frequency ω_{LO}:

$$v_{LO}(t) = V_{LO}\sin(\omega_{LO}t) \tag{11.60}$$

Suppose the IF output voltage $v_{IF}(t)$ is given by the product

$$v_{IF}(t) = K_C v_{RF}(t) v_{LO}(t), \tag{11.61}$$

where the constant K_C has the dimensions of V^{-1} in order to make the dimensions come out correctly. Inserting Equations 11.59 and 11.60 into Equation 11.61 and applying the trigonometric identity

$$\sin A \ \sin B = \frac{1}{2}[\cos(A-B) - \cos(A+B)] \tag{11.62}$$

give the following:

$$v_{IF}(t) = \frac{K_C V_{RF} V_{LO}}{2}\{\cos[(\omega_{RF} - \omega_{LO})t] - \cos[(\omega_{RF} + \omega_{LO})t]\} \tag{11.63}$$

The IF output contains two frequency components: one at the **sum frequency** $\omega_{RF}+\omega_{LO}$ and one at the **difference frequency** $\omega_{RF}-\omega_{LO}$. In the case that the LO frequency is higher than the RF, the fact that the frequency difference is negative is of no consequence, because the cosine function is indifferent to the sign of its argument: cos (X) = cos (−X). So as real frequencies are always positive, the difference frequency ω_{IF} is really the *absolute value* of the algebraic difference between ω_{RF} and ω_{LO}:

$$\omega_{IF} = |\omega_{RF} - \omega_{LO}| \tag{11.64}$$

While it is true that both the sum and difference frequencies are produced by an ideal multiplying mixer, in downconversion applications, the sum-frequency band is usually much higher than either the RF or LO frequencies and can be easily filtered out with a simple lowpass filter at the IF output, leaving only the desired difference frequency to be used for further processing. It is easy to see, however, that if one desires to do **upconversion**—in a transmitter, for example—then the circuit can be arranged to pass the sum frequency and not the difference frequency at the output.

Mixers are useful components, but their application must be accompanied by proper filtering in order to prevent various **spurious responses** from occurring. One of the simplest of these false (undesirable) responses is called the **image response**. Many FM broadcast-band receivers operate by downconverting a signal in the 88–108-MHz broadcast band to an IF of 10.7 MHz. Suppose the desired signal's unmodulated carrier is at 88.5 MHz, for example. Using low-side downconversion, the proper LO frequency to convert 88.5 MHz to 10.7 MHz is (88.5–10.7) = 77.8 MHz. However, if there happens to be a signal at (77.8–10.7) = 67.1 MHz present at the RF input of the mixer, it will also be downconverted to 10.7 MHz. This undesirable response is called an **image**, and the simplest way to reduce it is to put suitable RF filtering ahead of the mixer to reduce the amplitude of any out-of-band signals to the point that they will not interfere with the desired signals. Depending on how well a mixer circuit is designed, it can show numerous other spurious responses at harmonics of the LO frequency and elsewhere, so the design of an RF circuit using a mixer is not always a simple matter.

11.6.2 Phase Shifters and Modulators

It turns out that a multiplying-type mixer can also be used as a **modulator**. In general, a modulator is any circuit that impresses **modulation** on a **carrier** wave. A carrier is the signal or waveform a transmitter produces when it is not transmitting information in the form of modulation on the carrier. Two simple forms of modulation are **phase modulation** and **AM**.

Various types of phase modulation are popular for digital communications systems. Suppose a simple two-level binary waveform is to be transmitted via phase modulation. One way to achieve this is to transmit the carrier with no phase change when the digital input is 0 and to reverse the phase by 180° when the digital input is 1. This type of phase modulation is called **binary phase-shift keying** or **BPSK**. Figure 11.32 illustrates how an ideal multiplier-type RF mixer can be used to impose BPSK modulation on a carrier signal.

It turns out that certain types of mixers require any signal (input or output) containing a DC component to be connected to the IF port. So we show such a mixer being used to multiply a sine-wave LO carrier at the LO port by a binary waveform at the IF port that is held at either a positive voltage $+V$ (representing 1) or $-V$ (representing 0). When the sign of V changes, the output of the mixer changes phase by 180°, and a BPSK-modulated carrier results.

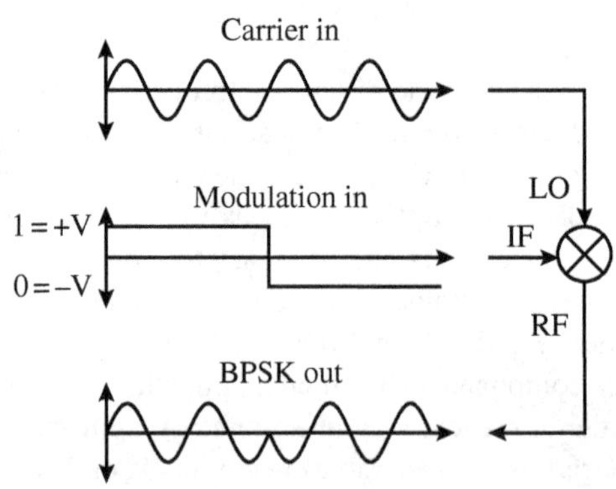

FIGURE 11.32 Sine-wave carrier applied to LO port phase modulated by binary modulating signal at IF port to produce binary phase-shift-keying modulated signal at RF port.

A multiplying-type mixer (analog multiplier) can also be used for AM, which varies the amplitude of the input carrier in accordance with the level of an AC modulating signal. AM can be used with analog signals, as in the AM broadcast band (535–1705 kHz), or with digital (two-level) signals, in which case it is termed **amplitude-shift keying** or **ASK**. In either case, the modulation can be carried out by multiplying the carrier signal function $\sin(\omega_c t)$ by 1 plus a scaled version of the modulating signal, which in the following example we will assume is a single-frequency sine wave at a radian frequency ω_m:

$$v_{AM}(t) = V_c[1 + p\sin(\omega_m t)]\sin(\omega_c t) \tag{11.65}$$

Analog modulation methods generally have a way to express the degree to which the carrier is modulated. In AM, this measure is the **percentage of modulation** or **modulation depth**, and in Equation 11.65, p is a fraction between 0 and 1 (or 0–100% in terms of percentage) that indicates the modulation depth. The time-domain and frequency-domain representations of an amplitude-modulated signal for modulation percentages of 0, 50, and 100% are shown in Figure 11.33.

With no modulating signal applied, the modulation percentage is 0% and the result is a pure sine-wave carrier at the frequency $f_c = \omega_c/2\pi$. For 50% modulation, the fraction $p = 0.5$, and at the peaks of the modulating waveform at a frequency $f_m = \omega_m/2\pi$, the instantaneous carrier amplitude is either reduced to 50% of its no-modulation value or increased to 150% of that value. The outline of the peaks of the amplitude-modulated carrier wave (e.g., on an oscilloscope trace) is known as the **envelope**, and with AM, the actual modulating signal is easily observed as the envelope of the carrier wave. This also means that AM signals can be demodulated with a very simple detector circuit called an **envelope detector**, which is basically a rectifier–filter circuit. The envelope faithfully reproduces the modulating signal for any percentage of modulation up to 100%, as shown in Figure 11.33c, in which the envelope varies from 0 to exactly twice the no-modulation amplitude on the positive peaks of the modulating waveform. However, if p exceeds 100%, the envelope no longer follows the modulating wave's shape, and envelope detection will lead to a distorted output. Other types of detectors such as **synchronous detectors** can overcome this problem but only at the cost of added complexity at the receiver.

In the frequency domain, the multiplication of the carrier by the modulating signal produces two **sidebands**, one on either side of the carrier. This is a specific example of a general result: modulation of any type on a single-frequency carrier wave produces sidebands at frequencies that are different from the carrier frequency. (It is impossible to convey information on a carrier wave without adding sidebands, so don't try to think of a way to do it!) The sidebands in this example appear spaced a distance f_m above and below the carrier frequency f_c as shown in Figure 11.33b and c. As the modulation percentage increases, the amplitude of the sidebands with respect to the carrier rises, reaching a limit of 6 dB below the carrier in the case of 100% modulation. Most digital modulation techniques can be considered as sophisticated combinations of amplitude and phase modulation, sometimes with the use of multiple carrier frequencies.

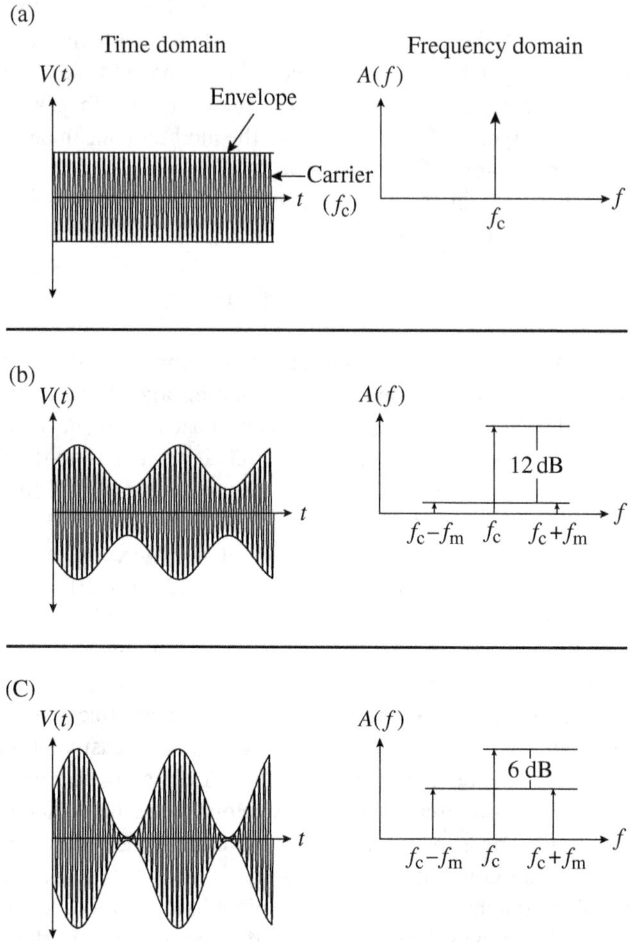

FIGURE 11.33 Time-domain and frequency-domain illustrations of amplitude modulation: (a) 0% modulation, (b) 50% modulation, and (c) 100% modulation.

A third type of modulation, called **frequency modulation** or **FM**, modulates or changes the instantaneous frequency of a carrier in accordance with the amplitude of the modulating signal. Because only the frequency changes, the amplitude of a frequency-modulated wave can remain constant and still convey the desired information. This is a great advantage both in amplifying the signal at the transmitter and receiving it at the receiver. At the transmitter, the power amplifiers need not be linear, because the amplitude of the carrier does not change. As with AM, FM produces sidebands, although the mathematical description of FM sidebands is quite complex. Basically, however, the wider the bandwidth of the modulating signal, the wider the RF bandwidth needed to transmit the information faithfully, as is the case with AM as well.

In an FM receiver, a special circuit called a **limiter** is used to eliminate changes in the amplitude of the received signal due to noise and interference from other stations.

If the received signal is strong enough and the degree of modulation (called **deviation** in FM) is chosen properly, limiting greatly improves the received signal-to-noise ratio, and the resulting system is almost as good for high-quality transmission as a digital communications system. Although frequency modulation can be achieved by starting with a constant-frequency carrier and phase-modulating it, this is not a good approach as compared to directly frequency-modulating the oscillator that produces the carrier.

11.6.3 RF Switches

Situations arise in many RF designs in which a common RF element must be time-shared between two different circuits. For example, there is usually room for only one antenna in a mobile-phone design, but both the transmitter and the receiver must use it (assuming the same frequency band is used for both). If a simple common connection is made that joins the transmitter, the receiver, and the antenna, part of the transmitter's energy would go into the receiver, wasting it and possibly damaging the receiver. And the transmitter's impedance, even if not operating, might interfere with the efficient transmission of energy from the antenna to the receiver. The solution to such a problem is an **RF switch**.

RF switches are designed to provide low-loss selectable pathways for RF energy while providing high loss along paths that are turned off. A good RF switch will transmit energy when turned on with losses of less than 1 dB while providing a loss exceeding 20–40 dB or more when turned off. Purely electronic RF switches can use three-terminal devices such as FETs for low-power applications, but for higher-power applications, a specialized semiconductor device called a **PIN diode** is often used. A PIN diode resembles an ordinary junction diode, except that between the P-type and N-type layers, a region with low doping is placed (*I* stands for **intrinsic** or undoped). When no DC current flows through the PIN diode, it behaves like a small capacitance and can be made to look like an open circuit with suitable external components. However, when appreciable DC current flows through it, the charge carriers in the *I* region turn it into a good conductor at radio frequencies, and it allows RF to pass through with low loss.

More recently, **MEMS** technology has been applied to the problem of RF switching, and MEMS-based switches are now available that have lower on-state loss and higher off-state isolation than many PIN-diode circuits. Early problems with reliability of MEMS switches have largely been overcome, and so you will find these types of devices used increasingly in RF switch applications in the future.

11.6.4 Oscillators and Multipliers

The topic of oscillators and oscillator design was dealt with at length in Chapter 7, so we will limit the discussion in this section to special considerations pertaining to RF oscillator design.

An oscillator whose frequency is determined exclusively by the components within the oscillator circuit itself and is not controlled by any external circuit is called a **free-running** oscillator. The only type of free-running oscillator that is generally suitable for operating radio transmitters in modern designs is a **crystal-controlled** oscillator, whose frequency-determining element is a quartz-crystal resonator, although **MEMS**-type resonators are becoming increasingly popular for RF designs as well. The **stability** of an oscillator (the constancy of its oscillation frequency) is one of the most important characteristics for RF applications, because if the transmitted frequency **drifts** outside of its allocated band, two bad things can happen: (1) the receiver may not be able to receive the signal, and (2) the FCC or other regulatory authority may take legal action if the drifting signal causes interference to other services. Good crystal-controlled oscillators have a stability measured in 1 part in 10^7 or better. For example, a 10-MHz crystal-controlled oscillator with a stability of 1 part in 10^7 will produce an output frequency of $10\,\text{MHz} \pm (10^7) \times (10^{-7}) = 10\,\text{MHz} \pm 1\,\text{Hz}$. This may or may not be sufficient for a specific application, so RF system designers should carefully investigate the frequency stability requirements of a given system design before specifying oscillator stability. Oscillators with better stability are more costly, especially if they must maintain their stability over a wide temperature range.

Almost any oscillator can be made controllable by an external DC control voltage with the addition of a **varactor** (also sometimes called a **varicap** or **variable-capacitance diode**), which is a semiconductor diode used as a voltage-variable capacitor. Figure 11.34 shows how a varactor can be used to shift the frequency of an RF oscillator in accordance with an externally imposed voltage V_C from a control circuit. Resistor R_1 isolates the control voltage source from the RF voltage present across the varactor. Over a limited range, the varactor's RF capacitance $C(V_C)$ can be expressed as a function of control voltage V_C with this linearized expression:

$$C(V_C) \approx C_0 + K_{VC}(V_C - V_{C0}), \tag{11.66}$$

where C_0 is the varactor's capacitance at a nominal control voltage of V_{C0}. Typically, a varactor must be reverse-biased with a constant DC bias voltage V_{C0}, and then it

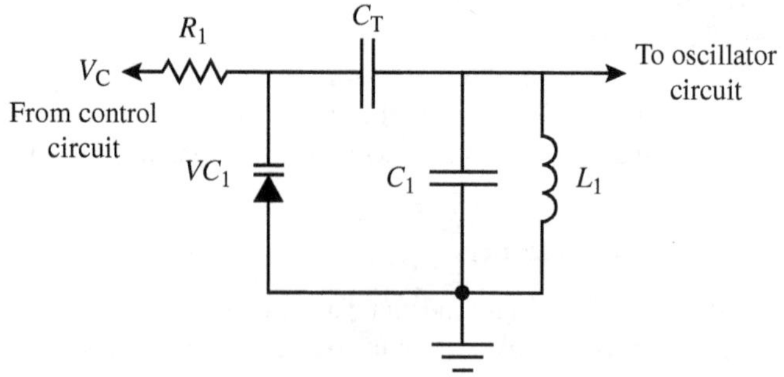

FIGURE 11.34 Oscillator resonator circuit (L_1–C_1) with varactor circuit (R_1–VC_1–C_T) connected to vary resonance frequency in accordance with control voltage V_C applied to varactor.

responds approximately linearly to small AC changes around the DC bias voltage. The constant K_{VC}, whose units are farads per volt, expresses how sensitive the varactor is to changes in bias voltage.

The tuned circuit L_1–C_1 is assumed to be the frequency-establishing part of an RF oscillator (e.g., a **Colpitts** oscillator circuit), which by itself would produce an output at a frequency

$$f_0 = \frac{1}{2\pi\sqrt{L_1 C_1}} \tag{11.67}$$

With the addition of the R_1–VC_1–C_T circuit, we can find the sensitivity of the oscillator frequency f_{OSC} to small changes in the capacitance of C_1:

$$\frac{df}{dC} = \frac{d}{dC}\left(\frac{1}{2\pi\sqrt{LC}}\right) = -\frac{f_0}{2C}, \tag{11.68}$$

and if the coupling capacitor C_T is much smaller than the varactor capacitance C_{VC}, you can show that

$$\frac{dC}{dC_{VC}} \approx \frac{2C_T}{C_{VC}}, \tag{11.69}$$

so combining Equations 11.68 and 11.69 with 11.66, we finally obtain this relation between the oscillator frequency change δf and the control voltage variation $\delta V_C = V_C - V_{C0}$:

$$\delta f = -\frac{f_0 C_T K_{VC}}{C_1 C_{VC}} \delta V_C \tag{11.70}$$

Within certain limits, then, a linear relationship between the varactor control voltage change δV_C and the oscillator frequency shift δf can be obtained. To a limited degree, such frequency adjustments can be obtained even with crystal-controlled oscillators, although the maximum frequency shift with high equivalent Q resonators such as quartz crystals is much smaller than with lower-Q resonators. Such frequency control can be used either directly or by incorporating the controlled RF oscillator in a **phase-locked loop** or **frequency-locked loop** that ensures the average frequency is stabilized to a reference frequency while allowing the instantaneous frequency of the oscillator to change in accordance with modulation.

If a very high RF output frequency is required for transmission, such as in the 2.4-GHz mobile-phone band, a problem arises in generating a very stable output. Highly stable oscillators using quartz or MEMS resonators usually have an upper frequency limit of a few hundred megahertz. If a higher output frequency is desired, either the stabilized oscillator frequency can be multiplied directly with a frequency-multiplier circuit or else used as a reference for a phase-locked loop. We have already discussed phase-locked loops and **frequency synthesizers** in Chapter 9, so only a brief mention of frequency multipliers is in order here.

Suppose an RF amplifier's input signal frequency is f_{IN} and the level of the input signal is gradually raised, so that the amplifier starts in a **class A** mode but moves gradually to **class B** and then to **class C**. If a spectrum analyzer is connected directly to the amplifier output with no filtering, the spectrum will show increasing amounts and numbers of **harmonics** of the input signal at $2f_{IN}$, $3f_{IN}$, $4f_{IN}$, etc. By convention, the harmonic $2f_{IN}$ at twice the fundamental frequency is called the **second harmonic**, the one at $3f_{IN}$ is the **third harmonic**, and so on.[5] Normally in class C amplifiers, the output circuit contains a filter or tuned circuit that passes the fundamental frequency and reduces or eliminates the harmonics. But a **frequency multiplier** can be designed by simply retuning the output circuit of a class C amplifier so that it passes the desired harmonic instead of the fundamental. So, for example, a times-three multiplier might havc an input frequency of 50 MHz and an output circuit tuned to the input's third harmonic at 150 MHz. For a given drive level and active device, the efficiency of a frequency multiplier falls off as the order of the harmonic increases, roughly as $1/f$. So a sixth-harmonic multiplier will be only half as efficient as a third-harmonic multiplier, for example. Frequency multipliers are not as commonly used now that inexpensive phase-locked-loop circuits can synthesize microwave-frequency outputs, but they are still found in some systems, especially high-power transmitters.

11.6.5 Transducers for Photonics and Other Applications

Besides the direct radiation of RF signals into space, there are other common applications that use electronic signals that must be dealt with using RF-style techniques. The modulation of lasers and optical signals requires wideband circuits that often process signals with frequency components well into the microwave range, above 1 GHz. Solid-state **laser diodes** can be directly amplitude modulated at VHF, but the drive circuit requirements are quite critical, because a fairly large amount of power is involved and heat dissipation and temperature control become challenging issues. Once produced, optical signals are transmitted over fiber-optic links to receivers that often employ **photodiodes** that convert the optical impulses to electrical ones. Again, because of the wide bandwidths involved, RF circuit techniques are often used to amplify the received signals and process them to be suitable for digital demodulation circuitry.

Other applications for RF systems that are not, strictly speaking, communications uses include medical imaging systems that use either **ultrasonic** transducers or **magnetic resonance imaging** (MRI) systems. Many ultrasonic imaging systems in medical applications employ frequencies in the megahertz range, and sophisticated RF and DSP circuits are needed to synthesize the images from arrays of ultrasonic transducers. MRI machines use extremely strong and precise magnetic fields to induce magnetic resonance in the nuclei of certain molecules. As a result, pulses of magnetic fields in the RF range produce RF signals that can be interpreted in terms of MRI, and complex RF generators and receivers are also employed in these systems.

[5] Do not confuse the EE term **harmonic** with the musical term **overtone**: an electrical engineer will identify as the **second harmonic** what a musician will identify as the **first overtone**.

Another special use of RF systems is in **radio astronomy**, which is the reception of naturally occurring radio waves from the sun, from other planets, or from sources in interstellar space. These sources can be either stars or interstellar gas, dust, or molecules. Signals received by radio astronomy receivers are extraordinarily weak, and so extreme measures such as **cryogenically cooled** receivers to reduce noise are used. The wavelengths received by radio astronomers cover the entire radio spectrum, from frequencies below a few megahertz up to the **terahertz** (THz, 10^{12} Hz) region. However, the atmosphere of the earth blocks many of these wavelengths, which means that the most effective platform for radio astronomy is a space probe outside the atmosphere. An unmanned spacecraft called the **Wilkinson Microwave Anisotropy Probe** (WMAP) was launched in 2001 and investigated the unevenness, or anisotropy, of the **cosmic microwave background** at frequencies ranging from 23 to 94 GHz. The cosmic microwave background is radiation left over from the Big Bang. Data from this and similar space probes is used to determine things like the estimated age of the universe—13.7 billion years is the current estimate—and its composition, which seems to be mostly stuff that we can't see or detect directly (**dark matter** and **dark energy**). Fortunately, the microwave background *can* be detected with the large antennas on the WMAP probe (see Fig. 11.35), and the subject of antennas is the last one we will take up in this chapter.

FIGURE 11.35 The Wilkinson Microwave Anisotropy Probe (WMAP) spacecraft was launched in 2001 and made fundamental discoveries about the nature of the universe by means of its microwave-dish receiving antennas (Photo courtesy of NASA).

11.6.6 Antennas

A radio **antenna** (sometimes called an **aerial** in the United Kingdom) converts an electromagnetic wave in space into an electric wave that travels along a transmission line to an electronic system for further processing. The same physical structure can usually be used either to **transmit** waves into space or to **receive** them. There are many different types and sizes of antennas, ranging from nanometer-size structures used for optical-wavelength **metamaterials** up to the 1000-foot-diameter (305-m) **Arecibo radiotelescope** dish built into a natural depression in Puerto Rico. Virtually all antennas have a few basic features in common, which we will now describe.

An antenna design can be characterized by its **frequency range** and **bandwidth** over which it performs its function with acceptable efficiency. Antennas such as the **yagi** shown in Figure 11.36 (named for Hidetsugu Yagi, who with Shintaro Uda developed the design in the 1920s) typically have narrow bandwidths of 10% or less, although they can be easily designed for any frequency from a few megahertz up to the microwave range. Other types of antennas such as the **parabolic dish** antenna are inherently broadband and can be used over a 2:1 frequency range or larger, depending on the design. The US National Aeronautics and Space Administration (NASA) operates a 70-m-diameter steerable parabolic dish antenna (Fig. 11.37) at Goldstone, California, as part of its Deep Space Network that receives the extremely weak signals from space probes such as the WMAP in Figure 11.35.

Another important property of an antenna is **gain**. Because most antennas are **passive** devices that simply pass through a percentage of whatever RF energy they receive, the term does not mean amplification in the same sense that is used for an active-device amplifier that provides gain. Instead, an antenna's gain measures its ability to focus or direct RF power applied to its terminals in a given direction. The only way that antenna gain in a passive structure can be obtained in one direction is by making the **radiated power density** from the antenna smaller in other directions.

FIGURE 11.36 Yagi antenna used to produce moderately directional beam of radio waves. The entire antenna is about 0.3 m (1 ft) long.

FIGURE 11.37 Seventy-meter Goldstone steerable parabolic dish antenna used in NASA's Deep Space Network. (Photo courtesy of NASA.)

Figure 11.38 shows how this works. In Figure 11.38a, we show a (hypothetical) **isotropic** (also called an **omnidirectional**) antenna. Such an antenna radiates equally well in all directions, although strictly speaking, a perfectly isotropic antenna is a mathematical fiction! By convention, the gain of an ideal (lossless and perfectly impedance-matched) isotropic antenna is exactly 1. Suppose the transmitter attached to such an antenna provides $P_{RAD} = 1$ W of power to it. If this power is radiated equally in all directions, we can calculate the **isotropic radiated power density** $U_{ISO}(R)$ (in W m^{-2}) at a distance R meters by dividing P_{RAD} by the area of a sphere with radius R:

$$U_{ISO}(R) = \frac{P_{RAD}}{4\pi R^2} \tag{11.71}$$

For example, if we measure the power density in free space (no other structures nearby) at a distance $R = 1$ km away from an isotropic radiator emitting 1 W evenly in all directions, the radiated power density at that distance will be

$$U_{ISO}(R) = \frac{1\text{ W}}{4\pi(1\text{ km})^2} = 79.58\times10^{-9}\text{ W m}^{-2}, \tag{11.72}$$

or 79.58 nW m^{-2}. The fact that the isotropic antenna in Figure 11.38a radiates equally well in all directions is symbolized by equal-length vectors radiating outward from it, although in reality you need to imagine the vectors radiating equally in a

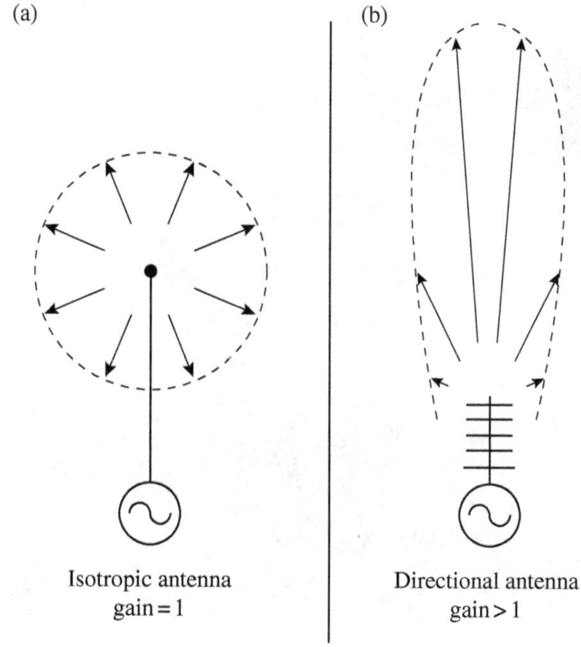

FIGURE 11.38 Vector lengths indicate *radiated power density* in a given direction. (a) Hypothetical *isotropic* antenna with gain = 1 (equal power density radiated in all directions). (b) Yagi antenna with gain >1 in vertical direction (up).

three-dimensional sphere (the figure captures only a two-dimensional cross section of this **antenna pattern**, as it is called).

Contrast the pattern of the isotropic antenna of Figure 11.38a with the antenna pattern of the directional yagi antenna shown in Figure 11.38b. Clearly, the radiated power density in one direction (upward) is greater for the same power P_{RAD} applied to the directional antenna, as indicated by the longer vectors in the upward direction. But this greater-than-unity gain has been obtained by reducing the radiated power density in other directions, as shown by the shorter vectors pointing in other directions. If U_{MAX} is the maximum radiated power density from the directional antenna and U_{ISO} is the power density (with the same applied power P_{RAD}) provided by the isotropic antenna, the **antenna gain** G of the directional antenna is defined as

$$G \equiv \frac{U_{MAX}}{U_{ISO}} \tag{11.73}$$

As with other power ratios in electronics, antenna gain can be expressed either as a numeric ratio or as dB, and the context will usually make clear which form is used.

Most RF system designers are not so much interested in the gain of an individual antenna, but in the total **link loss** that exists between the terminals of a transmitting antenna and the terminals of a receiving antenna designed to pick up part of the

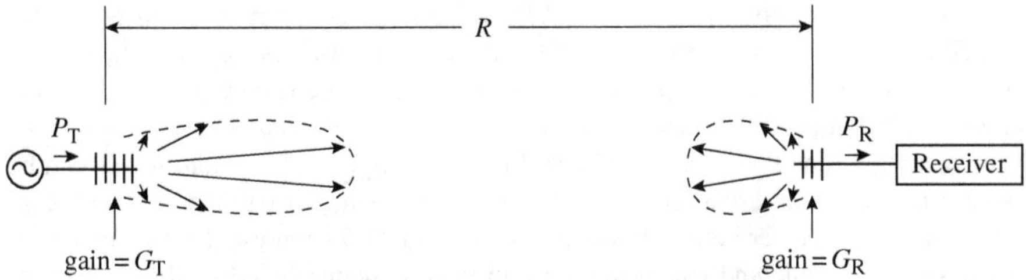

FIGURE 11.39 Generic radio link consisting of transmitter producing power P_T, transmitter antenna with gain G_T, link of distance R, receiving antenna with gain G_R, and receiver.

signal emitted from the transmitter. Now, we can combine the concept of **free-space loss** L_{FS} introduced in Equation 11.48 with the concept of antenna gain to find the link loss L_L. To do this, we will use only numeric ratios, not dB values.

Figure 11.39 illustrates a generic radio link. A transmitter produces power P_T, which goes to a (lossless, perfectly matched) transmitting antenna with gain G_T. Again, assuming the ideal free-space situation (no other objects nearby), the receiver at a distance R uses a second antenna with gain G_R to receive a power P_R, which goes to the receiver circuit. The quantity we wish to find is the link loss

$$L_L = \frac{P_T}{P_R} \tag{11.74}$$

(Because we are solving for a *loss*, not a gain, the larger number P_T is in the numerator and P_R is in the denominator, making the link loss a number >1.) Recalling Equation 11.48, we know that if both antennas were isotropic (if $G_T = G_R = 1$), the link loss L_L would simply equal the free-space loss L_{FS}:

$$L_L\big|_{G_T=G_R=1} = L_{FS} = \left(\frac{4\pi R}{\lambda}\right)^2 \tag{11.75}$$

But if we allow the receiver and transmitter gains to be greater than 1, the link loss will be *reduced* by these factors, so in general we will have the following equation for the link loss with antenna gains G_T and G_R:

$$L_L = \frac{L_{FS}}{G_T G_R} = \frac{(4\pi R/\lambda)^2}{G_T G_R} \tag{11.76}$$

So expressed, the link loss is a power-ratio number greater (usually much greater) than 1. Most textbooks discuss the problem in terms of a link "gain," which is the inverse of link loss, namely,

$$\frac{P_R}{P_T} = \frac{1}{L_L} = G_T G_R \left(\frac{\lambda}{4\pi R}\right)^2 \tag{11.77}$$

Expressed in this way, Equation 11.77 is called the **Friis transmission equation**, named for the same H. T. Friis that came up with the definition of noise figure. You should know that in these simple derivations, we have ignored many details of antenna engineering that need to be considered in a realistic case: antenna efficiency, impedance matching, and the issue of **polarization**. (Like light waves, radio waves can be polarized in a given direction, and mismatched polarizations can lead to significant additional loss.) Nevertheless, Equations 11.76 and 11.77 represent a best-case scenario for a radio link, and you cannot do significantly better than the value they yield, although you can easily do worse. We can show their use by calculating the free-space distance over which a simple data link could work using isotropic antennas and then showing how much farther it could operate with higher-gain antennas.

Suppose a 2.4-GHz **WLAN** uses a $P_T = 100$-mW transmitter and omnidirectional transmitting and receiving antennas and under these conditions can achieve a maximum free-space range of $R_1 = 60\,\text{m}$, beyond which the link abruptly fails to operate. We can use this information and Equation 11.77 to estimate what the minimum received signal power $P_R(\text{min})$ must be:

$$P_R(\text{min}) = \frac{P_T}{L_{FS}} = P_T\left(\frac{\lambda}{4\pi R}\right)^2 \tag{11.78}$$

At 2.4 GHz, the free-space wavelength $\lambda = 12.5\,\text{cm}$, and inserting these values gives for the minimum received power using isotropic antennas

$$P_R(\text{min}) = (0.1\,\text{W})\left(\frac{0.125\,\text{m}}{4\pi(60\,\text{m})}\right)^2 = 2.75\,\text{nW} \tag{11.79}$$

Now, suppose we procure two identical antennas, one for the transmitter and one for the receiver, each of which has a gain of 10 dB when pointed and aligned properly. The numerical gain ratio corresponding to 10 dB of power gain is therefore

$$G_T = G_R = 10^{(10\,\text{dB}/10)} = 10 \tag{11.80}$$

If we install these antennas at the transmitter and receiver site, what is the new longer range R_2 we can expect to obtain for line-of-sight (approximately free-space) transmission? Solving Equation 11.77 for R, we find

$$R_2 = \frac{\lambda}{4\pi}\sqrt{G_T G_R \frac{P_T}{P_R(\text{min})}} = \frac{0.125\ \text{m}}{4\pi}\sqrt{(10)(10)\frac{10^{-1}\ \text{W}}{2.75\times 10^{-9}\ \text{W}}} = 600\ \text{m} \tag{11.81}$$

We could have guessed this from the form of Equation 11.81 without doing any math, because the fact that both gains are under a square root means that if you raise both the receiver and transmitter gains by the same factor, the usable range increases by that same factor. So multiplying both receiver and transmitter gains by 10 increases the range by a factor of $10 \times 60\,\text{m} = 600\,\text{m}$.

11.7 RF DESIGN TOOLS

Because RF design is a hybrid discipline that uses conventional circuit theory, electromagnetic theory, and other borrowings from the science of physics, special design tools have been developed to assist the RF engineer in completing a design without extensive "cut-and-try" hardware breadboarding, which used to be a standard part of RF development projects. One of the earliest design tools developed especially for RF designers was the **Smith chart**, first published in 1939 by Philip Smith, an engineer working on antenna problems for Bell Telephone Laboratories. In the 1930s, Smith was engaged in research on HF antennas and needed to plot antenna impedances for a range of frequencies. Many antennas and resonant circuits used in RF systems show impedances that vary over an extremely wide range, from zero to values approaching infinity. On a conventional rectangular Cartesian *x–y* chart, only part of the **impedance locus** (set of impedance points) can be plotted on a chart with reasonably scaled axes. For example, if we try to plot the complex impedance of the series *L–C–R* circuit shown in Figure 11.40 on a rectangular-coordinate chart, the region of the locus near resonance at f_0, where $Z(f_0) = 50\,\Omega$, can be plotted as shown in Figure 11.41. But far away from resonance at frequencies lower or higher than the 3-dB-down frequencies, the imaginary part of the impedance becomes so large that the locus line falls off the chart, no matter what scales are used for the imaginary (*y*) axis.

By contrast, when the same impedance locus is plotted on the Smith chart shown in Figure 11.42, the entire locus fits on the chart, including all frequencies from DC to infinity. Impedances that go to infinity fit on the Smith chart because infinite impedance itself is a point at the extreme right-hand edge of the chart. The Smith chart plots a simple mathematical transformation of the complex impedance $Z(f)$ called the **reflection coefficient** $\Gamma(f)$:

$$\Gamma(f) = \frac{Z(f) - Z_0}{Z(f) + Z_0}, \tag{11.82}$$

in which Z_0 is called the **normalizing impedance** and appears in the center of the chart. Typically, Z_0 is set to the characteristic impedance of the transmission line used

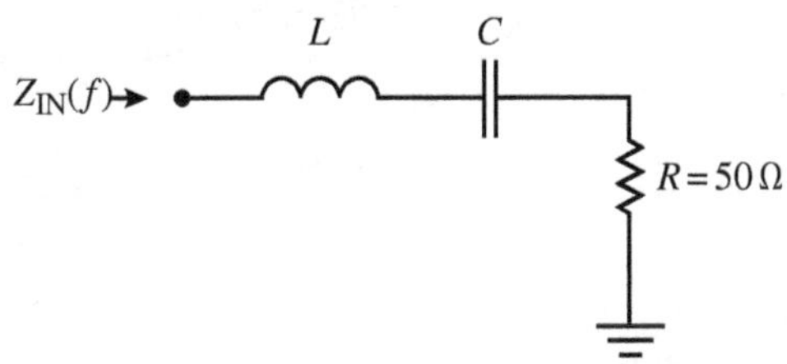

FIGURE 11.40 Series *L–C–R* circuit whose impedance versus frequency is to be plotted on both conventional and Smith charts.

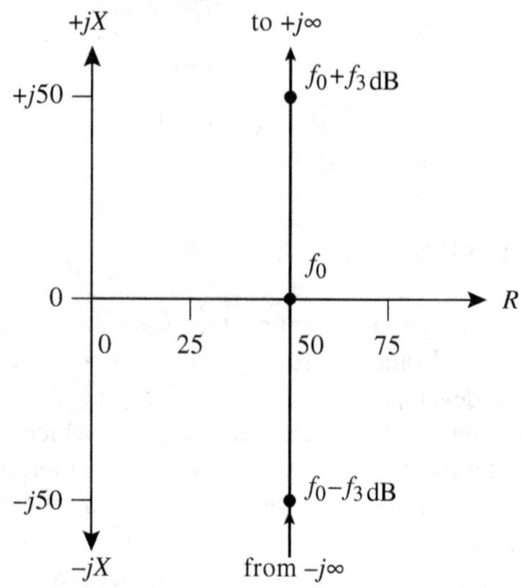

FIGURE 11.41 Only a portion of the impedance locus of the series-resonant circuit of Figure 11.40 can be plotted on this conventional rectangular Cartesian x–y axis chart.

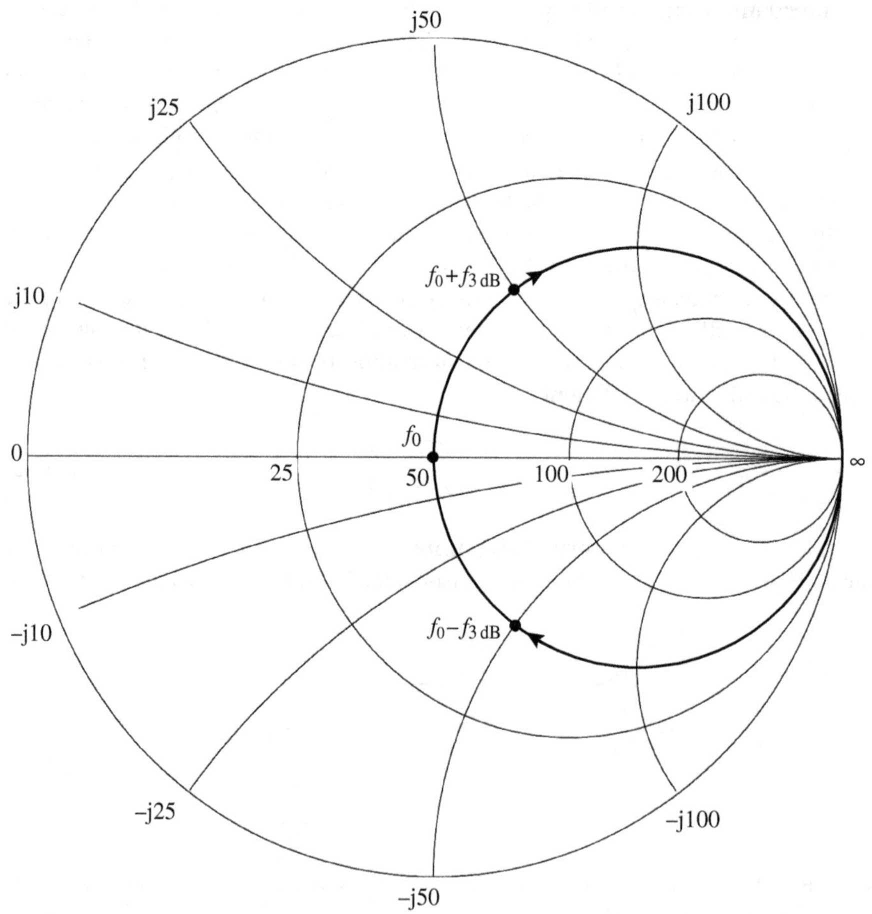

FIGURE 11.42 Smith-chart plot of $Z(f)$ for circuit in Figure 11.40. Note that entire frequency range from DC to infinity fits on chart.

in the RF system, typically 50 Ω. Smith charts are very useful in dealing with impedance problems on transmission lines, because when the transmission-line characteristic impedance is chosen to be Z_0, changing the length of the transmission line simply moves the impedance at its end in a circle around the chart. Smith charts are also useful for lumped-element circuit problems in RF designs as well.

With the advent of modern digital computers, numerous software packages were developed to assist in the often complex calculations needed to design filters, amplifiers, and matching networks in RF circuits and systems. Many lumped-element RF circuits can be analyzed and synthesized with network-analysis software (e.g., National Instruments Multisim™ or OrCAD™ PSpice), but evaluating the RF behavior of various physical structures such as connectors, transmission lines, and antennas requires specialized applications such as Agilent's EESof electronic design automation software, ANSYS's high-frequency structure simulator (HFSS), and COMSOL's finite-element "multiphysics" software. While the use of these advanced tools is beyond the scope of this book, you should be aware that they are widely used and have largely eliminated the need for the sort of intelligent guesswork that guided the labor-intensive hardware prototyping of RF circuits and systems in the past. However, no amount of design software can substitute for an engineer's good understanding of how a system works and what its critical design characteristics are.

BIBLIOGRAPHY

Coleman, C. *An Introduction to Radio-Frequency Engineering*. Cambridge, UK: Cambridge University Press, 2004.

Hagen, J. B. *Radio-Frequency Electronics: Circuits and Applications*, 2nd Edition. Cambridge, UK: Cambridge University Press, 2009.

Pozar, D. M. *Microwave Engineering*, 4th Edition. Hoboken, NJ: Wiley, 2012.

Rutledge, D. B. *The Electronics of Radio*. Cambridge, UK: Cambridge University Press, 1999.

Ruthroff, C. L. Some Broad-Band Transformers *Proceedings of the IRE*, (1959), 47:1337–1342.

PROBLEMS

Note: The problems below that involve noise calculations are simplified to the extent that all noise figure and noise temperature characteristics are assumed to be measured with a 50-Ω source impedance, which is perfectly matched to the amplifier input. This ideal condition is not possible to achieve in general, which means that actual noise calculations and measurements are more complex than these simplified problems indicate. However, the approach to noise outlined in this chapter is useful for approximate system designs where high accuracy is not required.

Problems of above-average difficulty are marked with an asterisk (*).

11.1. *Horizon distance for line-of-sight radio link.* While it is only an approximation that radio waves travel in straight lines, you can conservatively estimate the range of a microwave signal by calculating the distance to the horizon on a perfectly spherical earth. Assuming the earth is a perfect sphere with a radius R_e of 6371 km, calculate the straight-line distance d to the horizon (in km) when viewed from a height h of (a) 10 m, (b) 30 m, (c) 100 m.

11.2. *Maximum range of weather radar.* One use of radio waves not explored in this chapter is in **radar**. One type of radar set sends short intense radio-wave pulses to a target and uses the energy reflected from the target to deduce the target's position, distance, and other information. Suppose a weather-radar antenna is placed on a tower that is 20 m high. Assuming the radiation from the antenna barely grazes the horizon and then travels to a storm cloud at a height of 12 km, what is the maximum distance d the storm cloud can be from the radar and still be visible at the horizon? Use an idealized perfectly spherical earth with a radius R_e of 6371 km.

11.3. *Phase shift due to finite length of circuit conductors.* On a particular type of circuit board, assume a signal travels at 2/3 of the speed of light. Suppose a high-speed system using this type of board can tolerate a phase shift in a sine-wave signal due to propagation delay of only 15° before failing to work. Calculate the maximum length l_{MAX} (in m, or a suitable subdivision such as cm or mm) that a circuit-board trace carrying this signal can have at each of the following frequencies without encountering excessive phase shift due to the propagation delay along the trace: (a) $f = 10$ MHz, (b) $f = 30$ MHz, (c) $f = 500$ MHz, and (d) $f = 2.5$ GHz.

11.4. *Coaxial cable impedance and propagation speed.* To carry the high-power outputs of VHF FM and TV broadcast transmitters to tower-mounted antennas, a large-diameter copper pipe is often used as the outer conductor of a coaxial line to make a low-loss transmission line that can handle the high voltages involved without breaking down.

(a) Assuming that the dielectric in the transmission line is air ($\varepsilon_r = 1$) and the inner diameter $2b$ of the copper pipe used is 7.39 cm, find the outer diameter $2a$ (in cm) of the inner conductor needed to make the cable's impedance $Z_0 = 50.0\,\Omega$.

***(b)** Any real coaxial cable needs mechanical supports to keep the inner conductor centered in the outer conductor. One way of doing this with large-diameter coaxial lines is by mounting dielectric disks with holes

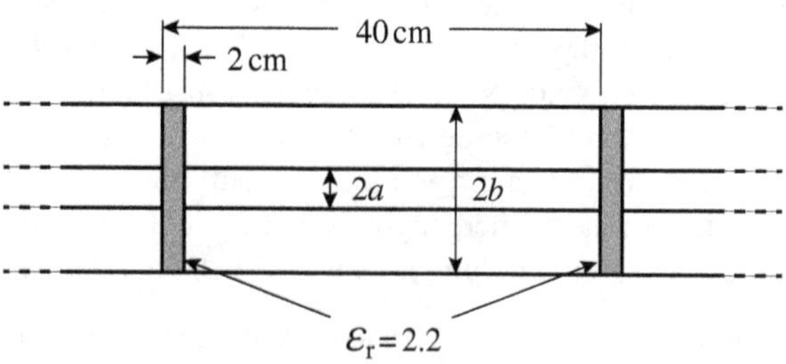

FIGURE 11.43 Dielectric disks separating inner and outer conductors of high-power coaxial transmission line.

through the center between the inner and outer conductors as shown in Figure 11.43. In the line shown, the disks are made of PTFE having a relative permittivity $\varepsilon_r = 2.2$ and are 2 cm wide and spaced 40 cm apart (center to center or left edge to left edge). As long as the wavelength of operation is much longer than 40 cm, you can estimate the effective capacitance per unit length of such a line by adding two capacitances in parallel: the capacitance C_d of a 2-cm-long line with $\varepsilon_r = 2.2$ and the capacitance C_a of a 38-cm-long line with $\varepsilon_r = 1.0$. Using the dimensions for a and b you found in part (a), find the impedance $Z_{0(D)}$ of this transmission line with the dielectric supports added. Is it much lower than 50 Ω?

11.5. *Fourier transform of single pulse*. In signal theory, the *Fourier transform* is the integral-based limit of the periodic infinite series known as the Fourier series. While a Fourier series necessarily produces a periodic function, the Fourier transform can be performed on a single isolated (nonperiodic) waveform, such as the one shown in Figure 11.44. The mathematical definition of the Fourier transform $F[V(t)]$ we will use here is

$$F[V(t)] = F(f) \equiv \int_{t=-\infty}^{+\infty} V(t)e^{-2\pi jft}\,dt \tag{11.83}$$

In general, the Fourier transform of a function is complex, having both real and imaginary parts. In order to find the amount of energy at different frequencies, we can define the **energy spectral density** $S(f)$ for this single (nonperiodic, noncontinuous) pulse as

$$S(f) \equiv |F(f)|^2, \tag{11.84}$$

which, because it is a squared magnitude, is always real and positive.

(a) Calculate and plot the energy spectral density of the pulse shown in Figure 11.44 for the frequency range $f = -5\pi/t_w$ to $+5\pi/t_w$. Your x-axis should be in dimensionless (normalized) units from $ft_w = -5\pi$ to $+5\pi$, and your y-axis should be in normalized units of $S(f)/(V_0 t_w)^2$ from 0 to 1. (Hint: You will be plotting the square of something called the "sinc function," and you may have to do a hand calculation for $f = 0$ because a software calculation may blow up.) Although negative frequencies do not exist for the baseband pulse as shown, if it is used to amplitude-modulate an RF carrier,

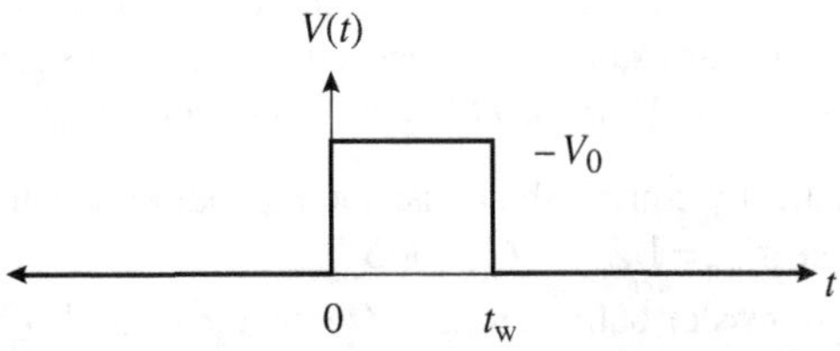

FIGURE 11.44 Single rectangular pulse of amplitude V_0 and duration t_w.

sidebands will appear both above and below the carrier frequency, as though the baseband frequencies were both positive and negative.

(b) In terms of the pulse width t_w, at what frequency f_{NULL} does the energy spectral density first go to zero? A system that passes frequencies up to f_{NULL} with no attenuation and gradually reduces frequencies above f_{NULL} will often reproduce an acceptable version of the pulse shown in Figure 11.44, though it will be rounded off some.

11.6. *Balanced and unbalanced transmission lines*. Suppose 1 mW of RF power is generated by a transmitter whose output is an unbalanced 75-Ω coaxial transmission line. However, the transmitting antenna requires a 300-Ω balanced input, which is fed by a short section of balanced transmission line with a characteristic impedance of 300 Ω.

(a) Calculate the RMS voltage V_{UN} and RMS current I_{UN} in the 75-Ω coaxial line if it is carrying 1 mW with no reflections (perfect matching).

(b) Calculate the RMS voltage V_{BAL} and RMS current I_{BAL} in the balanced 300-Ω transmission line for 1 mW of power.

(c) Suppose the unbalanced line is connected to the balanced line with a transformer having a turns ratio of N_1:N_2, where the N_2-turn winding faces the 300-Ω balanced line. If $N_1 = 100$ turns, about how many turns is N_2?

11.7. *Impedance-matching problem with L networks*. Suppose the output impedance Z_{OUT} of a certain antenna at 25 MHz is real: $Z_{OUT} = 15 + j0\ \Omega$. This (unbalanced) output impedance needs to be transformed up to $Z_0 = 50\ \Omega$ in order to match a 50-Ω transmission-line input impedance.

(a) Using an *L* network consisting of a series capacitor C_1 followed by a shunt inductor L_2 to ground, find values for C_1 and L_2 that will create a perfect impedance match at 25 MHz.

(b) Next, design another *L* network for the same impedance-matching problem from 15 to 50 Ω, but this time, use a series inductor L_1 followed by a shunt capacitor C_2 to ground, and find the values that provide a match at 25 MHz.

(c) Which design requires the smaller value of inductance? (Large inductors are generally more difficult to obtain than large capacitors.)

***11.8.** *Impedance-matching problem with pi network.* Suppose the source of RF energy in a circuit is a transistor whose output impedance Z_{OUT} at 25 MHz is a small resistance in series with a capacitive reactance: $Z_{OUT} = 5 - j20\ \Omega$. Following the steps described in the following, design a *pi*-network matching circuit to transform the transistor's output impedance to perfectly match a real load impedance of $Z_0 = 50\ \Omega$. Use the transistor's output capacitance as the input capacitor for your *pi* network. This will establish the *Q* of your circuit and leads to a unique solution.

(a) Mathematically convert the transistor's output impedance Z_{OUT} to an output admittance $Y_{OUT} = 1/Z_{OUT} = G_{OUT} + jB_{OUT}$.

(b) Find the transistor output circuit's *Q* with the formula $Q = B_{OUT}/G_{OUT}$.

(c) Using the L-network design formulas for $R_L < R_S$ in Figure 11.13, set $R_S = 1/G_{OUT}$ and $Q_D = Q$ and solve for the intermediate resistance $R_L = R_{IP}$.

(d) Design an L network with $X_1 = 1/B_{OUT}$ and X_2 being an inductive reactance. This network should match the transistor output to the resistance R_{IP}.

(e) Using the Figure 11.13 design formulas for $R_L > R_S$, design a second L network with a series inductance (reactance X_3) and shunt output capacitor (reactance X_4) to transform intermediate resistance R_{IP} up to the output resistance $R_L = Z_{OUT} = 50\,\Omega$.

(f) Finally, combine the inductive reactances X_2 and X_3 into one reactance, and draw a diagram of your final circuit. It should look like a single L network with a series inductor followed by a shunt output capacitor. But the input capacitor of the circuit is provided by the transistor, making it actually a *pi* network.

***11.9.** *Pi network with prescribed intermediate resistance. Vacuum tubes* are still used for some high-power RF applications, including amateur radio services. Most vacuum tubes are high-impedance devices with effective output impedance levels of $1000\,\Omega$ or more. In this problem, you will design a *pi*-network matching circuit for a vacuum-tube RF transmitter operating at $f = 7.20$ MHz (in what is known as the **40-meter band**). Design the network in two L-network stages that you connect together in cascade as the final design step, in accordance with the procedure shown in Figure 11.18. (*Hint*: To obtain answers that will check mathematically, carry at least four decimal places in all your calculations without rounding off for this problem.)

(a) Using the design formulas in Figure 11.13 for the step-down L network ($R_L < R_S$) with a shunt C_1 and series L_2, find values for C_1 and L_2 that will match a source resistance $R_S = 10\,\text{k}\Omega$ to an intermediate load resistance $R_{IP} = R_L = 25\,\Omega$. Do this by first calculating Q_D from $R_S = 10\,\text{k}\Omega$ and $R_L = 25\,\Omega$, and then find C_1 and L_2 from the formulas given.

(b) Next, design a second L network to transform the intermediate impedance of $25\,\Omega$ to the final desired load resistance of $50\,\Omega$. Use the step-up formulas in Figure 11.13 ($R_L > R_S$) to design this network, starting by calculating Q_U using $R_S = 25\,\Omega$ and $R_L = 50\,\Omega$. Find values for L_3 and C_4.

(c) Finally, connect the output of the step-down L network to the input of the step-up L network to produce a final *pi* network, drawing the circuit with two capacitors and a single combined inductor and showing all component values. Find the total value L_{TOT} of the (single) *pi*-network series inductor via $L_{TOT} = L_2 + L_3$. Calculate the Q_{TOT} of the combined circuits via $Q_{TOT} = Q_D + Q_U$. Check your solution by calculating the impedance "looking into" the *pi* network from the source side when it is loaded with $50\,\Omega$. Your result should be very close to $10{,}000 + j0$.

11.10. *Impedance ratios of autotransformers.* An **autotransformer** is a kind of transformer with only one winding and one or more **taps** (connections to intermediate points in the winding) brought out in addition to the two ends of

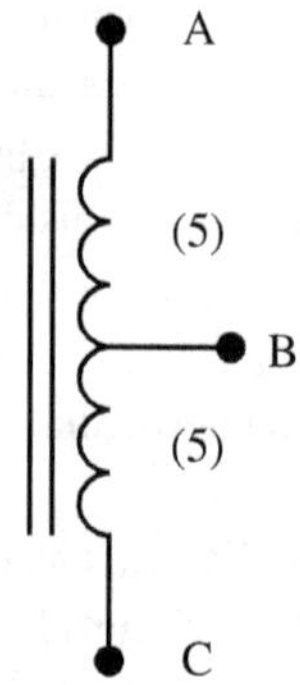

FIGURE 11.45 Schematic diagram of autotransformer shown in pictorial of Figure 11.20.

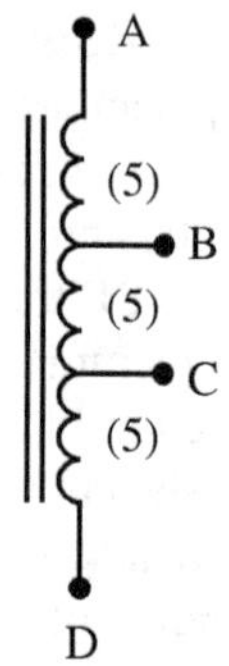

FIGURE 11.46 Two-tap autotransformer with three 5-turn windings.

the entire winding. The bifilar transformer pictorially illustrated in Figure 11.20 is an autotransformer, and its schematic diagram is shown in Figure 11.45. The numbers in parentheses in Figure 11.45 indicate the number of turns in each part of the winding. In calculating the impedance ratios of autotransformers, the number of turns between any pair of terminals can be treated as though they were turns on a separate winding.

(a) If a 300-Ω resistor is connected between terminals *A* and *C* in Figure 11.45, what impedance is seen between terminals *B* and *C*? (*Hint*: The effective turns ratio is 2:1.)

(b) Now, suppose a third 5-turn winding is added to make a two-tap autotransformer as shown in Figure 11.46. If a 300-Ω resistor is connected between terminals *A* and *D*, what is the equivalent resistance seen across terminals *B* and *D*? Across terminals *C* and *D*?

11.11. *Power-added efficiency of RF amplifier.* A certain solid-state device used as an RF amplifier is found to consume 2.2A at 24 VDC from its DC power supply when it is providing RF output power of $P_{OUT(RF)} = 39\,W$. To produce this amount of output power, its RF input power must be $P_{IN(RF)} = 4.2\,W$. Calculate the power-added efficiency η_{PA} of this amplifier. How does this

TABLE 11.2 Small-signal Parameters for 2N2222 with $I_C = 500\,\mu A$

Parameter	Value
Input capacitance C_i	25 pF
Input resistance R_i	50 Ω
Output capacitance C_o	8 pF
Output resistance R_o	28 kΩ

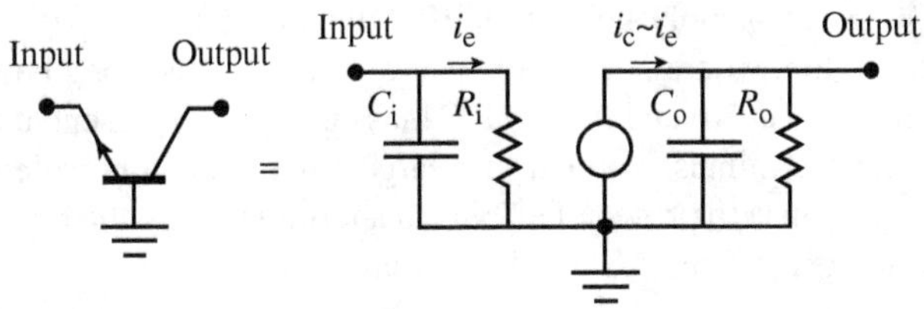

FIGURE 11.47 Approximate equivalent circuit of 2N2222 NPN BJT in common-base configuration for high-frequency amplifier.

figure compare to the conventional efficiency η, which neglects the RF input power required?

11.12. *Design of matching networks for 4-MHz BJT common-base amplifier.* This problem involves the design of a narrowband 4-MHz BJT RF amplifier driven by a 50-Ω source and driving a 50-Ω load. Unlike the common-emitter and common-collector configurations, the common-base connection for BJT RF amplifiers is almost always stable regardless of input and output impedances, because the grounded base provides good isolation between the input and output circuits. This common-base amplifier design problem uses an NPN BJT type 2N2222, which has the approximate common-base characteristics for a DC (bias) collector current of $I_C = 500\,\mu A$ shown in Table 11.2:

The approximate equivalent circuit is shown in Figure 11.47.

For the particular bias current chosen, no input matching network is required because the small-signal resistance of the base–emitter diode is 50 Ω. (The reactance of the input capacitance of 25 pF is so high at 4 MHz in comparison to 50 Ω that it can be neglected.) The real part i_e of the emitter current appears essentially unmodified as the collector current i_c, but the amplifier can produce power gain because the same current entering a low impedance in the emitter circuit encounters a higher impedance at the collector circuit, producing power gain.

(a) Match the collector output resistance to the load by designing an L network to step the 28-kΩ impedance of the output circuit down to a 50-Ω output load. Absorb the 8-pF output capacitance in a capacitor C_T in your matching network.

(b) Calculate the maximum theoretical power gain G of this design at the center frequency of 4 MHz, using the power-gain definition G=(power delivered to load)/(power absorbed from source). Use a real Thévenin equivalent source impedance of 50 Ω and load impedance of 28 kΩ, and ignore all reactances (presumably your matching networks have tuned them out).

*(c) Finish your design with a complete schematic diagram of the amplifier, including a common-emitter-type DC bias circuit (such as the one in Fig. 11.23) to provide a DC collector current of 500 μA with a DC power-supply voltage of 12 V. Include 10-nF bypass and coupling capacitors where needed to ensure that the transistor's base terminal is grounded at RF and that no DC bias currents are present at the input and output terminals. You can use large-value inductors called **RF chokes** (e.g., 1 mH) to provide DC bias to a terminal (e.g., the input and output) while leaving the AC and RF circuit undisturbed.

*(d) Finally, model your entire circuit with a circuit simulation software package such as Multisim™. You may need to adjust the matching-circuit values slightly in order for the output circuit to resonate at 4 MHz, or else the overall circuit will not show power gain. Such **manual tuning** is often needed in RF amplifier circuits to compensate for slight differences in device characteristics or wiring capacitance and inductance, although it is undesirable from a manufacturing point of view. What is the maximum power gain (in dB) and 3-dB-down bandwidth of your amplifier simulation?

11.13. *Current and impedance levels of push–pull RF amplifier*. Solid-state RF power amplifiers that provide more than 50 W or so must deal with very large currents, because the typical silicon RF BJT or FET cannot withstand a DC collector voltage higher than about 40 V. (Newer devices using materials such as **silicon carbide** can exceed this voltage limitation and provide greater power for the same current level.) Suppose a push–pull RF amplifier using a DC collector voltage of 36 V must provide a total peak RF output power of 200 W. Typically, the AC "swing" (peak-to-peak AC excursion) of voltage at the collectors cannot exceed twice the DC power-supply voltage.

(a) If each of the two devices in the push–pull amplifier delivers 100 W and has a peak-to-peak voltage of $(2\times36)=72$ V across its output terminals, what is the *peak* current $I_{C(PK)}$ at the collector of each device, assuming sinusoidal voltage and current?

(b) At this current $I_{C(PK)}$, how much series inductance L_S at a frequency of $f=30$ MHz will cause a voltage drop of 5 V (peak) in series with the collector, seriously compromising performance? This exercise shows why it is so important to minimize series inductance in the output circuits of high-power solid-state RF amplifiers.

11.14. *Threshold SNR calculation.* A certain digital receiver system requires a SNR of −3 dB (power ratio) to be above threshold for proper operation. The system's bandwidth B is 80 kHz and the effective input noise temperature of the system is $T_{SYS} = 1000$ K. What is the minimum signal power P_{MIN} at the system's input needed for the system to operate above threshold? Express your answer in both watts (numeric form) and dBm (dB relative to 1 mW).

11.15. *Power output from noise temperature calculation.* A certain low noise RF amplifier has a bandwidth $B = 100$ MHz and is connected to a source resistance at room temperature ($T_0 = 290$ K). The amplifier's average power gain over its rated bandwidth is $G_{dB} = 55$ dB. If the amplifier has an effective noise temperature of $T_{amp} = 100$ K, what is the total noise power P_{OUT} (internal amplifier noise plus amplified noise from the room-temperature load) measured at the amplifier output? What is the resulting output voltage V_{OUT} (RMS) across a 50-Ω load? (*Note*: Remember to convert decibel gains into numeric gains before using them in noise calculations.)

11.16. *System noise temperature of two amplifiers in cascade.* Two amplifiers, A_1 and A_2, that have the same bandwidth B are connected in cascade (output of A_1 goes to input of A_2). Amplifier A_1 has an effective noise temperature $T_{e1} = 35$ K and a gain $G_1 = 30$ (numeric ratio). Amplifier A_2 also has a gain $G_2 = G_1 = 30$, but its noise temperature is much higher: $T_{e2} = 500$ K. Assume for the purposes of this problem that the source is a 50-Ω resistor at 0 K, so that the only noise to be considered originates in the amplifiers themselves.

- **(a)** Calculate the system noise temperature $T_{SYS}(a)$ of the system of two cascaded amplifiers with A_1's output feeding A_2's input. (*Hint*: Use Eq. 11.42 with $T_{e3} = 0$.)
- **(b)** Now, reverse the order of the amplifiers in the cascade so that A_2's output feeds the input of A_1. What is the system noise temperature $T_{SYS}(b)$ of the cascade with the higher-noise amplifier placed at the front end? Is this a significant difference?

11.17. *Noise figure and noise temperature.* Table 11.3 allows conversion between noise *temperature* (in K) and noise *figure* (in dB). Both types of noise performance measurements are still in common use. Fill in the blanks where numbers are missing.

***11.18.** In the mobile-phone uplink shown in Figures 11.28 and 11.29 and described in the accompanying text, the low-noise amplifier was mounted on top of the tower so it could be connected immediately to the receiving antenna output. Recalculate the system noise temperature T_{SYS} for a system consisting of the coaxial cable followed by the amplifier, and recalculate the maximum acceptable link loss L_S(max) if the low-noise amplifier is instead mounted at the *base* of the tower, at the output end of the transmission line between the transmission line and the receiver input. Does this change degrade the system performance appreciably?

TABLE 11.3 Noise Figure–Noise Temperature Conversion Table

Noise figure (dB)	Noise temperature (K)
0	0
0.1	6.75
—	20.8
0.4275	—
1	—
—	250
3.01	—
—	628
10	—
—	5000

11.19. *LO frequency and IF for RF downconversion.* In the type of radio receiver called a **superheterodyne**, the incoming RF is converted by means of a mixer circuit to a lower frequency termed the intermediate frequency (*IF*). Filtering and other signal processing steps are often easier at the lower frequency, and tuning (varying the input frequency of the receiver) can be achieved by varying the local oscillator (LO) frequency used by the mixer circuit. Typically, the IF frequency is 10–30% of the highest RF frequency received. For the following questions, assume the FM receiver has an IF of 10.7 MHz and is designed to receive the range of frequencies from 88 MHz to 108 MHz (the FM broadcast band in the U. S.)

(a) What is the range of LO frequencies needed if high-side injection is used (LO frequency > RF frequency)?

(b) What is the range of LO frequencies needed if low-side injection is used (LO frequency < RF frequency)?

(c) What is the range of frequencies for which an image response can occur with high-side injection?

(d) What is the range of frequencies for which an image response can occur with low-side injection? Which type of injection will allow an image frequency range that includes part of the adjacent aircraft radio band (108–137 MHz)?

(e) What is the LO frequency range if l

11.20. *Peak and average power for AM.* While some types of modulation produce an RF waveform with a constant-amplitude envelope, amplitude modulation (AM) does not. This means that both the peak and average power for an AM signal will vary considerably with modulation, as you will discover in the following exercise.

(a) Suppose an AM transmitter delivers an RF waveform with amplitude of 100 V (peak) into a 50-Ω load with 0% modulation. What is the average power $P_{AVG}(0\%)$ delivered to the load with 0% modulation?

(b) With 100% modulation, what is the peak voltage $V_{PEAK}(100\%)$?

(c) With 100% modulation, what is the peak delivered power $P_{PEAK}(100\%)$?

(d) With 100% modulation, what is the average delivered power $P_{AVG}(100\%)$?

***11.21.** *Coupling capacitor for varicap frequency control.* The circuit of Figure 11.34 is used to vary the frequency of a (nominal) 50-MHz oscillator whose frequency is equal to the resonant frequency of C_1 in parallel with L_1. The value of L_1 is 470 nH. The varicap VC_1 used provides a capacitance of $C_{MIN}=10$ pF with maximum reverse bias and $C_{MAX}=30$ pF with zero bias. What value of coupling capacitor C_T will allow a tuning range of at least $\Delta f=100$ kHz ($f_{LOW}=49.95$ MHz and $f_{HI}=50.05$ MHz) when connected between the varicap and the L_1–C_1 tuned circuit?

11.22. *Friis transmission equation.* During a mountain-climbing expedition, it is necessary to communicate between adjacent mountain tops a distance $R=8$ km apart with two-way radios that use a frequency $f=462$ MHz, which is in a commercial VHF two-way radio band. The propagation path between the radios can safely be considered as a free-space path with no reflections.

(a) If both the radios have omnidirectional antennas, what is the link loss L_L (ratio >1) for the path between the two radios? State your result in dB.

(b) What is the link loss L_L if the citizens-band frequency of 27 MHz is used instead?

(c) Suppose the link frequency $f=462$ MHz and both radios now use yagi antennas, each with a gain of 12 dB. What is the link loss including gain of these antennas?

For further resources for this chapter visit the companion website at
http://wiley.com/go/analogmixedsignalelectronics

12

ELECTROMAGNETIC COMPATIBILITY

12.1 WHAT IS ELECTROMAGNETIC COMPATIBILITY?

The following is a true story. It happened to some senior electrical engineering students of my acquaintance who decided to enter an **IEEE robotics competition**. The particular challenge involved steering the robot accurately over a distance of a few meters. The students chose a two-motor robot with tank-type treads, each tread driven by one motor, on opposite sides of the robot. The motor shafts also had position readouts that produced a certain number of pulses per shaft revolution. In order for the robot to move in a straight line, both motors had to turn at the same rate. To make sure this happened, the students designed a pulse-counting routine in their control software to monitor each motor's speed and minimize the speed differences between the two motors. When they tried the routine with pulses from a lab-bench pulse generator, it worked fine, so they loaded it into the robot's software and instructed it to run in a straight line.

To their surprise, the robot had a fit—jerking this way and that, twisting around, and, in general, doing anything other than moving smoothly in a straight line. The contest was less than a week away, so the students had little time to diagnose and fix the problem. But diagnose it they did. After hours of investigations, they figured out that the DC motors on the robot used **brushes** (as most inexpensive DC motors do). The brushes make intermittent contact with the **commutator** (a segmented set of contacts) on the rotating armature, and as the brush contacts make and break fairly large currents, they create energetic impulses that contain a wide range of frequency

Analog and Mixed-Signal Electronics, First Edition. Karl D. Stephan.

Companion Website: http://wiley.com/go/analogmixedsignalelectronics

components. The motor-brush impulses were strong enough to interfere with the operation of the motor position readout circuits, causing them to produce spurious pulses. In effect, the position circuits were telling the microprocessor that the motors were doing crazy and impossible things, and the control software was trying to compensate for these nonexistent motions with the series of jerks and twitches that the students witnessed.

Due to lack of time, the students had to eliminate their planned feedback control for the tread motors and relied instead on manual adjustments of motor speed, which worked adequately for the short distances involved in the competition. They also learned a valuable lesson about a subject that is rarely treated in the usual electrical engineering curriculum: **electromagnetic compatibility** or **EMC**.

EMC is the study of how electromagnetic energy can cause unintended effects in electronic systems and how to eliminate these effects. The general name for these unintended effects is **electromagnetic interference** or **EMI**, and a subset of EMI that deals with such interference at radio frequencies is called **radio-frequency interference** or **RFI**. Electromagnetic theory is thus the scientific foundation of engineering for EMC, but many people who encounter an EMC-related problem may not have the background in electromagnetics necessary for a thorough understanding of a given situation. Nevertheless, the purpose of this chapter is to provide an easily understood introduction to the basic principles of EMC for readers who may not have had a course in electromagnetic theory. Accordingly, this chapter will concentrate on how to recognize problems that are related to EMC. More than other types of design difficulties, EMC problems tend to be system-level problems that require an understanding of the big picture of a given electronic system in its environment. As you might expect, solving EMC problems requires a good system-level understanding, as well as the requisite background in electromagnetics. Reading this chapter by itself will not make you an EMC expert, but it should point you in the right direction if you encounter an EMC problem and give you an idea of what steps to follow to fix it, including consultation with a qualified EMC expert if necessary.

The rest of the chapter is organized into four sections. In the first section, we distinguish between the two main categories of EMC problems: those related to communications systems and everything else. The communications type of EMC problem arose in the very early days of radio around 1900, before tuned circuits were used to discriminate between transmitter wavelengths. The noncommunications type of EMC problems had to await the application of electronics to fields other than radio, which began around 1920.

In the second section, we describe the four ways that EMI can be transferred: conduction, electric fields, magnetic fields, and electromagnetic radiation. There are not hard and fast boundaries between these ways or modes of transfer, but usually, one of them is dominant in any given situation, and knowledge of the transfer mode is critical to both diagnosing an EMC problem and in finding a solution.

In the third section, we discuss the basics of various approaches to EMC problem solving. The main methods in use are bypassing and filtering, proper grounding, and shielding, or a combination of these methods.

In the fourth section, we show that the best way to avoid EMC problems in a system design is to design with EMC in mind, rather than ignoring EMC issues until after a prototype is built and then doing the often expensive and time-consuming diagnostics and modifications that are sometimes necessary to fix EMC problems. The intelligent use of a few good EMC practices early in the design process can pay off in freedom from EMC problems later.

12.2 TYPES OF EMI PROBLEMS

12.2.1 Communications EMI

As mentioned earlier, EMC became an issue with the advent of the earliest radio communications systems. In Chapter 11, we described how the electromagnetic spectrum of radio wavelengths is a finite resource that must be exploited in a coordinated way. The radio transmission of information requires a nonzero amount of **bandwidth** in the radio spectrum. Very roughly speaking, the bandwidth that a signal requires is proportional to the number of bits per second transferred. Signals with low data rates, such as on-and-off signals for a remote-control device, can be transmitted in bandwidths only a few Hz wide, in principle. But high-speed data links such as streaming video transmit many **megabytes** (MB) per second and require much wider bandwidths of several MHz or greater, depending on the type of modulation and other factors.

With analog types of modulation (e.g., AM and FM), a radio link requires more or less exclusive use of a **channel**, which is the term for the band of radio frequencies allocated for that signal. In a properly designed radio communications system using allocated channels, the locations, transmitter powers, transmitter bandwidths, and receiver characteristics are all chosen so that each transmitter–receiver pair can operate independently of the others. This is illustrated schematically in Figure 12.1, which shows a conceptual drawing of a simple transmitter–receiver radio link. Transmitter T_1 emits a band of RF energy centered on frequency f_1. Receiver R_1 must

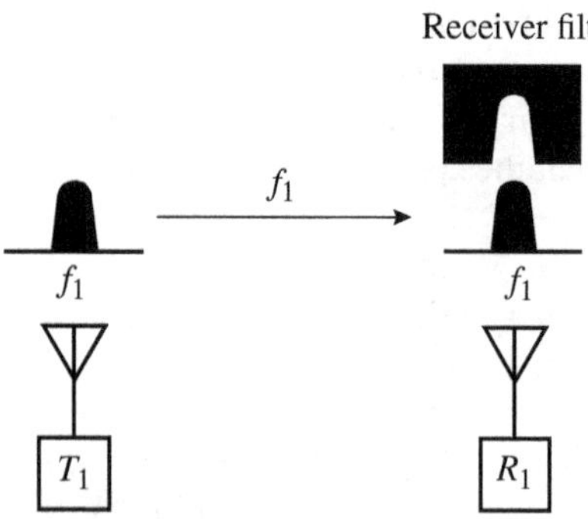

FIGURE 12.1 A transmitter T_1 sending a band of signals centered on frequency f_1 to receiver R_1, which filters out all received energy except for the transmitted signal near f_1.

be within transmitter T_1's **range** (meaning the geographic area where the signal strength from T_1 is strong enough to be received with adequate quality by R_1). An ideal receiver performs a **matched-filter** operation on the signal received from T_1, passing essentially all the significant frequency components from T_1 near f_1 while rejecting all other signals (from other transmitters, for instance). As the figure indicates, the spectral shape of the transmitted signal fits perfectly into the filter at the receiver, while other signals fit poorly or not at all. Real receivers do not perform quite as well as this ideal receiver, although they can approach the ideal quite closely.

If a second transmitter–receiver pair T_2–R_2 operates physically near the first pair so that each receiver is within range of both transmitters and a different channel at frequency f_2 is used by the second pair, no interference (EMI) results as long as the transmitter and receiver frequency characteristics are chosen properly, as shown in Figure 12.2. As the figure indicates, the signal at frequency f_2 from transmitter T_2 is out of the band of frequencies near f_1 passed by receiver R_1. Transmitter T_2's signal is therefore called an **out-of-band** signal with respect to the bandwidth of receiver R_1. Similarly, the signal at f_1 from transmitter T_1 is out of band with respect to receiver R_2's filter that receives only the channel at f_2. This shows how communication links in which receivers are within the range of more than one transmitter can nevertheless operate properly when the frequencies are appropriately allocated and used.

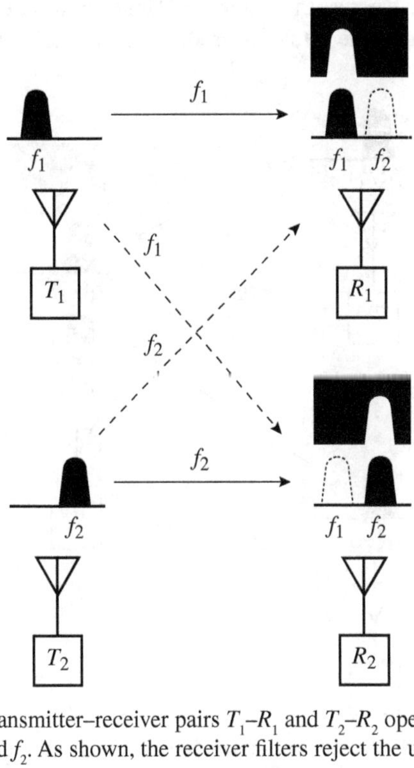

FIGURE 12.2 Two transmitter–receiver pairs T_1–R_1 and T_2–R_2 operating on different channels at frequencies f_1 and f_2. As shown, the receiver filters reject the undesired signal from the out-of-band transmitter.

The simplest case of communications EMI occurs when two or more transmitters emit signals in the same, or nearly the same, frequency band. For example, suppose both transmitters T_1 and T_2 try to use the *same* channel (frequency band f_1) to send signals to receiver R_1, which is set up to receive f_1, as shown in Figure 12.3. If the receiver is within the effective range of both transmitters, both transmitted signals will arrive at the receiver. Depending on the type of modulation, the stronger signal may overwhelm the weaker signal, but propagation phenomena such as **fading** and **multipath reception** may change the ratio of received signals so that the receiver picks up each signal alternately or possibly nothing at all. Or in the case of AM, both signals may be received along with distortion and undesirable interference. In any case, if the two interfering signals have comparable strengths at the receiver, the result of the situation in Figure 12.3 is EMI.

This is the most fundamental type of communications system EMI and is called **co-channel interference**, because it happens when signals on the same channel from two or more transmitters arrive at the receiver. For fixed-location stations such as broadcast transmitters, co-channel interference is prevented by careful allocation of frequencies with attention paid to providing enough geographical separation between stations that use the same channel and close regulation of the maximum power emitted by each transmitter. For example, in the AM broadcast band, certain channels are reserved for low-power stations, and the limited geographic range of such stations means that dozens of them across the continental United States can use the same channel, at least in the daytime. (The situation is different at night, when the **ionosphere**

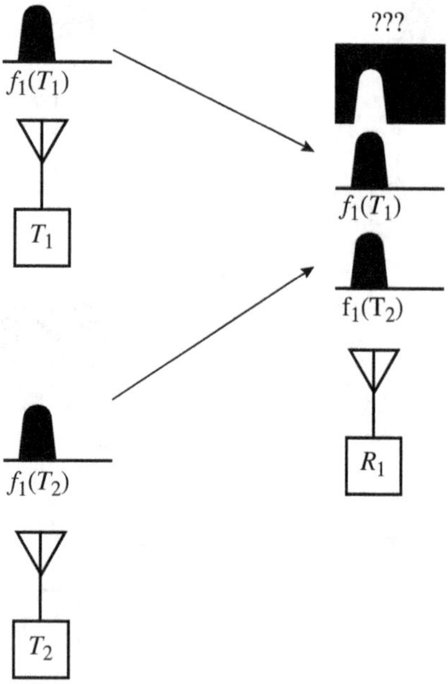

FIGURE 12.3 Two different transmitters attempting to use the same frequency f_1 produce co-channel interference at the receiver.

effectively reflects certain wavelengths back to earth for distances of hundreds or thousands of miles.) For mobile services such as mobile phones (cell phones), the range of the base-station transmitters is intentionally limited to a region called a cell, so that the same set of frequencies can be reused in other cells that are separated by enough distance to prevent interference.

More and more radio services now use various forms of digital modulation rather than analog modulation. With digital modulation, the interference situation is more complex, because many types of digital modulation produce signals that cover a wide range of frequencies that can be used simultaneously by other transmitters and receivers. The details of how digital signals interfere with each other is beyond the scope of this book, but while digital modulation techniques can achieve significant improvements in system performance compared to analog modulation using the same frequency allocation, there is a limit to the **channel capacity** of digitally modulated RF signals as well. As we mentioned in Chapter 11, signals transmitted with digital modulation tend to have an "all-or-nothing" character. As the strength of an interfering signal increases, the quality of reception shows no change at first and then abruptly degrades to zero. Most mobile phones now use digital modulation, but most people are familiar with what happens when a mobile phone becomes out of range—the signal becomes intermittent and eventually disappears. In densely populated areas, such signal degradation can be caused by interfering signals as much as by the relative weakness of the desired signal. In short, EMI can be a problem for digitally modulated signals as well as those using analog modulation.

A less common type of communications EMI problem that can nevertheless be very difficult to solve is communications EMI due to **intermodulation** in a nonlinear element in the transmitter or receiver, or (rarely) elsewhere. As mentioned in Chapter 11, when two sine-wave signals at frequencies f_1 and f_2 combine in a nonlinear circuit element, it is possible to produce a sum signal at a frequency $f_3=f_1+f_2$, as well as a difference frequency and other sums and differences of multiples of both f_1 and f_2. In communications systems terms, these are called **intermodulation products**. (One particular type of intermodulation, in which the amplitude modulation on one carrier appears on the carrier of a second amplitude-modulated signal, is called **cross modulation**.) Suppose that a pair of transmitters emitting frequencies f_1 and f_2 intermodulate in some nonlinear element as shown in Figure 12.4 to produce a sum frequency $f_3=f_1+f_2$. Depending on its amplitude, the intermodulation product at f_3 can interfere with, or even overwhelm, the legitimate signal at f_3 from transmitter T_3 that receiver R_3 is designed for.

The output stages of high-power amplifiers (especially solid-state ones) are very nonlinear. Because tower space is expensive, a typical radio transmitting tower will support a large number of transmitting antennas, each of which is operating at one or more of a number of different frequencies. If sufficient care is not taken to isolate the output of one transmitter from another, the high-power output of one transmitter can go backward through adjacent transmitting antennas into the output stage of a second transmitter and produce undesirable intermodulation products. This is one reason why such transmitters are often protected by special low-loss filters to keep such intermodulation from occurring.

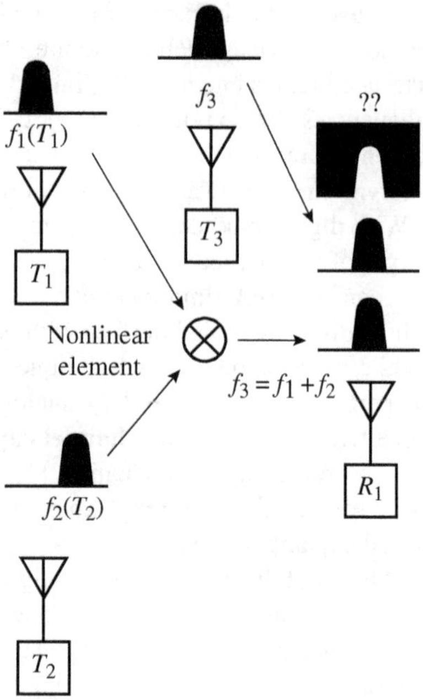

FIGURE 12.4 Example of two signals at frequencies f_1 and f_2 combining in a nonlinear element to produce spurious sum frequency f_3 that interferes with intended signal from transmitter T_3.

A poorly designed receiver **front end** (input stages) can also produce undesirable intermodulation products unless the circuit is carefully designed to be very linear and is adequately protected from out-of-band signals by filtering.

Finally, intermodulation sometimes occurs even with properly designed and isolated transmitters and receivers. The magnetic and electric fields near a transmitting antenna can be so large that the currents in passive objects such as antenna elements, dish reflectors, or even rusty hardware nearby can produce nonlinear intermodulation products. As you can imagine, such EMI problems are very difficult to track down, but when the usual sources of problems have been eliminated and high-power transmitters are involved, this so-called passive intermodulation may be the culprit.

Fortunately, the vast majority of communications systems EMI problems are due to a combination of poor transmitter or transmitter system design, poor receiver design, or improper geographic location of the transmitter or receiver. The solutions to such problems are relatively straightforward, and for certain classes of communications systems such as military equipment, elaborate EMC specifications and performance criteria have been developed. Following these specifications will not always mean that no EMI will occur, but the rules are very effective at minimizing such problems.

12.2.2 Noncommunications EMI

Radio communications systems are designed to emit and receive electromagnetic radiation, so it stands to reason that EMI problems were first identified in the field of radio. But as electronic systems began to be used for noncommunications purposes—for example, instrumentation, controls, and radar—problems arose in which sources of EMI interfered with the proper operation of systems that were not designed to receive electromagnctic radiation in the first place. While the hardware involved in noncommunications EMI is very different in function from communications hardware, the EMI process is similar. Just as in a communications link one has a transmitter, a propagation medium, and a receiver, every EMI situation in a noncommunications setting has a **source** of EMI (in the role of transmitter), a **transfer path** (in the role of propagation medium), and the system interfered with (in the role of receiver), colloquially called the **victim**, as shown in Figure 12.5. The main difference between the communications-type EMI situation and a noncommunications EMI problem is that the noncommunications source is not designed to be a transmitter, nor is the victim designed to be a receiver. But physical devices are unaware of the designer's intentions, so if the physics permits EMI to happen, it will happen. The fact that EMI problems are nearly always accidental is one reason why EMI problems can be so unexpected and hard to track down. In the example we began the chapter with, the students designing the robot were not trained to view a DC motor as a transmitter nor a microprocessor board as a receiver, but these components accidentally functioned as such to produce the EMI problem we described.

All EMI problems have these three common elements: source, transfer medium, and victim. Because the source and the victim are not usually optional elements in a design, most EMC efforts concentrate on preventing the method of transfer from allowing the EMI produced by the source to reach the victim. In the next section, you will learn how the four ways or modes of EMI transfer—conduction, electric fields, magnetic fields, and radiation—tend to arise in different circumstances and which mode is more likely to occur in a given situation.

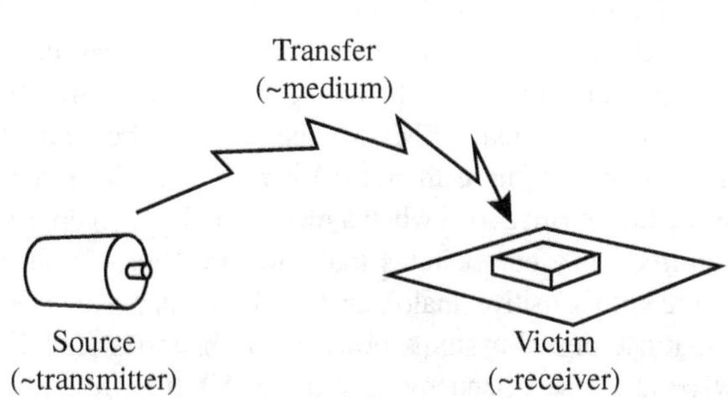

FIGURE 12.5 Basic EMI situation in which an EMI source (a DC motor as transmitter) sends interfering energy through a transfer method (as propagation medium) to a victim system (a microprocessor board as receiver).

12.3 MODES OF EMI TRANSFER

Because EMI is *electromagnetic* interference, it can travel from a source to a victim only by electromagnetic means. EMC specialists divide these means into four practical categories: (1) conduction, meaning a path provided by wires, cables, circuit-board traces, or other good electrical conductors; (2) electric fields, which can be modeled as capacitive coupling; (3) magnetic fields, which can be modeled as inductive coupling; and (4) electromagnetic radiation, which involves both electric and magnetic fields traveling together as a radio wave. Each category has its own characteristics, and identifying which of the transfer modes is dominant in a given EMI situation takes the designer a long way toward arriving at a solution.

12.3.1 Conduction

When EMI is carried by means of conductive pathways, the designer can often use circuit theory to analyze it and to find ways to prevent it. The most common type of conductors implicated in conduction-transfer EMI are power-supply lines, followed by ground leads and ground conductors. The reason for this is not hard to figure out. Every part of an electronic system with active components requires power and ground connections, and therefore, power-supply conductors and ground return paths are typically shared by every part of a system. Just as an impure city water supply can spread disease germs throughout an entire city, a poorly regulated or inadequately bypassed power supply can furnish EMI transfer paths that effectively interconnect many parts of a system that should not be interconnected.

We have already discussed one classic type of conduction-transfer EMI in Chapter 6: the situation in which a high-gain amplifier's output stage couples part of its output back through the power supply lead to its own input, leading to undesirable oscillation. This situation was shown in Figure 6.6, along with the solution: insertion of power-supply bypass capacitors and isolation resistors in the power-supply leads. The *R–C* filters thus formed increase the transfer loss through the power-supply circuit for the frequencies amplified by the amplifier stages, reducing the feedback below the point at which it causes problems.

In general, the sources involved in conduction-transfer EMI draw considerable power, and the current they consume has a significant AC component. Amplifier output stages are one common source of conduction EMI. Mechanical switches and relays that control significant amounts of power (more than 1 W) can also produce conduction-transfer EMI, because the waveforms that arise when a mechanical switch opens and closes contain abundant high-frequency components that can travel throughout a power-supply network and interfere with sensitive analog or digital circuitry. This is especially true in portable devices that use digital systems powered by batteries. The HI and LO logic levels in such systems are separated by as little as 500 mV, which means the **noise margin** (tolerance of a digital input for noise voltage added to the desired digital signal) is very small, making these devices unusually sensitive to conducted EMI in the form of pulses or high-frequency noise. Also, batteries can have rather high **internal resistance**, which increases the power-supply voltage fluctuation for a given change in current drain.

Another common conduction path for conduction-transfer EMI in AC-powered equipment is the AC power line itself. Poorly designed equipment can produce non-periodic **transients** and periodic current waveforms with high-frequency components that go backward into the connection between the equipment and the AC power source. Once EMI gets into a building's power wiring, it can go almost anywhere and can cause problems that are extremely difficult to solve. By the same token, AC power is not always a pure, well-regulated sine wave. Lighting equipment using fluorescent tubes and their ballasts, heating equipment, and **universal motors** that use brushes and commutators can all produce abundant amounts of EMI that can travel through a building's power wiring to whatever is plugged into it. Unless a system's AC-operated power supply is designed to prevent power-line EMI from entering the system, it may allow outside interference to disrupt its operation, but only when such interference is present. This can lead to peculiar situations such as finding that a particular product works well in some factories, but not in others where the power-line noise is higher.

Other types of conductors along which EMI can travel include signal cables (both shielded and unshielded); instrumentation cables used for industrial sensors, thermostats, fire and burglar alarms, and so on; and even ground wires, such as the third (green) wire used for a safety enclosure ground in AC-powered equipment. Grounding is discussed later in the section on alleviating EMI, and intelligent grounding is one of the most important steps a designer can take to prevent conduction-transfer EMI.

To diagnose conduction-transfer EMI, one must demonstrate that the interference causing the problem is carried by identifiable conductors from an identifiable source to an identifiable victim. If the conduction pathways are power-supply lines, one cannot simply cut the lines to see if the interference goes away: the functions of the powered system go away too! One diagnostic test that can sometimes be performed to establish the nature of conduction-transfer EMI is to substitute a temporary separate test power supply for the normal system power supply. This test is shown in Figure 12.6.

In Figure 12.6a, the system's common power supply, the EMI source, and the EMI victim are shown along with the EMI pathway from the source to the victim through the power-supply wiring that is shared between the source and victim. (The smoke lines rising from the victim are a figurative way of showing that EMI is preventing the victim from working properly, though the problem need not issue in literal smoke!) If the designer suspects that the power-line conductor is carrying the EMI responsible for the problem, the temporary setup in Figure 12.6b can be tried. The suspected power lines are cut, and a temporary laboratory supply that can handle the suspected victim's power-supply needs is substituted for the system supply. As Figure 12.6b shows, this operation blocks the EMI path and will allow the victim to operate normally, if the power-supply connection is truly the main pathway for EMI. Normal operation in the diagnostic test indicates that the power-supply line is indeed carrying EMI from the source to the victim. Diagnosing the problem is only the first step toward solving it but a very important one. Solutions to this and other EMI difficulties will be discussed in Section 12.4.

Power-supply lines are not the only type of conductive pathway that can carry EMI, although they are frequently found to be at fault. Any type of conductor

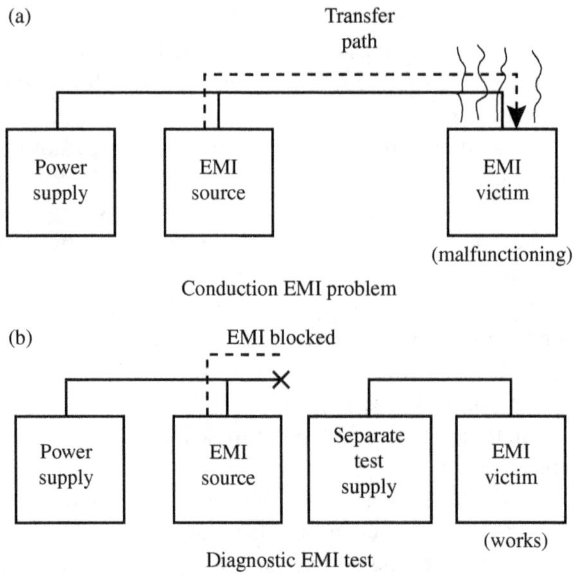

FIGURE 12.6 (a) System with conduction-transfer EMI problem caused by EMI transfer from source through power-supply line to victim. (b) Diagnostic test using separate test power supply to victim, demonstrating that EMI is conducted through power-supply line.

or conductive material—a ground lead, a metallic chassis, or even a nonelectrical conductive component such as a water pipe—can potentially conduct EMI from a source to a victim. If an electrical connection is not necessary for the proper function of the system, the solution can be as simple as inserting an insulator to interrupt the conductive pathway. If the electrical connection is a necessary part of the system, such as the USB cable shown in Figure 11.8, cutting the connection is not an option, and other solutions such as the ferrite choke shown in the figure can reduce or prevent RFI and other types of EMI.

12.3.2 Electric Fields (Capacitive EMI)

If you comb your hair on a dry day with a plastic comb, the comb will acquire an electrostatic charge. It will then pick up small pieces of paper, and if you listen carefully while combing, you may hear a crackling sound, because the high electric fields (measured in units of volts per meter) can exceed the dielectric strength of air (about $3\,\mathrm{MV\,m^{-1}}$) causing **dielectric breakdown** and producing tiny sparks. These effects are due to the electric field created around the comb due to the net charge on it. The field can be high enough to attract bits of paper from a distance of a few cm, but sensitive sensors can detect an electrostatic field like this at a distance of many meters.

If there is a voltage difference between any pair of conductors, an electric field can exist between the conductors and produce a net charge Q proportional to the voltage difference V. The ratio of charge to voltage (Q/V) has the dimensions of coulombs per volt, or capacitance (farads). A capacitor is simply a component designed to hold a certain amount of charge Q when a voltage V is applied across its terminals. But capacitance can be defined and measured between any two conductors in an electronic system. And if two conductors share what is called **mutual capacitance**, meaning that a voltage difference between them produces a measurable change in charge, this capacitance can couple energy between the conductors and thus furnish a path for EMI. Coupling via electric fields in this way is the second of the four categories of EMI transfer and often one of the most significant ones.

The most typical situation in which capacitive EMI can occur is in **high-impedance** circuits, where one conductor carrying a large AC voltage is placed near a second conductor connected to a sensitive part of the circuit. Examples of high-impedance circuits are instruments that measure extremely low currents, such as the ionization-current detectors in some types of smoke detectors, and amplifiers for weak high-impedance signals such as those from **piezoelectric** transducers, which are used in some types of motion sensors and actuators. The same circuit element can act as both source and victim in some types of EMI, as shown in the pictorial sketch in Figure 12.7.

Suppose a signal from a high-impedance source enters a **printed circuit board** (**PCB**) at the point labeled "input" in Figure 12.7. It travels along a conductive trace to the input of an integrated-circuit amplifier, where it is amplified and sent via another trace to an output terminal labeled "output." The EMI problem arises from the fact that the input and output traces are close enough together on the board to share a common electric field when a potential difference (voltage) appears between them. This mutual electric field gives rise to a mutual capacitance C_M, symbolized by the small capacitors in dashed lines connecting the input and the output. Depending on the amplifier's gain and phase response, the mutual capacitance between output and input may provide a **loop gain** magnitude greater than 1 and a phase shift of zero degrees at one or more frequencies. If this happens, the oscillator theory developed in Chapter 7 tells us that the circuit will be **unstable** and there is a chance it will **oscillate**. Even if the capacitive coupling is insufficient to cause oscillation, it can be

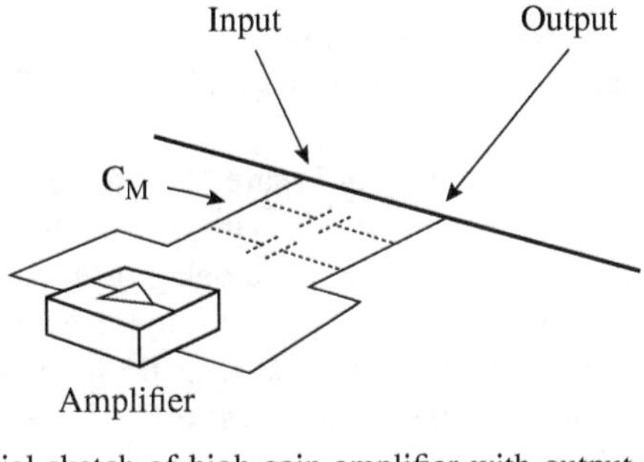

FIGURE 12.7 Pictorial sketch of high-gain amplifier with output circuit-board trace near input trace, leading to mutual capacitance and electric-field EMI.

enough to result in severe **frequency distortion** such as peaks and dips of several dB in the amplifier's frequency response.

In this particular example, the same circuit played a dual role of both EMI source and EMI victim, but a similar situation involving two separate signal paths can result in **crosstalk** due to mutual capacitance. Crosstalk refers to the undesired transfer of a signal from one path to another path. For example, in a stereo music amplifier, two separate audio signals—left and right—are amplified independently. If the high-voltage output of the left-channel amplifier, for example, is situated so that it is close to the input leads of the right channel, the right channel can pick up some of the left channel's signal, leading to interchannel crosstalk and a loss of **stereo separation**. Good layout practices that help to prevent electric-field EMI include separating high-level conductors from low-level conductors and incorporating extra preventive measures such as shields when necessary, as we will discuss later.

12.3.3 Magnetic Fields (Inductive EMI)

As the word implies, EMI can take place by means of either electric or magnetic fields (or both). We have seen how differences in electric potential (voltage) can produce changes in charge—that is, current—in a conductor that is in the electric field of another conductor. In that case, the thing that starts the ball rolling is a voltage difference. If there is no voltage difference, that means there will be no EMI transferred by means of an electric field.

EMI transferred by magnetic fields begins with currents, not voltages, because changing magnetic fields in electronic systems arise from changing currents. A typical magnetic-field EMI problem occurs when a conductor carrying a large varying current produces a magnetic field around itself. A complete understanding of the different ways electric and magnetic fields behave requires knowledge of electromagnetic theory, but some intuitive geometrical ideas about how they differ can be gathered from pictures.

In electromagnetism, a **field line** describes a path that an imaginary charged particle called a **test charge** would move along if placed in the field. This is simple enough to understand with electric fields. Harking back to the example of a comb with, say, a positive charge on it, you can see that if a negatively charged bit of paper is placed near the comb, it moves toward the comb because opposite charges attract. And if, for example, a conducting wire carries a negative charge, a negatively charged particle such as an electron will tend to move straight away from the wire, because like charges repel each other.

This fact is reflected in the diagram of Figure 12.8, which shows the electric and magnetic fields present around a long straight wire. You are looking at a cross section of the wire—imagine that the wire's length is along your line of sight and the black dot in the center is the wire itself, where it goes through the page from front to back. In this example, the wire is both maintained at a positive voltage above ground (which is too far away to show in the picture) and carries a current. The electric-field lines are straight and point away from the surface of the wire, which is the direction a positive test charge would move if placed near the wire. And the strength of the

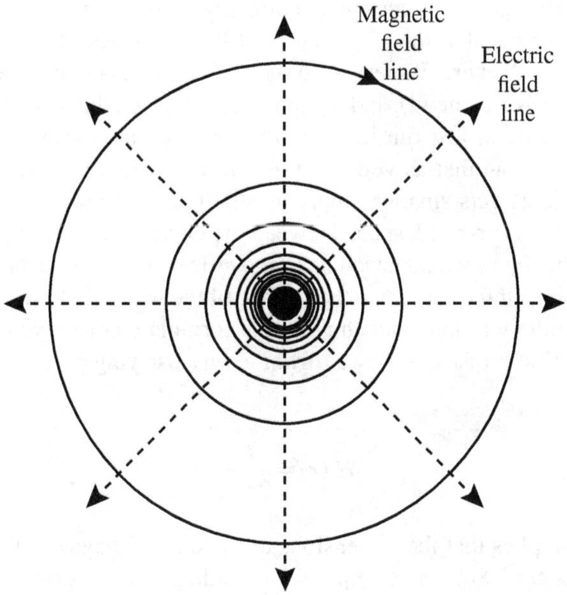

FIGURE 12.8 Electric fields (dashed lines) and magnetic fields (solid lines) surrounding a current-carrying wire (black circle) at a positive voltage above ground (ground not shown).

electric field (measured in volts per meter) is proportional to the wire's voltage with respect to ground. Roughly speaking, the field strength in a diagram like this is proportional to how close the field lines are to each other. So near the wire, the electric-field lines are closer together, implying that the electric field is higher there. For the special case of an infinitely long round wire in free space that has a uniformly distributed charge q on it (in coulombs per meter), the electric field at a distance r (meters) from the axis of the wire is (for r greater than the wire's radius)

$$E(r) = \frac{q}{2\pi\varepsilon_0 r} \quad (12.1)$$

in which the constant ε_0 is called the **permittivity of free space** and has a value of $8.854\times10^{-12}\,\mathrm{F\,m^{-1}}$. So, for example, an infinitely long wire with 1 nC of charge on each meter's length will produce an electric field at 1 m away from the wire of

$$E(1\mathrm{m}) = \frac{1\,\mathrm{nC\,m^{-1}}}{2\pi\,(8.854\times10^{-12}\,\mathrm{F\,m^{-1}})(1\,\mathrm{m})} = 17.97\,\mathrm{V\,m^{-1}}. \quad (12.2)$$

This is a very small field but nonetheless capable of affecting sensitive high-impedance circuits adversely.

Now, consider the magnetic-field lines. If the wire were at a high voltage but carried no current, there would be *no* magnetic field, because (except for permanent magnets) all magnetic fields in electronic systems are created by currents. But this wire happens to be carrying a current in a direction going away from the reader (into the page). It turns out that the magnetic-field lines produced by this current form *circles* centered on the wire. If there were such a thing as a positive **magnetic charge** (there isn't, as far as we know!), and you put this charge on the outer circle and let go of it, the charge would just run in a circular orbit around the wire as long as the current persisted. Note that as you get closer to the wire, the spacing between the magnetic-field lines gets smaller, implying that the magnetic field increases near the wire too. (There is a blank space between the closest magnetic-field line and the wire, because the figure would have gotten too crowded otherwise, but the field does keep on increasing all the way up to the surface of the wire.) To be quantitative, if the wire is (again) infinitely long and straight, the formula that tells you the **magnetic-field intensity** H at a distance r away from a wire carrying a current of I amperes is simply

$$H(r) = \frac{I}{2\pi r}. \tag{12.3}$$

Equation 12.3 implies that the dimensions of the unit of magnetic-field intensity H is **amps per meter**.[1] So, for example, if a welding cable carries a direct current of 200A, the magnetic-field intensity 1 m away from it is $H=(200\text{A})/(2\pi \times 1\,\text{m})=31.8\,\text{A}\,\text{m}^{-1}$. This field intensity is comparable to the earth's magnetic field, when expressed in terms of amperes per meter ranges between about 20 and $50\,\text{A}\,\text{m}^{-1}$ on the earth's surface.

The significant thing about magnetic fields with regard to EMI is that a changing magnetic field near a conductor *induces a voltage* in that conductor. So if two wires lie side by side (as in Fig. 12.7), and one is carrying AC current, a magnetic field will appear around the current-carrying wire (call it wire No. 1). If some of this magnetic field is present at the second wire and the orientation is correct (basically, if the two wires are not perfectly perpendicular), the changing magnetic field from wire No. 1 at the wire No. 2 will produce a changing voltage in No. 2. The second wire will behave like there is a tiny voltage source in series with it, and the voltage produced will be proportional to the amplitude of the changing current in the first wire.

This is a long way of saying that conductors near each other can couple magnetically. This basic principle is how all transformers work. A transformer is simply two or more separate conductors coupled by means of a magnetic field. And in fact, transformers are often the source of magnetic-field EMI, because not all the magnetic field that couples the transformer windings stays in the transformer. Some field always *leaks* into the space around the transformer, and if a conductor carrying a signal to a sensitive circuit passes near the transformer, there is a good chance that magnetic-field EMI can occur.

[1]It turns out that the magnetic quantity that has real effects is the **magnetic flux density** B, measured in units called **teslas** (T). In free space, the relationship between B and H is $B=\mu_0 H$, where $\mu_0=4\pi\times 10^{-7}\,\text{H}\,\text{m}^{-1}$ is called the **permeability of free space**. But because B and H are usually proportional except inside some magnetic materials, we have chosen to discuss only H in this chapter for simplicity.

You can now see why magnetic-field EMI is current driven, while electric-field EMI is voltage driven. While high-impedance circuits are most prone to have electric-field EMI problems, low-impedance circuits tend to have magnetic-field EMI problems, because the same amount of power carried at a low impedance will involve a current that is higher and a voltage that is lower than the high-impedance case. Types of circuits that tend to use high currents and low impedances include audio amplifiers that drive speakers, power circuits that operate **solenoids** (electromagnets) and electric motors, some types of lighting and display circuits that produce high currents, and many types of power supplies. If a system you are designing involves any high-current AC inputs or outputs, you should bear in mind that magnetic-field EMI is a possibility.

As we mentioned, transformers can be a source of magnetic-field EMI and can also be the victim if an external magnetic field couples to the transformer's windings. The same is true of inductors that can also produce and couple magnetic fields by virtue of the coils of wire inside them. Some types of inductors are more prone to make external magnetic fields than others. Inductors whose windings are wound on a ring-shaped **toroidal core** tend to leak less magnetic field into the environment than other types of inductors whose coils are basically cylindrical. Also, the orientation of coils in inductors and transformers can have a great effect on how much magnetic-field EMI they pick up. Regardless of the details, you should remember that inductors and transformers are especially likely to pick up magnetic-field EMI and be wary of using them in sensitive low-impedance circuits such as amplifier inputs without considering carefully the possibility that they will pick up interfering magnetic fields.

12.3.4 Electromagnetic Fields (Radiation EMI)

Up to this point, we have been purposely vague about the distances involved in electric-field and magnetic-field EMI. The reason is that the transfer distance—how far the source is from the victim—can depend on a great many factors, such as the voltages or currents involved at the source, the sensitivity of the victim, the orientations of circuit components, and so on. Nevertheless, these two types of EMI tend to be localized in comparison to either conducted EMI or the topic we will now take up: radiated EMI.

The fundamental difference between radiated EMI and magnetic- or electric-field EMI is the rate at which the different types of fields decrease with distance r from the source. (This r is a free-space distance—imagine the source is isolated and floating in outer space by itself.) No matter what type of field source is involved in producing EMI at a single frequency, fundamental principles of electromagnetism tell us that the fields it produces can be sorted out into two categories: electric and magnetic fields near the source and radiated fields farther away.

The behavior with distance r of all the fields depends on the dimensionless ratio $(2\pi r/\lambda)$. This number expresses the distance r by **normalizing** it to the wavelength λ of the source's emission, which is assumed to be at a single frequency. (If more than one frequency is involved, you simply analyze the waveform in question into its component frequencies and figure out what each one does by itself.)

Certain types of electric and magnetic fields are significant mostly in a region where $(2\pi r/\lambda)$ is less than one—that is, within a sixth of a wavelength or less of the source. Note that for realistic problems, the source cannot be infinitely long in any dimension, unlike the hypothetical wire in Figure 12.8. So in this case, the distance r is measured from the point of observation of the field, to the center of a source of finite extent in all three dimensions.

The region in which $(2\pi r/\lambda)$ is less than one is called the **near field**, and the near field is where most electric-field and magnetic-field EMI problems occur. In fact, these fields are generally referred to simply as either electric fields or magnetic fields, precisely because they do not radiate a long distance and they are found mostly in the near field.

As the distance r from the (finite-sized) source increases, the amplitudes of the magnetic and electric fields in the first (near-field) category fall off very fast, at rates of either $1/r^2$ or $1/r^3$. These rates of decrease are shown in Figure 12.9. When you realize that power transfer depends on the square of the field amplitudes, you can understand why a few wavelengths away from the source the electric and magnetic fields in the near-field category are usually negligible. For example, a field whose amplitude falls off as $1/r^2$ will transfer power at a rate that falls off as $(1/r^2)^2 = 1/r^4$. So if we double the distance r, the power transfer falls off by a factor of $1/2^4 = 1/16$. And the $1/r^3$ fields fall off even faster. But that is not true of the second category of fields, the **radiated field**.

The important characteristic of the radiated field is that it decreases more slowly with distance, as $1/r$, even when the source is finite in extent. (The electric and magnetic fields in Fig. 12.8 also decreased as $1/r$, but that was for a fictional infinitely long source that is never encountered in reality.) As Figure 12.9 shows, at a

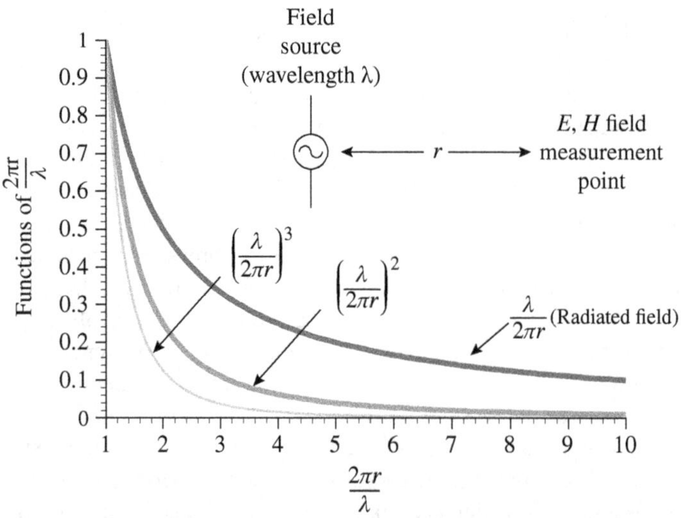

FIGURE 12.9 Amplitudes of fields versus distance r from source at wavelength λ. Localized electric and magnetic fields vary as $1/r^3$ or $1/r^2$; radiated field varies as $1/r$.

distance where $(2\pi r/\lambda)=10$, the near-field electric and magnetic fields have decreased in amplitude by factors of 100 or 1000, but the radiated field has decreased only by a factor of 10. The radiated fields are the type of fields used in radio communications, precisely because they show such a slow decrease of amplitude with distance. A radiated field's **power density** in $\mathrm{W\,m^{-2}}$ goes as $1/r^2$, which is the well-known **inverse-square law** that applies to light waves and all other types of electromagnetic radiation, as well as acoustic waves and many other types of waves. Radio waves are referred to as electromagnetic radiation because, in free space, both electric and magnetic fields travel together in a kind of lockstep arrangement. Wherever you find a radiated electric field, you will also find a radiated magnetic field and vice versa. In the simplest kind of radiated electromagnetic wave (technically called a **transverse electromagnetic plane wave**), the electric- and magnetic-field lines are at right angles to each other and are also at right angles to the direction the wave is traveling in, called the **direction of propagation**. Following our convention of showing more intense fields as field lines more closely spaced together, we have illustrated a small section of such a plane electromagnetic wave in Figure 12.10.

There are several things to note about this picture. It is an instantaneous snapshot of the fields of a traveling wave, showing their configuration at a single moment in time. First, the wave's **wavelength** is twice the distance from the negative peak of the electric field to the positive peak, just as you would expect in a water wave, sound wave, or any other kind of wave. Next, the intensities of the electric field (in $\mathrm{V\,m^{-1}}$) and the magnetic field (measured in **amps per meter** or $\mathrm{A\,m^{-1}}$) vary together: where the electric field is the highest, the magnetic field is also. This fact leads to an interesting quantity called the **impedance of free space** η_0, which is about 377 Ω. The impedance of free space is the ratio of the electric-field intensity to the magnetic-field intensity of a free-space radiated wave and is constant for most types of radiated waves in air or a vacuum. (If the wave propagates through a medium other than air, this impedance changes.)

The next thing to notice about the traveling radiated electromagnetic wave is that both the electric and the magnetic fields are perpendicular to the direction of propagation. In dealing with radiated EMI, you need to know how the radiated fields tend to be oriented so that you can understand how the radiation is emitted by the source and received by the victim. If we "unfroze" the picture and let the wave continue on its way, you should imagine the entire pattern moving uniformly to the right at the speed of light. So at a given stationary point, a victim would experience both an alternating magnetic field and an alternating electric field at the frequency f of the radiated wave. The direction of the electric field is called the wave's **polarization**, and in general, this direction can be any vector that is perpendicular to the wave's direction of travel.

Most radiated EMI is produced by conductors carrying a current whose frequency f (assuming it is sinusoidal) is related to the wave's wavelength λ by the familiar equation $\lambda = c/f$, where c is the speed of light. For appreciable radiation to take place, typically, a conductor's length has to be at least a few percent of the free-space wavelength λ. Otherwise, the conductor cannot radiate energy efficiently, and the electric

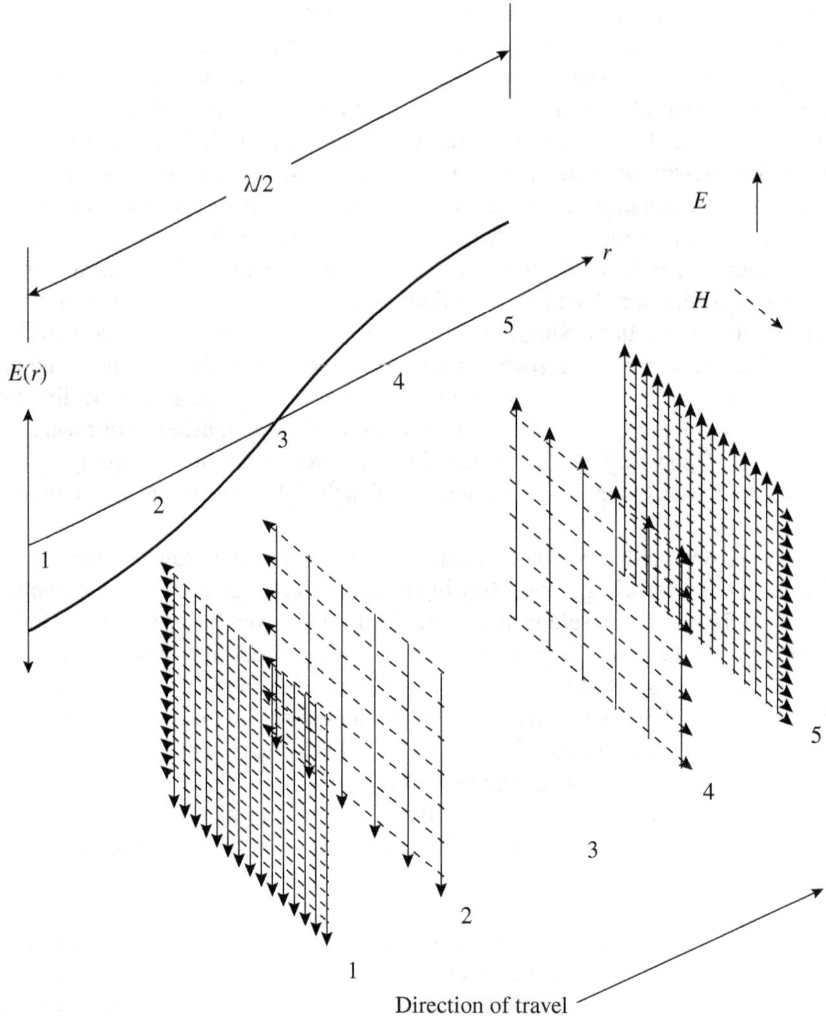

FIGURE 12.10 Sections of a frozen-in-time radiated plane electromagnetic wave showing electric-field lines (solid) and magnetic-field lines (dashed), along with a plot of electric-field intensity $E(r)$ versus distance r in the direction of propagation.

and magnetic fields around it merely store energy and give it back to the circuit twice each cycle. But energy that is radiated in the form of an electromagnetic wave such as the one in Figure 12.10 is forever lost to the circuit, which means that such radiation involves a **loss** in the radiating circuit. This loss is modeled by including a resistive component called the **radiation resistance** in the equivalent-circuit model.

Radiated EMI is often emitted by high-frequency circuits that are physically large enough to radiate some of their energy efficiently. Depending on the frequency and the power level involved, radiated EMI can interfere with a wide variety of other systems. For example, suppose an RF heating unit is installed in a factory to seal plastic bags. Such units typically use a frequency such as 27.12 MHz in an ISM band. If the unit's power supply accidentally couples some of the generated energy to a long extension cord leading to the unit, this cord could form a radiator if it is an appreciable fraction of 11 m (= c/27 MHz) long. Depending on the wave's intensity, this 27-MHz energy could interfere with the operation of computers, control systems, communications devices, and other important electronics. And because it is a radiated field, not simply a localized electric or magnetic field, it could travel many meters before finding its victim—even outside the factory building!

Now that you know about the four different ways that EMI is transferred from a source to a victim, we will briefly describe some of the main techniques used to prevent EMI in electronic systems.

12.4 WAYS TO REDUCE EMI

The problems caused by EMI can take many forms: mysterious intermittent malfunctions, interfering signals that seem to come from nowhere, and other difficulties that can be very hard to track down once a system is built. That is why good designers routinely include suitable precautions in designs to prevent EMI from happening in the first place, at least under most conditions. It is much easier to prevent EMI by building in preventive measures than it is to ignore the EMI issue during the design phase and then be forced to undertake a lengthy and expensive diagnostic campaign and redesign once EMI problems show up in a prototype system. Measures that can reduce or eliminate EMI should be a part of every designer's toolkit. While one can go overboard in EMI reduction, the experienced designer will learn which types of systems are more prone to EMI and where to use preventive measures. But first you need to learn what these measures are.

Each of the four transfer modes of EMI—conduction, electric fields, magnetic fields, and radiated fields—needs to be dealt with in a way that is appropriate for it. Some cases of EMI involve more than one transfer mode, but usually one is dominant, and when you eliminate the worst offender, the next-worst one shows up, and so on. The first set of methods to prevent EMI—bypassing and filtering—apply primarily to EMI that is conducted along system wiring. The other methods, namely, grounding and shielding, apply more to EMI that is transferred by means of fields. However, the boundaries between these categories are fuzzy. For example, the way a bypass or filter circuit is grounded and shielded may have a large impact on how well it performs. At any rate, a suitable combination of the techniques described in the following text should go a long way toward reducing or eliminating EMI problems.

12.4.1 Bypassing and Filtering

Bypassing means connecting power-supply lines (and other lines that should be at a constant potential) to ground via a suitable capacitor or capacitors. **Filtering**, in the context of EMI, means using filters designed to pass desirable energy—DC or AC power or signals—while blocking conducted EMI that is traveling along the same pathway. Viewed this way, bypassing is a subset of filtering, because a grounded bypass capacitor forms a lowpass filter with whatever impedance it is connected to. But the design approaches to the two types of circuits are different, so we will treat them separately here.

Not all bypass capacitors prevent EMI. Some are used simply to maintain an internal circuit junction at a nearly constant potential so that an amplifier will not lose gain. For example, in the common-emitter amplifier in Figure 10.26 (reproduced here as Fig. 12.11a), the **emitter bypass capacitor** C_1 keeps the emitter terminal of the transistor at a nearly constant voltage. If C_1 were removed, the AC emitter current caused by the AC base voltage would produce an AC emitter voltage because of the voltage drop across the emitter resistor. The AC emitter voltage thus developed is in phase with the original base voltage, leaving less of the total input signal available to do useful work across the emitter–base junction and lowering the stage's gain. A similar role is performed by capacitor C_2 in the common-base RF amplifier shown in Figure 12.11b, taken from Figure 11.23. If C_2 were not present, AC base current would cause an AC base voltage to appear because of the voltage drop across the base-bias voltage divider. Again, the AC base voltage would reduce the portion of the signal voltage available to appear across the base–emitter junction, thus reducing the stage's gain. To obtain the maximum gain from these circuits, it is important to include the bypass capacitors noted, but they do not have much effect on potential EMI problems.

However, bypass capacitor C_3 can reduce EMI as well as improve the amplifier's performance. Unless an ideal power supply is connected directly to the $+V_{CC}$ terminal of the RF amplifier, the AC collector current passing through the output transformer will produce a voltage across the equivalent series impedance of the power supply, whatever it happens to be. For example, if the power supply is at the far end of a long power-supply lead with considerable inductance, the AC voltage drop across this inductance can be an appreciable fraction of the DC supply voltage. The AC voltage

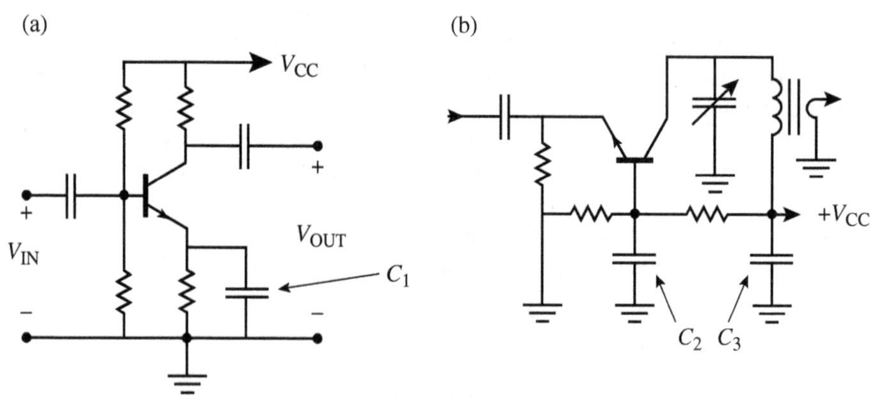

FIGURE 12.11 Bypass capacitors C_1, C_2, and C_3 used in examples of amplifier circuits: (a) common-emitter amplifier, (b) common-base RF amplifier.

across the power-supply impedance reduces the voltage available at the transformer for usable output power, and this reduces the stage's gain. From an EMI point of view, the power-supply impedance also causes a portion of the amplifier's RF output to be present in the power-supply voltage of any other circuit connected to that same power-supply lead. For example, if a high-gain front-end RF amplifier stage is connected to the same power-supply lead as the common-base amplifier shown, the front end may receive the RF signal from the common-base amplifier through the power supply, and this could cause interference or oscillation—in other words, EMI. So the installation of bypass capacitor C_3 can benefit both the individual amplifier stage's performance and reduce the chances that the system will encounter an EMI problem as well.

In general, any time a circuit draws current from a power supply and the current contains high-frequency components, it is a good idea to install bypass capacitors as physically close to the circuit drawing power as possible. Examples of these types of circuits include switching power supplies, high-speed digital circuits of all types, RF amplifiers, audio-amplifier output stages, and motor, LED, and display drivers. The term "high power" is relative to the power source and the scale of the system. In a 500-watt public-address-system amplifier, a circuit that draws 2 W is not that big a deal. But in a system powered by a 1.5-V hearing-aid battery with a high internal resistance, a circuit that intermittently draws 50 mW from the battery can be regarded as high power and in need of bypassing and other conducted-EMI prevention measures.

From the discussion in Chapter 6, you should know that a **filter** is a circuit that transmits some frequencies with little or no loss while attenuating others by a specified amount. Figure 12.12, modified from Figure 6.7, showed how the usual circuit configuration of a **two-port** filter is arranged to receive a signal from an equivalent source with open-circuit voltage v_s in series with source impedance Z_S and transfers part of the signal to the load impedance Z_L as output voltage v_o. When filters are used in analog signal paths, the source and load impedances Z_S and Z_L are usually well-defined resistances, and this makes the filter design relatively straightforward. If the filter circuit's characteristics are known (e.g., in terms of its **impedance matrix** $[Z]$ as discussed in Chapter 3), one can easily calculate the ratio v_o/v_s and predict how the filter will attenuate any given frequency component of the signal. If the output of a circuit or system contains undesirable frequency components that will cause EMI, a filter can be designed to pass the desired frequencies while attenuating the

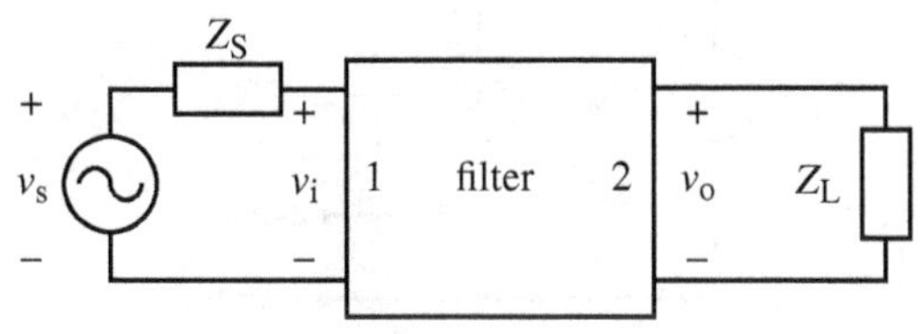

FIGURE 12.12 Generic two-port filter circuit with equivalent voltage source v_s in series with source impedance Z_S, and output voltage v_o across load impedance Z_L.

undesired ones so that they are no longer troublesome, and this is a common motivation for the use of conventional analog filters in RF systems.

However, if EMI appears on power-supply wiring, the filter situation is different. Suppose we wish to design an EMI filter to be used on power-supply leads instead of signal leads. An ideal power supply has a source impedance Z_S of approximately zero, and the AC impedance of most power-supply loads is not well known either. If the power supply in question is the AC utility supply, the impedances involved are extremely low (often 1 Ω or less) and can vary in their resistive or reactive parts from minute to minute, depending on other loads connected to nearby building wiring. The unpredictability of source and load impedances in typical power-supply situations means that EMI filters must be designed to deal with a wide variety of circumstances and their performance in blocking EMI cannot always be guaranteed without prior knowledge of the impedance environment.

A further complication involved in AC utility-supply EMI filters is that the EMI to be filtered can be of either the **common-mode** type or the **differential-mode** type. Common-mode and differential-mode voltages were explained in Chapter 5. Taking an AC power-line cord set as an example, there are three conductors involved: the so-called "hot" lead (usually color-coded black), the neutral lead (color-coded white), and the safety ground lead (usually colored green). To express the voltage relationships among three conductors requires at least two voltage variables. While the most obvious pair of voltage variables is hot to ground and neutral to ground, the same amount of information is also conveyed by expressing these voltages in terms of a **common-mode voltage** v_{CM} and a **differential-mode voltage** v_{DM}. These voltages are produced by the set of voltage sources shown in Figure 12.13. As you can tell from the figure, if only a common-mode voltage is present, you will measure the *same* voltage v_{CM} between each line conductor (either hot or neutral) and ground. A voltmeter connected between the hot and neutral leads will read zero. On the other hand, if only a differential-mode voltage is present, a voltmeter connected between hot and neutral will register the full differential-mode voltage v_{DM}, while only half this voltage will appear from each conductor to ground. These relations are expressed

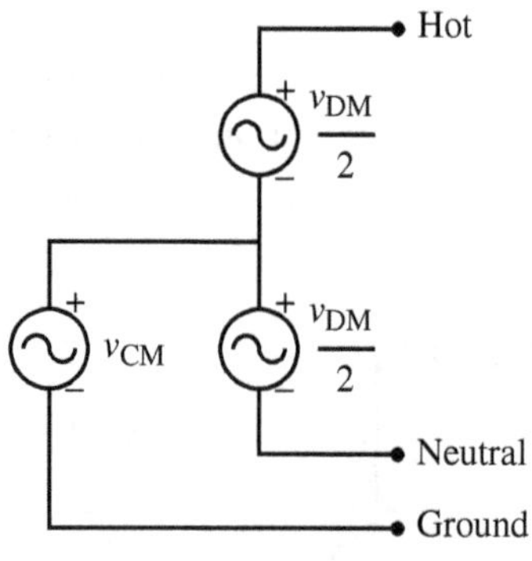

FIGURE 12.13 Voltage sources for common-mode and differential-mode EMI voltages on AC power-line pair.

by the following conversion equations, where v_{HG} is the voltage measured from hot (+) to ground (−) and v_{NG} is measured from neutral (+) to ground (−):

$$v_{HG} = v_{CM} + \frac{v_{DM}}{2} \tag{12.4}$$

$$v_{NG} = v_{CM} - \frac{v_{DM}}{2} \tag{12.5}$$

Many sources of EMI tend to produce either mainly differential-mode or common-mode interference, and an effective EMI filter must be able to deal with both differential-mode and common-mode interference waveforms.

A typical EMI filter for use on an AC utility-supply line is shown in Figure 12.14. It is housed in a solid-sheet-metal case that keeps electric fields within the filter housing and also tends to prevent EMI currents within the filter from passing outside or through it. (More details about shielding for EMI are presented later.)

A typical AC power EMI filter circuit is shown in Figure 12.15. Capacitors C_1 and C_2 are connected between the hot and neutral leads and thus reduce differential-mode EMI energy. Capacitors C_3 and C_4 bypass each power-line lead to the ground lead and will reduce common-mode EMI. Additional common-mode isolation is provided by the pair of coupled inductors L_1–L_2, which pass differential-mode energy (including the low-frequency power-line energy at 50 or 60 Hz) but are wound in such a way as to present a substantial inductive impedance to common-mode EMI current. The resistor R does not improve the filter's EMI performance, but provides a discharge path to ground in the event that the filter's capacitors retain a dangerously high voltage if the filter is disconnected from the power line.

The performance of these types of EMI filters is typically evaluated in terms of **insertion loss** in a 50-Ω system. (Insertion loss is the increase in loss through a transmission line when a component such as a filter is inserted in the line.) Good EMI

FIGURE 12.14 AC power-line EMI filter packaged in shielded metal case.

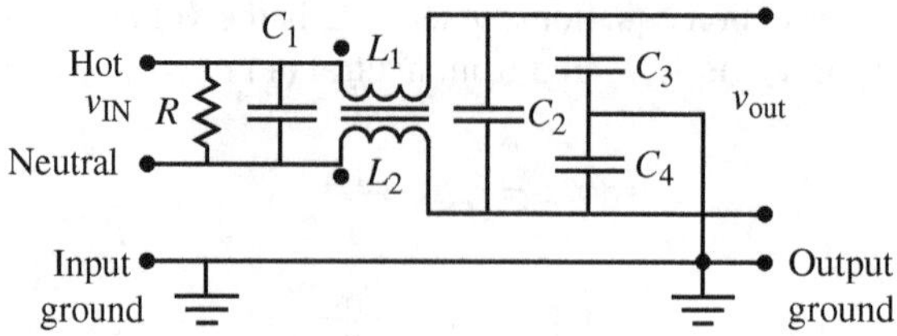

FIGURE 12.15 Typical circuit of AC power-line EMI filter.

filters can provide insertion loss in the range of 50 to 70 dB or more at frequencies that are often of concern for EMI problems. For example, many switching power supplies operate at frequencies in the 20–500 kHz range, so high insertion loss in this frequency range is a good feature of an EMI filter.

12.4.2 Grounding

The word **ground** in electrical engineering theory means a node whose potential is zero by definition. While it is mathematically convenient to have such an ideal voltage reference for theoretical designs, real physical circuits never provide grounds that behave in an ideal way. An ideal ground would have exactly zero impedance at all frequencies, and this is impossible for anything larger than a mathematical point. The technique of **grounding** involves designing a system so that all ground connections perform their functions with acceptable quality and without causing EMI or other problems. No ground is perfect, but most grounds can be made good enough so that they do not cause EMI problems on their own.

Some of the most baffling problems that improper grounding can cause arise from the situation known as a **ground loop**. Ground loops take various forms, but one of the most common ones is shown in Figure 12.16, which depicts an actual EMI situation that the author dealt with. In a performance venue, a video signal from the stage needed to be sent some distance away to be projected on a screen for viewing by overflow crowds. This involved transmitting an analog video signal from a camera about 50 m (150 ft) to a video display unit inside an auditorium.

The typical analog video signal source produces a waveform with an amplitude of about 1 V peak to peak when connected to a load impedance of 75 Ω. The camera was connected to the video amplifier input of the display unit through a coaxial cable with an impedance of 75 Ω. Besides the video camera, other electrical equipment was used at the remote location and provided with 120-VAC 60-Hz power through an extension cord. Both the camera and the video amplifier were properly grounded to the AC power-supply ground at their respective locations.

When the system was first installed, an EMI problem showed up. The video signal as received at the display was very distorted, showing slowly moving horizontal bars that made the image almost unusable. The reason for this EMI was that a portion (about 0.5 V) of the 60-Hz power was getting into the video circuit. This probably happened in the following way.

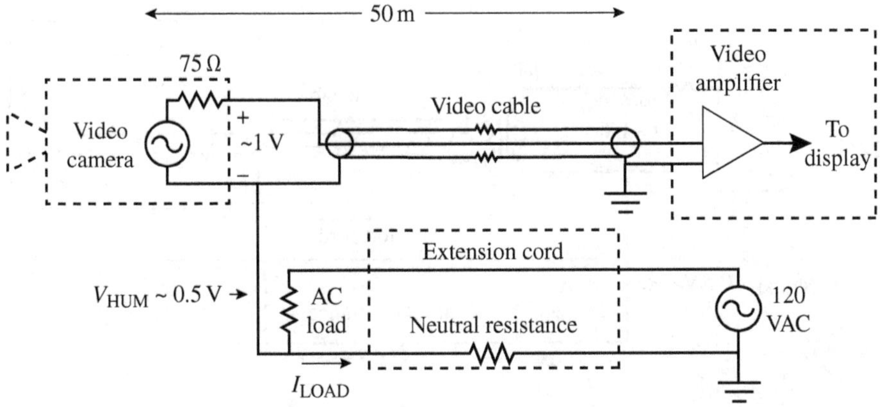

FIGURE 12.16 Video display system with EMI problem caused by ground loop.

All ground connections of any length have some resistance. The extension cord used had both neutral and ground leads, but both of them had some resistance and we have lumped both leads together in a single return line in Figure 12.16, representing this resistance as the neutral resistance shown in the figure. When a substantial AC load current I_{LOAD} flows through such a resistance, an AC voltage develops across this resistance. If we take the true ground reference to be the entry point of the AC utility supply to the building, then this means that all other "grounds" in the AC power system are not really at ground, but above ground by some small AC voltage. And the larger the load on the system, the larger these stray AC voltages can become because of the larger voltage drops across the ground resistances.

In accordance with good engineering practice, the video camera's ground was connected to the power-system ground, either directly or through circuit capacitance, and the stray 60-HzAC voltage present on the power-system ground appeared in series with the desired 1-V video signal. From the point of view of the video amplifier, its input terminal saw the sum of the desired 1-V video signal and the undesired 60-Hz EMI voltage, which led to the distortion in the picture. Note that despite efforts to ground all appropriate points, including the coaxial cable's ground shield, the cable's ground also has resistance, and any stray 60-Hz currents flowing along that ground will contribute to the EMI problem as well. In such a situation, you can usually trace a closed loop through all the grounds, in this case from the video-amplifier ground to the AC power-system ground, through the extension-cord neutral and the camera ground, and back through the coaxial cable ground lead to the amplifier ground. This is why the connection is termed a ground loop. Ground loops do not always lead to EMI problems, but the potential for problems is usually there.

This problem was solved with a **ground-loop isolator**, which is a specialized type of **isolation transformer**. (In audio-signal work, such devices are sometimes called **direct boxes**.) A video isolation transformer transmits video frequencies (typically 30 Hz to 4 MHz) but blocks the passage of any low-frequency current that might otherwise pass from one winding's ground terminal to the other winding's ground

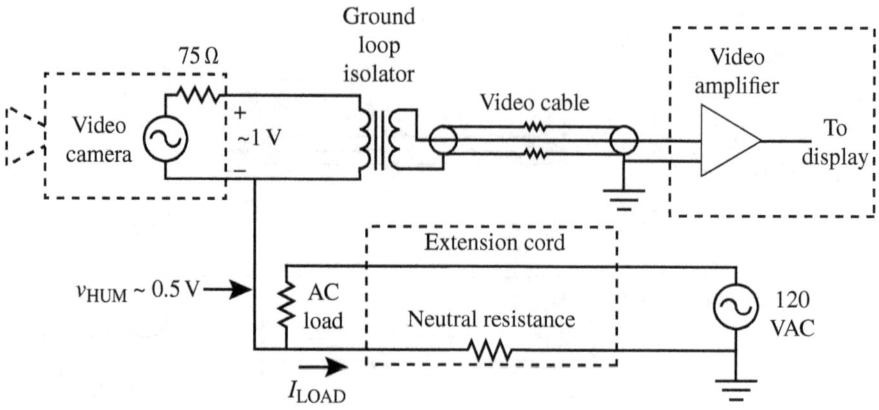

FIGURE 12.17 Ground-loop EMI problem solved with ground-loop isolator.

terminal. The isolator transmits only the *voltage difference* presented to its input terminals. When the isolator was installed properly as shown in Figure 12.17, the only signal it transmitted to the coaxial cable was the 1-V video signal. Although the entire camera was also at the 60-Hz stray-voltage potential, this was a *common-mode* voltage, and the way the isolator is wired makes it insensitive to common-mode voltages. No 60-Hz stray current can travel along the coaxial cable's shield in Figure 12.17, and so the signal that appeared at the video amplifier was clean, with virtually no 60-Hz EMI present. Note that you can no longer trace a closed loop through the ground paths in Figure 12.17, because the ground-loop isolator has broken the loop.

In general, any time there are different parts of a system that must share the same ground through lengthy interconnected ground leads, it is a good idea to avoid ground loops. One way to do this is with a **star** type of grounding method (also referred to as the **one-point** method). Suppose there are five power-using units in a system, in which unit 1 uses the least power and unit 5 uses the most. Suppose the five units are situated roughly as shown in the pictorial-schematic diagram of Figure 12.18, along with the AC power source for the units. From a cost point of view, it is probably most economical to connect the system's grounds in a daisy-chain fashion as Figure 12.18 shows, going from unit 5 at the left to the next closest unit (unit 2) and so on to the AC supply. The problem with this cost-saving approach is that the large current drawn by unit 5 must flow through the entire ground system wiring, from 5 to 2, to 1, to 4, to 3, and finally to the system's power-supply ground. In doing so, the current will cause small but nonzero voltage drops across the ground conductors interconnecting each unit. The result is that the "ground" potential that each unit sees is slightly different. If these units attempt to exchange low-level signals, the signals are almost certain to be contaminated by voltages produced in the ground-lead resistances by power current from unit 5, and EMI will result.

Applying the star grounding approach to this system results in the connections shown in Figure 12.19. The AC power source is physically moved to be as close as possible to the unit using the most power, unit 5. Then *independent* ground leads are

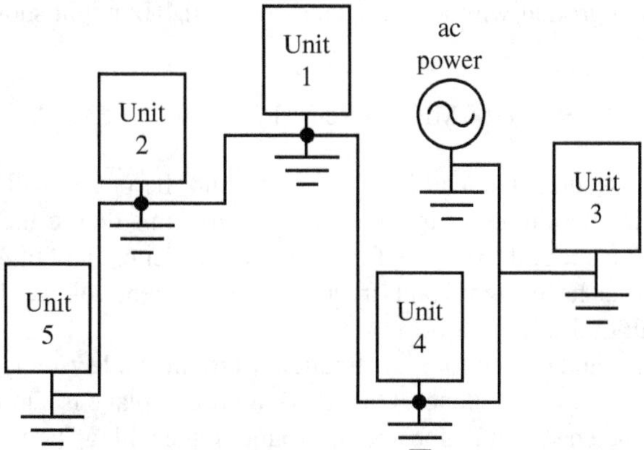

FIGURE 12.18 One way to connect the grounds of five power-using electronics units to an AC power-supply ground. In this pictorial drawing, ground symbols indicate ground terminals, not literal ground.

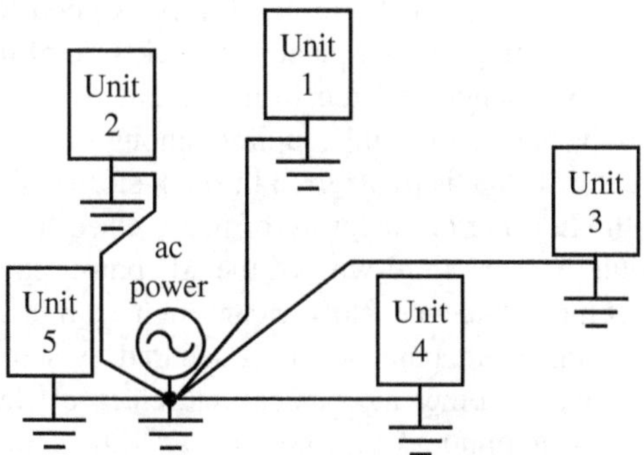

FIGURE 12.19 The system of Figure 12.18 with ground connections rewired in accordance with the star or one-point method of grounding.

connected between the AC power's ground terminal, which is the center of the star, and the ground terminals of every other unit. In this way, the ground lead of each unit carries only the ground current consumed by that unit. If possible, the star's center should be arranged that the ground leads carrying more current are shorter than leads carrying less current. In this way, voltage drops along the ground leads are lowered compared to the wiring approach of Figure 12.18, and less chance for harmful ground-potential EMI results.

Good grounding techniques are especially important with high-frequency and high-speed electronic systems. Wires a few cm long that have almost no impedance at low frequencies become small inductors at higher frequencies. Depending on the wire's diameter and its proximity to a **ground plane** (a grounded sheet-like conductor), a single isolated wire has an inductance of approximately 10–20 nH per cm.

So a 2.5-cm-long ground wire at a frequency of 100 MHz might show an inductive reactance X_L of

$$X_L = 2\pi(100\,\text{MHz})(2.5\,\text{cm})(15\,\text{nH}\,\text{cm}^{-1}) = 23.5\,\Omega. \tag{12.6}$$

If the designer assumes this lead is a perfect ground, he or she will be in error by about 23.5 Ω! Digital circuits can show ground problems due to the inductance of ground connections too, if the clock frequency or the rising and falling edges of the digital waveforms have significant frequency components above 10 MHz or so, which almost all digital signals do.

Good **PCB** ground design for high-frequency circuits and systems often involves the extensive use of ground planes. On a PCB, a ground plane is simply a large area of copper left unetched and connected to ground. Other things being equal, a wide conductor will have less inductance than a narrow conductor, and a large-area ground plane can be considered as a wide, low-inductance conductor interconnecting all grounds attached to it. Good RF PCB layouts usually leave ground planes everywhere that it is possible to have them, isolating terminals and leads by gaps between the ground plane and the other conductors. In **multilayer PCBs**, it is possible to make **vias**, which are connections between the layers. A good RF PCB layout will have numerous vias connecting ground planes on each side of the PCB, so that no ground plane becomes an island, isolated from the others. The effect of all these ground planes is to prevent electric-field coupling among the remaining PCB leads, which helps to prevent EMI that is transferred through electric fields.

Finally, the metallic housing or enclosure of many electronic systems is usually grounded to the ground (green) third wire of the AC power source, when present. While a large piece of metal like an enclosure can often act as a good ground, this is not always the case. Hinges, attachment points secured by screws, layers of paint applied before assembly, and other mechanical and electrical factors can make an apparently solid steel or aluminum enclosure appear to potential EMI more like a Swiss cheese full of holes. For example, aluminum and aluminum alloys become covered with a very thin but tough layer of **aluminum oxide** when exposed to air. This oxide coating is a nonconductor, and although it can be ruptured with enough mechanical force and good grounds can therefore be made to aluminum components, in some situations, the oxide coating remains intact between two aluminum parts and creates a high-resistance path for ground currents, causing EMI problems. More about enclosures and how to use them to prevent EMI is covered in the next section, on **shielding**.

12.4.3 Shielding

The word **shield** has acquired a new meaning in electronics in recent years. Originally, it meant a component or structure that blocks electric, magnetic, or electromagnetic fields for such purposes as reducing EMI. More recently, the term has also been used to refer to an auxiliary circuit board that stacks onto a main microprocessor board so as to make pin-to-pin interconnections conveniently. Just to be clear, we are not

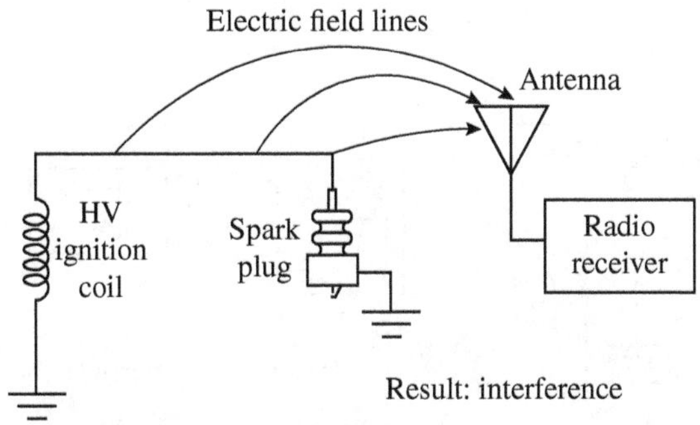

FIGURE 12.20 Automotive ignition system with unshielded high-voltage lead transferring EMI to radio receiver via electric-field lines.

talking about the microprocessor-board add-on type of **shield** in this section, although shields (in the second sense) may sometimes benefit from shields in the first sense.

An **electromagnetic shield** acts to reduce the influence of fields that would otherwise interfere with the proper operation of the circuit or system that is shielded. Shields behave differently with regard to low-frequency electric fields, low-frequency magnetic fields, and high-frequency electromagnetic fields, so we will begin by describing the basics of how a shield works. We will start with an example of electric-field shielding, because that is usually the simplest and most effective type of shielding.

When radios began to be installed in automobiles in the 1920s, severe problems were encountered with interference from the engine's ignition system. Then as now, the typical internal-combustion ignition system consists of a high-voltage **ignition coil** (a type of transformer) that produces pulses of voltages in excess of 20 kV (see Fig. 12.20). These pulses are sent to the **spark plugs**, where a small plasma discharge (spark) provides enough heat energy to ignite the combustible mixture of fuel and air in each cylinder at the proper instant in the power cycle. The wires from the ignition coil to each spark plug are called **ignition wires** and are insulated to withstand the high-voltage pulses without breaking down. But insulation by itself does nothing to stop **electric-field lines** from appearing between the high-voltage conductor and any other conductors in the vicinity, including the radio's antenna. If you like, you can consider the high-voltage wire, the insulator and air between the wire and the antenna, and the antenna as forming a small-value capacitor. Even though this capacitor's value is well below 1 pF, the high-voltage pulses are large enough to drive significant high-frequency currents into the sensitive input of the radio receiver, causing interference in the form of pops that coincide with the engine's revolutions.

The most effective way to prevent this type of electric-field-transferred EMI is with an **electrostatic shield**. In the case of automotive ignition wires, the shield can take the form of a conductive enclosure similar to the ground or outer conductor of a coaxial cable. (In fact, coaxial cables are **self-shielded** from this type of interference, which is why they are often used in sensitive systems.) Installation of an electrostatic shield that completely surrounds the ignition system is shown in Figure 12.21 as a dashed line connected to the system's ground at the base of the ignition coil. Now, the

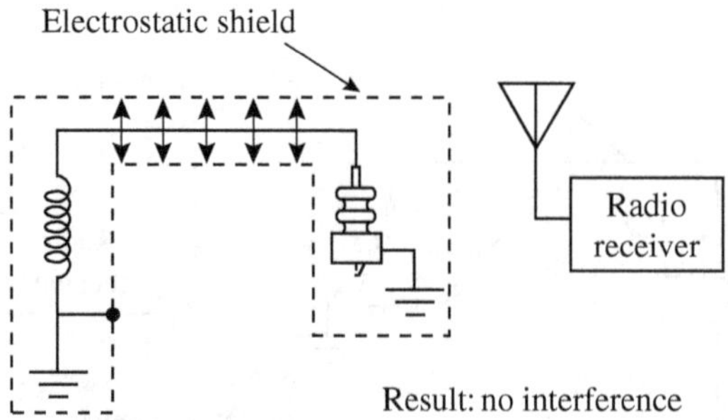

FIGURE 12.21 The automotive ignition system of Figure 12.20 with a conductive shield (dashed lines) to prevent EMI transferred via electric fields.

electric field lines leaving the high-voltage wire terminate on the inside of the shield conductor instead of the radio antenna, and the EMI problem has been solved.

Notice that in Figure 12.21, the shield completely encloses the entire high-voltage circuit. Strictly speaking, this is impossible, because it is necessary to make openings in the shield to provide power, ground, and signal leads to the ignition system. However, these conductors can be adequately protected from causing EMI with filters, bypass capacitors, and other EMI prevention techniques suitable for conduction-transfer EMI. This does not change the fact that the ideal shield to prevent electric-field EMI is a solid conductor that entirely encloses the source of EMI in question. If you think of the EMI field as water or smoke that could escape through a small hole or slit, the best shield structure will be watertight without any possibility for leakage through openings of any size or shape.

While it is hard to achieve this ideal in most practical situations, it is relatively easy to come close with sheet-metal boxes, circuit-board ground plane patterns, and even conductive metal or paint evaporated or sprayed onto plastic. All these measures and more have been used to reduce electric-field-transferred EMI, and once the main electric-field pathway for EMI has been identified, it is often a simple matter to find space where a grounded conducting electric-field shield can be installed.

Magnetic-field EMI is much more difficult to deal with compared to electric-field EMI. The reason is that, at least for DC and low frequencies, magnetic fields are not affected significantly by shields made of nonmagnetic conducting material. As the frequency of interference rises and the field gradually behaves more like a radiated electromagnetic wave, conducting shields increase in effectiveness, but this is generally not the case until the frequency exceeds 10 MHz or more. Magnetic fields in the 50–60 Hz power-line range and audio range (20 Hz–20 kHz) usually penetrate thin nonmagnetic conductive shields with little or no attenuation.

Consequently, a stubborn case of magnetic-field EMI may call for the use of **magnetic shielding materials**. Most magnetic shields consist of an alloy such as **permalloy** or **mu metal** that has extremely high **magnetic permeability**, symbolized by μ. A thorough understanding of magnetic permeability requires a knowledge of electromagnetic theory, but basically, if a current-carrying wire is placed near a metal with high μ, the current will induce a very high magnetic field in the metal.

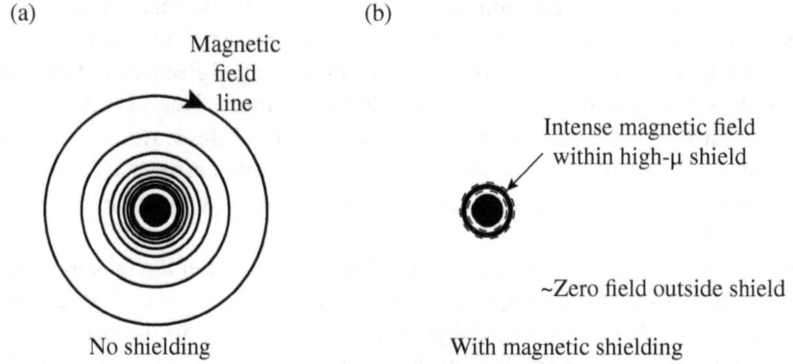

FIGURE 12.22 (a) Magnetic-field lines surrounding cross section of current-carrying conductors with no magnetic shielding. (b) Same current-carrying conductor surrounded by magnetic shield of high-μ material.

Very roughly speaking, the high-permeability material attracts nearly all of the available magnetic-field lines inside itself, leaving very little magnetic field to pass through the material to regions outside the shield.

In Figure 12.22a, we show the cross section of a current-carrying conductor with circular magnetic-field lines as they appear in the absence of other conductors or any type of shielding. As shown in the figure, the magnetic-field lines grow in intensity as you approach the surface of the wire, but some amount of field extends for a long distance away from the wire. Now suppose we surround the wire with a tubular magnetic shield made of high-μ material such as mu metal, as shown in Figure 12.22b. Within the shield itself, the current will induce an extremely high field—as much as 100,000 times or more the field that was present before the shield was added. But the useful side effect of this extremely high field within the shield is that almost no magnetic field is left to appear outside the shield. So any other conductors that were formerly receiving magnetic-field EMI from the current-carrying conductor shown will no longer pick up magnetic-field interference in this way, once the magnetic shield is added. Note that effective magnetic shielding requires special high-permeability materials. Ordinary magnetic metals such as iron, nickel, and certain steels do not have enough permeability to provide much in the way of magnetic shielding and may even make things worse.

No shielding is perfect, and there are several disadvantages and limitations of magnetic shielding material. Most types of high-permeability material are rather costly and heavy and require special fabrication techniques to preserve their desirable magnetic properties. They also have frequency limitations, becoming less effective as frequency increases. Nevertheless, certain types of components such as **cathode-ray tubes (CRTs)**, **photomultiplier tubes (PMTs)**, and some types of **magnetic memories** such as hard disk drives can be sensitive to external magnetic fields, and magnetic shields may be needed to prevent EMI in these cases.

If the EMI in question is transferred by means of radiated electromagnetic fields, the same type of nonmagnetic conductive shield used for electric fields is usually effective against radiated electromagnetic fields as well. Because the electric and magnetic fields in an electromagnetic wave travel together, a good conductor tends to reflect the whole wave, which includes both its magnetic and electric fields. However, you should know that an opening in a shield that is more than a small fraction of a wavelength in its longest dimension will allow part of the wave to leak through the shield. So long slits, cracks, or nonconducting joints between two conducting plates will compromise the performance of a shield used to reduce EMI caused by radiated fields.

As long as the largest opening in the shield is much smaller than a wavelength in every dimension—a small hole, say—a conductive shield with such openings can work nearly as well as if it were solid. A familiar example of this type of shield is the window in a microwave oven. The interior of a typical microwave oven has several hundred watts of 2.45-GHz energy inside it. If this energy were allowed to leave the oven chamber, it would cause all kinds of problems ranging from breaking the law (regulatory agencies set the maximum allowable RF emission for such devices) to setting fires and killing people with heart pacemakers. So how does the window let you to see into the oven without allowing the microwaves to escape?

The wavelength λ of a free-space 2.45-GHz wave is

$$\lambda = \frac{c}{f} = \frac{3.10^8\,\mathrm{m\,s^{-1}}}{2.45\,\mathrm{GHz}} = 122.4\,\mathrm{mm} \tag{12.7}$$

The window in a typical microwave oven is covered by a metal sheet that is perforated with a regular pattern of holes. Each hole is about 1.5 mm in diameter and is spaced about 2.5 mm away from the adjacent holes (center to center). In terms of a 2.45-GHz wavelength, this diameter is about (1.5/122.4) = 1.2% of a wavelength. The amount of electromagnetic radiation that penetrates through a hole that small is very low, even if there are many of them side by side. So a small round hole is allowable in an electromagnetic shield, as long as its diameter is only a few percent of the highest wavelength involved and no conductor passes through it. Obviously, if an electronic system is to be entirely enclosed in such a shield, signal and power conductors often have to pass through it, but preventing EMI from conductors is a matter of filtering and bypassing.

Another useful technique to reduce radiated EMI is the use of **absorbing material**. Certain types of semiconducting substances such as carbon-doped foam plastic tend to absorb incoming electromagnetic radiation rather than reflect or transmit it. Sometimes, radiation within a shielded enclosure becomes a problem when one part of a circuit inside the shield interferes with another part by means of electric or magnetic fields. Inserting a block of absorbing material at an appropriate location inside the enclosure can reduce or eliminate this type of EMI. Although the military field of **electronic countermeasures** is outside the scope of this book, **stealth aircraft** designed to be invisible to radar beams are covered with sophisticated absorbing materials to minimize the amount of radar waves that the aircraft reflects.

We have provided little in the way of quantitative data or equations to help you deal with EMI problems. The reason is that, except for conducted EMI that can be handled with filters and bypass capacitors, EMI is not just a circuit problem, but an electromagnetics problem. There is not room in this book to provide the electromagnetics background you would need to understand most of the formulas used in quantitative analysis of field-transferred EMI problems. So instead, we have used pictures and qualitative descriptions to characterize the various kinds of EMI and some measures that can be used to deal with them. The references at the end of this chapter should be consulted for specific quantitative measures one can take to evaluate a given design for potential EMI before it is built and to analyze an EMI problem once it occurs.

12.5 DESIGNING WITH EMI AND EMC IN MIND

In this concluding section, we will discuss some of the legal requirements for EMC that many electronic products must meet and how to include EMC considerations in all stages of a design project.

12.5.1 EMC Regulators and Regulations

Because EMI can involve radiated electromagnetic waves and the electromagnetic spectrum is a public resource, regulatory bodies in many nations have issued directives and laws that apply to the EMC performance of electronic systems sold or used in their respective regions. In addition, certain international bodies such as the **International Electrotechnical Commission** (**IEC**) and the **International Standards Organization** (**ISO**) have issued guidelines that, while not always having the force of law, are nevertheless widely observed. In the United States, EMC regulations are issued primarily by the **Federal Communications Commission** (**FCC**). While details of the various rules differ, most of them follow a similar pattern.

Before a product falling under the regulations can be legally sold, it must pass a series of specified EMC tests that are designed to measure the level of fields it emits in various wavelengths. These tests require specialized equipment such as **spectrum analyzers** and **anechoic chambers**, and all but the largest manufacturers typically hire a specialized EMC-testing firm to do them. Often, the equipment is found to radiate excessively under some conditions, and so the design has to be modified until the EMC test is reliably passed. Once the equipment passes EMC testing, the manufacturer receives a certificate verifying this fact. The certificate sometimes confers on the manufacturer the privilege of placing a **declaration of conformity label** on the product nameplate, which indicates that the design has passed all applicable tests. Similar procedures apply in most other countries, although the details of the tests differ.

12.5.2 Including EMC in Designs

Regulations aside, EMI problems can disable a design at a very inconvenient stage in the design process. As you have seen, EMI tends to be a system-level problem, involving at least three factors: the source, the transfer medium, and the victim. Tests of smaller subsystems in the early stages of a project will often show no EMI problems initially. For example, an RF transmitter may work fine by itself on a test bench. An RF receiver destined for use in the same system as the transmitter may operate within its specifications when it is tested alone, separately from the transmitter. But when the two units are combined into a system, the transmitter as an EMI source may send EMI through a shared power-supply line to the receiver as victim, and EMI shows up. By that time, a lot of effort has been spent on the project already. This is not the best way to proceed.

In many areas of engineering, **simulation software** has reduced or eliminated the need for lab-based prototype testing of the "cut-and-try" type. This is also true in the EMC field. In the last few decades, a number of software products have become available that can be used to model entire systems in order to identify EMI problems and meet EMC specifications. As is true with any computer model, the simulations are only as good as the data provided to them. Because no software can simulate absolutely everything that goes on in an electronic system, the user of EMC software must develop a sense of judgment about how to translate a given actual design into a form that will be compact enough to run efficiently on EMC software but detailed enough to give realistic results. The learning curve for using EMC software is rather steep, and EMC simulation work tends to be done by specialists with enough experience to run simulations intelligently. But everyone doing electronic designs should be aware that EMI and EMC problems can be simulated, and much unnecessary lab work avoided, when the right software is available and properly used.

Even in the absence of EMC software, designers who are not EMC specialists can nevertheless build good EMC practices into their designs. Power-supply lines should be routinely bypassed in many locations with capacitors whose values are chosen to provide maximum bypassing for the range of frequencies that is most likely to cause trouble. Audio and low-speed digital circuitry (clock speeds below 50 MHz) benefit most from bypass capacitors in the range of 10 nF to 1 μF or larger. Bear in mind that electrolytic capacitors (usually with values greater than 1 μF) do not function well above 1 MHz or so. High-speed digital circuits and RF circuits should use additional bypass capacitors with values below 1 nF.

If any leads carry signals whose maximum level is less than 100 mV, the designer should seriously consider using coaxial cable or a comparable type of shielded conductor for any such leads that are longer than a few cm. This is true especially for high-quality audio systems and other equipment that covers a wide **dynamic range** (ratio of maximum signal to minimum noise level). It's better to install coaxial cables and their accompanying connectors at the beginning of a design than to leave them out at first, only to find later that you need them but don't have room for them.

If the analog system being designed has a high gain at a single frequency or range of frequencies, choosing the best physical layout for the amplifier stages is a vital part of a good design. The best layout is a straight line to provide maximum physical separation between the sensitive input stages and the high-level output stages. Sometimes, design restrictions or other factors prevent this, but keeping as large a separation as possible between the input and output of a high-gain amplifier cascade can prevent oscillation-type EMI.

Power supplies can both produce EMI and prevent it. EMI prevention is almost always necessary for switching power supplies, because all such supplies involve the rapid turn on and turn off of large currents, and most such circuits include inductors that can produce both conducted EMI and magnetic-field EMI. EMI prevention in switching power supplies must be applied both to the power-supply input, typically connected to the AC power line, and to the outputs to prevent the switching frequency or its harmonics from interfering with the operation of the electronics that forms the power supply's load.

While a poorly-regulated and bypassed power supply can provide a pathway among subsystems for conducted EMI, a well-regulated and well-bypassed power supply can prevent such troubles.

Even if a completed system operates well and passes required EMC testing on its own, the designer cannot assume that it will work in all environments it is likely to be used in. Once a unit is installed in a user's location, it is subject to EMI from anything in the vicinity, even radiation from a source that can be many meters or even kilometers away. EMI can come from as far away as outer space! Satellite dish antennas aimed at **geosynchronous satellites** can point directly at the sun at certain times of year as the sun appears to pass behind the satellite. Depending on the antenna's sensitivity and the satellite's signal strength, the naturally occurring radio emission from the sun can overwhelm the signal, causing the link to fail for a few minutes. Fortunately, this type of failure is predictable, but it shows that the EMI-conscious designer should be familiar with the environment in which the design will be used. You should use some imagination in thinking of ways that the system could fail because of EMI originating outside its boundaries. And if such problems are likely, it will pay in the long run to head them off before they happen.

BIBLIOGRAPHY

Kaiser, K. L. *Electromagnetic Compatibility Handbook.* Boca Raton, FL: CRC Press, 2005.

Paul, C. R. *Introduction to Electromagnetic Compatibility*, 2nd Edition. Hoboken, NJ: Wiley-Interscience, 2006.

Weston, D. A. *Electromagnetic Compatibility: Principles and Applications.* New York, NY: Marcel Dekker, 1991.

PROBLEMS

Note: Because EMC problems often involve many aspects of electronics, some of these problems require you to use material from previous chapters. References to these chapters are noted when needed.

Problems of above-average difficulty are marked with an asterisk (*).

12.1. *Bandwidth estimation of various signals.* The following list of electronic signals used for various purposes is in no particular order. Rearrange them so that they are in order of increasing bandwidth required, from the least bandwidth to the most bandwidth (there may be more than one right answer to this question):

(a) one-way mobile telephone text message;
(b) one-way mobile telephone call;
(c) watching a movie on a 360-by-480 pixel window on an Internet browser;
(d) loading a webpage over a connection that takes 20 s to load;
(e) data link transmitting 10 temperature readings per second, each accurate to 3 decimal digits;
(f) loading a webpage over a connection that takes 2 s to load;
(g) garage door opener;
(h) watching a movie transmitted over an analog cable channel;
(i) sending data over an intercontinental undersea fiber-optic cable; and
(j) watching a movie transmitted over a digital cable channel.

12.2. *Intermodulation problem.* In Chapter 4, the problem of intermodulation due to third-order nonlinearities in amplifiers was considered. Suppose a high-power amplifier amplifies two narrowband RF signals of equal amplitude. One is at frequency $f_1 = 162.40$ MHz and the other is at $f_2 = 162.55$ MHz. If the amplifier has any third-order nonlinearity in its transfer function, it will produce two additional signals besides the intended ones, one at a frequency f_L below f_1 and the other at a frequency f_H above f_2.

(a) Based on the analysis in Chapter 4 of third-order intermodulation, find f_L and f_H.
(b) If the amplifier has a third-order intercept of +35 dBm (defined at the output) and the two desired signals are each at an output level of +5 dBm, what is the level of the 3rd-order intermodulation products?

12.3. *Identification of EMI source, transfer medium, and victim.* In each of the following EMI situations, describe (1) the source or transmitter, (2) the transfer medium, and (3) the victim or receiver. Also state which (if any) of these three items is most responsible for the EMI problem.

(a) A truck driver buys an illegal 2-kW amplifier for his 27-MHz **citizens-band (CB)** radio. (The legal limit for a transmitter in this band is only 5 W.) He uses the transmitter to contact a girlfriend and expresses himself in indelicate language just as he is driving by a funeral home. The mourners inside, listening to the eulogy, are surprised to hear the trucker's words blasting over the PA system.
(b) You have invited your friends over to watch the final World Cup football (soccer in the United States) game on your direct-broadcast satellite TV receiver. As the game gets down to its final minutes, a heavy rainstorm begins outside your home and your TV screen goes blank.
(c) You have spent the last three years of your life working on a book, and the only complete copy is contained on an external magnetic hard drive in your

den. Your 5-year-old nephew receives the gift of a powerful samarium–cobalt magnet at a birthday party at your home. As you criticize the gift giver for giving what is potentially a dangerous object to such a young person, you see the 5-year-old walk over to your desk with the magnet, which snaps to the side of your hard drive. Your manuscript, it turns out, is toast.

12.4. *EMI filter problem.* The EMI filter circuit shown in Figure 12.15 has components designed to attenuate both common-mode and differential-mode EMI. Typically, a passive lowpass filter circuit's attenuation increases at the rate of 20 dB/decade for every pole it contains. The circuit in Figure 12.15 is basically a series-shunt *ladder network*, in which a series element is followed by a shunt element (connected to ground) then a series element and so on. In a lowpass ladder network, every series inductor (counting two or more components in parallel as one element) provides a pole, as does every shunt capacitor (again, counting two or more in parallel as one element).

(a) Considering only those elements that are significant for the common-mode voltage applied to the circuit, estimate the rate at which the filter's attenuation falls off above its cutoff frequency. Express your answer in dB per decade. Assume there is a small but nonzero series resistance R in the power supply's equivalent circuit.

(b) Now considering the elements significant to the differential-mode voltage applied to the filter, estimate the attenuation rate of increase, expressing your answer in dB per decade.

***(c)** If $R = 1$ ohm and $C_1 = C_2 = C_3 = C_4 = 50\,\text{nF}$, what is the filter's 3-dB-down cutoff frequency for differential-mode signals? Assume there is no load connected to the V_{OUT} terminals for this analysis.

12.5. *Electric-field EMI coupling problem.* Suppose two circuit traces on a printed circuit board (PCB) run parallel to each other for a distance of 30 mm. Also suppose that a calculation based on dimensions and dielectric material surrounding the traces has shown that the mutual capacitance between these two traces is $40\,\text{fF}\,\text{mm}^{-1}$ ($1\,\text{fF} = 10^{-15}\,\text{F}$).

(a) Suppose trace A carries a low current at a sine wave voltage of 250 V (RMS) at a frequency $f = 53\,\text{kHz}$ and the impedance of the circuit connected to trace B is $250\,\text{k}\Omega$. Approximately, how much EMI voltage (RMS) $V_{EMI}(a)$ will be induced in trace B by the proximity of trace A?

(b) Now, assume the same power in trace A is carried by a larger current at a voltage of only 25 V and the impedance of the trace connected to circuit B is only $600\,\Omega$. What is the RMS EMI voltage $V_{EMI}(b)$ under these low-impedance conditions? This example shows how high-impedance circuits are more prone to electric-field-coupled EMI than low-impedance circuits are.

12.6. *Magnetic-field EMI coupling problem.* A certain Hall-effect sensor on a circuit board is sensitive to stray magnetic fields and will malfunction if the total magnetic-field strength at its location exceeds $100\,\text{A}\,\text{m}^{-1}$. A long, straight, isolated wire near the sensor carries a DC current of 8A and is $r=5$ cm away from the sensor. Approximating the long wire as infinitely long (which is not a bad assumption as long as it extends several times r either way from its closest point to the sensor), use Equation 12.3 to estimate the magnitude of the magnetic field H at the sensor. Will the sensor malfunction?

12.7. *Radius of near-field zones for various frequencies.* For each frequency given in the list below, calculate the distance r_{NF} for which $2\pi r/\lambda=1$. This distance is the approximate boundary between the area where magnetic and electric fields are significant, and the region where primarily radiated electromagnetic fields are present.

(a) $f=1.2$ MHz (in AM broadcast band)
(b) $f=2$ GHz (computer clock rate)
(c) $f=60$ Hz (AC power line)
(d) $f=116$ MHz (aircraft communications band).

12.8. *Power density and field intensity for radiated EMI.* The **power density** U of a radiated electromagnetic wave is the total power W carried by a cross section of the wave divided by the cross-sectional area A and is measured in units of watts m^{-2}. In Chapter 11, we explained how to calculate the **isotropic power density** $U_{ISO}(r)$ as a function of distance r from a source that radiates uniformly in all directions. Equation 11.71 shows that the power density from such a source at a distance r is simply the total power divided by the area of a sphere of radius r. While actual power density in a given terrestrial situation will generally be different from U_{ISO}, this value is sometimes a good starting point for approximate calculations. Given a power density U of a single-frequency electromagnetic plane wave and the impedance of free space $\eta_0=376.7\,\Omega$, it can be shown that you can calculate the *peak* value of the oscillating electric field $E_{PK}(U)$ as $E_{PK}(U)=\sqrt{2\eta_0 U}$. The peak value of the magnetic field associated with the plane wave is $H_{PK}(U)$, which is $H_{PK}(U)=\sqrt{2U/\eta_0}$. Assume a VHF two-way radio emits a power of 1W isotropically at a frequency of 157.5MHz. Calculate the peak electric and magnetic fields resulting from this isotropic radiator (ignoring any reflection, refraction, or absorption by surroundings) at a distance of

(a) $r=10$ m,
(b) $r=100$ m, and
(c) $r=1$ km.

12.9. *Orientation of victim and electromagnetic-wave EMI.* The rule that an antenna must be a few percent of a wavelength long to operate effectively becomes a problem with long-wave signals (below 3MHz), where the wavelengths exceed 100m. As a result, some radios receive only the magnetic field associated with the radiated wave, because magnetic-field pickup devices can be much smaller than a wavelength and still operate reasonably well. One way this is done is with a type of antenna called a **loopstick**, which is basically a long cylindrical magnetic core inside a coil of wire. The loopstick antenna is sensitive only to the component of the magnetic field that is parallel to the axis

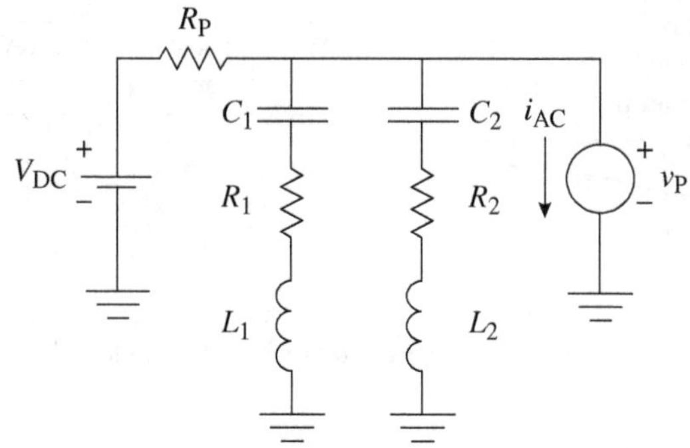

FIGURE 12.23 Equivalent circuits of battery power supply, two bypass capacitors, and power-supply load for Problem 12.10.

of the loopstick. Most long-wave signals are oriented (**polarized**) so that the electric field is vertical (perpendicular to ground) and the magnetic field is horizontal (parallel to the ground). Suppose you wish to receive a signal from an AM broadcast-band station *A* that is located due north of your receiver. An interfering station *B* on the same frequency produces a much stronger wave at your location, and station *B* is located due east of you.

(a) How should you orient the axis of the receiver's loopstick to maximize the signal from station *A* and minimize it from station *B*?

(b) If you were lost on a large lake surrounded by radio stations at known locations, how could you use a loopstick-equipped radio (no GPS!) as a direction finder?

***12.10.** *Effectiveness of bypass capacitors and frequency.* As explained in Chapter 3, physical capacitors always have some parasitic inductance and resistance associated with them. These parasitic impedances restrict the useful frequency range of a real capacitor so that bypassing a wide range of frequencies effectively often requires the use of two or more capacitors of different values in parallel. Figure 12.23 shows the equivalent circuits of a battery power supply with equivalent series resistance $R_P = 1\,\Omega$. The power supply's load is modeled by an ideal current sink that draws a current $i_{AC} = (500\,\text{mA})\sin[2\pi(200\,\text{Hz})t] + (100\,\text{mA})\sin[2\pi(30\,\text{MHz})t]$. Two bypass capacitors can be used. Capacitor C_1, an electrolytic type, has these values for its equivalent circuit: $C_1 = 1200\,\mu\text{F}$, $R_1 = 50\,\text{m}\Omega$, and $L_1 = 8\,\mu\text{H}$. Capacitor C_2, a monolithic ceramic type, has these equivalent-circuit values: $C_2 = 470\,\text{pF}$, $R_2 = 4\,\text{m}\Omega$, and $L_2 = 60\,\text{nH}$.

(a) Assuming neither C_1 nor C_2 is present, calculate the RMS voltages present at v_{AC} across the load at 200 Hz (call it V_{LF}) and at 30 MHz (call it V_{HF}). This calculation simply involves multiplying each frequency component of the current by $1\,\Omega$ and converting peak voltage to RMS

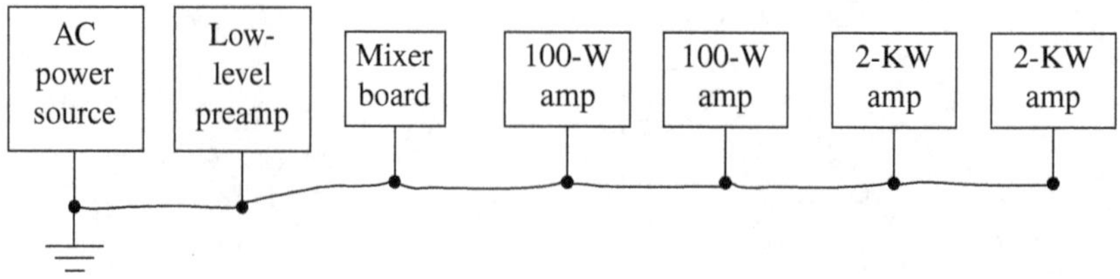

FIGURE 12.24 Pictorial drawing of original power-supply ground-lead wiring of audio-equipment rack for Problem 12.12, showing actual locations and lengths.

voltage. These voltages can be considered as EMI voltages that will appear on the power-supply line.

(b) Next, assume that the electrolytic capacitor C_1 is present but C_2 is not present. By how many dB does the V_{LF} EMI voltage decrease when this capacitor is installed? What is the dB improvement in V_{HF} when C_1 is installed?

(c) Next, assume only C_2 is installed (and not C_1), and calculate the dB improvement (decrease) in EMI levels for V_{LF} and V_{HF}.

(d) Finally, include both C_1 and C_2, and calculate the dB improvement in EMI levels for V_{LF} and V_{HF}. This problem shows why two different capacitors are sometimes needed to deal with widely separated EMI frequencies.

12.11. *Common-mode and differential-mode voltages*. Suppose the following voltages are measured between the designated terminals of an AC line power source shown in Figure 12.13: voltage between hot (+) and ground (−), $(121\,\text{V})\sin[2\pi(60\,\text{Hz})t] + (30\,\text{mV})\sin[2\pi(1\,\text{MHz})t] - (500\,\text{mV})\sin[2\pi(40\,\text{kHz})t]$, and voltage between neutral (+) and ground (−), $(1\,\text{V})\sin[2\pi(60\,\text{Hz})t] - (30\,\text{mV})\sin[2\pi(1\,\text{MHz})t] + (400\,\text{mV})\sin[2\pi(40\,\text{kHz})t]$. For each frequency component (60 Hz, 40 kHz, and 1 MHz),

(a) State whether the frequency component is purely common mode, purely differential mode, or a combination.

(b) Express each mode in terms of a sum of a common-mode voltage v_{CM} and a differential-mode voltage v_{DM}, after solving Equations 12.4 and 12.5 for v_{CM} and v_{DM} in terms of v_{HG} and v_{NG}.

12.12. *Rewiring of grounds in star pattern.* A technician with no knowledge of EMI grounding techniques saved copper by wiring a rack full of audio equipment with AC power-supply ground leads as shown in Figure 12.24. As you can see, the ground leads are "daisy chained" with the sensitive low-level preamp closest to the AC power source and the high-power amplifiers farthest away. While keeping the components in a straight line, rearrange the components and redraw the ground leads so that the new drawing reflects a good "star" type of independent ground lead wiring with higher-power units having shorter leads to the AC power source.

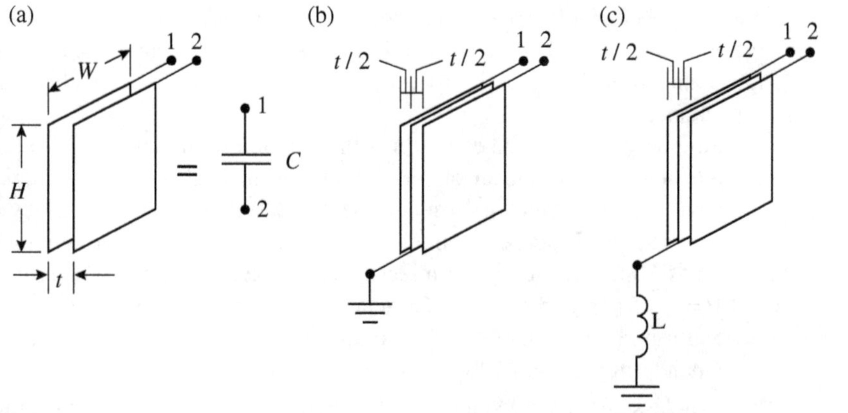

FIGURE 12.25 (a) Parallel -plate capacitor and equivalent circuit. (b) Parallel-plate capacitor with electrostatic shield plate inserted between original two plates. (c) Shield plate with inductance L of ground lead connecting it to ground.

12.13. *Perforated shield opening in HF transmitter housing.* An interesting characteristic of electromagnetic fields is that they obey the principle of "electrodynamic similitude."[2] This principle says that if you have a physical structure made of a perfect conductor and you solve for the electromagnetic fields around it at a frequency f_1, you can use the same solution for a larger structure S times as big if you also lower the frequency by the same factor, so that the new frequency f_2 is related to the original frequency f_1 by $f_2 = f_1/S$. So, for example, if we know that a perforated screen of a certain size and spacing of holes will work well as a shield at $f_1 = 2.45$ GHz, we can figure out what dimensions of screen will work just as well for an HF transmitter working at $f_2 = 24.5$ MHz. Using the typical microwave-oven window-shield dimensions given in Section 12.4., calculate the dimensions for a similar perforated screen that will work well as a shield for a HF transmitter operating at 24.5 MHz.

12.14. *Effectiveness of electrostatic shield in parallel-plate capacitor.* A **parallel-plate capacitor** is formed when two conducting plates of height H and width W are placed parallel to each other at a spacing t, which is much smaller than either W or H. Figure 12.25a illustrates such a capacitor. It can be shown that if the dielectric material between the plates is air, the value C of this capacitor is

$$C = \frac{\varepsilon_0 HW}{t}. \tag{12.8}$$

[2]The principle of electrodynamic similitude is discussed at more length in J. Stratton, *Electromagnetic Theory* (New York: McGraw-Hill, 1941), pp. 488–490. If the original conductor has finite conductivity, the conductivity must also be scaled as frequency squared to maintain exact similitude with scaling.

where $\varepsilon_0 = 8.854 \times 10^{-12}\,\mathrm{F\,m^{-1}}$ and the dimensions H, W, and t are all in meters. This formula neglects the **fringing fields** that appear at the edges of the conductors, but usually, these fields do not contribute much to the capacitance.

(a) Suppose plate 1 of the capacitor is the output terminal of a high-voltage piezoelectric-transducer driver circuit in a medical ultrasonic imaging device. The voltage on terminal 1 is $v_D = 200\,\mathrm{V}$ (rms) at a frequency of $f = 1.3\,\mathrm{MHz}$. Suppose terminal 2 is the input of a sensitive amplifier, whose equivalent circuit is a resistance of $2\,\mathrm{k\Omega}$ to ground. Also assume $H = W = 5\,\mathrm{mm}$ and $t = 1\,\mathrm{mm}$. Draw an equivalent circuit of this situation, and use it to calculate the EMI voltage v_{EMI}(a) that appears at terminal 2 when terminal 1 has 200 V on it.

(b) Now assume a third shielding plate is inserted in between the original HV plate 1 and amplifier plate 2, as shown in Figure 12.24b. Assume the third plate is connected to ground by means of an ideal zero-impedance ground connection. The third plate forms two capacitors C_1 and C_2 of equal value. Neglecting the fringing fields, draw an equivalent circuit of the situation in Figure 12.24b, and calculate the magnitude (rms) of the new v_{EMI}(b) that results when the electrostatic shield is inserted.

(c) Finally, suppose the ground lead is not perfect, but has an equivalent circuit consisting of an inductor L whose value is $L = 10\,\mu\mathrm{H}$. Draw the equivalent circuit of the resulting situation and calculate the rms magnitude of v_{EMI}(c). If the proper signal present at terminal 2 is only $50\,\mu\mathrm{V}$ rms, will the voltage v_{EMI}(c) be large enough to cause significant interference?

For further resources for this chapter visit the companion website at
http://wiley.com/go/analogmixedsignalelectronics

APPENDIX

TEST EQUIPMENT FOR ANALOG AND MIXED-SIGNAL ELECTRONICS

A.1 INTRODUCTION

As electronic systems have grown more complex, so have the items of test and measurement equipment needed to develop and repair them. There is a bewildering variety of test gear available today, from inexpensive **digital voltmeters** (**DVMs**) given away as "swag" at technical conferences to sophisticated custom-designed factory test systems costing millions. The purpose of this appendix is to describe the functions and specifications of the most common types of analog and mixed-signal test equipment commonly found in electronics laboratories.

Test equipment for use with electronic systems falls into one of three broad categories: stimulus equipment, response equipment, and stimulus/response equipment. As the word "stimulus" implies, **stimulus equipment** actively provides energy of some form to the device or circuit under test: power, in the case of **laboratory power supplies**, or various waveforms in the cases of **function generators** and **signal generators**. **Response equipment** passively receives energy from the system under test and analyzes it in some way, producing data for the user in the form of a screen display or a data file. An **oscilloscope** is one type of response equipment. Finally, **stimulus/response equipment** both provides a signal to the system under test and analyzes its response to the stimulus. The **ohmmeter** function of a DVM both stimulates a resistor with a small voltage and measures its response to the stimulus—the current drawn—to calculate resistance.

Analog and Mixed-Signal Electronics, First Edition. Karl D. Stephan.

Companion Website: http://wiley.com/go/analogmixedsignalelectronics

We begin with the most common type of stimulus equipment found in electronics laboratories: the laboratory power supply.

A.2 LABORATORY POWER SUPPLIES

Nearly every electronic system needs electric power, and in developing **prototype** systems (early experimental and developmental versions of a system), it is often convenient to use **laboratory power supplies** for the DC power needed. A lab power supply can be as simple as a dry-cell battery or as complex as desired. Obviously, the supply must be capable of delivering more power than the system under test will require. A supply that can handle a small battery-powered consumer device will not suffice for powering a system that delivers a kilowatt to a load. But power capability is not the only important power-supply specification.

This discussion is restricted to *DC* power supplies. (While power supplies capable of delivering AC waveforms are available, they are specialty items.) The simplest type of DC power supply is the *single-unit* supply, whose equivalent circuit is shown in Figure A.1a. Most lab power supplies have an output voltage adjustment that allows continuous variation of the output voltage between a low value or zero and the maximum output voltage specified for the supply. This variability is indicated by the slantwise arrow across the battery symbol. The positive and negative terminals of most lab power supplies are **floating**, meaning that there is *no connection* between either terminal and the safety (and chassis) ground or any other power-supply terminal. The safety ground in all test equipment is connected between exterior conducting parts of the case and the green ground wire in the AC-power **cord set** that plugs into the wall. The third prong of the AC power plug connects the chassis ground to the building ground and bypasses any dangerous leakage current to ground rather than allowing it to pass through a user's body! (Chapter 10 contains more information on safety grounding.)

To allow for maximum flexibility in connecting power supplies to loads, most power-supply output terminals are intentionally not connected to the power-supply chassis or safety ground. Sometimes, the safety ground is brought out to the front

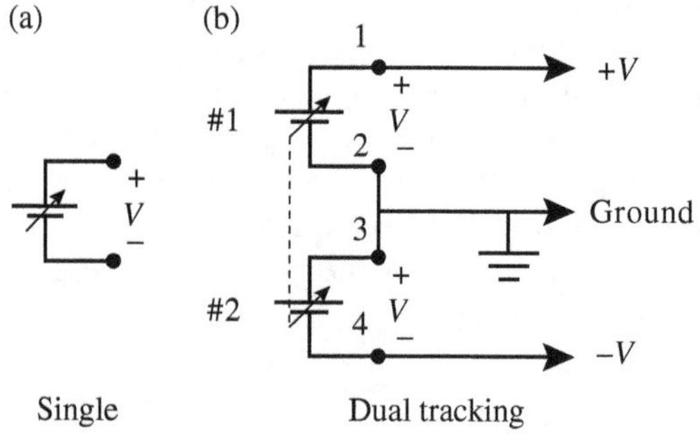

FIGURE A.1 (a) Single-unit power supply. (b) Dual tracking power supply wired to provide $\pm V$ and ground.

panel in the form of a green terminal, but it is up to the user whether to use this terminal, and most systems will work fine without it. Floating terminals are also often featured in **dual power supplies** that can provide equal and opposite DC power-supply voltages with respect to a third ground terminal. The connections for this use are shown in Figure A.1b.

Many analog circuits that use **op amps** require symmetrical power-supply voltages so that the op amps can source both positive and negative voltages to their loads. Students are often baffled as to how to connect dual supplies to provide balanced DC voltages for such circuits. The proper connection diagram is shown in Figure A.1b. Terminal 1 is the positive terminal of supply #1. (In most power supplies, the positive terminals are colored red, and the negative terminals are colored black.) To provide $+V$ to the load, terminals 1 and 2 are connected between the $+V$ line of the load and ground. This requires grounding the negative terminal of supply #1 to the ground lead of the load at terminal 2, as shown. If only the $+V$ supply were needed, that would complete the power-supply wiring.

However, the $-V$ supply must also be connected, and in order to provide a negative voltage to ground, terminal 3 (the positive lead of supply #2) must be connected to the load's ground. Connecting a positive terminal to ground bothers some people, but it is strictly necessary in the case of balanced dual supplies. The negative lead of supply #2 (terminal 4) therefore becomes the supply source for the $-V$ required by the load and is connected as shown. Note that *both* terminals 2 and 3 are connected to each other *and* to the load's ground lead. If *any* of these leads are left out—especially the ground lead from the interconnected terminals 2 and 3 to the load's ground—the circuit will either not operate at all or will behave very strangely. Leaving out the common ground lead in a dual-supply connection is a common mistake to make, but if you have read this far, you have no excuse for making it!

The dashed line between the arrows on supplies #1 and #2 indicates that these are **tracking power supplies**, meaning that the adjustment of a single voltage control will affect the output of both supplies simultaneously. Usually, some means of monitoring the supply's output voltages is provided, and before the load is connected, you should be sure that the voltage is adjusted to the value required by the load.

The total power delivered by a power supply is the sum of the power delivered by each pair of terminals. So if a dual supply is providing $\pm 15\,\text{V}$ at 2A each, its total delivered power is $(15\,\text{V} \times 2\,\text{A}) \times 2 = 60\,\text{W}$. Most power supplies are rated by their total maximum power output, which is of course less than the maximum power they draw from the AC line because no power supply is 100% efficient. The best efficiency is shown by **switching power supplies**, discussed in Chapter 10, because the semiconductor devices in them are switched between fully on and fully off, leading to minimum power dissipation and better efficiency. However, the noise produced by switching can be a problem for some sensitive circuits, so **analog power supplies** may be a better alternative in some cases. Analog supplies regulate output voltage by dissipating excess power in transistors operating at DC. This leads to almost noise-free output voltage, but inevitably reduces efficiency compared to the switching-type supply. Further details on power-supply specifications, including the definitions of **line regulation** and **load regulation**, are given in Chapter 10.

One feature of many lab supplies not discussed in Chapter 10 is that of **current limiting**. For the protection both of the circuit being powered and of the power supply itself, it is desirable to limit the maximum current the supply provides. This limit is usually settable by the operator with a panel control and can be set anywhere from the maximum output current the supply can handle down to zero. (Mistaken setting of the current-limit control to zero can lead the user to suspect a malfunctioning power supply, but always check the setting of the current-limit control first before concluding that a power supply is defective.) If the load tries to draw a current in excess of the current-limit setting, some supplies simply lower the output voltage so as to maintain the output current at the limit setting. Other supplies enter a permanent current-limit mode that the user must manually reset. In any case, the user of a current-limited power supply does not need to be too concerned that a short circuit will cause a high overcurrent that could lead to damage or a fire, although setting the current limit too high can still allow such problems to occur. In general, one should set the current limit to a value slightly higher than the maximum current expected to be drawn by the circuit, but no higher.

While power-supply manufacturers make every reasonable attempt to design supplies so that they are a good imitation of an ideal voltage source, no power supply is perfect, and the user must bear this fact in mind in connecting a power supply to a system. Every power supply has a small but nonzero **output impedance**, and this fact should be borne in mind when connecting loads that require very nearly constant voltages. Even if a power supply functions like an ideal voltage source, its good performance can be compromised by poor wiring practices. In Chapters 6 and 12, we discuss the problems that can occur if good power-supply wiring and grounding practices are not followed. Problems can arise from the resistance of long, thin lead wires between a power supply's terminals and a load circuit that attempts to draw high current from it. If more than 1 A or so is drawn by the system under test, you should consider using thick, short, low-resistance wires in all power-supply and ground connections. The resistance of copper wires of various diameters is tabulated in **wire tables** available on the Internet, and it is a simple matter to calculate the resistance of a given length of wire of a specified size. For example, #22 **American wire gauge** (**AWG**) copper wire has a diameter of 0.644 mm (0.0253 inches) and a resistance of 52.96 mΩ m^{-1}. So a 2-m length of such wire (about 6 ft) has a resistance of over 0.1 Ω. That may not sound like much, but if your circuit attempts to draw 10 A at a nominal voltage of 5 V from the supply, the voltage dropped in a pair of 6-ft wires leading to the power supply will be (0.2 ohm)(10 A) = 2 V, leaving only 3 V for the circuit, which will obviously not be happy! Even if the delivered DC voltage is adequate, the resistance and inductance of long power-supply leads can cause undesirable coupling among analog circuits, leading to oscillation and **electromagnetic interference** (**EMI**). Further details about power-supply-related EMI are given in Chapter 12.

A.3 DIGITAL VOLT-OHM-MILLIAMMETERS

Meters capable of measuring fundamental electrical quantities such as voltage, current, and resistance have been available since the dawn of electrical engineering. For most of that time, meters took the form of **electromechanical** devices. For many decades, the heart of most electrical meters was a sensitive electromechanical mechanism called the **D'Arsonval** meter movement, named after its inventor Jacques-Arsène d'Arsonval (1851–1940). In the D'Arsonval movement, a rectangular coil of fine wire is mounted between bearings and placed in a magnetic field. Current through the coil exerts a twisting force (**torque**) on the coil, which is balanced by a small spring so that a pointer attached to the coil moves across a graduated scale through a certain angle proportional to the current. The D'Arsonval movement is used in most pointer-style analog meters and although digital displays have replaced it for most applications, it is still found in some inexpensive types of test equipment. It is quite sensitive, requiring as little as 50 μA or less for a **full-scale** deflection, and can be made reasonably rugged.

However, mechanical meters have limitations, and so both analog and digital electronics were applied to the basic meter idea to yield the modern **DVM** or **digital volt-ohm-milliammeter** (**DVOM**). Most such instruments combine the functions of a DC voltmeter, an ammeter, and an ohmmeter and often throw in other functions such as continuity testing, AC voltage and current measurements, and frequency counting.

The most important specification of any measurement test equipment is its **accuracy**, which is not the same as **precision** or **resolution**. (Accuracy, precision, and resolution are discussed at more length in Chapter 8.) Accuracy is the degree to which a given measured quantity agrees with the true or exact value of that quantity, as established by an agreed-upon **measurement standard**. For example, if you weigh a glass of water on a scale and the scale reads 1.0 kg, the accuracy of the scale amounts to how closely the true weight of the glass of water approximates the standard mass of the **international prototype of the kilogram**, a physical lump of metal kept by the International Bureau of Weights and Measures in Sèvres, France. There are standards for every common electrical quantity such as the volt, the ampere, and the ohm, and the accuracy of measurements of these quantities is simply how close the indicated value is to the true value, usually stated in terms of a percentage. If a meter is rated at 2% accuracy, that means the manufacturer guarantees its reading will be within ±2% of the true value.

Another important feature of a DVM is its **range** or **ranges**. Typical DVMs have a variety of selectable ranges that set the maximum value the device can measure. For example, a 3 1/2-digit DVM with automatic polarity display can show a reading from −1999 to +1999 (the 0 or 1 in the leading decimal place is the half digit). If the lowest voltage range is given as 200 mV, that means the meter at that setting can display voltages from −199.9 to +199.9 mV. Any voltage outside that range will result in an overload indication of some type. The **number of display digits** determines the **resolution** of a DVM, which is usually related to the device's accuracy. For example, a meter with only 5% accuracy does not need more than a 2 1/2-digit display, because any digits beyond the second decimal place are meaningless. On the other hand, a calibration-laboratory-quality voltmeter with 0.01% accuracy needs to have a 5-digit display, at a minimum.

The current measurement in a DVM is usually performed with a different set of input terminals than the ones used for voltage, because a low-resistance **shunt** is usually permanently connected between the instrument's current terminals, and it would be inconvenient to switch this shunt internally. Most DVMs can measure currents ranging from as low as a few μA up to 10 A or more. As mentioned earlier, DVMs measure resistance by imposing a small voltage across the device being measured and using the resulting current to calculate resistance. The reading is meaningful only for **linear** components such as resistors, capacitors, and inductors and indicates only the component's DC resistance. Some DVMs have a **diode** function that tells the user whether a diode is good or bad, but that is all.

AC voltages and currents can also be measured by most DVMs, but caution is advised here. Unless the instrument is stated to be a **true RMS** voltmeter, the AC voltage ranges are calibrated under the assumption that the input voltage waveform is a sine wave. That is approximately true in the case of AC power-line voltages, but complex audio and digital signals are not sine waves, and the voltage readings of these waveforms will be incorrect when compared to their true RMS values. Also, most DVMs have a **frequency response** limited to the audio range (about 20 kHz) and will become increasingly inaccurate above that frequency. If AC voltage or current measurements above the audio range are necessary, the user will need to procure a radio-frequency (**RF**) **voltmeter** or **power meter**, which are specialized instruments that are designed to maintain their accuracy at frequencies in the MHz or GHz ranges.

Finally, one should be aware that a DVM loads the circuit it is connected to. The type of loading depends on whether the instrument is being used to measure current or voltage.

Because voltage measurements are always made with the instrument in **parallel** with the two points of the circuit whose voltage difference is desired, the ideal voltmeter would draw no current at all from the circuit. This is impossible in practice, but real DVMs have a finite **input resistance** that is very high. A typical standard value for the input resistance on DC voltage ranges is 10 MΩ, although the manufacturer's literature should be consulted if one needs to know this value. Specialized high-impedance voltmeters called **electrometers** are available with much higher input resistances, but special techniques are needed to use them.

When used to measure current, a DVM is always connected in **series** with the conductor whose current is being measured. This requires physically breaking the conductor somehow and inserting the ammeter in series with the break. (The only ways to avoid breaking the circuit are by using a **clamp-on ammeter**, which measures the magnetic field produced by the current, or by including in the original circuit design a small resistance called a **current shunt** and measuring the voltage across it.) The ideal ammeter's series resistance is zero, but DVMs always have a small but nonzero resistance when measuring current. This is usually less than 1 Ω for the high-current ranges but can be as high as 1 kΩ or more for low-current ranges. Again, the manufacturer's specifications should be consulted if problems will arise from a voltage drop across the ammeter terminals.

A.4 FUNCTION GENERATORS

If a circuit is designed to process an input signal or voltage, it is not always convenient to use an actual signal source such as a transducer, microphone, or antenna. For this reason, the stimulus-type instrument known as a **function generator** was developed.

The early form of the function generator, the **sine-wave oscillator**, was developed in the 1930s for use in testing audio equipment such as telecommunications gear, studio mixer boards, and public-address systems. As such, it originally covered only the audio-frequency range (approximately 10 Hz to 20 kHz) and produced only sine waves. Later versions added capabilities such as the generation of **square waves** (more generally, rectangular waves) and **triangle waves**, and featured extended frequency ranges up to 2 MHz or higher.

Inexpensive function generators use primarily analog circuitry to produce sine, square, and triangle waves with acceptably low **total harmonic distortion** in the range of 5% or less. (Equation 4.14 provides a definition of total harmonic distortion.) Besides maximum waveform distortion, other important specifications for a function generator include **output impedance** and **maximum peak-to-peak output voltage**. An output impedance of 50 Ω is fairly standard, and a good function generator should be able to deliver at least a 5 V peak-to-peak waveform of any type into a 50-Ω load, with more being desirable. Other desirable features are good **frequency accuracy** (how well the frequency-dial reading agrees with the actual frequency of the output) and **frequency stability** (how nearly constant the frequency is over time). Because of inherent limitations in analog circuitry, frequency accuracy and stability better than 5% or so is rarely achieved in analog function generators, although some instruments feature a **digital frequency counter** readout that shows the user what the actual frequency is to better than 1% accuracy.

Many function generators provide for a **DC offset**, which is a constant DC voltage added to or subtracted from the AC output. The DC offset is helpful in generating a **transistor–transistor logic (TTL)-compatible square wave**, for example, or square waves for use as clock or other signals for digital circuitry in general. In Figure A.2, we show how the DC offset of a function generator can be used to obtain a HI voltage of +5 V and a LO voltage of 0 V. The basic idea is to set the AC square-wave output with its peak-to-peak value equal to the peak-to-peak level required (5 V in the case of TTL) and then adjust the DC offset so that the LO logic level is at 0 V. These are the desirable logic levels to be used with the type of digital logic circuitry known as **TTL**. Newer types of logic families use lower HI voltages such as 3.6 V or less, but these can also be produced by suitable adjustment of the peak-to-peak level of the square-wave output of a function generator combined with adjustment of the DC offset to make the LO level 0 V.

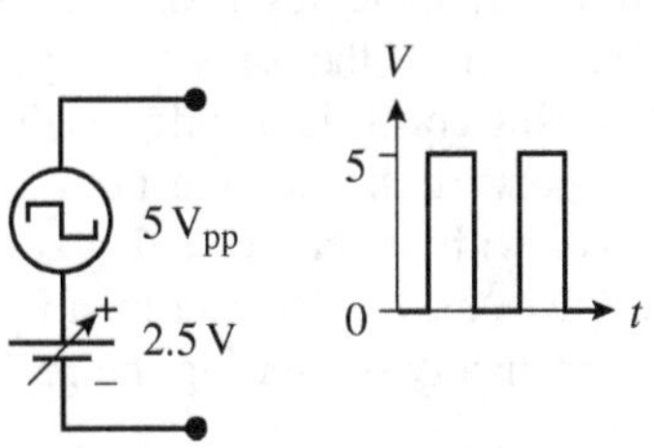

FIGURE A.2 AC square wave of 5 V (peak to peak) added to DC offset of +2.5 V to produce a TTL-compatible square wave in which HI = +5 V and LO = 0 V.

A.5 OSCILLOSCOPES

The portrayal of a changing voltage or current on a Cartesian x–y graph with voltage or current as the vertical (y)-axis and time as the horizontal x-axis is a useful function for many types of analog and digital electronic design, development, and troubleshooting work. Early attempts to achieve this type of measurement used electromechanical means such as electromagnetically deflected pens that inked moving strips of paper or small mirrors that reflected beams of light onto photosensitive paper strips. These devices had serious limitations that were overcome by the **cathode-ray oscillograph** invented by Karl Ferdinand Braun in 1897. But it was not until the 1950s that the **oscilloscope** in its modern form became generally available for use by television repair technicians and other electronics workers. While such **analog oscilloscopes** are still available, they are limited by the fact that no permanent record of their display can be conveniently made, so when **digital storage oscilloscopes** became available at reasonable prices, most users turned to them because of the convenience of downloading their digital data outputs for analysis and reporting.

The most important specification of an oscilloscope is its **bandwidth**, which expresses the highest sine-wave frequency that the instrument can display accurately. For example, a 100-MHz **scope** will display a 100-MHz sine wave reasonably well, but will not accurately portray a 100-MHz square wave, because the harmonics of the 100-MHz square wave extend well beyond the instrument's maximum bandwidth. Because most waveforms contain harmonics, caution is advised when measuring waveforms whose fundamental frequency is greater than 30% or so of the scope's rated bandwidth, because considerable **waveform distortion** may occur.

Most scopes have controls for **vertical sensitivity**, which expresses the change in voltage of the input signal that corresponds to one vertical screen division. Typical ranges of vertical sensitivity go from 5 mV per division up to 50 V per division or higher. Sensitivities much greater than this (below 5 mV per division) are not useful, because the **noise floor** of scopes of reasonable bandwidth is a few mV. In reading or measuring voltage with a scope, the user should take into account any **voltage division** provided by the **scope probe**. The probe is the device used to connect the scope input to the circuit under test. Some probes provide a direct connection and are known as **X1** probes because the indicated reading on the scope screen is multiplied by 1 to give the actual reading. Other probes have switch-selectable division ratios of **X1** and **X10**. The advantage of the *X*10 setting is that it provides a greater bandwidth and less loading on the circuit under test than the *X*1 setting does. However, for low-bandwidth and low-level signals, the *X*1 setting provides greater sensitivity. Many scopes have a **probe-setting control** that informs the instrument of the probe's scale factor. If this control is present, the user must ensure that the actual setting of the switch on the probe agrees with the probe-setting control setting on the scope. Otherwise, the displayed voltages may be in error by a factor of 10 or more. (Some scopes take care of this automatically by sensing the probe's setting, but not all of them do.)

The **time base** system of the scope establishes the horizontal scale in terms of the elapsed time per horizontal division of the display. (In older analog scopes, this function was achieved by "sweeping" an electron beam across the screen, so this setting is sometimes referred to as **sweep speed**.) Controls are available to set the sweep speed to a range of values as low as 10 s per division up to as high as 100 ns per division or faster, depending on the scope's maximum bandwidth. The best setting of this control produces a display with two or three fundamental-frequency cycles of the waveform of interest. If the fundamental frequency is known, it is easy to calculate the appropriate sweep speed that will yield a usable waveform. If the signal's frequency is not known in advance, one way to set the sweep speed control is to start at a very low speed (a long time per division) and slowly advance it to higher speeds (lower times per division) until a usable waveform results.

Even when this is done, the display may still be erratic and move around irregularly. This is probably because the **trigger controls** have not yet been adjusted properly to provide **synchronization** between the waveform and the horizontal sweep operation. The function and use of the trigger controls is shown in Figure A.3. In order to display a stationary view of a periodic waveform, the scope's horizontal sweep circuits must **trigger** (initiate) the horizontal sweep at the same point at each period (or multiple of periods if more than one period is displayed). This is done by means of a trigger system that has two primary controls: **trigger level** and **trigger polarity**. The trigger level control establishes the level (position on the display screen) that the waveform must cross in order to trigger the sweep. With analog scopes, the trigger event typically occurs before the sweep starts and so cannot usually be seen on the screen. But in digital scopes, the display is not a real-time one, so usually the trigger point is displayed at the horizontal center of the screen (unless the user selects another position). The trigger polarity determines whether it is a positive-going or negative-going part of the waveform that triggers the sweep.

In Figure A.3a, the trigger level is set higher than any point of the waveform, and so no synchronization is possible. Under these conditions, the display may be blurry or partially recognizable, though in rapid apparent motion. Adjusting the trigger level downward in this situation will bring it into the range of levels occupied by the waveform. In Figure A.3b, a positive-going trigger set at +2 vertical divisions has been selected. This places the upward-going part of the triangle wave so that it crosses the +2-division mark at the horizontal center of the screen, as shown by the small circle in the figure. Moving the trigger level downward to −1 division while keeping a positive-going trigger polarity leads to the waveform shown in Figure A.3c. As the user adjusts the level, the waveform is seen to "slide" horizontally so as to cross the center line at the varying vertical level. If the trigger level is kept at −1 division but the trigger polarity is switched from positive going to negative going, the waveform shown in Figure A.3d will result. In this way, any part of the waveform of interest can be centered on the screen or even enlarged for more detailed inspection with special horizontal-scaling functions on some scopes.

If the waveform being examined is not periodic—thermal noise, for example—no stable display will be possible, because the waveform does not repeat itself in time. Nevertheless, some estimates can be made of average amplitude and frequency ranges even with nonperiodic waveforms, although scopes are most useful when the waveform in question is reasonably periodic so that a nearly stable display can be examined.

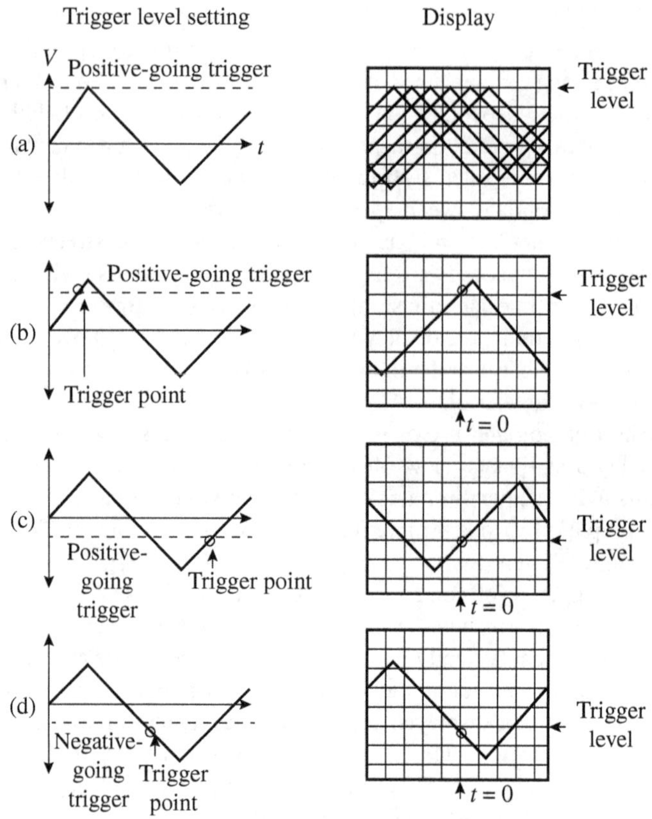

FIGURE A.3 Four settings of trigger level and polarity for a triangle wave. (a) Trigger level set too high—waveform not synced. (b) Positive-going trigger with level set at +2 vertical divisions. (c) Positive-going trigger with level set at −1 vertical division. (d) Negative-going trigger with level set at −1 vertical division.

Some digital scopes allow the user to do a **single-shot** capture of an intermittent or one-time waveform. In this mode, the trigger circuit does not operate until the proper trigger voltage is received. Then exactly one sweep is made and the waveform captured is preserved for examination or downloading.

Other advanced features of some digital scopes include an **autorange** function, which automatically sets the trigger, vertical, and horizontal controls to display a usable waveform in many cases; **digital measurement** options that calculate and display values of various waveform features such as peak and RMS amplitude, frequency, period, and other quantities of interest; **multiple-channel inputs** that allow more than one simultaneous waveform to be displayed and measured; and **data storage functions** that can store not only an image of the waveform but a detailed table of the voltage versus time and all instrument settings to a variety of digital storage media or to an Internet connection. Furthermore, some digital scopes can perform **Fourier transforms** of a signal waveform, which converts the instrument into a rudimentary **spectrum analyzer**. Spectrum analyzers are discussed briefly in Section A.7.

A.6 ARBITRARY WAVEFORM GENERATORS

Many electronic signals in use today are quite complex, especially in the case of communications systems that use sophisticated digital modulation techniques. Analog function generators cannot produce such waveforms, so in recent years, a mixed-signal instrument called an **arbitrary waveform generator** (**AWG**) has become available at prices that are within the range of well-equipped educational electronics laboratories. An AWG can do everything an analog function generator can do, and much more besides, because the output waveform is an analog version of a digitally synthesized waveform produced by the digital portion of the instrument. An AWG uses the signal-generation technique called **direct digital synthesis** (**DDS**), which produces a waveform by generating a digital version of it and then sending the resulting sequence of binary words to a **digital-to-analog converter** (**DAC**).

A simplified block diagram of an AWG is shown in Figure A.4. The user selects from a wide variety of signals with a set of controls, which can be either hardwired controls or **softkeys** on an input screen. Some types of AWGs will accept digital input in the form of a table of voltage values versus time, so the user can cause the production of a truly *arbitrary* waveform, within certain limits. Once the parameters of the waveform to be produced are presented to the digital system, digital signal processing software generates the desired sequence of binary words that represent the waveform called for. An internal clock (usually a highly accurate crystal-controlled oscillator) establishes the timing and frequency for all waveforms produced, which means that the frequency accuracy of the output waveform is as good as the internal clock's accuracy, typically 1 part in 10^7 or better. (This level of frequency accuracy is much better than the best analog function generators.) The binary words representing the output waveform are sent in sequence to a high-quality DAC, which produces an analog version of the waveform that appears at the output.

Because the waveform is produced digitally, almost any type of modulation can be synthesized with an AWG. The only limitation is that the signal must be periodic. Every AWG has a maximum memory storage capacity, which limits the longest period that a waveform can have. So it is difficult to produce true random noise with an AWG, although a **pseudorandom** noise sequence can be generated. Most AWGs

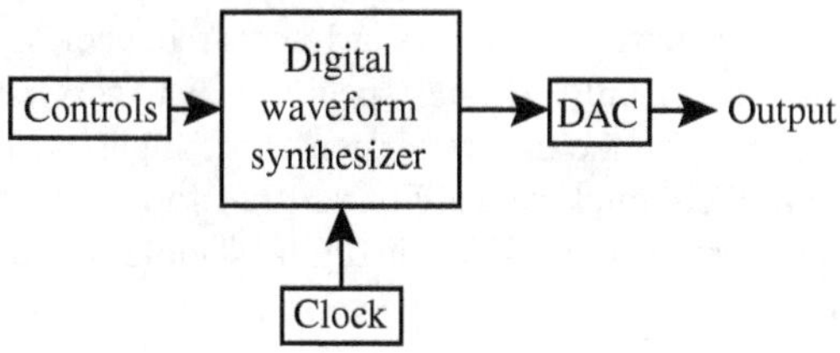

FIGURE A.4 Simplified block diagram of arbitrary waveform generator (AWG).

can synthesize waveforms that have amplitude, frequency, and phase modulation, as well as combinations of these. The maximum conversion speed of the DAC imposes a maximum output frequency limit, but reasonably priced AWGs can generate waveforms at frequencies up to 20 MHz or higher.

A.7 OTHER TYPES OF ANALOG AND MIXED-SIGNAL TEST EQUIPMENT

Most routine experimental and developmental work involving analog and mixed-signal circuits can be performed with the equipment described in Sections A.1–A.6. Amplifiers, oscillators, and analog-signal-processing circuits can be tested with stimulation provided by a function generator or an AWG, and responses measured with a DVM or scope. More complex circuits and systems, especially in the areas of communications and RF, may require more sophisticated test gear, so we will briefly describe a few of the more advanced types of test equipment that are often useful for analog and mixed-signal design.

A.7.1 Spectrum Analyzers

Strictly speaking, the **spectrum** of a signal is its Fourier[1] transform. As you probably know, the **Fourier transform** is a mathematical operation that takes a finite time sequence of a voltage or current function as its input and transforms it into a function of *frequency*. Although in general, the Fourier transform of a real function is a complex function, the squared *magnitude* of the Fourier transform at a given frequency measures the energy at that frequency in the original signal. So a display of the magnitude of the Fourier transform of a sample of a time-varying signal will give the viewer a good idea of how the signal's energy is distributed over a range of frequencies.

This type of information is especially of interest in communications systems, where **bandwidth** limitations imposed by RF spectrum regulations or the limited bandwidth of a transmission system makes it important to know exactly what frequency ranges are occupied by a given signal.

The earliest spectrum analyzers used exclusively analog circuitry and resembled **superheterodyne** radio receivers, in that they used local oscillators and narrowband filters to produce a display on an oscilloscope-like screen. The display was not a true Fourier transform but approximated it, and elaborate measures were necessary to prevent spurious responses from interfering with the display. As digital technology advanced, more of the **back-end** functions were taken over by digital signal processing, and present spectrum analyzers at lower frequencies perform nearly all their internal operations digitally. Digital spectrum analyzers perform a **discrete Fourier transform** (**DFT**) on the digital version of the input waveform. A DFT is a discrete version of a true complex-valued Fourier transform and as such contains all the information that the original time-domain digital sample contained. In addition to allowing the display of the DFT's magnitude versus frequency to indicate the **power**

[1]*Fourier* is pronounced "four-*yay*," not "four-ee-er."

spectrum of the signal, the DFT can be used in **digital filtering** operations and other signal processing work as well. The DFT is how many digital scopes perform spectrum analysis, although a custom-designed spectrum analyzer will often perform better than a general-purpose digital scope with a spectrum analysis feature.

A.7.2 Logic Analyzers

In the development and maintenance of digital systems, one often needs to get a picture of what is going on in the system in terms of certain data streams and the timing relationships among them. Because many digital systems use busses with widths of 32–64 bits or more, it is impossible to observe the sequence of digital words on such a bus with an oscilloscope having only a few channels. For these types of problems, **logic analyzers** have been developed that can sense and capture large blocks of digital data and decode it for display in a recognizable form.

Early logic analyzers were equipped with a number of small clip-on probes that had to be individually attached to each of the data lines to be examined. While some lower-end analyzers still operate this way, since 2000, a new method of attaching a logic analyzer to a digital circuit has been developed: the **compression probe**. Compression probing requires cooperation on the part of the board designer to provide **surface-mount technology** (**SMT**) pads that interface with a removable mechanical **retention module**. When it is desired to observe the digital bus with a logic analyzer, the user mechanically attaches the retention module, which aligns the analyzer's compression interconnect with the circuit's SMT pads and allows the electrical connections between the digital bus and the analyzer. The compression interconnect is specially designed with internal resistors to produce minimum loading on each line—often less than 1 pF. This allows the analyzer to monitor the digital circuit's operation without affecting the electrical environment significantly, which was difficult with the earlier manual-attachment probes.

Once the digital data has been captured by the logic analyzer, many such instruments can perform timing, display, and even decoding operations on the data, even to the extent of displaying only data on an Ethernet bus that meets certain criteria set by the user. While the details of digital-circuit analysis and troubleshooting are beyond the scope of this text, engineers dealing with systems that include extensive digital circuitry (which is nearly every system of any size today) should be aware of the powerful diagnostic resource that logic analyzers represent.

A.7.3 Network Analyzers

A **network analyzer** is a stimulus/response instrument that both provides a stimulus to a component or circuit, usually in the form of an AC sine wave, and measures the response of the circuit to that stimulus. In a sense, the **ohmmeter** function of a DVM is a type of network analyzer, because it stimulates the resistor under test with a known voltage (or current) and measures its response in terms of the resulting current (or voltage). Generally, however, the term **network analyzer** is reserved for devices that use AC signals as stimuli.

Network analyzers can be either the **scalar** type, which records only the magnitude of the response, or the **vector** type, which records both magnitude and phase. One early form of scalar network analyzer was called a **sweep generator**, which in combination with an oscilloscope provided a visual display of a circuit's **frequency response**. These instruments were useful in **aligning** the **intermediate-frequency (IF)** amplifiers of analog radio and TV receivers and continue to be used in special applications for RF circuit design and manufacturing.

However, more information can be obtained from the vector type of network analyzer. A network analyzer can have one, two, or more than two **ports**. A port consists of a pair of wires with respect to which a voltage or current is defined. A **one-port** vector network analyzer can measure the complex ratio between the voltage across a component and the current through it. This ratio is the complex **impedance**, and if the stimulus frequency is varied, a plot of the component's impedance versus frequency can be obtained. Another name for a one-port vector network analyzer is an **LCR meter**, and the display of such meters can be set to read out impedance in terms of the inductance and **Q** (**quality factor**) of a coil or the equivalent *R–C* series circuit of a capacitor at a specified frequency. Such instruments are available with various specified accuracies and frequency ranges, although the ones with better accuracy and higher frequency limits tend to be quite costly.

Two-port vector network analyzers can be used to measure a variety of components and systems ranging from simple **three-terminal devices** such as transistors up to entire amplifier systems. Another term for a two-port network analyzer is a **Bode plotter**, because the vector data obtained by stimulating one port with a sine wave of a known magnitude and phase and measuring the magnitude and phase at the second port can be graphed using the familiar **Bode-plot** format that displays magnitude in dB and phase in degrees, both plotted against the logarithm of frequency. If properly performed, a Bode-plot laboratory measurement can be plotted on the same graph along with the same measurement made in network-analysis software such as Multisim™, which allows direct comparisons between theory and experimental results. For example, the circuit simulation response shown in Figure 6.28 for the guitar-amp example can be compared to a laboratory measurement made with a Bode plotter, and if the theoretical model properly captured all the significant features of the design, the theoretical and experimental gain and phase curves will agree pretty closely. The usefulness of such a measurement is obvious.

At higher frequencies in the RF and microwave range, conventional voltage and current measurements become increasingly difficult to make accurately, and vector network analyzers using a **standard characteristic impedance** (typically 50 Ω) are the only show in town, so to speak. They can be used to measure the performance of both systems and individual devices such as RF transistors, as long as proper **biasing** means are provided. Such measurements are challenging to perform with high precision (repeatability) and accuracy and require costly and specialized equipment and calibration techniques. Nevertheless, it is possible to make precision impedance and gain measurements at virtually any frequency in use in electronics today, so the old days when educated guesses were the only way to design a circuit have largely passed from the scene.

INDEX

Note: Page numbers in *italics* refer to Figures; those in **bold** to Tables.

Analog and Mixed-Signal Electronics, First Edition. Karl D. Stephan.

Companion Website: http://wiley.com/go/analogmixedsignalelectronics